Progress in Probability and Statistics
Vol. 3

Edited by
Peter Huber
Murray Rosenblatt

Birkhäuser
Boston · Basel · Stuttgart

[Progress in Probability and Statistics.]

S. Asmussen
H. Hering

Branching Processes

1983

Birkhäuser
Boston • Basel • Stuttgart

Authors:

Søren Asmussen
Institute of Mathematical
Stochastik
University of Copenhagen
5, Universitetsparken
2100 Copenhagen Ø, Denmark

Heinrich Hering
Institut für Mathematische
Statistik
Universität Göttingen
Lotzestr. 13
3400 Göttingen, West Germany

CIP-Kurztitelaufnahme der Deutschen Bibliothek

Asmussen, Søren:
Branching processes / S. Asmussen ; H. Hering.
- Boston ; Basel ; Stuttgart ; Birkhäuser, 1983.
(Progress in probability and statistics ;
Vol. 3)
ISBN 3-7643-3122-4

NE : Hering, Heinrich: ; GT

Library of Congress Cataloging in Publication Data

Asmussen, Søren.
Branching processes.

(Progress in probability and statistics ; v. 3)
Bibliography:
1. Branching processes. I. Hering, H. (Heinrich), 1940- . II. Title. III. Series.
QA274.76.A78 1983 519.2'34 82-22704
ISBN 3-7643-3122-4 (Switzerland)

ISBN: 3-7643-3122-4

Printed in USA

TABLE OF CONTENTS

PART C: MULTIGROUP BRANCHING DIFFUSIONS ON BOUNDED DOMAINS

Chapter V: Foundations

Chapter VI: Limit theory for subcritical and critical processes

Chapter VII: Basic limit theory for supercritical processes

Chapter VIII: More on the limiting behaviour of linear functionals

FIRST APPEARANCE OF SPECIALLY MARKED RELATIONS:

PREFACE

Branching processes form one of the classical fields of applied probability and are still an active area of research. The field has by now grown so large and diverse that a complete and unified treatment is hardly possible anymore, let alone in one volume. So, our aim here has been to single out some of the more recent developments and to present them with sufficient background material to obtain a largely self-contained treatment intended to supplement previous monographs rather than to overlap them.

The body of the text is divided into four parts, each of its own flavor. Part A is a short introduction, stressing examples and applications. In Part B we give a self-contained and up-to-date presentation of the classical limit theory of simple branching processes, viz. the Galton-Watson (Bienayme-G-W) process and its continuous time analogue. Part C deals with the limit theory of Markov branching processes with a general set of types under conditions tailored to (multigroup) branching diffusions on bounded domains, a setting which also covers the ordinary multitype case. Whereas the point of view in Parts A and B is quite pedagogical, the aim of Part C is to treat a large subfield to the highest degree of generality and completeness possible. Thus the exposition there is at times quite technical. Part D concerns simple branching diffusions on unbounded domains, age-dependent processes and their generalizations to models in which the individuals have completely unrestricted reproduction patterns, and finally models in which different individuals cooperate in producing offspring. For easier reference a number of auxiliary results and tools we use are compiled in an Appendix.

To each chapter there is a section with bibliographical comments. They record our immediate sources without giving a complete history of the subject. Similarly, the list of references at the end of the book is not encyclopedic, but is restricted to those publications we explicitly quote in the text or the comments.

Topics not covered could easily fill a second volume. Among them are processes with random or deterministically varying environment, decomposable processes (except for some results on immigration), the whole complex of diffusion approximations, continuous-state or, more generally, Borel measure-valued processes, models with infinitely many particles, or the scarcely developed field of controlled branching. References up to 1972 are to be found in the book of Athreya and Ney (1972). Of more recent date are Kaplan (1973,1974). and Tanny (1976,1977) on random environment, Fearn (1972), Jagers (1973), Goettge (1975), Cohn and Hering (1981) on varying environment, Ogura (1975), Polin (1976,1977), Foster and Ney (1978) on decomposable processes, Kurtz (1978), Papanicolaou (1978) on diffusion approximations, and Kallenberg (1979) on simple continuous-state processes. Concerning Borel measure-valued processes and branching random fields we only mention the review by Dawson and Ivanoff (1978), for processes with infinitely many particles also the book by Matthes, Kerstan, and Mecke (1978).

Thanks go first of all to Arletta Havlik and Cathy Stevens for efficiently typing the bulk of the manuscript, To Tony Augustine for administrative assistance, to Klaus Enderle for help with the corrections, and to Dominique Grard and Marrie Powell for some last-minute typing jobs. H.Hering would also like to thank the Mathematics Department of Cornell University, the Institute of Mathematical Statistics of the University of Copenhagen, as well as the Statistics Department of the University of Melbourne for their hospitality, and he gratefully acknowledges the financial support given by Stiftung Volkswagenwerk, the Danish Natural Science Research Council, and the University of Melbourne.

Copenhagen, 1981 — S.A.

Göttingen, 1981 — H.H.

PART A

INTRODUCTION

CHAPTER I

BRANCHING PHENOMENA AND MODELS

1. SIMPLE BRANCHING PROCESSES

In a wide sense, a branching process could be described as a stochastic model for the development of a population, the members of which we call individuals or particles, and which independently reproduce (and possibly in addition evolve in some other manner).

The prototype of a branching process, modelling the phenomenon in its purest form, is the simple Galton-Watson or, historically more correct, Bienaymé-Galton-Watson processes $\{Z_n\}_{n\in\mathbb{N}}$. We refer to the parameter $n\in\mathbb{N}$ as time and to $Z_n\in\mathbb{N}$ as the size of the nth generation (though sometimes other interpretations are relevant as is seen below). The individuals reproduce independently, each according to the same offspring distribution F (concentrated on $\mathbb{N}$). That is, if $X_{n,i}$ represents the number of offspring of the ith individual of the nth generation, then

$$Z_{n+1} = \sum_{i=1}^{Z_n} X_{n,i}$$

and the $X_{n,i}$ are independent, each with distribution F. The realizations of the process are often depicted in a family tree:

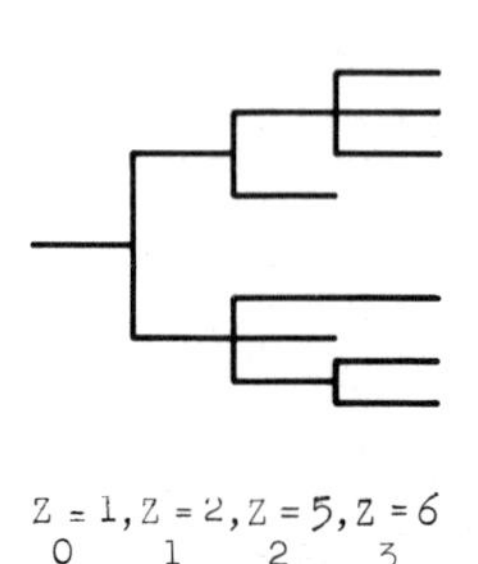

$Z_0 = 1, Z_1 = 2, Z_2 = 5, Z_3 = 6$

Examples compatible with this setting are species of plants or insects which have a well-defined reproduction cycle of one or several years, tied to the natural calendar, such that no generation lives to see the final stage of the next generation. If there are two separate sexes, consider, e.g., the female line.

Another example is the <u>electron multiplier</u>, an instrument to detect and measure weak-current beams of charged elementary particles, particularly electrons. The incoming particle hits a metal plate and strikes loose a number of electrons, which by means of an electric field are directed towards a second plate. Here they generate more electrons, which in turn are made to hit a third plate, and so on. The process ends at the last plate.

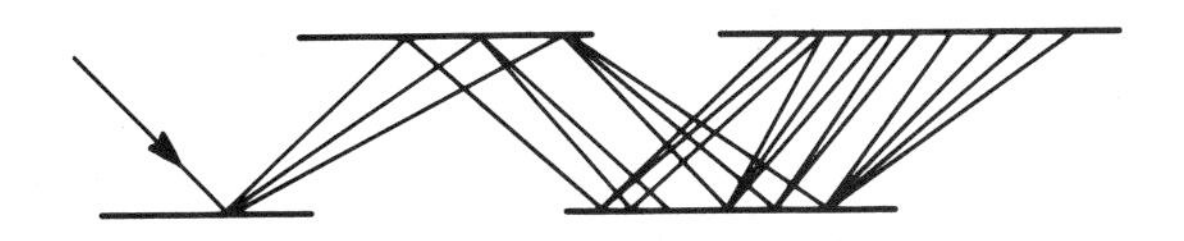

A more notorious example, some simple aspects of which fit in at this point, is the neutron chain reaction in nuclear reactors and atomic bombs. In a medium containing fissionable material such as Uranium 235, Plutonium 239 or Thorium 232, a neutron (possibly produced by instantaneous fission) may collide with a nucleus and either be scattered or absorbed. When absorbed, it may cause the nucleus to split and release several new neutrons, which in turn may induce other nuclei to split.

Further examples, which will not be discussed here, are certain chemical chain reactions and cascades of defects in complicated networks such as electronic computers.

Even if generations overlap in time, as is the case in the nuclear chain reaction which really has to be studied as a continuous time process, it is sometimes of interest to record successive generation sizes $Z_0, Z_1, \ldots$. Thus the original example of Galton and Watson was survival of surnames in English peerage: The ancestor (corresponding to $Z_0 = 1$) is a man with some given surname, Z_1 the number of his sons, Z_2 of his grandsons and so on. The problem is to study the <u>extinction probability</u> $q = P(Z_n = 0$ eventually). Similar examples arise in population genetics and epidemiology. In epidemics, the birth of the ancester corresponds to the occurence of some mutant gene and Z_1 is then the number of offspring which themselves carry the mutation.

In epidemics an individual's direct descendents are the healthy individuals it infects. It has been suggested to use the model, e.g., to approximate the probability of a major epidemics by $1-q$. Also for neutron chain reactions, many calculations have been done for the generation sequence.

Still, with generations overlapping, one will more often be interested in the size of the population at subsequent units of time than at subsequent generations. The offspring distribution then becomes the distribution of the descendants of all generations of the ancestor which are alive after one unit of time. In such examples the sequence of physical generations is usually called the imbedded generation process. Even if this is a Galton-Watson process, the population recorded at subsequent units of time need not be so. In fact, some conditions like independence of the reproduction of the age are needed.

These considerations lead in a natural way to a continuous time version $\{Z_t\}_{t\geq 0}$ of our process. Let τ be the time from birth to reproduction. Then by age independence,

$$P(\tau > t+s \mid \tau > t) = P(\tau > s).$$

It is well-known that this equation characterizes the exponential distribution so that $P(\tau > s) = e^{-\lambda s}$. In other words, for any individual alive at time t the probability of reproducing in the time interval $[t, t+dt]$ is λdt. Thus, if n independently evolving individuals are present, the probability of one of them reproducing is $n\lambda dt$ and that of two or more is $O(dt^2) = 0$. That is, if reproduction results in $k = 0,1,2...$ individuals with probability p_k, then

$$P(Z_{t+dt} = Z_t+k-1 \mid Z_s;\ 0 \leq s \leq t) = Z_t \lambda p_k dt, \quad k \neq 1,$$

$$P(Z_{t+dt} = Z_t \mid Z_s;\ 0 \leq s \leq t) = 1 - Z_t \lambda (1-p_1) dt.$$

The process $\{Z_t\}_{t\geq 0}$ is called the simple Markov branching process.

By construction, any discrete skeleton $Z_0, Z_\delta, Z_{2\delta}, \ldots$ of $\{Z_t\}$ is a Galton-Watson process. Not every Galton-Watson processes can, however, be constructed this way, for example, no process whose F has finite support. As in the discrete time model, reproduction need not coincide with the physical death of the parent, who may survive as possible part of the offspring. An important example is the pure birth process (sometimes called also the Yule or Yule-Furry process) where $p_2 = 1$, all other $p_k = 0$. One usually thinks of the individuals as having infinite lifelengths and giving birth at various times during the life, and birth times of any individual being a poisson process with intensity λ.

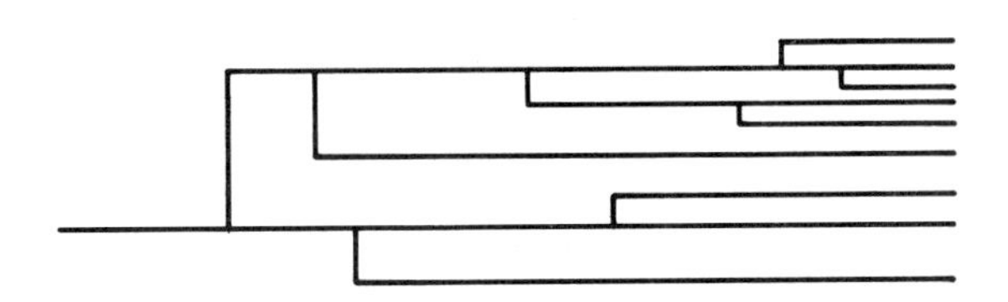

A slightly different interpretation is the binary split modelling, say the growth of cell populations, for example, in tumors.

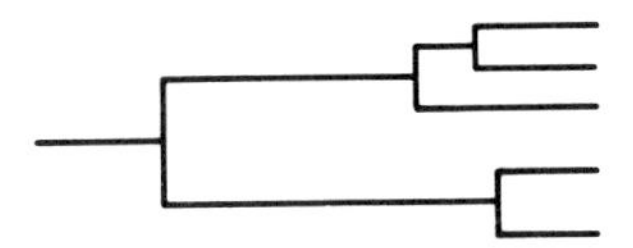

In the particular case when $p_k > 0$ only if $k = 0$ or $k = 2$, one speaks of the birth-death process.

Before proceeding to more complicated branching models, let us briefly discuss the limitations of the processes considered so far when modelling the described physical and biological phenomena.

Applying the model to species with a reproduction cycle parallel to the calendar, we are tacidly assuming that the relevant environment does not change significantly from generation to generation, due to outside influence or extreme population densities. In the two-sex case the existence of a sufficiently densely distributed male population is part of the environmental condition.

In the case of an electron multiplier we are neglecting that the number and energy distribution of the electrons struck loose depends on the impuls and energy of the incoming particle.

In a neutron chain reaction the probabilities of collision and absorption also depend on the energy of the neutron. Scattering of a neutron does in general lead to a change in energy. That is, at this point we would have to restrict ourselves to a homogeneous medium containing only heavy isotopes, like the core of an atomic bomb, where scattering leads to almost no change in the neutron's energy and consideration of just one energy is meaningful (one-group theory). Another problem is the finiteness of the medium. Unless the boundary is reflecting, as is the case in some smaller versions of the bomb, or the dimensions of the medium are very large compared with the mean free path of the neutrons (which can only be the case in a reactor), the larger probability of a neutron near the boundary to get lost to the outside cannot be neglected. The changes of the medium due to the consumption of fissionable material or change of temperature are usually slow in the time scale set by the sequence of neutron generations. A similar remark applies to the early stages of an epidemic.

Finally, no matter how one interprets the time of reproduction, the models considered so far allow only for physical lifetimes which are constant or geometrically distributed in discrete time, or exponentially distributed in continuous time. For example in cell proliferation, one has a well-defined lifelength distribution which is not of any of these types.

In the next sections, we introduce some of the basic modifications of simple branching processes needed to describe such phenomena in a satisfactory way.

2. P-TYPE PROCESSES

One of the simplest and yet most versatile extensions of the model of Section 1 is to allow for several types of particles. We consider here the case of a finite number p of types. To describe the process mathematically, suppose the types are $1, 2, \ldots, p$. The population at time n is then specified by the p-vector $Z_n = (Z_n(1), \ldots, Z_n(p))$, where $Z_n(i)$ is the number of type i individuals at time n. The individuals still reproduce independently, but the offspring distribution depends on the type of the parent. That is, in discrete time we can write

$$Z_{n+1} = \sum_{i=1}^{p} \sum_{k=1}^{Z_n(i)} X^i_{n,k}$$

where the $X^i_{n,k}$ are independent, with values in $\mathbb{N}^p$ and distribution F^i depending only on i. Different components of $X^i_{n,k}$ may be dependent. In continuous time the time from birth to reproduction of a type i individual is exponential with mean $1/\lambda_i$ depending on i, and the vector $(z(1), \ldots, z(p))$ of offspring results with probability $p^i_{z(1), \ldots, z(p)}$.

The setting of p-type processes is very flexible from the application point of view, in particular when used as approximation. When modelling the evolution of a species, we can now account for several alleles, admitting differentiation and mutation. For an epidemic we can allow for several periods of the illness, such as incubation, latent, and infectious phases. Suppose, for example, we want to distinguish between the infectious phase $i = 1$ and the non-infectious phase $i = 2$ and the between times t and $t+dt$ a non-infectious individual becomes infectious with probability $\lambda_1 dt$ and that an infectious one dies with probability $\mu_1 dt$ and infects a healthy individual with probability $\mu_2 dt$. To fit this into the above framework, one only has to let $p^1_{01} = 1$, all other $p^1_{ij} = 0$, $\lambda_2 = \mu_1 + \mu_2$, $p^2_{00} = \mu_1/\lambda_2$, $p^2_{11} = \mu_2/\lambda_2$. Other examples of multitype processes in epidemiology are veneral diseases, where each of the two types , produce the other, carrier models (for example tuberculosis) where the carriers appear healthy but may transmit the disease to others, and host-vector diseases. E.g., in malaria the hosts are the humans and the vectors the mosquitos. Here in addition the parasites has life-cycles within both hosts and vectors, with well-defined phases (so that, e.g., only parasites of one particular phase can be transmitted from host to vector), and thus one could subdivide both hosts and vectors into several types, according to which phases are present.

A somewhat different approach in continuous time leads to the so-called multi-phase birth process. Here every particle moves through the types $1 \to 2 \to \dots \to p$ and when leaving type p splits into two type 1 particles. If, as is usually assumed, all $\lambda_i = \lambda$, the physical lifelength becomes Erlangian (i.e., gamma with p degrees of freedom). The types often have no concrete interpretation, but are introduced to formally include models with such lifelength distributions in the theory of Markov branching processes.

We can now also take care of the fact that at least in most biological cases lifetime distributions are not exponential and that reproduction probabilities may depend on age. Suppose for simplicity that the individuals are distinguished only by age. Divide the age scale into intervals of equal length ("age-groups"), and use the discrete-time p-type process as approximation. The set of types is the set of the first p age-groups. An individual of the ith age-group at time n moves to the $(i+1)^{th}$ age-group, or dies, and may in either case produce some offspring in the first age-group. For $i = p$ death occurs almost surely. Thus F^i is concentrated on $\{x: x(i+1) = 0$ or $1,\ x(j) = 0$ for $j \neq 0, t+1\}$. While with a one-type process we could model survival of a birth event by the parent only by interpreting a member of the offspring as parent, not being able to distinguish it from the other offspring, the parent is now labelled by its age-group.

Next we return to the neutron chain reaction. In a classical reactor the energy of the neutrons arising from a split is deliberately reduced by a sequence of collisions with the light nuclei of a non-fissionable component, the so-called moderator, built into the reactor core. Consequently, we divide not only the medium but also the energy scale into a finite number of intervals ("energy-groups"). Our set of types is now the product of the set of energy groups and the set of space cells. More precisely than by its position and energy a neutron is characterized by its position and impuls vector. Thus we can go further and divide the corresponding six-dimensional product space ("phase space") into cells.

Another class of examples are cosmic ray showers. When passing through matter, elementary particles of high energy, known to occur in cosmic radiation, or artifically produced by accelerators, initiate cascades. One distinguishes a hard and a soft component. The first is based on strong nuclear interaction and involves mainly nucleons and mesons. The second is based on electromagnetic interaction, involving photons and electrons of positive and negative charge.

Taking only first and second order effects, the electron-photon cascade is composed of the following events: (1) An electron looses energy by radiating a photon, (2) a photon of sufficiently high energy is converted into a positron-negatron pair, (3) a positron of the cascade and a negatron of the medium annihilate each other, producing a pair of photons, (4) an electron looses energy by ionizing an atom along its path.

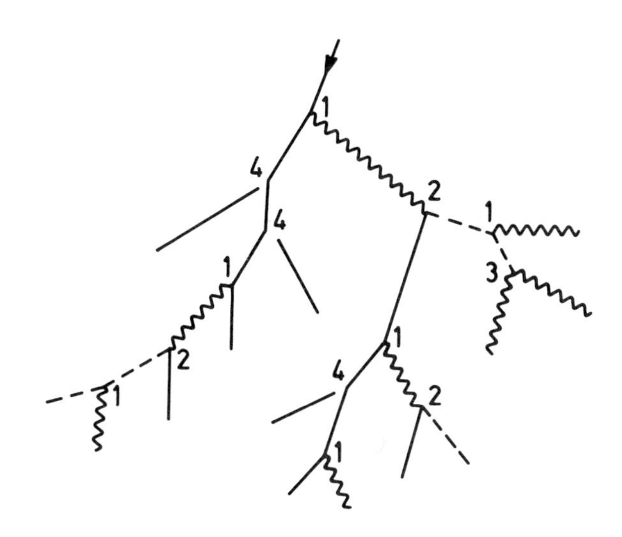

When approximating such a cascade by a p-type process, we again divide the energy scale into a finite number of intervals and define as our set of types the product of the set of energy groups and a set of three points representing photons, positrons, and negatrons.

Higher order effects are scattering with production of several electrons and photons and ionization cascades, that is, cascades initiated by electrons struck loose from atoms in the medium by an electron of the primary electron-photon cascade. At high energies the loss of energy by ionization is low. Otherwise it is fairly realistic to model the ionization as a continuous non-random loss of energy, in which case the approximation by p-type processes makes sense only with a discrete time parameter. The continuous, non-random energy loss corresponds to reversed aging: Particles are created at an energy at most equal to the energy of the parent and progress towards zero energy.

3. AGE-DEPENDENCE

The discrete approximations of continuous variables such as age, energy, location, etc. may suffice for practical purposes, but are not always the most elegant or natural ways to set up the model. We next consider some more direct constructions of age dependent models.

A simple example is the Bellman-Harris process, specified by the lifetime distribution G and the offspring distribution F. The offspring is produced at the time of death and independent of the life-length. The model extends the Markov branching process in a similar way as the renewal process extends the Poisson process. In fact, following any branch in the family tree of the imbedded generation process yields a Poisson process in case of a Markov branching process and a renewal process in case of a Bellman-Harris process.

The cell split model fits particularly well into this framework (with F degenerate at 2). From the point of view of demography, the Bellman-Harris process does, however, not provide an adequate description: Reproduction may take place several times during life and interpreting the parent as possible part of the offspring leads to not very natural restrictions on the reproduction mechanism. Instead, the age-dependent birth-death process has been suggested. Here an individual of age a dies within $[t, t+dt]$ with probability $\lambda(a)dt$ and gives birth (and survives) with probability $\mu(a)dt$. If $\int_0^\infty \lambda(a)da = \infty$, then lifetimes are finite and the lifetime distribution G has the density

$$\frac{dG(a)}{da} = \lambda(a)e^{-\int_0^a \lambda(b)db} .$$

More generally the residual lifetime of an individual aged x has distribution G_x with density

$$\frac{dG_x(a)}{da} = \lambda(a+x)e^{-\int_0^a \lambda(b+x)db} .$$

Apart from the demographic applications, one could also use this model in say epidemiology, where the age-dependence of μ would model the different degrees of infectiveness during the illness.

There are still some phenomena left, which cannot be described by the age-dependent birth-death process. There is a minimal waiting time between successive births, and in addition to age also the number of earlier births and their placement in life will matter. A model allowing for these features, and in fact the most general one-type age-dependent process constructed so far, is the Crump-Mode process, associated equally much with the names of Ryan and Jagers. Here the basic parameters are the time τ of death and a point process ξ on $[0,\infty)$, whose epochs represent the times at which the individual gives birth. Thus $\xi(a)$ is the number of children before age a and $\xi(\infty)$ the total number of children. The process is then characterized by the joint distribution of τ and ξ, and the point is that no restrictions are put on the particular form of this distribution. The Bellman-Harris process comes out as a special case by demanding that $\tau < \infty$, $\xi(\infty) < \infty$, that all epochs of ξ are at τ, and that τ and $\xi(\infty)$ are independent.

In the same way as for simple branching processes, the applicability of these models is greatly improved by allowing for several types. The precise formulations of the models suggest themselves. As an example, consider the cell split, where the one-type Bellman-Harris process provides an adequate description of the evolution of the total population size. The lifespan of one particular cell is, however, divided into four biologically well-defined phases G_1, S, G_2, M (e.g. M is the mitotic phase). Assuming independence of the duration of the phases, one could use a four-type Bellman-Harris process to represent information on say also the proportion of cells in the mitotic phase. Of course, the lifetime distribution in the one-type process is the convolution of the four lifetime distributions of the four-type process.

4. GENERAL PROCESSES

In the last section we have discussed the direct modelling of age as a continuous variable. Just as well one can model continuous motion in, say, a geographic region, or an energy range, instead of motion along the age-scale. A formally unified setting for such processes and indeed any process modelling the development of a finite population, whose members behave independently, is provided by the so-called <u>general Markov branching process</u>.

The idea is simply to replace the finite set of types in Section 2 by an arbitrary set X of types. That is, the state of the process at any given time is now given by the total number n of particles present and their respective types $x_1,\dots,x_n \in X$. The development of the process $\{x_t\}$ in the state space X of all finite populations of particles with types in X is Markovian, and different particles and their lines of descent evolve independently. If both, the time parameter t and the set of types X, are continuous, the type of a particle may change continuously with time. In fact, one may think of the motion of a particle up to the time of reproduction as a Markov process on X. Occurence of a branching event and the law governing the resulting offspring depend only on the local path behaviour.

As a main case, suppose the Markovian motion is described by $\{x_t\}_{t\geq 0}$, the motion is stopped between t and $t+dt$ with probability $k(x_t)dt$, and this stopping results in an offspring population in X with law $\pi(x_{t-},\cdot)$, where π is a transition kernel from X to X. The distribution of the time τ of the first occurence of branching ("life-time distribution"), conditioned on the path of the ancestor, is then simply

$$P(\tau > t\,|\,x_s,\ s \geq 0) = e^{-\int_0^t k(x_s)ds} .$$

If all offspring arises at the point of death of the ancestor, π is called a <u>local branching law</u>.

In the age-dependent birth-death process, $\{x_t\}$ is linear motion to the right on $X = [0,\infty)$, $k(x) = \lambda(x) + \mu(x)$, and a branching event at age x results in death with probability $\lambda/(\lambda+\mu)$ and in creation of an additional particle with age 0 with probability $\mu/(\lambda+\mu)$. Thus in this case the branching law is <u>non-local</u>.

Setting the Crump-Mode process up in this framework would require a very large set of types. For example, the type of an individual could be the time τ of death together with the whole point process on $[0,\tau]$ describing the reproduction. In cases such as this, however, the practical value of the formulation as a general Markov branching process may be questioned.

An important class of branching processes, which we shall treat at some length later, is given by branching diffusions. Diffusions have been used as approximate descriptions of various types of motion, from the motion of a neutron in a reactor to that of a member of a biological population. For a simple branching diffusion the set of types is a connected region in a Euclidean space or a differentiable manifold. The motion of a particle on this region is determined by a transition probability density which is the fundamental solution of a parabolic differential equation, often called the diffusion or heat equation, with some boundary conditions. In addition to possible absorption at the boundary the motion is stopped with a termination density k, which in general is a function of the particle position. Particles produced in a branching event may arise at the point of death of their immediate ancestor or somewhere else, not necessarily all at the same point.

More useful than a simple branching diffusion is, of course, a *multigroup branching diffusion*. This is a special case of a branching diffusion with a disconnected diffusion domain on a non-local branching law: The connected components of X are all congruent. Particles may in particular arise only at points which are pictures under the given congruence of the point of death of their immediate ancestor ("*quasi-local branching law*"). Using this model for a neutron chain reaction, each connected component corresponds to one energy group. A fission produces particles in the higher energy groups, and the subsequent branching events result either in absorption, fission, or simply descent into a lower energy group, if there still is one.

Still fitting into the general framework is a modification which takes into account the phenomenon of retarded branching: A fission caused by a slow neutron (thermal energy) is not instantaneous but is delayed by an approximately exponentially distributed time. For simplicity, consider the one-group case. Add to the original set of types X a set Y' congruent to a subset X' of X containing all the points at which stopping can result in fission.

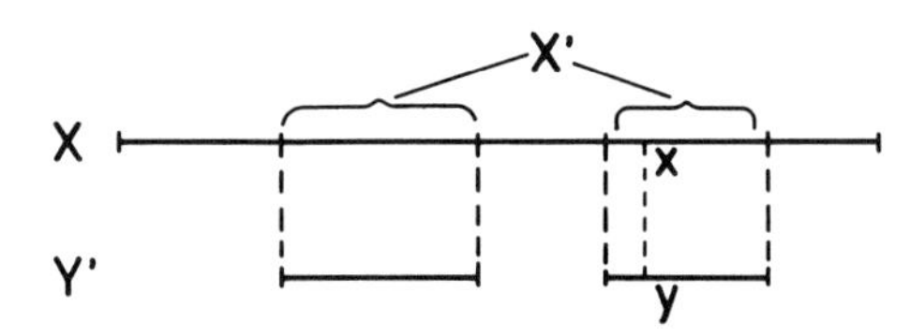

An event of retarded branching can now be modelled as composition of two ordinary branching events: A particle stopped at $x \in X'$ disappears and is replaced by a particle at the picture $y \in Y'$ of x. This particle stays at y for an exponentially distributed time, then disappear and is replaced by a population of particles in X, in the reactor example at x.

For biological applications it will in general be better to work with age-dependent branching diffusions. In the simplest case we have just one lifetime distribution, independent of the particles path, but not necessarily exponential, which is used instead of the density k to stop the motion. The set of types is now the product of the diffusion domain and the age-scale. It is obvious how to proceed to more elaborate models involving general lifetime distributions, possibly also dependent on the particle path.

Modelling the first and second order effects of the electron-photon cascade at moderate energies, we consider motion along the energy scale towards zero. The set of types is now $X = \{1,2,3\} \times [0,\infty)$, where $1,2,3$ represent photon, negatron, and positron, or since all energy values occuring in the cascade are at most equal to the initial value, $X = \{1,2,3\} \times [0,C^*]$ with a finite E^*. The process is characterized as follows:

(1) A photon of energy E at least equal to twice the rest energy E_r of an electron stays at this energy level for an exponentially distributed time, after which it is converted into a negatron-positron pair with kinetic energies $(E-E_r)u$ and $(E-E_r)(1-u)$, where u is distributed on $[0,1]$ with density $p_1(u)$. A photon of energy $E > 2E_r$ cannot be converted.

(2) A negatron moves along the energy scale, continuously loosing kinetic energy (ionization loss) according to $dE = -\alpha k(E)dt$, where $k(E)$ is a positive function tending to zero as E tends to zero. A negatron of kinetic energy E is stopped with density $\beta k(E)$, and the results are a photon with energy Eu and a negatron with kinetic energy $E(1-u)$, where u is distributed on $[0,1]$ with density $p_2(u)$.

(3) A positron continuously looses kinetic energy as does a negatron and is stopped with density $\gamma k(E)$, whereupon it is replaced with probability δ by a photon and a positron, with energy distributions as in the case of the negatron, and with probability $(1-\delta)$ by a pair of photons with energies $(E+E_r)u$, $(E+E_r)(1-u)$, where u is distributed on $[0,1]$ with density $p_3(u)$.

BIBLIOGRAPHICAL NOTES

Most of the models and examples given are standard, see Harris (1963), Ikeda, Nagasawa, and Watanabe (1968,1969), and Jagers (1975). an exception is the retarded branching model, cf.Asmussen and Hering (1977).

PART B

SIMPLE BRANCHING PROCESSES

CHAPTER II

THE GALTON-WATSON PROCESS: PROBABILISTIC METHODS

1. INTRODUCTION

Let $\{Z_N\}_{N\in\mathbb{N}}$ be a Galton-Watson process with offspring distribution F. That is (with the usual convention $\Sigma_1^0 = 0$),

$$(1.1) \qquad Z_{N+1} = \sum_{i=1}^{Z_N} X_{N,i}$$

where the $X_{N,i}$ $(N = 0,1,\ldots,i = 1,2,\ldots)$ are i.i.d. with common distribution F. Being concentrated on $\mathbb{N}$, F is specified by the point probabilities $p_0, p_1, \ldots$ at $0,1,\ldots$.

Formally $\{Z_N\}$ is a time-homogeneous Markov chain with state space $\mathbb{N}$. The transition probabilities can be expressed in terms of the convolution powers of F in an obvious manner. Similarly, the dependence on the initial distribution of F reduces simply to that given $Z_0 = m$, the process evolves (due to the independence of the different lines of descent) as the sum of m independent copies of a process with $Z_0 = 1$. In this manner, the offspring distribution becomes the main parameter of the process.

To avoid trivialities, assume throughout that F is non-degenerate (i.e. $p_k < 1$ for all k) and that $P(Z_0 = 0) < 1$. We use the notation

$$\mathfrak{F}_N := \sigma(X_{n,i};\ n=0,\ldots,N-1,\ i = 1,2,\ldots),\ \mathfrak{F}_0 := \sigma(Z_0),$$

$$m := EX_{n,i} = E(Z_1 | Z_0 = 1) = \int_0^\infty x dF(x) = \sum_{k=0}^{\infty} kp_k$$

and let $E := \{Z_N = 0 \text{ eventually}\}$ be the set of extinction, $q := P(E | Z_0 = 1)$ the extinction probability.

The usefulness of the Markov chain set-up seems quite limited, since the main result in that direction is

1.1. PROPOSITION. *The state 0 is absorbing and all other states $1,2,\ldots$ transient. That is, $Z_N \to \infty$ a.s. on E^c.*

PROOF. That 0 is absorbing is inherent in the definition. To prove transience, we must show that $r_k := P(Z_N = k \text{ for some } N = 1,2,\ldots | Z_0 = k) < 1$ for $k = 1,2,\ldots$ But

$$r_k \leq P(Z_1 > 0 | Z_0 = k) = 1 - p_0^k < 1 \quad \text{if} \quad p_0 > 0,$$

$$r_k = P(Z_1 = k | Z_0 = k) = p_1^k < 1 \quad \text{if} \quad p_0 = 0$$

since in the latter case $Z_0 \leq Z_1 \leq \cdots$ □

Clearly, $P(E|\mathfrak{F}_0) = q^{Z_0}$. Thus if $q = 1$, then $PE = 1$, while otherwise $P(Z_N \to \infty) = 1 - Eq^{Z_0} > 0$. It is clear that $q > 0$ if and only if $p_0 > 0$. The criterion determining whether $q = 1$ or $q < 1$ is

1.2. PROPOSITION. *If* $m \leq 1$, *then* $q = 1$, $P(Z_N \to \infty) = 0$ *while if* $1 < m \leq \infty$, *then* $q < 1$, $P(Z_N \to \infty) > 0$.

Various proofs will come out from the discussion below.

If the offspring distribution is a mixture of an atom at 0 and a geometric distribution on $\{1,2,\ldots\}$, i.e.

$$p_k = cb^{k-1} \quad k = 1,2,\ldots, \quad p_0 = 1 - p_1 - p_2 - \cdots \tag{1.2}$$

then it is possible to explicitly calculate the distribution of Z_N given $Z_0 = 1$, which is again of the form (1.2) (with $c = c_N$ and $b = b_N$ depending on N). This fact (and recursive formulas for the c_n, b_n) could, of course, be derived by elementary calculations with the point probabilities, using the fact that convolution powers of the geometric distribution are negative binomial. We omit the details since the analytic method of Chapter III is shorter and more elegant. No such explicit formulas are known in other non-trivial examples, e.g. of a Poisson offspring distribution. In Chapter III we shall, however, compute the generating functions $f_N(s) = E(s^{Z_N}|Z_0 = 1)$ and we give here some expressions for the first and second moments. It is obvious from (1.1) that (irrespective of whether the mean m or the variance σ^2 in the offspring distribution are finite)

$$E(Z_{N+1}|\mathfrak{F}_N) = mZ_N, \quad \mathrm{Var}(Z_{N+1}|\mathfrak{F}_N) = \sigma^2 Z_N, \tag{1.3}$$

$$E(Z_{N+1}|Z_0 = 1) = mE(Z_N|Z_0 = 1) = \cdots = m^{N+1}, \tag{1.4}$$

$$\sigma_{N+1}^2 := \mathrm{Var}(Z_{N+1}|Z_0 = 1) = \begin{cases} \dfrac{\sigma^2 m^{N+1}(m^{N+1}-1)}{m^2-m}, & m \neq 1 \\ (N+1)\sigma^2 & m = 1 \end{cases}, \tag{1.5}$$

$$E(Z_{N+k}|\mathfrak{F}_N) = m^k Z_N, \quad \mathrm{Var}(Z_{N+k}|\mathfrak{F}_N) = \sigma_k^2 Z_N. \tag{1.6}$$

Indeed, (1.5) follows by iterating

$$\sigma_{N+1}^2 = E\mathrm{Var}(Z_{N+1}|\mathfrak{F}_N) + \mathrm{Var}\,E(Z_{N+1}|\mathfrak{F}_N) = m^N\sigma^2 + m^2\sigma_N^2.$$

(1.4) gives the first proof of $q = 1$ if $m \leq 1$. Indeed, if $P(Z_N \to \infty | Z_0 = 1) > 0$, then $m^N = E(Z_N|Z_0 = 1) \to \infty$ so that $m > 1$. If $m < 1$, then $q = P(Z_N \to 0|Z_0 = 1) = 1$ could also be derived from A 1.4 and $\Sigma_0^\infty E(Z_N|Z_0 = 1) < \infty$.

Combining with the martingale convergence theorem A 2.2, we get from (1.6)

1.3. PROPOSITION. *If* $m < \infty$, *then* $\{W_N\} := \{Z_N/m^N\}$ *is a non-negative martingale w.r.t.* $\{\mathfrak{F}_N\}$. *Hence a.s.* $W := \lim_{N\to\infty} W_N$ *exists,* $0 \leq W < \infty$ *and* $E(W|\mathfrak{F}_0) \leq W_0$, $E(W|Z_0 = 1) \leq 1$.

This result seems to have its main interest in the *supercritical* case $1 < m < \infty$. As an application to also the *subcritical case* $m < 1$ and the *critical case* $m = 1$ we can, however, prove once more that $q = 1$ if $m \leq 1$. Indeed, since then m^{-N} is bounded away from 0, Z_N/m^N cannot have a finite limit on the set $\{Z_N \to \infty\}$ which hence has probability 0.

It follows that $P(W > 0) = P(0 < W < \infty) > 0$ can only occur in the supercritical case. It is natural to ask whether here

$$\{W > 0\} = \{Z_N \to \infty\} = E^c \quad \text{a.s.,} \tag{1.7}$$

i.e. whether m^N is the proper normalization of Z_N on E^c. Indeed this is not always the case and this phenomenon motivates some of the investigations of the next sections. We give here some preliminary estimates:

1.4. PROPOSITION. *Either* $W = 0$ *a.s. or* (1.7) *holds.*

1.5. PROPOSITION. *If* $1 < m < \infty$, $\sigma^2 < \infty$, *then* (1.7) *holds and furthermore* $W_N \overset{\mathcal{L}^2,\mathcal{L}^1}{\to} W$, $E(W|Z_0 = 1) = 1$ *and* $\mathrm{Var}(W|Z_0 = 1) = \sigma^2/(m^2-m)$.

1.6. COROLLARY. *If* $1 < m \leq \infty$, *then* $q < 1$ *and for any* $m_1 < m$, $\lim_{N\to\infty} Z_N/m_1^N = \infty$ *a.s. on* E^c.

In the proof of 1.4, we need

1.7. LEMMA. Let $W_{N,i}$ be the W-variable associated with the Galton-Watson process initiated by the ith individual alive at time N. Then the $W_{N,i}$ $(i = 1,\dots,Z_N)$ are independent conditionally upon $\mathfrak{F}_N$, $P(W_{N,i} \leq w|\mathfrak{F}_N) = P(W \leq w|Z_0 = 1)$ and a.s.

$$W = m^{-N} \sum_{i=1}^{Z_N} W_{N,i}. \tag{1.8}$$

The proof is an immediate consequence of the branching property, viz. the independence of the $i=1,\dots,Z_N$ lines of descent starting at time N. Letting $Z_{N,i,k}$ be the population size in line i at time $N+k$, (1.8) follows from $Z_{N+k} = \Sigma_1^{Z_N} Z_{N,i,k}$, $Z_{N,i,k}/m^k \to W_{N,i}$ as $k \to \infty$.

PROOF OF 1.4. Suppose $\tilde{q} := P(W = 0|Z_0 = 1) < 1$. Then by A 2.5 and 1.7, we have a.s.

$$\begin{aligned} I(W = 0) &= \lim_{N\to\infty} P(W = 0|\mathfrak{F}_N) \\ &= \lim_{N\to\infty} P(W_{N,i} = 0 \ i = 1,\dots,Z_N|\mathfrak{F}_N) = \lim_{N\to\infty} \tilde{q}^{Z_N} \\ &= \lim_{N\to\infty}\{\tilde{q}^{Z_N} I(Z_N \to \infty) + \tilde{q}^{Z_N} I(E)\} = 0 + I(E). \qquad \square \end{aligned}$$

PROOF OF 1.5. Suppose $Z_0 = 1$. Then by (1.5), $\sup_N \operatorname{Var} W_N = \sigma^2/(m^2-m) < \infty$ and the convergence theorem for $\mathcal{L}^2$-bounded martingales yields the $\mathcal{L}^2$-statements as well as $EW = 1$. But $EW = 1$ clearly excludes $W = 0$ a.s. so that (1.7) must hold. $\square$

If $1 < m < \infty$, $\sigma^2 < \infty$ one could obtain the existence of W and parts of 1.5 without reference to martingales, by elementary calculations using (1.5), (1.6) to show that $\{W_N\}$ is Cauchy in $\mathcal{L}^2$. We omit the details.

PROOF OF 1.6. Let a be chosen so large that $\tilde{m} := E(X_{N,i} \wedge a) > m_1$, $\tilde{m} > 1$, let k be fixed for a while and define

$$\tilde{Z}_{N+1} := \begin{cases} Z_{N+1} & N < k \\ \sum_{i=1}^{\tilde{Z}_N} X_{N,i} \wedge a & N \geq k \end{cases},$$

Let $\tilde{F}$ be the distribution of $X_{N,i} \wedge a$ (i.e. $\tilde{F}(x) = F(x)$ $x < a$, $\tilde{F}(x) = 1$ $x \geq a$) and $\tilde{q}$ the extinction probability in a Galton-Watson process with offspring distribution $\tilde{F}$. Since $\tilde{F}$ has finite variance and $\{\tilde{Z}_{N+k}\}_{N\in\mathbb{N}}$ is a Galton-Watson process with offspring distribution $\tilde{F}$, it follows from 1.5 that

$$(1.9) \qquad \lim_{N\to\infty} \tilde{Z}_N/\tilde{m}^N > 0 \quad \text{a.s.}$$

on $A_k := \{\tilde{Z}_N \to \infty\}$. As k increases, so does $\tilde{Z}_N$ and hence A_k. Furthermore

$$PA_k = E(1-\tilde{q}^{Z_k}) = E(1-\tilde{q}^{Z_k})I(E) + E(1-\tilde{q}^{Z_k})I(Z_N \to \infty)$$

tends to $P(Z_N \to \infty)$ as $k \to \infty$ since $\tilde{q} < 1$. Since $A_k \subseteq \{Z_N \to \infty\}$, (1.9) must hold on $\{Z_N \to \infty\}$. Clearly, (1.9) implies $Z_N/m_1^N \to \infty$ and $q < 1$. □

The argument is of a model to be frequently used, i.e. to use truncation to approximate by $\mathcal{L}^2$-bounded quantities (here the $\tilde{Z}_N$), to which elementary methods apply.

2. THE KESTEN-STIGUM THEOREM

We consider from now on in Sections 2-5 the supercritical case $1 < m < \infty$. In order to refine and make more precise the estimates of the growth of Z_n on $\{Z_n \to \infty\}$, which were obtained in Section 1, the first obvious question is when $W = \lim Z_n/m^n$ is non-zero, cf. 1.4. The answer is provided by the theorem of Kesten and Stigum (1966a) which asserts that $P(W > 0) > 0$ if and only if

$$\text{(X LOG X)} \qquad \int_0^\infty x \log^+ x \, dF(x) = \sum_{k=2}^\infty k \log k \, p_k < \infty$$

while otherwise $W = 0$ a.s. The scope of this result is even somewhat wider than just to deal with the non-degeneracy of W. In fact, we have:

2.1. THEOREM. *Suppose* $Z_0 = 1$. *Then the following assertions are mutually equivalent and equivalent to* (X LOG X):

(2.1) $P(W > 0) > 0$

(2.2) $EW = 1$

(2.3) $W_n \xrightarrow{\mathcal{L}^1} W$, i.e., $E|W-W_n| \to 0$

(2.4) The sequence $\{W_n\}_{n\in\mathbb{N}}$ is uniformly integrable.

(2.5) $E \sup_n W_n < \infty$

The hardest parts of this theorem turn out to be (X LOG X) $\Rightarrow$ (2.2) and (2.1) $\Rightarrow$ (X LOG X). Also the establishment of (2.5) requires a non-trivial argument. For the rest of the theorem, we note that (2.2), (2.3) and (2.4) are equivalent by general results on uniform integrability A 3.3 and that the implications (2.2) $\Rightarrow$ (2.1) and (2.5) $\Rightarrow$ (2.4) are obvious.

In the present chapter we shall give three (somewhat related) probabilistic proofs of the Kesten-Stigum theorem, while an analytic proof comes out as a special case of Chapter III.

The first proof is based on the expansion $W = 1 + \Sigma_0^\infty \{W_{n+1} - W_n\}$ of W in a sum of martingale increments

$$W_{n+1} - W_n = \frac{1}{m^{n+1}} \sum_{i=1}^{Z_n} \{X_{n,i} - m\} \, .$$

We approximate $W_{n+1} - W_n$ by a $\mathcal{L}^2$-bounded martingale increment

$\widetilde{W}_{n+1}-W_n+R_n$ defined by truncating the $X_{n,i}$ at m^n,

$$\widetilde{W}_{n+1} := \frac{1}{m^{n+1}} \sum_{i=1}^{Z_n} X_{n,i} I(X_{n,i} \leq m^n)$$

$$R_n := E(W_n-\widetilde{W}_{n+1}|\mathfrak{F}_n) = E(W_{n+1}-\widetilde{W}_{n+1}|\mathfrak{F}_n)$$

$$= E(\frac{1}{m^{n+1}} \sum_{i=1}^{Z_n} X_{n,i} I(X_{n,i} > m^n)|\mathfrak{F}_n) = \frac{W_n}{m} \int_{m^n}^{\infty} x \, dF(x).$$

We first compute three series similar to those of Kolmogorov A 4.1:

2.2. LEMMA. Without conditions beyond $1 < m < \infty$,

$$\sum_{n=0}^{\infty} P(\widetilde{W}_{n+1} \neq W_{n+1}) < \infty, \quad \sum_{n=0}^{\infty} \text{Var}\{\widetilde{W}_{n+1}-W_n+R_n\} < \infty,$$

while $\Sigma_0^\infty ER_n = m^{-1}\Sigma_0^\infty \int_{m^n}^{\infty} x \, dF(x) < \infty$ if and only if (X LOG X) holds.

PROOF. We give the calculations in some detail, since similar computations come up in a number of other proofs. They read:

(2.6)
$$\sum_{n=0}^{\infty} P(\widetilde{W}_{n+1} \neq W_{n+1}) = \sum_{n=0}^{\infty} P(X_{n,i} > m^n \text{ for some } i = 1,\ldots,Z_n)$$

$$= \sum_{n=0}^{\infty} EP(X_{n,i} > m^n \text{ for some } i = 1,\ldots,Z_n|\mathfrak{F}_n) \leq \sum_{n=0}^{\infty} EZ_n \int_{m^n}^{\infty} dF(x)$$

$$= \sum_{n=0}^{\infty} m^n \int_{m^n}^{\infty} dF(x) = \int_0^{\infty} \sum_{n=0}^{\infty} m^n I(x > m^n) dF(x) = \int_0^{\infty} O(x) dF(x) < \infty,$$

(2.7)
$$\sum_{n=0}^{\infty} \text{Var}\{\widetilde{W}_{n+1}-W_n+R_n\} = \sum_{n=0}^{\infty} E \, \text{Var}(\widetilde{W}_{n+1}|\mathfrak{F}_n)$$

$$= \sum_{n=0}^{\infty} E \frac{Z_n}{m^{2n+2}} \text{Var} \, X_{n,i} I(X_{n,i} \leq m^n) \leq m^{-2} \sum_{n=0}^{\infty} m^{-n} \int_0^{m^n} x^2 dF(x)$$

$$= m^{-2} \int_0^{\infty} x^2 \sum_{n=0}^{\infty} m^{-n} I(x \leq m^n) dF(x) = \int_0^{\infty} x^2 O(\tfrac{1}{x}) dF(x) < \infty,$$

(2.8)
$$\sum_{n=0}^{\infty} \int_{m^n}^{\infty} x \, dF(x) = \int_0^{\infty} x \sum_{n=0}^{\infty} I(x > m^n) dF(x) = \int_0^{\infty} x \, O(\log x) dF(x). \quad \square$$

2.3. LEMMA. Without conditions beyond $1 < m < \infty$, $\Sigma_0^\infty\{\tilde{W}_{n+1}-W_n+R_n\}$ converges a.s. and in $\mathcal{L}^1$, and $\tilde{W}_{n+1} = W_{n+1}$ eventually.

PROOF. Combining with 2.2, the last assertion follows by the Borel-Cantelli lemma and the first by the martingale convergence theorem, since the term of the series $\Sigma_0^\infty\{\tilde{W}_{n+1}-W_n+R_n\}$ are martingale increments by definition and the martingale is $\mathcal{L}^2$-bounded by 2.2. □

PROOF OF (X LOG X) $\Rightarrow$ (2.2). Since $\Sigma_0^\infty ER_n < \infty$ and $R_n \geq 0$, $\Sigma_0^\infty R_n$ converges in $\mathcal{L}^1$. Thus $\Sigma_0^\infty\{\tilde{W}_{n+1}-W_n\}$ does so by 2.3 and hence $E(\Sigma_{N+1}^\infty\{\tilde{W}_{n+1}-W_n\}) \to 0$, $N \to \infty$. But clearly,

$$EW = E(1 + \sum_{n=0}^{N}\{W_{n+1}-W_n\} + \sum_{n=N+1}^{\infty}\{W_{n+1}-W_n\})$$

$$\geq 1 + 0 + E(\sum_{n=N+1}^{\infty}\{\tilde{W}_{n+1}-W_n\})$$

so that indeed $EW \geq 1$ and (by 1.3) $EW = 1$. □

PROOF OF (2.1) $\Rightarrow$ (X LOG X). Let $\underline{W} := \inf_n W_n$ and note that $P(\underline{W} > 0) = P(W > 0) > 0$. By 2.3, $\Sigma_0^\infty\{W_{n+1}-W_n+R_n\}$ converges a.s. and hence $\Sigma_0^\infty R_n$ does so, since $\Sigma_0^\infty\{W_{n+1}-W_n\}$ exists (viz. equals $W-1$). Thus a.s.

$$\infty > \sum_{n=0}^{\infty} R_n \geq m^{-1}\,\underline{W}\sum_{n=0}^{\infty}\int_{m^n}^{\infty} x\,dF(x)$$

implying that $\Sigma_0^\infty \int_{m^n}^\infty x\,dF(x) < \infty$ and hence (by 2.2)(X LOG X). □

We next give the second proof, which is more similar to the original one by Kesten and Stigum. The idea is here to approximate $\{Z_n\}_{n\in\mathbb{N}}$ by a time inhomogeneous $\mathcal{L}^2$-bounded Galton-Watson process $\{\tilde{Z}_n\}_{n\in\mathbb{N}}$ defined by

$$\tilde{Z}_0 := 1,\quad \tilde{Z}_{n+1} := \sum_{i=0}^{\tilde{Z}_n} X_{n,i}I(X_{n,i} \leq Am^n).$$

2.4. LEMMA. As $A\uparrow\infty$, $P(\tilde{Z}_n = Z_n$ for all $n)\uparrow 1$.

PROOF. Since $\tilde{Z}_n \leq Z_n$,

$$P(\tilde{Z}_n \neq Z_n \text{ for some } n) \leq \sum_{n=0}^{\infty} P(\tilde{Z}_n \neq Z_n) \leq \sum_{n=0}^{\infty} E\,\tilde{Z}_n\int_{Am^n}^{\infty} dF(x)$$

$$\leq \sum_{n=0}^{\infty} m^n \int_{Am^n}^{\infty} dF(x) = \int_A^\infty \sum_{n=0}^{\infty} m^n I(x > Am^n)\,dF(x) = \int_A^\infty O(\tfrac{x}{A})\,dF(x).$$

The r.h.s. clearly tends to zero as $A\uparrow\infty$. □

2.5. LEMMA. Define $\tilde{W}_n := \tilde{Z}_n/m^n$. Then $\tilde{W}_n$ is a $\mathcal{L}^2$-bounded supermartingale and hence there exists a random variable $\tilde{W}$ such that $0 \leq \tilde{W} < \infty$, $\tilde{W}_n \to \tilde{W}$ a.s. and in $\mathcal{L}^1$.

PROOF. The supermartingale property follows from

$$(2.9) \qquad E(\tilde{Z}_{n+1}|\mathfrak{F}_n) = \tilde{Z}_n \int_0^{Am^n} x\, dF(x) \leq m\, \tilde{Z}_n .$$

Furthermore,

$$\mathrm{Var}(\tilde{Z}_{n+1}|\mathfrak{F}_n) = \tilde{Z}_n \mathrm{Var}\, X_{n,i} I(X_{n,i} \leq Am^n) \leq \tilde{Z}_n \int_0^{Am^n} x^2 dF(x),$$

$$\mathrm{Var}\, \tilde{Z}_{n+1} = \mathrm{Var}\, E(\tilde{Z}_{n+1}|\mathfrak{F}_n) + E\, \mathrm{Var}(\tilde{Z}_{n+1}|\mathfrak{F}_n) \leq m^2 \mathrm{Var}\, \tilde{Z}_n + m^n \int_0^{Am^n} x^2 dF(x).$$

It follows by iteration and the calculation in (2.7) that

$$\sup_n \mathrm{Var}\, \tilde{W}_{n+1} \leq \sum_{n=0}^{\infty} m^{-n} \int_0^{Am^n} x^2 dF(x) < \infty,$$

proving $\mathcal{L}^2$-boundedness since $0 \leq E\, \tilde{W}_{n+1} \leq 1$. □

It follows from (2.9) and 2.5 that

$$(2.10) \qquad E\tilde{W}_{n+1} = E\tilde{W}_n m^{-1} \int_0^{Am^n} x\, dF(x) = E\tilde{W}_n (1 - m^{-1} \int_{Am^n}^{\infty} x\, dF(x)),$$

$$(2.11) \qquad E\tilde{W} = \lim_{n\to\infty} E\tilde{W}_{n+1} = \prod_{n=0}^{\infty} (1 - m^{-1} \int_{Am^n}^{\infty} x\, dF(x)).$$

PROOF OF (X LOG X) $\Longrightarrow$ (2.2). By 2.4 and monotone convergence, $E\tilde{W}\uparrow EW$ as $A\uparrow\infty$. Now consider the series

$$(2.12) \qquad \log E\tilde{W} = \sum_{n=0}^{\infty} \log(1 - m^{-1} \int_{Am^n}^{\infty} x\, dF(x)).$$

The nth term is bounded in absolute value by a constant (independent of $A \geq 1$ and n) times

$$\int_{Am^n}^{\infty} x\, dF(x) \leq \int_{m^n}^{\infty} x\, dF(x).$$

The sum of the r.h.s. is finite by (2.8). Letting $A\uparrow\infty$ in (2.12) and using dominated convergence, it follows that $\log E\tilde{W} \to 0$ and hence that $\log EW = 0$. □

PROOF OF (2.1) $\Longrightarrow$ (X LOG X). Since $\widetilde{W}\uparrow W$ as $A\uparrow\infty$, it suffices to prove that $\widetilde{W} = 0$ for any fixed A if (X LOG X) fails. But using (2.12), the inequality $\log(1-x) \leq -x$ and a calculation similar to (2.8), it follows that

$$\log E\widetilde{W} \leq -m^{-1} \sum_{n=0}^{\infty} \int_{Am^n}^{\infty} x \, dF(x) = \infty. \quad \square$$

The third probabilistic proof of the Kesten-Stigum theorem comes out in Section 5 in connection with a closer study of the case where $m < \infty$, but (X LOG X) fails. To complete the proof of 2.1, it remains to establish (2.5) under one of the equivalent conditions (X LOG X), (2.1), (2.2), (2.3), (2.4). This comes out easily from

2.6. LEMMA. *Suppose* (2.1) *holds and let* $M := \sup W_n$. *Then there exists constants* $A, B \in (0,\infty)$ *such that for all* $x \geq 1$ *and all* $n \in \mathbb{N}$,

$$P(W \geq Ax) \geq BP(M \geq x). \tag{2.13}$$

In fact, it follows upon integration by parts from (2.13) that

$$EM = \int_0^{\infty} P(M \geq x)dx \leq 1+B^{-1} \int_1^{\infty} P(W \geq Ax)dx \leq 1 + \frac{EW}{BA} < \infty .$$

PROOF OF 2.6. Let $W^{(1)}, W^{(2)}, \ldots$ be i.i.d. each distributed as W. Since $0 < EW < \infty$, it follows by the law of the large numbers that if $0 < A^* < EW$, then for some $B > 0$

$$P\left(\frac{W^{(1)}+\ldots+W^{(n)}}{n} \geq A^*\right) \geq B$$

for all $n = 1,2,\ldots$ Define for some fixed $x \geq 1$, $F_n := \{W_n \geq x, W_k < x, k = 0,1,\ldots,n-1\}$. Letting [ ] denote integer part, $[xm^n] \geq 1$ for all n and thus by 1.7,

$$P(W \geq Ax \mid F_n) \geq P\left(m^{-n} \sum_{i=1}^{[xm^n]} W_{n,i} \geq Ax\right)$$

$$\geq P\left(\sum_{i=1}^{[xm^n]} W_{n,i} \geq A[xm^n]\right) \geq B.$$

It follows that

$$BP(M \geq x) = B \sum_{n=0}^{\infty} PF_n \leq \sum_{n=0}^{\infty} P(W \geq Ax \mid F_n)PF_n \leq P(W \geq Ax). \quad \square$$

3. FINER LIMIT THEOREMS: FINITE OFFSPRING VARIANCE

The question of refining the basic results for the Galton-Watson process splits roughly in two. In the present section, the aim is to obtain the typical behaviour not hesitating imposing regularity conditions (the appropriate one turns out to be $\sigma^2 < \infty$). In the next section, the consequences of weakening the assumptions are then studied.

For simplicity, we assume throughout that $p_0 = 0$ so that $q = 0$.

A main problem is, of course, the rate of convergence of W_N to W. Here we have

3.1. THEOREM. *Suppose that* $1 < m < \infty$ *and that* $0 < \sigma^2 < \infty$ *and define* $\tau^2 := \text{Var}(W|Z_0 = 1) = \sigma^2/(m^2-m)$,

$$U_N := \left(\frac{m^N}{\tau^2 W_N}\right)^{1/2}(W-W_N) .$$

Then: (i) *For all* $y \in \mathbb{R}$,

$$\lim_{N\to\infty} P(U_N \leq y|\mathfrak{F}_N) = \Phi(y) \text{ a.s.} \tag{3.1}$$

In particular, the limiting distribution of U_N *exists and is standard normal.*

(ii) *It holds a.s. that*

$$\varlimsup_{N\to\infty} \frac{U_N}{(2 \log N)^{1/2}} = 1, \quad \varliminf_{N\to\infty} \frac{U_N}{(2 \log N)^{1/2}} = -1. \tag{3.2}$$

We first give a proof based upon the additivity property already exploited in 1.7, the set-up of which we now generalize:

3.2. PROPOSITION. *Let* $Y := Y(Z_0, Z_1, \ldots)$ *be some functional of the process and* $Y_{N,i}$ *the same functional evaluated in the Galton-Watson process initiated by the* i^{th} *individual at time* N. *Then the* $Y_{N,i}$ $(i = 1,\ldots,Z_N)$ *are independent conditionally upon* $\mathfrak{F}_N$, *with* $P(Y_{N,i} \leq y|\mathfrak{F}_N) = P(Y \leq y|Z_0 = 1) := F(y)$. *Define* $S_N := \Sigma_1^{Z_N} Y_{N,i}$ *and suppose that* $1 < m < \infty$, $E(Y|Z_0 = 1) = 0$, $0 < \tau^2 := \text{Var}(Y|Z_0 = 1) < \infty$. *Then*: (i) *For all* $y \in \mathbb{R}$

$$\lim_{N\to\infty} P\left(\frac{S_N}{(\tau^2 Z_N)^{1/2}} \leq y|\mathfrak{F}_N\right) = \Phi(y) \text{ a.s.} \tag{3.3}$$

In particular, the limiting distribution of $S_N/(\tau^2 Z_N)^{1/2}$ *exists and is standard normal.*

(ii) *It holds a.s. that*

$$(3.4)\qquad \underline{\lim}_{N\to\infty} \frac{S_N}{(2\tau^2 Z_N \log N)^{1/2}} \geq -1, \quad \overline{\lim}_{N\to\infty} \frac{S_N}{(2\tau^2 Z_N \log N)^{1/2}} \leq 1,$$

with the inequalities replaced by equalities if for some $k < \infty$ Y *depends on* $Z_0, \dots, Z_k$ *only* (i.e., is $\mathfrak{F}_k$-measurable).

3.3. REMARK. The normalizings in 3.1, 3.2 can be rewritten in a number of ways. E.g. Z_N could be replaced by $m^N W$.

The first CLT (3.3) in 3.2 is immediate from $Z_n \to \infty$, $F^{*n}(y\sqrt{\tau^2 n}) \to \Phi(y)$, so that

$$(3.5)\qquad P(S_N \leq y\sqrt{\tau^2 Z_N} \,|\, \mathfrak{F}_N) = F^{*Z_N}(y\sqrt{\tau^2 Z_N}) \to \Phi(y).$$

The second CLT follows by taking expectations in (3.5). Some more work is required for the

PROOF OF THE LIL IN 3.2. Write for notational convenience $EY_{N,i}$ instead of $E(Y_{N,i}|\mathfrak{F}_N)$ etc. and define

$$Y'_{n,i} := Y_{n,i} I(|Y_{n,i}| \leq m^{n/2}), \quad \tilde{Y}_{n,i} := Y'_{n,i} - EY'_{n,i},$$

$$\tilde{S}_n := \sum_{i=1}^{Z_n} \tilde{Y}_{n,i}, \quad \tilde{\omega}_n^2 := \mathrm{Var}(\tilde{S}_n|\mathfrak{F}_n), \quad T_n := \tilde{S}_n/\tilde{\omega}_n .$$

By a standard moment inequality,

$$E|\tilde{Y}_{n,i}|^3 \leq E|Y'_{n,i}|^3 + 3E|Y'_{n,i}|(E|Y'_{n,i}|)^2 + 3E|Y'_{n,i}|EY'^2_{n,i} + |Y'_{n,i}|^3$$

$$\leq 8E|Y'_{n,i}|^3 = 8\int_{|y|\leq m^{n/2}} |y|^3 dF(y).$$

Letting C be the Berry-Esseen constant, we get

$$\Delta_n := \sup_{y\in R} |P(T_n \leq y|\mathfrak{F}_n) - \Phi(y)| = 8C\tilde{\omega}_n^{-3} Z_n \int_{|y|\leq m^{n/2}} |y|^3 dF(y).$$

Since one can easily check that

$$(3.6)\qquad \lim_{n\to\infty} \tilde{\omega}_n^2/Z_n = \tau^2,$$

it follows that the assumption A(7.3) $\Sigma_0^\infty \Delta_n < \infty$ of A 7.2 holds since

$$\sum_{n=0}^{\infty} m^{-n/2} \int_{|y|\leq m^{n/2}} |y|^3 dF(y) = \int_{-\infty}^{\infty} |y|^3 \Sigma m^{-n/2} I(|y| \leq m^{n/2}) dF(y)$$

$$\leq \int_{-\infty}^{\infty} |y|^3 O(|y|^{-1}) dF(y) = \int_{-\infty}^{\infty} O(y^2) dF(y) < \infty.$$

Thus by A 7.2 it suffices to verify

$$\overline{\lim_{n\to\infty}}\, S_n/(2\tau^2 Z_n \log n)^{1/2} = \overline{\lim_{n\to\infty}}\, T_n/(2 \log n)^{1/2} .$$

Recalling (3.6) and the explicit definitions of $Y'_{n,i}$, $\tilde{Y}_{n,i}$, T_n it suffices that

$$\sum_{i=1}^{Z_n} \{Y_{n,i} - Y'_{n,i}\} = o(m^{n/2} (\log n)^{1/2}),$$

$$\sum_{i=1}^{Z_n} EY'_{n,i} = o(m^{n/2} (\log n)^{1/2}),$$

or, appealing to Kronecker's lemma A(5.2) that

$$\sum_{n=0}^{\infty} m^{-n/2} (\log n)^{-1/2} \sum_{i=1}^{Z_n} |Y_{n,i} - Y'_{n,i}| < \infty,$$

(3.7)

$$\sum_{n=0}^{\infty} m^{-n/2} (\log n)^{-1/2} \sum_{i=1}^{Z_n} |EY'_{n,i}| < \infty.$$

Noting that $Z_n = O(m^n)$ and that

$$|EY'_{n,i}| = |E(Y'_{n,i} - Y_{n,i})| \leq E|Y'_{n,i} - Y_{n,i}|$$

$$= E|Y_{n,i}| I(|Y_{n,i}| > m^{n/2}) = \int_{|y|>m^{n/2}} |y| dF(y),$$

it suffices for both assertions of (3.7) that

$$(3.8) \qquad \sum_{n=0}^{\infty} m^{-n/2} (\log n)^{-1/2} m^n \int_{|y|>m^{n/2}} |y| dF(y) < \infty$$

(for the first, take the mean). And (3.8) certainly holds since even

$$\sum_{n=0}^{\infty} m^{n/2} \int_{|y|>m^{n/2}} |y| dF(y) = \int_{-\infty}^{\infty} |y| \Sigma m^{n/2} I(|y|>m^{n/2}) dF(y) = \int_{-\infty}^{\infty} O(y^2) dF(y) < \infty. \quad \square$$

PROOF OF 3.1. Letting $Y := W-1$ in 3.2, it follows from (1.6) that $S_n = m^n(W-W_n)$. Therefore the central limit theorem in 3.1 and the $\overline{\lim} \geq -1$, $\underline{\lim} \leq 1$ parts of the law of the iterated logarithm follow at once from 3.2. Let next $Y := W_k - 1$, $\tau_k^2 := \text{Var } W_k$. Then $S_n = m^n(W_{n+k}-W_n)$,

$$\frac{m^n(W-W_n)}{(2\tau^2 Z_n \log n)^{1/2}} = \frac{m^{n+k}(W-W_{n+k})}{(2\tau^2 Z_{n+k}\log(n+k))^{1/2}} \cdot \left(\frac{Z_{n+k}\log(n+k)}{Z_n \log n}\right)^{1/2} m^{-k}$$

$$+ \frac{S_n}{(2\tau_k^2 Z_n \log n)^{1/2}} \cdot \frac{\tau_k}{\tau}$$

so that the $\overline{\lim}$ in 3.1 is at least

$$-\lim_{n\to\infty}\left(\frac{Z_{n+k}\log(n+k)}{Z_n \log n}\right)^{1/2} m^{-k} + \frac{\tau_k}{\tau} = -m^{-k/2} + \frac{\tau_k}{\tau}$$

which tends to one as $k \to \infty$. $\underline{\lim} \leq -1$ is proved in a similar manner.

□

As another example of just the same type, we shall consider the obvious empirical estimator

$$\hat{m}_N := \frac{Z_1 + \ldots + Z_N}{Z_0 + \ldots + Z_{N-1}}$$

of m based upon all parent-offspring combinations in the first N generations (in fact, $\hat{m}_N$ is known to be the maximum likelihood estimator of m in the statistical model specified by all offspring distributions).

3.4. PROPOSITION. (a) *As* $N \to \infty$, $\hat{m}_N \to m$ a.s.; (b) *The limiting distribution as* $N \to \infty$ *of* $[W(1+m+\ldots+m^{N-1})]^{1/2}(\hat{m}_N - m)$ *is normal with mean zero and variance* σ^2; (c)

$$\overline{\lim_{N\to\infty}}\left[\frac{W(1+m+\ldots+m^{N-1})}{2\sigma^2 \log N}\right]^{1/2}(\hat{m}_N - m) = 1 \text{ a.s.}, \quad \underline{\lim} = -1.$$

PROOF. Part (a) is an easy consequence of $Z_n/m^n \to W \in (0,\infty)$ a.s. For parts (b), (c), define

$$U_N := \Sigma_0^{N-1}\{Z_{n+1} - mZ_n\} = (Z_0 + \ldots + Z_{N-1})(\hat{m}_N - m),$$

$$c_N^2 := \text{Var}(U_N | Z_0 = 1) = \sigma^2(1+m+\ldots+m^{N-1}).$$

Then part (b) is equivalent to that $U_N/W^{1/2}c_N$ is asymptotically standard normal. Let $Y := U_k$ in 3.2. Then $U_{N+k} = U_N+S_N$ and by 3.2, $S_N/(Wm^N c_k^2)^{1/2}$ is asymptotically standard normal. Hence since $c_k^2 m^N/c_{N+k}^2 \to 1-m^{-k}$,

$$P(S_N/(Wc_{N+k}^2)^{1/2} \leq y) \to \Phi(y/(1-m^{-k})^{1/2})$$

as $N \to \infty$. Furthermore, using Chebycheff's inequality,

$$P(U_N/(Wc_{N+k}^2)^{1/2} < -\varepsilon) \leq P(W < w)+P(U_N/(wc_{N+k}^2)^{1/2} < -\varepsilon)$$

$$\leq P(W < w) + \frac{c_N^2}{\varepsilon^2 wc_{N+k}^2} .$$

Given $\delta > 0$, we can choose w so small and k so large that this is $< \delta$ for all N. Then

$$\overline{\lim_{N\to\infty}} P(U_N/(Wc_N^2)^{1/2} \leq y) = \overline{\lim_{N\to\infty}} P(U_{N+k}/(Wc_{N+k}^2)^{1/2} \leq y)$$

$$\leq \overline{\lim_{N\to\infty}} \{P(U_N/(Wc_{N+k}^2)^{1/2} < -\varepsilon) + P(S_N/(Wc_{N+k}^2)^{1/2} \leq y+\varepsilon)\}$$

$$\leq \delta+\Phi((y+\varepsilon)/(1-m^{-k})^{1/2}).$$

Letting $k \to \infty$, $\delta \downarrow 0$ and $\varepsilon \downarrow 0$, it follows that the $\overline{\lim}$ is at most $\Phi(y)$. $\underline{\lim} \geq$ is obtained in a similar manner, proving (b).

For part (c), letting $Y := Z_1-m$ yields $S_n = Z_{n+1}-mZ_n$ and it follows that

$$\sup_n |Z_{n+1}-mZ_n|/(m^n \log n)^{1/2} < \infty \text{ a.s.},$$

$$(3.9) \qquad U_N \leq \sum_{n=0}^{N-1} \frac{|Z_{n+1}-mZ_n|}{(m^n \log n)^{1/2}}(m^n \log n)^{1/2} = O([m^N \log N]^{1/2}).$$

Letting $Y := U_k$ as in (b) and $A_N := U_N/(2c_N^2 W \log N)^{1/2}$, it follows from $U_{N+k} = U_N+S_N$ that

$$\overline{\lim_{N\to\infty}} A_{N+k} \leq \overline{\lim_{N\to\infty}} \frac{U_N}{(2c_{N+k}^2 W \log(N+k))^{1/2}} + \overline{\lim_{N\to\infty}} \frac{S_N}{(2c_{N+k}^2 W \log(N+k))^{1/2}} .$$

By (3.9), the first term is $O(m^{-k/2})$ as $k \to \infty$, and by 3.2, the second is

$$\overline{\lim_{N\to\infty}} \frac{S_N}{(2c_k^2 W m^N \log N)^{1/2}} \left(\frac{c_k^2 m^N}{c_{N+k}^2}\right)^{1/2} = 1\cdot(1-m^{-k})^{1/2}.$$

Similarly

$$\overline{\lim_{N\to\infty}} A_{N+k} \geq -O(m^{-k/2})+(1-m^{-k})^{1/2}$$

so that $\overline{\lim}\ A_N = 1$, which is equivalent to the lim sup-part of (c). The $\underline{\lim}$-part is similar.

Though useful and widely applicable, 3.2 does not cover all CLT's coming up. As an example, consider a weighted sum of the form

$$U_N := \sum_{n=0}^{N-1} a_{N-n}\{Z_{n+1}-mZ_n\}.$$

Using 3.2 with $Y := Z_1-m$, it follows that the nth term is asymptotically normal conditionally upon $\mathfrak{F}_n$ with variance $\sigma_{N,n}^2 := a_{N-n}^2 Z_n \sigma^2$. Hence one might expect U_N to be asymptotically normal with variance

$$(3.10) \qquad \sum_{n=0}^{N-1} \sigma_{N,n}^2 = \sigma^2 \sum_{n=1}^{N} Z_{N-n} a_n^2 \cong W\tau_N^2 := \sigma^2 W m^N \sum_{n=1}^{N} m^{-n} a_n^2$$

provided that the contribution of the nth term to U_N becomes negligible (as $N \to \infty$ with n fixed), i.e., that $\sigma_{N,n}^2 = o(\tau_N^2)$ for n fixed, which is equivalent to

$$(3.11) \qquad a_N^2 = o(m^N \sum_{n=1}^{N} m^{-n} a_n^2).$$

Now if the last k terms in U_N dominate for large k in the sense that

$$(3.12) \qquad \lim_{k\to\infty} \underline{\lim_{N\to\infty}} \frac{\sum_{n=N-k}^{N-1} \sigma_{N,n}^2}{\tau_N^2} = \lim_{k\to\infty} \underline{\lim_{N\to\infty}} \frac{\sum_{n=1}^{k} m^{-n} a_n^2}{\sum_{n=1}^{N} m^{-n} a_n^2} = 1$$

(which implies (3.11), then just the same argument as in the proof of 3.4 (where $a_n = 1$) yields the CLT for $U_N/W^{1/2}\tau_N$. But if, e.g., the terms in U_N are of equal magnitude, for example if $a_n = m^{n/2}$, then (3.11) holds but not (3.12), and another approach is required. This example may be of limited intrinsic interest, but relevant examples along the same lines will come up in Chapter VIII. We specialize to the present situation the main tool used there, which is essentially an adaptation of the martingale CLT A 8.1 to martingale difference triangular arrays indexed by the total set of all individuals ever alive:

3.5. THEOREM. *Let (for each fixed* N*)* $X^N_{n,i}$ *be r.v. such that* $X^N_{n,i}$ *is* $\mathfrak{F}_{n+1}$*-measurable and that conditionally upon* $\mathfrak{F}_n$ *the* $X^N_{n,i}$ $(i = 1,\dots,Z_n)$ *are independent with* $E(X^N_{n,i}|\mathfrak{F}_n) = 0$. *Define*

$$S_N := \sum_{n=0}^{\infty}\sum_{i=1}^{Z_n} X^N_{n,i},\quad s_N^2 := \sum_{n=0}^{\infty}\sum_{i=1}^{Z_n}\operatorname{Var}(X^N_{n,i}|\mathfrak{F}_n)$$

and suppose that S_N *converges in* $\mathcal{L}^2$*, i.e., that* $Es_N^2 < \infty$*, that* $s_N^2 \xrightarrow{P} W$ *and that the conditional Lindeberg condition*

$$(3.13)\qquad L_N := \sum_{n=0}^{\infty}\sum_{i=1}^{Z_n} E(X^{N\,2}_{n,i}I(|X^N_{n,i}| > \epsilon)|\mathfrak{F}_n) \xrightarrow{P} 0 \quad \forall\, \epsilon > 0$$

holds. Then the limiting distribution of $S_N/W^{1/2}$ *exists and is standard normal.*

We defer the proof to Chapter VIII, since no simplifications occur in the present situation. The above martingale approach will prove useful also for more complicated LIL's.

3.6. EXAMPLES. Consider first the CLT for $W-W_N$ in 3.1. Let $\tau^2 := \sigma^2/(m^2-m)$ and

$$S_N := \tau^{-1}m^{N/2}(W-W_N) = \tau^{-1}m^{N/2}\sum_{n=N}^{\infty}\{W_{n+1}-W_n\}$$

$$= \tau^{-1}\sum_{n=N}^{\infty}\sum_{i=1}^{Z_n}\frac{m^{N/2}}{m^{n+1}}\{X_{n,i}-m\}.$$

This is of the form in 3.5, with $X^N_{n,i} := m^{N/2}\{X_{n,i}-m\}/m^{n+1}$ for $n \geq N$ and $X^N_{n,i} := 0$ otherwise. Clearly a.s.,

$$s_N^2 = \tau^{-2}\sum_{n=N}^{\infty} Z_n\frac{m^N}{m^{2n+2}}\sigma^2 = (1-\frac{1}{m})m^N\sum_{n=N}^{\infty}W_n m^{-n} \to W.$$

If $\widetilde{F}$ is the common distribution of the $X_{n,i}-m$, the Lindeberg condition reduces to

$$\sum_{n=N}^{\infty} Z_n\frac{m^N}{m^{2n}}\int_{|x|>\epsilon m^n/m^{N/2}} x^2 d\widetilde{F}(x) \xrightarrow{P} 0 \quad \forall\, \epsilon > 0,$$

which holds because of

$$\sum_{n=N}^{\infty} m^{N-n} \int_{|x|>\epsilon m^{n-N/2}} x^2 d\tilde{F}(x) \leq \int_{|x|>\epsilon m^{N/2}} x^2 d\tilde{F}(x) \sum_{n=0}^{\infty} m^{-n} \to 0.$$

We next verify the conjecture above that the condition (3.11) is sufficient for the asymptotic normality of $W^{-1/2}S_N$ where

$$S_N = \tau_N^{-1} U_N = \sum_{n=0}^{N-1} \sum_{i=1}^{Z_n} \frac{a_{N-n}}{\tau_N}\{X_{n,i}-m\},$$

(as pointed out above, the CLT for $\hat{m}_N - m$ in 3.4 is a special case of this). Here

$$s_N^2 = \sum_{n=0}^{N-1} Z_n \frac{a_{N-n}^2}{\tau_N^2} \sigma^2 = m^{-N} \frac{\sum_{n=1}^{N} a_n^2 Z_{N-n}}{\sum_{n=1}^{N} m^{-n} a_n^2} = \frac{\sum_{n=1}^{N} m^{-n} a_n^2 W_{N-n}}{\sum_{n=1}^{N} m^{-n} a_n^2}$$

tends a.s. to W since (3.11) makes it permissible to approximate W_{N-n} by W. The quantity to be inspected in the Lindeberg condition becomes

$$(3.14) \qquad L_N' := \sum_{n=0}^{N-1} Z_n \frac{a_{N-n}^2}{\tau_N^2} \int_{|x|>\epsilon\tau_N/|a_{N-n}|} x^2 d\tilde{F}(x).$$

Now by (3.11), $a_N^2 \leq \delta^2\tau_N^2$ when $N \geq N_1$ and hence, since $\tau_N^2 \uparrow$, $a_{N-n}^2 \leq \delta^2\tau_N^2$ when $N-n \geq N_1$. Since $\tau_N^2 \uparrow \infty$, also $a_k^2 \leq \delta^2\tau_N^2$ for $k = 1,\ldots,N_1$ when $N \geq k_2$. Hence all $a_{N-n}^2 \leq \delta^2\tau_N^2$ when $N \geq k_2$ so that

$$EL_N' \leq \int_{|x|>\epsilon/\delta} x^2 d\tilde{F}(x) \sum_{n=0}^{N-1} m^n \frac{a_{N-n}^2}{\tau_N^2} = \int_{|x|>\epsilon/\delta} x^2 d\tilde{F}(x).$$

As first $N \to \infty$ and next $\delta \downarrow 0$, it follows that indeed $EL_N' \to 0$, $L_N \xrightarrow{P} 0$ [we take this opportunity to point out that the verification of the Lindeberg condition in Asmussen and Keiding (1977) Th.2.2 is in error and should be replaced by the above argument].

4. FINER LIMIT THEOREMS: INFINITE OFFSPRING VARIANCE

We assume throughout $q = 0 = p_0$, $1 < m < \infty$ and that (X LOG X) holds so that $W > 0$ a.s. We shall consider mainly two problems, strong laws for $W-W_N$ (i.e., a.s. estimates) and the relation between properties of the offspring distribution F and the distribution of W. We remark, however, that results concerning the last of these problems may lead to weak laws for $W-W_N$, using the decomposition 1.7 in ways related to the proof of the CLT in Section 3, but we shall not go into this deeply.

In the case of finite variance, the relevant strong law for $W-W_N$ is the LIL of Section 3. In the strong law of large numbers one could think of the usual LIL as a convergence rate result, and a counterpart for infinite variance is then the Marcinkiewicz - Zygmund version (Neveu (1965, pp.152-155) stating that the sample mean tends to the mean at a certain rate if and only if an associated moment exist. We shall give some results of similar type:

4.1. THEOREM.
(i) *Let* $1 < p < 2$, $1/p + 1/q = 1$. *Then* $W-W_n = o(m^{-n/q})$ *if and only if*

$$\int_0^\infty x^p \, dF(x) < \infty.$$

(ii) *Let* $\alpha > 0$. *Then* $W-W_n = o(n^{-\alpha})$ *if and only if*

$$\int_y^\infty x[\log x - \log y] dF(x) = o([\log y]^{-\alpha}). \tag{4.1}$$

(iii) *Let* $\alpha > 0$. *Then* $\Sigma_{n=0}^\infty n^{\alpha-1}\{W-W_n\}$ *converges if and only if* $\mu_{\alpha+1} < \infty$, *where*

$$\mu_\beta := \int_0^\infty x[\log^+ x]^\beta dF(x)$$

[to get a feeling for (4.1), note that

$$\mu_{\alpha+1} < \infty \Longrightarrow \int_y^\infty x \log x \, dF(x) = o([\log y]^{-\alpha}) \Longrightarrow (4.1) \tag{4.2}$$

$$\Longrightarrow \mu_{\alpha+1-\varepsilon} < \infty \; \forall \, \varepsilon > 0$$

as is easily seen upon integration by parts].

To estimate $W-W_N$ we write $W-W_N = \Sigma_N^\infty \alpha_n$ where $\alpha_n := W_{n+1}-W_n$ and use

4.2. LEMMA. Let $\{\alpha_n\}$, $\{\beta_n\}$ be series of real numbers such that $0 < \beta_n \uparrow \infty$. Then

$$(4.3) \qquad \sum_{n=0}^{\infty} \alpha_n\beta_n \text{ converges} \implies \sum_{n=N}^{\infty} \alpha_n = o(1/\beta_N).$$

This is a tail sum analogue of Kronecker's lemma A (5.2) and follows in a similar manner from Abel's Lemma A (5.1).

The proof of 4.1 has a number of features in common with the first proof of the Kesten-Stigum theorem in Section 2.

We first consider part (ii), the proof of which is particularly well suited to demonstrate the ideas. We let $\beta_n := n^\alpha$ and instead of studying $\Sigma\, \alpha_n\beta_n = \Sigma\, n^\alpha\{W_{n+1}-W_n\}$, we approximate by $\Sigma\, n^\alpha\{\tilde{W}_{n+1}-W_n+R_n\}$ defined as in Section 2, only with $X_{n,i}$ truncated at m^n/n^α rather than m^n. Calculations similar to (2.6), (2.7) yield

$$\sum_{n=0}^{\infty} P(\tilde{W}_{n+1} \neq W_{n+1}) \simeq \sum_{n=0}^{\infty} \mathrm{Var}[n^\alpha\{\tilde{W}_{n+1}-W_n+R_n\}] \simeq \mu_\alpha$$

and as in the proof of Lemma 2.2, we have immediately

4.3. LEMMA. Let $\alpha > 0$ and suppose $\mu_\alpha < \infty$. Then $\Sigma\, n^\alpha\{\tilde{W}_{n+1}-W_n+R_n\}$, $\Sigma\, n^\alpha\{W_{n+1}-W_n+R_n\}$ converges a.s.

PROOF OF (ii). Suppose first $\mu_\alpha < \infty$ (which is substantially weaker than (4.1), cf. (4.2). Combining 4.2 and 4.3 yields

$$o(N^{-\alpha}) = \sum_{n=N}^{\infty} \{W_{n+1}-W_n+R_n\} = W-W_N + \sum_{n=N}^{\infty} R_n.$$

Therefore $W-W_N = o(N^{-\alpha})$ is equivalent to

$$o(N^{-\alpha}) = \sum_{n=N}^{\infty} R_n = \sum_{n=N}^{\infty} m^{-1}W_n \int_{m^n/n^\alpha}^{\infty} x\, dF(x)$$

or, since

$$(4.4) \qquad 0 < \inf_n W_n \leq \sup_n W_n < \infty \text{ a.s.},$$

to

$$(4.5) \qquad o(N^{-\alpha}) = \sum_{n=N}^{\infty} \int_{m^n/n^\alpha}^{\infty} x\, dF(x).$$

Define $y_n := m^n/n^\alpha$, $N(x) := \sup\{n : y_n \leq x\}$. Then (4.5) can be rewritten as

$$(4.6) \qquad o([\log y_N]^{-\alpha}) = \int_{y_N}^{\infty} x(N(x)-N)dF(x), \quad N \to \infty.$$

Apparently (4.6) is weaker than

$$(4.7) \qquad o([\log y]^{-\alpha}) = \int_{y}^{\infty} x(N(x)-N(y))dF(x), \quad y \to \infty$$

but if (4.6) holds, so does (4.7) since for $y_N \leqq y < y_{N+1}$ then

$$\int_{y}^{\infty} x(N(x)-N(y))dF(x) \leqq \int_{y_N}^{\infty} x(N(x)-N)dF(x) = o([\log y_N]^{-\alpha})$$

$$= o([\log y]^{-\alpha}).$$

Now from the definition of $N(x)$ it can be verified that

$$N(x) = \frac{\log x}{\log m} + \frac{\alpha}{\log m} \log \log x + O(1).$$

As $x,y \to \infty$, the mean value theorem for the log yields $\log \log x - \log \log y = o(\log x - \log y)$ so that the right-hand side of (4.7) is

$$\int_{y}^{\infty} x(\log x - \log y)\left(\frac{1}{\log m} + o(1)\right)dF(x) + \int_{y}^{\infty} x\, O(1)dF(x).$$

Since $\mu_\alpha < \infty$, the last term is $o([\log y]^{-\alpha})$ and therefore conditions (4.7) and (4.1) are equivalent, completing the proof when $\mu_\alpha < \infty$.

Suppose next $\mu_\alpha = \infty$. Then by (4.2), certainly (4.1) fails and we have to prove that $W-W_n = o(n^{-\alpha})$ must fail too. Since we assume (X LOG X), we can find β such that $1 \leq \beta < \alpha$ and that $\mu_\beta < \infty$, $\mu_{\beta+1/2} = \infty$. Then from (4.2) and the first part of this proof it follows that $W-W_n = o(n^{-\beta})$ fails and the proof is complete since $\beta < \alpha$. □

PROOF OF (iii). Let $\beta_n := \Sigma_1^n k^{\alpha-1}$. Then

$$\sum_{n=1}^{N} \alpha_n \beta_n = \sum_{k=1}^{N} k^{\alpha-1} \sum_{n=k}^{\infty} \alpha_n - \beta_N \sum_{n=N+1}^{\infty} \alpha_n$$

and from (4.3) it follows by letting $N \to \infty$ that

$$(4.8) \qquad \sum_{n=1}^{\infty} \alpha_n \beta_n \text{ converges} \implies \sum_{k=1}^{\infty} k^{\alpha-1} \sum_{n=k}^{\infty}{}' \alpha_n \text{ converges.}$$

Let $\tilde{W}_{n+1}$, R_n be defined as above. Using $\beta_n \simeq n^\alpha$ one obtains

$$\sum_{n=0}^{\infty} P(\tilde{W}_{n+1} \neq W_{n+1}) \simeq \sum_{n=0}^{\infty} \mathrm{Var}\{\beta_n\{\tilde{W}_{n+1}-W_n+R_n\}\} \simeq \mu_\alpha .$$

Thus if $\mu_\alpha < \infty$, $\Sigma\, \beta_n\{W_{n+1}-W_n+R_n\}$ converges a.s. and from (4.8), we get the a.s. convergence of

$$\sum_{k=0}^{\infty} k^{\alpha-1} \sum_{n=k}^{\infty} \{W_{n+1}-W_n+R_n\} = \sum_{k=0}^{\infty} k^{\alpha-1}(W-W_k+\sum_{n=k}^{\infty} R_n).$$

Thus the convergence of $\Sigma\, k^{\alpha-1}\{W-W_k\}$ is equivalent to that of

$$\sum_{k=0}^{\infty} k^{\alpha-1} \sum_{n=k}^{\infty} R_n = m^{-1} \sum_{k=0}^{\infty} k^{\alpha-1} \sum_{n=k}^{\infty} W_n \int_{m^n/n^\alpha}^{\infty} x\; dF(x)$$

or, appealing to (4.4), to that of

$$\sum_{k=0}^{\infty} k^{\alpha-1} \sum_{n=k}^{\infty} \int_{m^n/n^\alpha}^{\infty} x\; dF(x) = \sum_{n=0}^{\infty} \beta_n \int_{m^n/n^\alpha}^{\infty} x\; dF(x).$$

Using $\beta_n \simeq n^\alpha$, this precisely reduces to $\mu_{\alpha+1} < \infty$ and the proof is complete when $\mu_\alpha < \infty$.

If $\mu_\alpha = \infty$, then of course $\mu_{\alpha+1} = \infty$. Assuming (X LOG X) we can choose β, $1 \leqq \beta < \alpha$, such that $\mu_\beta < \infty$, $\mu_{\beta+1} = \infty$. Then the first part of the proof excludes the convergence of $\Sigma\, n^{\beta-1}\{W-W_n\}$ and Abel's criterion (Bromwich (1908), pg.48) that of $\Sigma\, n^{\alpha-1}\{W-W_n\}$. □

PROOF OF (i). We let $\beta_n := m^{n/q}$ and truncate $X_{n,i}$ at $m^{n/p}$. Then

$$\sum_{n=0}^{\infty} P(\tilde{W}_{n+1} \neq W_{n+1}) \simeq \sum_{n=0}^{\infty} \mathrm{Var}[m^{n/q}\{\tilde{W}_{n+1}-W_n+R_n\}] \simeq \int_0^{\infty} x^p dF(x). \tag{4.9}$$

Assuming the right-hand side to be finite, we have a.s. convergence of $\Sigma\, m^{n/q}\{\tilde{W}_{n+1}-W_n+R_n\}$, $\Sigma\, m^{n/q}\{W_{n+1}-W_n+R_n\}$ and (4.3) gives

$$o(m^{-N/q}) = W-W_N+\sum_{n=N}^{\infty} R_n = W-W_N+m^{-1}\sum_{n=N}^{\infty} W_n \int_{m^{n/p}}^{\infty} x\; dF(x).$$

But the last term is $o(m^{-N/q})$, since

$$\sum_{n=0}^{\infty} m^{n/q} \int_{m^{n/p}}^{\infty} x\; dF(x) = \int_0^{\infty} O(x^p)dF(x) < \infty,$$

and it follows that $W-W_N = o(m^{-N/q})$, proving one way of the result.

For the converse, the method in the proofs of part (ii), (iii) and in Section 2 does not apply, because the condition for convergence in (4.9) is not weaker than that for the result and our proof is here totally different.

Suppose $W-W_N = o(m^{-N/q})$. In particular, $W_{n+1}-W_n = o(m^{-n/q})$ so that on $\{W > 0\}$

$$(4.10) \qquad m^{-n/p}\sum_{i=1}^{Z_n}\{X_{n,i}-m\} = m^{n/q}\{W_{n+1}-W_n\} \to 0.$$

Now consider i.i.d.r.v. $U_{n,i}, U'_{n,i}$ $(n = 0,1,2,\ldots, i = 1,2,\ldots)$ which are independent of $\mathfrak{F} := \sigma(X_{n,i})$ and each distributed as $X_{n,i}-m$. Define $U^s_{n,i} := U_{n,i}-U'_{n,i}$, $F^s(x) := P(|U^s_{n,i}| \leq x)$,

$$M_n := m^{-n/p}\max_{i=1,\ldots,Z_n}|U_{n,i}|,\quad p(n,\epsilon) := P(M_n > \epsilon|\mathfrak{F}),$$

$$(4.11) \qquad q(n,\epsilon) := P(m^{-n/p}|\sum_{i=1}^{Z_n}U_{n,i}| > \epsilon|\mathfrak{F})$$

$$= P(m^{-n/p}|\sum_{i=1}^{Z_n}\{X_{n,i}-m\}| > \epsilon|\mathfrak{F}_n)$$

and M^s_n, $p^s(n,\epsilon)$, $q^s(n,\epsilon)$ by replacing $U_{n,i}$ by $U^s_{n,i}$. Applying the Borel-Cantelli lemma first to the M^s_n and next the $U^s_{n,i}$, it follows that

$$(4.12) \quad \sum_{n=0}^{\infty}p^s(n,\epsilon) < \infty \iff M^s_n \leq \epsilon \text{ eventually} \iff \sum_{n=0}^{\infty}Z_n\int_{\epsilon m^{n/p}}^{\infty}dF^s(x) < \infty.$$

Now from (4.10), (4.11) and the conditional Borel-Cantelli lemma, $\Sigma\, q(n,\epsilon) < \infty$ for all $\epsilon > 0$. Since $p^s(n,\epsilon) \leq 2q^s(n,\epsilon) \leq 4q(n,\epsilon/2)$ by A 6.4 , the l.h. sum in (4.12) must thus be convergent. Hence the r.h. sum must be so and appealing to (4.4), this reduces by the usual calculations to $\int_0^\infty x^p dF^s(x) < \infty$ and hence, by standard facts on symmetrization, $\int_0^\infty x^p dF(x) < \infty$. □

We next turn to the problem of relating the properties of the distribution of W to the offspring distribution. An example of such results in case of finite variance is the explicit expression $\sigma^2/(m^2-m)$ for Var W. We study here instead moments of the form $EWf(W)$, with $f(x) = o(x)$ as $x \to \infty$. For example:

4.4. THEOREM (a) *The* p^{th} *moment* $(1 < p < 2)$ *of* W *and* F *are finite at the same time. More precisely,*

$$(4.13)\qquad m^{-p}\int_0^\infty x^p dF(x) \leqq EW^p \leqq 1 + \frac{\int_0^\infty x^p dF(x)}{m^p - m}.$$

(b) $\mu_{\alpha+1} < \infty$ *implies that* $EW(\log^+ W)^\alpha < \infty$.

The basic tool is

4.5. LEMMA. *Let* $f:[0,\infty) \to [0,\infty)$ *be concave and let* $S := X_1 + \ldots + X_N$ *be a sum of* N *independent r.v.* $X_i \geqq 0$. *Then*

$$(4.14)\qquad ESf(S) \leqq ESf(ES) + \sum_{i=1}^{N} EX_i f(X_i).$$

PROOF. The assumptions on f imply subadditivity, $f(a+b) \leqq f(a)+f(b)$, $a,b \geqq 0$. Thus

$$ESf(S) = \sum_{i=1}^{n} EX_i f(S) \leqq \sum_{i=1}^{n} \{EX_i f(\sum_{j\neq i} X_j) + EX_i f(X_i)\}$$

$$\leqq ESf(ES) + \sum_{i=1}^{n} EX_i f(X_i),$$

since by Jensen's inequality

$$EX_i f(\sum_{j\neq i} X_j) = EX_i Ef(\sum_{j\neq i} X_j) \leqq EX_i f(E\sum_{j\neq i} X_j) \leqq EX_i f(ES). \qquad \Box$$

PROOF OF 4.4. Letting $N := Z_n$, $X_i := X_{n,i}/m^{n+1}$, $S := m^{-n-1}\sum_{i=1}^{Z_n} X_{n,i} = W_{n+1}$ yields

$$E(W_{n+1} f(W_{n+1})|\mathfrak{F}_n) \leqq E(W_{n+1}|\mathfrak{F}_n) f(E(W_{n+1}|\mathfrak{F}_n)) + \sum_{i=1}^{Z_n} E(\frac{X_{n,i}}{m^{n+1}} f(\frac{X_{n,i}}{m^{n+1}})|\mathfrak{F}_n)$$

$$= W_n f(W_n) + m^{-1} W_n \int_0^\infty x f(\frac{x}{m^{n+1}}) dF(x)$$

and it follows that

$$EWf(W) \leqq \underline{\lim}\, EW_{N+1} f(W_{N+1}) = \underline{\lim}\{f(1) + \sum_{n=0}^{N} \{EW_{n+1} f(W_{n+1}) - EW_n f(W_n)\}\}$$

$$\leqq f(1) + \sum_{n=0}^{\infty} m^{-1} EW_n \int_0^\infty x f(\frac{x}{m^{n+1}}) dF(x)$$

$$= f(1) + m^{-1} \int_0^\infty x \sum_{n=0}^{\infty} f(\frac{x}{m^{n+1}}) dF(x).$$

Letting $f(x) = x^{p-1}$, the r.h. side of (4.13) is immediate. The l.h. side follows by convexity since

$$\int_0^\infty x^p dF(x) = m^p EW_1^p = m^p E(E(W|\mathfrak{F}_1))^p \leqq m^p EW^p.$$

In (b), $[\log^+ x]^\alpha$ does not satisfy the assumption on $f(x)$, but so

$$f(x) = \begin{cases} c_1 x & 0 \leqq x \leqq x_0 \\ [\log^+ x]^\alpha + c_2 & x_0 \leqq x < \infty \end{cases}$$

if we chose first $x_0 > 1$ such that $d^2/dx^2 (\log x)^\alpha < 0$ when $x \geqq x_0$ and let

$$c_1 := \frac{d}{dx}(\log x)^\alpha \Big|_{x=x_0}, \quad c_2 := c_1 x_0 - (\log x_0)^\alpha .$$

(b) follows at once, since one easily checks $\Sigma\, f(x/m^{n+1}) = O([\log^+ x]^{\alpha+1})$.

□

The above methods and results could be exploited somewhat further, but we shall not go into this in detail. The converse of (b) is known to be true. Also moments of order $\geqq 2$ could be treated. The relevant generalization of (4.14) becomes

$$(4.15) \qquad ES^\nu f(S) \leqq ES^\nu f(ES) + \sum_{\mu=1}^{\nu} \binom{\nu}{\mu} ES^{\nu-\mu} \sum_{i=1}^{N} EX_i^\mu f(X_i), \quad \nu = 1,2,\ldots$$

With later applications in mind, we give

4.6. COROLLARY. Define $M := \sup_n W_n$. Then (a) the pth moment $(1 < p < 2)$ of M and F are finite at the same time, (b) $\mu_{\alpha+1} < \infty$ implies that $EM(\log^+ M)^\alpha < \infty$.

Computing the moments of M upon integration by parts, the proof is an obvious combination of 2.6 and 4.4.

5. THE SENETA-HEYDE THEOREM

The problem is to make more precise the upper bound $Z_n = o(m^n)$ of the Kesten-Stigum theorem in the case where (X LOG X) fails. The assertion of the Seneta-Heyde theorem is

5.1. THEOREM. *If $1 < m < \infty$, then there exists a sequence c_n of constants such that $c_{n+1}/c_n \to m$ and that Z_n/c_n has an a.s. limit vanishing precisely on the set of extinction.*

Of course, $c_n \cong m^n$ if and only if (X LOG X) holds.

As a by-product of the study of problems related to 5.1, we shall also obtain another proof of the Kesten-Stigum theorem as well as introduce a martingale associated with two independent copies of the process and of some independent interest. Assume for simplicity throughout in the following $p_0 = 0$ so that $P(Z_n \to \infty) = 1$.

We first need to strengthen the provisional estimate 1.6 of the growth of Z_n:

5.2. LEMMA. $\underline{\lim}_{n\to\infty} Z_{n+1}/Z_n > 1$ a.s.

5.3. LEMMA. *The result of 5.2 implies*

$$(5.1) \qquad \sum_{n=0}^{\infty} Z_n I(x > Z_n) = O(x), \quad \sum_{n=0}^{\infty} Z_n^{-1} I(x \leq Z_n) = O(\tfrac{1}{x}).$$

5.4. LEMMA. $\lim_{n\to\infty} Z_{n+1}/Z_n = m$ a.s.

PROOF OF 5.2. Choose first a such that $\tilde{m}: = EX_{n,i} \wedge a > 1$ and define $\tilde{Z}_{n+1}: = \Sigma_1^{Z_N} X_{n,i} \wedge a$. Then $E(\tilde{Z}_{n+1}|\mathfrak{F}_n) = \tilde{m}Z_n$ and by 1.6,

$$\sum_{n=0}^{\infty} \mathrm{Var}\left(\frac{\tilde{Z}_{n+1}}{Z_n}\Big|\mathfrak{F}_n\right) = \mathrm{Var}(X_{n,i} \wedge a) \sum_{n=0}^{\infty} Z_n^{-2} < \infty$$

a.s. Hence by Chebycheff's inequality and the conditional Borel-Cantelli lemma, $\tilde{Z}_{n+1}/Z_n \to \tilde{m}$ a.s. But clearly $Z_{n+1} \geq \tilde{Z}_{n+1}$. $\square$

PROOF OF 5.3. Define $N(x): = \sup\{n: Z_n < x\}$ and assume without loss of generality that $Z_{n+1}/Z_n \geq \tilde{m}$ for all n and some $\tilde{m} > 1$ (so that $Z_n \leq \tilde{m}^{-k} Z_{n+k}$). Then the sums in (5.1) are

$$\sum_{n=0}^{N(x)} Z_n = Z_{N(x)} \sum_{n=0}^{N(x)} \frac{Z_n}{Z_{N(x)}} \le x \sum_{n=0}^{N(x)} \tilde{m}^{n-N(x)} \le \frac{x}{1-1/\tilde{m}} ,$$

$$\sum_{n=N(x)+1}^{\infty} \frac{1}{Z_n} = \frac{1}{Z_{N(x)+1}} \sum_{n=N(x)+1}^{\infty} \frac{Z_{N(x)+1}}{Z_n} \le \frac{1}{x} \sum_{n=N(x)+1}^{\infty} \tilde{m}^{N(x)+1-n} \le \frac{x}{1-1/\tilde{m}} .$$

□

PROOF OF 5.4. Let $\tilde{Z}_{n+1} := \Sigma_1^{Z_n} X_{n,i} I(X_{n,i} \le Z_n)$, $m(n) := E(X_{n+1} I(X_{n,i} \le Z_n) | \mathfrak{F}_n)$. By 5.3,

$$\sum_{n=0}^{\infty} \mathrm{Var}\left(\frac{\tilde{Z}_{n+1}}{Z_n} \Big| \mathfrak{F}_n\right) \le \sum_{n=0}^{\infty} \frac{1}{Z_n} \int_0^{Z_n} x^2 dF(x) = \int_0^{\infty} O(x) dF(x) < \infty ,$$

so that it follows as in the proof of 5.2 that $\tilde{Z}_{n+1}/Z_n - m(n)$ tends to zero a.s. But since $Z_n \to \infty$, also $m(n) \to m$ and the proof is completed by observing that $\tilde{Z}_{n+1} = Z_{n+1}$ eventually since by 5.3,

$$\sum_{n=0}^{\infty} P(\tilde{Z}_{n+1} \neq Z_{n+1} | \mathfrak{F}_n) \le \sum_{n=0}^{\infty} Z_n \int_{Z_n}^{\infty} dF(x) = \int_0^{\infty} O(x) dF(x) < \infty . \qquad \square$$

5.5. THEOREM. (Grey) *Let $\{Z_n^*\}_{n\in\mathbb{N}}$ be a Galton-Watson process independent of $\{Z_n\}_{n\in\mathbb{N}}$ and with the same offspring distribution* F. *Then $\{Y_n\}_{n\in\mathbb{N}} := \{Z_n/(Z_n+Z_n^*)\}_{n\in\mathbb{N}}$ is a $(0,1)$-valued martingale, and the a.s. limit* $Y := \lim_n Y_n$ *satisfies* $Y \in (0,1)$ *a.s.*

We note that 5.5 contains the Seneta-Heyde theorem as a corollary. Indeed, from the existence of Y it follows that Z_n/Z_n^* has a limit U in $[0,\infty]$ and since $1/(1+U) = Y$, $Y \in (0,1)$ implies that $U \in (0,\infty)$. Thus as the c_n one could take Z_n^* for one of the typical realizations of the Z_n^*-process for which U exists and $\in (0,\infty)$ a.s.

PROOF OF 5.5. By symmetry,

$$E\left(\frac{X_{n,i}}{\sum_{i=1}^{Z_n} X_{n,i} + \sum_{i=1}^{Z_n^*} X_{n,i}^*} \Big| \mathfrak{F}_n, \mathfrak{F}_n^*\right) = E\left(\frac{X_{n,i}^*}{\sum_{i=1}^{Z_n} X_{n,i} + \sum_{i=1}^{Z_n^*} X_{n,i}^*} \Big| \mathfrak{F}_n, \mathfrak{F}_n^*\right) = \frac{1}{Z_n+Z_n^*} .$$

From this the martingale property is immediate since

$$E(Y_{n+1} | \mathfrak{F}_n, \mathfrak{F}_n^*) = E\left(\frac{\sum_{i=1}^{Z_n} X_{n,i}}{\sum_{i=1}^{Z_n} X_{n,i} + \sum_{i=1}^{Z_n^*} X_{n,i}^*} \Big| \mathfrak{F}_n, \mathfrak{F}_n^*\right) = \frac{Z_n}{Z_n+Z_n^*} = Y_n$$

(note the similarity of the argument with the proof of the reversed martingale property in the SLLN). From $Y_n \in (0,1)$ and the martingale convergence theorem it is immediate that a limit $Y \in [0,1]$ exists, and that $E(Y|Z_0,Z_0^*) = Y_0 = Z_0/(Z_0+Z_0^*)$. Replacing $\{Z_n^*\}$ by $\{Z_{n+k}^*\}$, it follows that $Y^k := \lim_{n\to\infty} Z_n/(Z_n+Z_{n+k}^*)$ exists for each k, and $E(Y^k|Z_0,Z_k^*) = Z_0/(Z_0+Z_k^*)$. Therefore, using 5.4 in the form $Z_{n+k}^*/Z_n^* \to m^k$, it follows that

$$P(Y = 1) = P(\frac{Z_n}{Z_n^*} \to \infty) = P(\frac{Z_n}{Z_{n+k}^*} \to \infty) = P(Y^k = 1) \leq EY^k = E\,\frac{Z_0}{Z_0+Z_k^*}.$$

Letting $k \to \infty$, we get $P(Y = 1) = 0$. By symmetry, $P(Y = 0) = 0$. □

The only insight provided by this proof is how the c_n behave is that $c_{n+k}/c_n \to m^k$. We next present some different approaches more along the lines of Sections 2 and 4, which lead to more explicit expressions:

5.6. THEOREM. *The c_n can be defined by choosing first c_0 such that $\int_0^{c_0} x dF(x) > 1$ and letting*

$$c_{N+1} := c_N \int_0^{c_N} x\, dF(x) = c_0 m^{N+1} \prod_{n=0}^{N} (1 - \frac{1}{m}\int_{c_n}^{\infty} x\, dF(x)). \tag{5.2}$$

5.7. COROLLARY. *If $\mu_{1/2} := \int_0^\infty x(\log^+ x)^{1/2} dF(x) < \infty$, then as the c_N one can take*

$$c_N := m^N \exp(-\frac{1}{m}\int_0^\infty x(N \wedge \log_m x) dF(x)). \tag{5.3}$$

We first give a direct proof of 5.7 (without reference to 5.6), which, however, contains the same estimates needed to derive 5.7 from 5.6. Both this proof and the proof of 5.6 will yield new variants of the proof of the Kesten-Stigum theorem.

In the proof of 5.7, we take $Z_0 = 1$ and use the identity

$$\frac{Z_{N+1}}{m^{N+1}} = \prod_{n=0}^{N} \frac{Z_{n+1}}{mZ_n} = \prod_{n=0}^{N} (1 + \frac{1}{m} A_{n+1}) \tag{5.4}$$

where $A_{n+1} := (Z_{n+1} - mZ_n)/Z_n$. Define

$$\tilde{Z}_{n+1} := \sum_{i=1}^{Z_n} X_{n,i} I(X_{n,i} \leq Z_n),$$

$$\tilde{A}_{n+1} := \frac{\tilde{Z}_{n+1} - mZ_n}{Z_n}, \quad \epsilon_n := -E(\tilde{A}_{n+1}|\mathfrak{F}_n) = \int_{Z_n}^\infty x\, dF(x).$$

Then by 5.3,

$$(5.5)\qquad \sum_{n=0}^{\infty} P(\tilde{Z}_{n+1} \neq Z_{n+1}|\mathfrak{F}_n) \leq \sum_{n=0}^{\infty} Z_n \int_{Z_n}^{\infty} dF(x) = \int_0^{\infty} O(x)dF(x) < \infty,$$

$$(5.6)\qquad \sum_{n=0}^{\infty} \mathrm{Var}(\tilde{A}_{n+1}|\mathfrak{F}_n) \leq \sum_{n=0}^{\infty} \frac{1}{Z_n} \int_0^{Z_n} x^2 dF(x) = \int_0^{\infty} O(x)dF(x) < \infty.$$

5.8. LEMMA. <u>If</u> $1 < m < \infty$, <u>then</u> $\Sigma_0^N A_{n+1} = -\Sigma_0^N \epsilon_n + B_n$ <u>where</u> B_n <u>has a a.s. limit</u> $B \in (-\infty, \infty)$. <u>If furthermore</u> $\mu_{1/2} < \infty$, <u>then</u> $\Sigma_0^\infty A_{n+1}^2 < \infty$ <u>a.s.</u>

PROOF. The series $\Sigma_0^\infty \{\tilde{A}_{n+1} + \epsilon_n\}$ converges since the terms are martingale increments and (5.6) holds. The first assertion is therefore immediate since (5.5) implies that eventually $\tilde{Z}_{n+1} = Z_{n+1}$ and $\tilde{A}_{n+1} = A_{n+1}$. For the second, note first that $Z_n \geq Bm_1^n$ with $B > 0$ and $1 < m_1 < m$. Therefore

$$\sum_{n=0}^{\infty} \epsilon_n^2 \leq \sum_{n=0}^{\infty} \int_{Bm_1^n}^{\infty} x\, dF(x) \int_{Bm_1^n}^{\infty} y\, dF(y)$$

$$= \int_0^{\infty}\int_0^{\infty} xy\, O(\log x \wedge \log y) dF(x)dF(y)$$

$$\cong \int_0^{\infty} x\, dF(x)\{\log x \int_x^{\infty} y\, dF(y) + \int_0^x y \log y\, dF(y)\}.$$

If $\mu_{1/2} < \infty$, then both terms in $\{\ \}$ are $O((\log x)^{1/2})$ so that $\Sigma_0^\infty \epsilon_n^2 < \infty$. Thus by (5.6),

$$\sum_{n=0}^{\infty} E(\tilde{A}_{n+1}^2|\mathfrak{F}_n) = \sum_{n=0}^{\infty} \{\epsilon_n^2 + \mathrm{Var}(\tilde{A}_{n+1}|\mathfrak{F}_n)\} < \infty,$$

implying that $\Sigma_0^\infty \tilde{A}_{n+1}^2 < \infty$ and hence $\Sigma_0^\infty A_{n+1}^2 < \infty$. ☐

Define

$$g_N(x) := \sum_{n=0}^{N} I(Z_n < x), \quad \delta_N := \frac{1}{m} \sum_{n=0}^{N} \epsilon_n = \frac{1}{m} \int_0^{\infty} x\, g_N(x) dF(x).$$

When, $\Sigma_0^\infty A_{n+1}^2 < \infty$, it follows by a first order Taylor expansion of $\log(1 + \frac{1}{m} A_{n+1})$ that (5.4) can be rewritten as

$$(5.7)\qquad \frac{Z_{N+1}}{m^{N+1}} := D_N e^{\frac{1}{m}\sum_{n=0}^{N} A_{n+1}} := E_N e^{-\frac{1}{m}\sum_{n=0}^{N}\epsilon_n} := E_N e^{-\delta_N}$$

where D_N, E_N have a.s. limits in $(0,\infty)$ and (by 1.3, 1.6) for some $\varkappa_1$, $\varkappa_2$

$$(5.8) \qquad \varkappa_1(N \wedge \log_m x) \leq g_N(x) \leq \varkappa_2(N \wedge \log_m x).$$

Without $\Sigma_0^\infty A_{n+1}^2 < \infty$, we can at least (using $\log(1+x) \leq x$ and 5.8) obtain

$$(5.9) \qquad \frac{Z_{N+1}}{m^{N+1}} \leq e^{\frac{1}{m}\sum_{n=0}^{N} A_{n+1}} = e^{B_N/m}\, e^{-\delta_N}.$$

THIRD PROOF OF THE KESTEN-STIGUM THEOREM. If (X LOG X) holds, the conditions for (5.7) are satisfied and (5.8) implies the existence of a limit of δ_N. Thus (5.7) yields $W \in (0,\infty)$. If (X LOG X) fails, then by (5.8) $\delta_N \to \infty$ and $W = 0$ follows by (5.9). □

PROOF OF 5.7. Define $N(x):=\sup\{n: Z_n < x\}$ so that $g_N(x) = N \wedge N(x)$. Since $Z_{N(x)+1}/Z_{N(x)} \to m$ as $x \to \infty$, $\log_m x - \log_m Z_{N(x)+1}$ is bounded and solving (5.7) for $N = N(x)$, we get

$$N(x) + 1 = \log_m Z_{N(x)+1} + \log_m e\,\delta_{N(x)} - \log_m E_N,$$

$$(5.10) \qquad N(x) + 1 = \log_m(x) + \log_m e\,\delta_{N(x)} + O(1).$$

Now since $\mu_{1/2} < \infty$,

$$\int_0^\infty x(N \wedge \log_m x)\,dF(x) = \int_0^{m^N} x \log_m x\, dF(x) + N\int_{m^N}^\infty x\, dF(x) = o(N^{1/2}),$$

so that, by (5.8), $\delta_N = o(N^{1/2})$. Thus by (5.10)

$$N(x) = O(\log_m x), \quad \delta_{N(x)} = o((\log_m x)^{1/2}),$$

$$(5.11) \qquad h(x):=N(x) - \log_m(x) = o((\log_m x)^{1/2}).$$

Define $F_N':=\{x: \log_m x + h(x) \leq N,\ \log_m x \leq N\}$, $F_N'':=\{x: \log_m x + h(x) > N,\ \log_m x \leq N\}$, $F_N''':=\{x: \log_m x + h(x) \leq N,\ \log_m x > N\}$. Then

$$m\delta_N - \int_0^\infty x(N \wedge \log_m x)\,dF(x) = \int_0^\infty x(N \wedge (\log_m x + h(x)) - N \wedge \log_m x)\,dF(x)$$

$$= \int_{F_N'} xh(x)\,dF(x) + \int_{F_N''} x(N - \log_m x)\,dF(x) + \int_{F_N'''} x(\log_m x + h(x) - N)\,dF(x)$$

$$\to \int_0^\infty xh(x)\,dF(x) + 0 + 0$$

since $F_N' \uparrow [0,\infty)$, $F_N'' \to \emptyset$, $F_N''' \to \emptyset$ and the integrals on F_N'', F_N'' are dominated by $x|h(x)|$ which is integrable because of (5.11). Combining with (5.7) the proof is complete. $\square$

In the proof of 5.6, we first give

5.9. LEMMA. *Let* $\{c_N'\}$, $\{c_N''\}$ *be defined by* (5.2), *corresponding to the starting values* c_0' *and* c_0'', *respectively. Then* c_N'/c_N'' *has a limit* $\gamma \in (0,\infty)$.

PROOF. Choose k, ℓ with $c_{\ell-1}' \leq c_k'' \leq c_\ell'$. Then since the mapping $c \to c\int_0^c x\, dF(x)$ is non-decreasing, $c_{N+\ell-1}' \leq c_{N+k}'' \leq c_{N+\ell}'$ for all N and

$$\frac{c_{N+k+1}''}{c_{N+k}''} = \int_0^{c_{N+k}''} x\, dF(x) \leq \int_0^{c_{N+\ell}'} x\, dF(x) = \frac{c_{N+\ell+1}'}{c_{N+\ell}'} .$$

Therefore $c_{N+k}''/c_{N+\ell}'$ decreases to a limit β and since $c_{N+\ell}'/c_{N+\ell-1}' \to m$, we have $1/m \leq \beta \leq 1$. Take $\gamma := \beta m^{k-\ell}$. $\square$

PROOF OF 5.6. Let c_N^A be defined by (5.2) with $c_0 := A$. It is immediate that

$$\frac{Z_{N+1}^A}{c_{N+1}^A} := \frac{1}{c_{N+1}^A} \sum_{i=1}^{Z_N^A} X_{n,i}\, I(X_{n,i} \leq c_N^A) \quad (Z_0^A := 1)$$

is a non-negative martingale and hence has a limit $W^A \in [0,\infty)$. From 5.9 it follows that $\{W^A = 0\}$ and $q^A := P(W^A = 0)$ are non-increasing in A. From $c_{N+1}^A/c_N^A \to m$, $EZ_N^A = c_N^A/A$ and 5.3 we get

$$\sum_{n=0}^{\infty} \operatorname{Var}\Bigl(\frac{Z_{N+1}^A}{c_{N+1}^A}\Bigr) \leq \sum_{n=0}^{\infty} \frac{EZ_N^A}{c_{N+1}^{A\,2}} \int_0^{c_N^A} x^2 dF(x) = \int_0^{\infty} O(x) dF(x) < \infty.$$

Therefore $EW^A = EZ_0^A/c_0^A = 1/A > 0$ and $q^A < 1$. Letting $F_A = \{\tilde{Z}_{N+1} \neq Z_{N+1}$ for some $N\}$, we get as in the proof of 5.3

$$PF_A \leq \sum_{n=0}^{\infty} EZ_N^A \int_{c_N^A}^{\infty} dF(x) \leq \frac{1}{A} \int_A^{\infty} \frac{x}{1-1/m^A}\, dF(x)$$

if $c_{N+1}^A/c_N^A > m^A$ for all N. As $A \uparrow \infty$, $m^A \uparrow m$ and hence $PF_A \downarrow 0$. The probability of Z_N/c_N^A having a limit $\tilde{W} \in [0,\infty)$ for some (and then by 5.9 any) A does not depend on A by 5.9 and must hence be one. Similarly $P(\tilde{W} = 0)$ is independent of A and as in the proof of 1.4

$$I(\widetilde{W} = 0) = \lim_{n\to\infty} P(\widetilde{W} = 0|\mathfrak{F}_n)$$

$$\leq \lim_{n\to\infty} P(W^A = 0|\mathfrak{F}_n) = \lim_{n\to\infty} (q^n)^{Z_n^A} = 0 \quad \text{on} \quad \{Z_n \to \infty\}.$$

As $A \uparrow \infty$, $\{\widetilde{Z}_n^A \to \infty\} \uparrow \{Z_n \to \infty\}$. $\square$

6. IMMIGRATION

Suppose that at time n Y_n new individuals immigrate into the population and start Y_n new independent Galton-Watson processes $\{Z_k^{n,\nu}\}_{k\in\mathbb{N}}$ $\nu = 1,\dots,Y_n$. The resulting superposition

$$(6.1) \qquad \hat{Z}_N := \sum_{n=0}^{N} \sum_{\nu=1}^{Y_n} Z_{N-n}^{n,\nu}$$

is then a _Galton-Watson process with immigration_.

We assume that, conditionally upon $\mathfrak{Y} := \sigma(Y_0, Y_1, \dots)$, the $\{Z_k^{n,\nu}\}$ are independent and each distributed as $\{Z_k\}$ (corresponding to the case $Z_0 = 1$). No assumptions at all are made for the moment on the immigration process $\{Y_n\}_{n\in\mathbb{N}}$.

We first consider the supercritical case $1 < m < \infty$. Intuitively, one feels that if the immigration is dominated by the branching in some appropriate sense, then the rate of growth of $\hat{Z}_N$ should be the same as of Z_N (i.e. m^N assuming (X LOG X)).

6.1. THEOREM. _Conditionally upon_ $\mathfrak{Y}$, $\{\hat{W}_N\} := \{\hat{Z}_N/m^N\}$ _is a non-negative submartingale and hence has an a.s. limit_ $\hat{W} \in [0,\infty)$ _on_ $\{\sup_N E(\hat{W}_N \mid \mathfrak{Y})\}$ $= \{S := \Sigma_0^\infty m^{-n} Y_n < \infty\}$. _If_ (X LOG X) _holds, then conversely_ $\hat{W}_N \to \infty$ _a.s. on_ $\{S = \infty\}$. _Finally, on_ $\{S < \infty\}$ _we have the identification_ $\hat{W} = \widetilde{W}$, _where_

$$\widetilde{W} := \sum_{n=0}^{\infty} m^{-n} \sum_{\nu=1}^{Y_n} W^{n,\nu}, \quad W^{n,\nu} := \lim_{k\to\infty} Z_k^{n,\nu}/m^k.$$

6.2. COROLLARY. _If the_ Y_n _are i.i.d., then_

$$(6.2) \qquad E \log^+ Y_1 < \infty$$

implies that $P(S < \infty) = 1$ _and hence the a.s. existence of_ $\hat{W}$. _If_ (6.2) _fails, then_ $\overline{\lim_N} \hat{W}_N = \infty$ _a.s._ (also if (X LOG X) fails) _and more generally, there is no sequence_ $\{c_n\}$ _of constants such that_ $\hat{Z}_N/c_N$ _converges a.s. to a r.v._ W^* _with_ $P(W^* = \infty) = 0$, $P(W^* > 0) > 0$.

PROOF OF 6.1. The submartingale property is immediate from

$$E(\hat{Z}_{N+1} \mid \hat{Z}_N, \mathfrak{Y}) = m\hat{Z}_N + Y_{N+1} \geq m\hat{Z}_N.$$

It follows also by iteration or directly from the definition that

$$E(\hat{W}_N | \mathfrak{F}) = \sum_{n=0}^{N} m^{-n} Y_n .$$

Here conditionally upon $\mathfrak{F}$, Doob's condition $\sup E(\hat{W}_N | \mathfrak{F}) < \infty$ holds on $\{S < \infty\}$ and thus $\hat{W}$ exists. It is clear that for any M

$$\underline{\lim}_{N\to\infty} \hat{W}_N \geq \sum_{n=0}^{M} m^{-n} \sum_{\nu=1}^{Y_n} W^{n,\nu}$$

so that $W \geq \tilde{W}$. If (X LOG X) holds, then $\tilde{W} = \infty$ on $\{S = \infty\}$ (and hence $\hat{W}_N \to \infty$) since on $\{\tilde{W} < \infty\}$, $\infty > \Sigma\Sigma E(m^{-n}W^{n,\nu} | \mathfrak{F}) = S$, as is seen by conditioning upon $\mathfrak{F}$ and using the criterion A <u>4.2</u> for convergence of weighted sums of i.i.d. finite mean r.v. It only remains to prove that $\hat{W} = \tilde{W}$ a.s. on $\{S < \infty\}$. From above, $\hat{W} \geq \tilde{W}$. Now just observe that $E(\tilde{W} | \mathfrak{F}) = S$ (by positivity), $E(\hat{W} | \mathfrak{F}) \leq \underline{\lim}\, E(\hat{W}_N | \mathfrak{F}) = S$. □

The following lemma needed for <u>6.2</u> will be useful also in the subcritical case

<u>6.3</u>. LEMMA. <u>Let</u> $0 < \delta < 1$ <u>and let the</u> Y_n <u>be i.i.d. Then</u> (6.2) <u>is equivalent to either of</u>

(6.3) $$\sum_{n=0}^{\infty} P(Y_n \geq \delta^{-n}) < \infty,$$

(6.4) $$Y_n \leq \delta^{-n} \quad \text{eventually},$$

(6.5) $$\sum_{n=0}^{\infty} \delta^n Y_n < \infty \quad \text{a.s.}$$

PROOF. The equivalence of (6.2) and (6.3) is by now routine and the equivalence of (6.3) and (6.4) just Borel-Cantelli. To show the equivalence of (6.2) and (6.5), note that (6.4) is clearly necessary for (6.5). The equivalence of (6.2) and (6.4) shows that if (6.2) holds, then $Y_n \leq \delta_1^{-n}$ eventually when $\delta < \delta_1 < 1$ and hence $\Sigma\delta^n Y_n < \infty$. Alternatively, one could use the three-series criterion. □

PROOF OF 6.2. If (6.2) holds, then by 6.3 with $\delta := 1/m$, $S < \infty$ a.s. Suppose next that (6.2) fails. Then by 6.3, $Z_N \geq Y_N \geq m_1^N$ i.o. for all $m_1 > m$ so that we have $\overline{\lim}\ \hat{Z}_N/m^N = \infty$. More generally, for the last assertion of 6.2 it suffices to prove that the conditions on c_N, W^* imply that $c_{N+1}/c_N \to m$ (since then $c_N \leq m_1^N$ eventually for $m_1 > m$). This follows already from the convergence in distribution of Z_N/c_N to W^*: Write

$$(6.6) \qquad \frac{\hat{Z}_{N+1}}{c_{N+1}} = \frac{c_N}{c_{N+1}} \frac{\hat{Z}_N}{c_N} \frac{1}{Z_N} \sum_{i=1}^{\hat{Z}_N} X_{N,i} + \frac{Y_{N+1}}{c_{N+1}}$$

with the $X_{N,i}$ i.i.d. conditional upon $(\ ,\hat{Z}_N)$ and each having the offspring distribution. By the WLLN, $\hat{Z}_N^{-1}\Sigma_1^{\hat{Z}_N} X_{N,i} \overset{P}{\to} m$. Since $P(\hat{Z}_N \to \infty) \geq P(Z_N \to \infty) > 0$, we must have $c_N \to \infty$ so that the last term in (6.6) tends in distribution to zero. Therefore the limiting distribution of $\hat{Z}_{N+1}/c_{N+1}$ can only be the same as of $\hat{Z}_N/c_N$ if $c_{N+1}/c_N \to m$. □

We next go to the case $m \leq 1$, where one could hope that the immigration might compensate the extinction so that $\hat{Z}_N$ would converge in distribution (while it is easy to see that no interesting a.s. convergence results will hold). Assuming that the Y_n are i.i.d., we show below that (under regularity assumptions which could safely be assumed in applications) this holds in the subcritical case $m < 1$ but not the critical case $m = 1$:

6.4. THEOREM. _If_ $0 < m < 1$, _then_ $\hat{Z}_N$ _converges in distribution to a proper law if and only if_ (6.2) _holds while otherwise_ $\hat{Z}_N \underset{D}{\to} \infty$ (i.e.

$P(\hat{Z}_N \leq a) \to 0$ for all $a < \infty$). _If_ $m = 1$ _and_ $\sigma^2 < \infty$, _then_ $\hat{Z}_N \overset{D}{\to} \infty$.

The usual proofs of this are analytic. We shall give a proof where calculus enters only in form of the rather elementary

6.5. LEMMA. _Define_ $q_n := P(Z_n > 0)$. _Then if_ $0 < m < 1$, $q_n \leq m^n$ _and for any_ $m_1 < m$, $q_n \geq m_1^n$ _eventually. If_ $m = 1$ _and_ $\sigma^2 < \infty$, _then_ nq_n _tends to a constant in_ $(0,\infty)$.

This is to be proved in Chapter 3. A.s. convergence then comes in via

6.6. LEMMA. _The distribution of_ $\hat{Z}_N$ _is the same as of_

$$\check{Z}_N := \sum_{n=0}^{N} \sum_{\nu=1}^{Y_n} Z_n^{n,\nu}.$$

In particular (since $\check{Z}_N \uparrow$), $\hat{Z}_N$ _converges in distribution if and only if_ $\check{Z}_\infty := \Sigma_0^\infty \Sigma_1^{Y_n} Z_n^{n,\nu} < \infty$ _a.s._

PROOF. Obvious since (by a change of variables)

$$\hat{Z}_N = \sum_{n=0}^{N} \sum_{\nu=1}^{Y_{N-n}} Z_n^{N-n,\nu}$$

and the distribution of $(Y_0,\ldots,Y_N)$ is the same as that of $(Y_N,\ldots,Y_0)$ and the distribution of $Z_k^{n,\nu}$ does not depend on n,ν. □

PROOF OF 6.4. In order that $\check{Z}_\infty < \infty$ a.s. it is necessary that $Z_n^{n,\nu} > 0$ for only finitely many n,ν with $\nu = 1,\ldots,Y_n$. Conditioning upon $\mathfrak{Y}$, this is equivalent to $\Sigma_0^\infty q_n Y_n < \infty$ by the Borel-Cantelli lemma. If $m = 1$, $\sigma^2 < \infty$ this can never hold since then no matter the distribution of Y_1, $\Sigma_0^\infty Y_n/n = \infty$. If $0 < m < 1$, then combining 6.3 and 6.5 shows that $\Sigma_0^\infty q_n Y_n < \infty$ if and only if (6.2) holds. Suppose conversely that (6.2) holds and that $0 < m < 1$. Define $Y_n^* := Y_n I(Y_n \leq \delta^{-n})$ with $m < \delta < 1$, $\check{Z}_\infty^* = \Sigma_0^\infty \Sigma_1^{Y_n^*} Z_{n,\nu}^n$. Then by 6.3, $Y_n^* = Y_n$ eventually so that $\check{Z}_\infty^*$ and $\check{Z}_\infty$ are finite at the same time. But $\check{Z}_\infty^* < \infty$ since

$$E(\check{Z}_\infty^* | \mathfrak{Y}) = \sum_{n=0}^{\infty} \tilde{Y}_n E Z_n \leq \sum_{n=0}^{\infty} \left(\frac{m}{\delta}\right)^n < \infty. \quad \square$$

BIBLIOGRAPHICAL NOTES

The results in this chapter are all known or have been published elsewhere, the main exceptions being the expressions in 5.6 and 5.7 for the Seneta constants. Many proofs are, however, not the standard ones. A preliminary survey of some of the ideas involved is given in Asmussen (1978). In particular, the first proof of the Kesten-Stigum theorem simply specializes the proof given in a more general setting in Asmussen and Hering (1976 a). The second proof is more similar to the original proof by Kesten and Stigum (1966 a), with the simplification becoming possible by admitting one type only. 3.1 is due to Heyde (1970 b), Heyde and Leslie (1971), while the LIL part of 3.2 has been given its present form in Asmussen (1977), see also Heyde and Leslie (1971). Parts (a) and (b) of 3.1 are standard in estimation theory (see, e.g., the survey by Dion and Keiding (1978) and the references therein), while (c) and the rest of Section 2 is based on Asmussen and Keiding (1978). Except for some details, Section 4 follows Asmussen (1978) almost verbatim. The Seneta-Heyde theorem has been standard for some time. 5.3 was suggested by Asmussen and Kurtz (1980). 5.5 is due to Grey (1977. The rest of Section 5 is new, though 5.6 is related to Theorem 1.3.1 of Schuh and Barbour (1977). Uchiyama (1976) has a weaker form of 5.7, not assuming $\mu_{\frac{1}{2}} < \infty$, but imposing a regular variation condition. The first part of Section 6 is from Asmussen and Hering (1976 b). Some ideas were developed independently by Athreya, Parthasarathy, and Sankaranarayanan (1974). 6.4 is due to Seneta (1970). The present proof seems new.

Chapter III

The Galton-Watson Process: Analytic Methods

1. SUBCRITICAL PROCESSES

Let $P_n(i,j)$ be the n-step transition function of a Bienaymé-Galton-Watson process, define the probability generating functions (p.g.f.'s)

$$f_n(s) := \sum_{j=0}^{\infty} s^j P_n(1,j), \quad n = 1,2,\ldots, \quad |s| \leq 1,$$

and set $f_0(s) := s$, $f(s) := f_1(s)$, $|s| \leq 1$. The branching property may then be written in the form

$$\sum_{j=0}^{\infty} s^j P_n(i,j) = (f_n(s))^i, \quad i,j = 1,2,\ldots .$$

Combining this with the Chapman-Kolmogoroff equation,

$$\sum_{j=0}^{\infty} P_n(i,j) P_\ell(j,k) = P_{n+\ell}(i,k),$$

we get the semigroup relation

$$f_n(f_\ell(s)) = f_{n+\ell}(s). \tag{1.1}$$

The first moment of the first generation is

$$m := f'(1-) \leq \infty.$$

Since $f_{n+\ell}(0) = f_n(f_\ell(0)) \geq f_n(0)$, the extinction probability

$$q := \lim_{n\to\infty} f_n(0) = \lim_{n\to\infty} P_n(1,0)$$

always exists. Excluding the trivial case that $p_1 = P_1(1,1) = 1$, $f(s)$ is strictly convex, so that the equation $f(s) = s$ has at most two solutions in $\mathbb{R}_+$. By definition $f(1) = 1$. If $m < 1$ the second solution, if existent, must be > 1, i.e., $q = 1$. If $m = 1$ $(p_1 \neq 1)$, there is no second solution, so that $q = 1$, and if $m > 1$, there always exists a second solution < 1, which will be seen to be equal to q.

1.1. EXAMPLE. Only very few examples are known, for which f_n has been calculated explicitly. Perhaps the most interesting ones are the fractional linear generating functions: Let $0 < c$, $p < 1$, $c \leq 1-p$,

$$p_k = cp^{k-1}, \quad k \geq 1,$$

$$p_0 = 1 - \sum_{k=1}^{\infty} p_k = \frac{1-c-p}{1-p} ,$$

$(p_k = P_1(1,k))$. Then

$$f(s) = 1 - \frac{c}{1-p} + \frac{cs}{1-ps} ,$$

$$m = \frac{c}{(1-p)^2} .$$

If $m = 1$, then $c = (1-p)^2$,

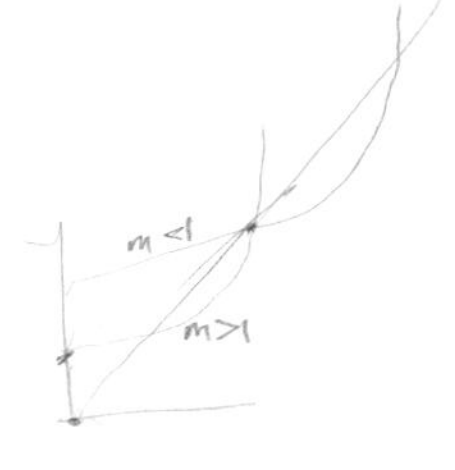

$$f(s) = \frac{p-(2p-1)s}{1-ps} ,$$

$$f_n(s) = \frac{np-(np+p-1)s}{1-p+np-nps} .$$

If $m \neq 1$, the second solution of $f(s) = s$ is

$$s_0 = \frac{1-c-p}{p(1-p)} ,$$

so that

$$\frac{f(s)-s_0}{f(s)-1} = \frac{1}{m} \frac{s-s_0}{s-1} .$$

Iterating this,

$$\frac{f_n(s)-s_0}{f_n(s)-1} = \frac{1}{m^n} \frac{s-s_0}{s-1} ,$$

which can be solved to yield

$$f_n(s) = 1 - m^n\left(\frac{1-s_0}{m^n-s_0}\right) + \frac{m^n\left(\frac{1-s_0}{m^n-s_0}\right)^2}{1 - \frac{m^n-1}{m^n-s_0} s} .$$

Fractional linear generating functions had occurred in work on functional iteration, cf. Schroeder (1871). They proved useful in the study of the extinction probability of male lines of descent, cf. Lotka (1931 a,b, 1939), and they play an important role in connection with the problem of embeddability of BGW processes in continuous time branching processes, cf. Karlin and McGregor (1968 a,b).

1.2. THEOREM. *If* $0 < m < 1$, *then*

$$\frac{f_n(s)-f_n(0)}{1-f_n(0)} \to g(s),\ n \to \infty,\quad s \in [0,1],$$

where $g(s)$ *is the unique p.g.f. solution of*

$$1 - g(f(s)) = m(1 - g(s)),\ g(0) = 0. \tag{1.2}$$

For convenience we now and then formulate results in terms of the process $\{Z_n, P\}$, $P(Z_0=1) = 1$, although the proofs rely only on the transition function, respectively, the generating function.

1.3. COROLLARY. (Yaglom's theorem) *If* $0 < m < 1$, *then*

$$P(Z_n = j | Z_n > 0) \to Q(j)\quad n \to \infty,$$

where Q *is a proper probability distribution*.

PROOF OF 1.2. Define

$$g_n(s) := \frac{f_n(s)-f_n(0)}{1-f_n(0)} = 1 - \frac{1-f_n(s)}{1-f_n(0)},$$

$$k(s) := \frac{1-f(s)}{1-s}.$$

Then

$$1 - g_n(s) = \frac{k(f_{n-1}(s))}{k(f_{n-1}(0))}(1-g_{n-1}(s)).$$

Since $f_n(s)\uparrow$, $s\uparrow$, and by convexity also $k(s)\uparrow$, $s\uparrow$, it follows that $1-g_n(s)\uparrow$, $n\uparrow$. Hence, the limit $g(s)$ exists and is a (possibly defective) p.g.f. From the definitions of g_n, k, and (1.1),

$$1-g_n(f(s)) = k(f_n(0))(1-g_{n+1}(s)).$$

Since $f_n(0)\uparrow 1$ and $k(s)\uparrow m$, as $s\uparrow 1$, this yields (1.2). Now let $s\uparrow 1$ in (1.2),

$$1-g(1-) = m(1-g(1-)).$$

As $m < 1$, $g(-1) = 1$. That is, g is non-defective.

Suppose there is a second p.g.f. solution $\tilde{g}$ of (1.2). Iterate (1.2), differentiate,

$$g'(f_n(s))f_n'(s) = m^n g'(s),$$

(1.3)

$$\tilde{g}'(f_n(s))f_n'(s) = m^n\tilde{g}'(s),$$

and note that for every $s \in [0,1)$ there exists an integer ℓ such that

$$f_\ell(0) \leq s \leq f_{\ell+1}(0).$$

Then by (1.1) and repeated use of (1.3)

$$\frac{\tilde{g}'(s)}{g'(s)} = \frac{\tilde{g}'(f_n(s))}{g'(f_n(s))} \leq \frac{\tilde{g}(f_{n+\ell+1}(0))\cdot g'(f_{n+\ell+1}(0))}{g'(f_{n+\ell+1}(0))\cdot g'(f_{n+\ell}(0))}$$

$$\leq \frac{\tilde{g}'(0)}{g'(0)} \cdot \frac{g'(f_{n+\ell+1}(0))}{g'(f_{n+\ell}(0))} = \frac{\tilde{g}'(0)}{g'(0)}\, m\, \frac{f'_{n+\ell}(0)}{f'_{n+\ell+1}(0)}$$

$$= \frac{\tilde{g}'(0)}{g'(0)} \cdot \frac{m}{f'(f_{n+\ell}(0))} \to \frac{\tilde{g}'(0)}{g'(0)}, \quad n \to \infty.$$

That is,

$$\frac{\tilde{g}'(s)}{g'(s)} \leq \frac{\tilde{g}'(0)}{g'(0)}.$$

Since $\tilde{g}$ and g are interchangeable in the argument, we have, in fact, equality. As $\tilde{g}(0) = g(0) = 0$, this implies $g \equiv \tilde{g}$. □

1.4. REMARK. Let T be the extinction time, i.e., $T = n$ if and only if $Z_{n-1} > 0$ $Z_n = 0$. If $q > 0$, the limits

$$Q(j) = \lim_{n\to\infty} P(Z_n = j \mid n < T < \infty), \; j = 1,2,\ldots$$

exist regardless of the value of $m < \infty$. For $m = 1$ the proof is the same as for $m < 1$, but Q is no probability distribution, in fact $Q \equiv 0$. If $m > 1$, note that the generating function of $\{P(Z_n = j \mid n < T < \infty)\}$ may be written in the form

$$g_n^*(s) = \frac{f_n^*(s) - f_n^*(0)}{1 - f_n^*(0)},$$

where $f_n^*(s)$ is the n-fold iterate of $q^{-1}f(qs)$. Now the same argument applies once more, and the limit generating function g^* is the unique p.g.f. solution of

$$1-g^*\left(\frac{f(sq)}{q}\right) = f'(q)(1-g^*(s)).$$

1.5. PROPOSITION. *The limit function* g *of Theorem 1.2 satisfies*

$$1-g(1-s) = sL(s), \tag{1.4}$$

where $L(s)$ *is slowly varying as* $s \downarrow 0$.

In the proof we refer to the expansion

$$\begin{aligned} &1-f(s) = (m-r(s))(1-s) \\ &0 = r(1) \leq r(s) \leq r(s') \leq m, \quad 0 < s' \leq s \leq 1. \end{aligned} \tag{1.5}$$

PROOF OF 1.5. Define

$$h_n(s) := m^{-n}(1-g(1-s(1-f_n(0)))).$$

As f is convex,

$$f(1-s(1-f_n(0))) \leq 1-s(1-f_{n+1}(0)).$$

Since g is non-decreasing, it follows by (1.2) that $h_n(s)\downarrow$, $n\uparrow$. Hence, the limit

$$h(s) := \lim_{n\to\infty} h_n(s)$$

exists. It is concave, non-decreasing, and $h(0) = 0$, $h(1) = 1$. By (1.1) and (1.5) there exists for every $\epsilon > 0$ an n_0 such that for all $n \geq n_0$

$$(1-\epsilon)m(1-f_{n-1}(0)) \leq 1-f_n(0) \leq (1+\epsilon)m(1-f_{n-1}(0))$$

and thus

$$h_{n-1}((1-\epsilon)sm) \leq mh_n(s) \leq h_{n-1}((1+\epsilon)sm).$$

Since h is continuous, this yields

$$h(ms) = mh(s).$$

Since h is concave with $h(0) = 0$, it follows that $h(s)/s$ is constant, and hence, as $h(1) = 1$, that $h(s) = s$. That is,

$$\frac{1-g(1-s(1-f_n(0)))}{1-g(1-(1-f_n(0)))} \to s, \; n \to \infty.$$

Now apply the lemma of Rubin and Vere-Jones, A.13.6 . □

The property (1.4) has some simple, explicit consequences. Let W be a random variable with distribution $\{Q(j)\}$. Then it follows from (1.4) by Karamata's Tauberian theorem, A 14.1, that

$$\int_0^s P(W > t)dt = L_1(s^{-1}), \; s > 0,$$

where $L_1(s) \sim L(s)$, $s \to 0$. Applying the density version A 14.2 of A 14.1,

$$P(W > s) = o(s^{-1}L_1(s^{-1})),$$

which implies

$$EW^{\alpha} < \infty, \; 0 < \alpha < 1.$$

1.6. THEOREM. *Suppose* $0 < m < 1$. *Then*

$$P(Z_n > 0) = m^n L_2(m^n), \tag{1.6}$$

where L_2 *is slowly varying at* 0.

PROOF. Define

$$\alpha(s) := 1-g(1-s), \; s \in [0,1].$$

From (1.2)

$$\alpha(1-f_n(0)) = m^n.$$

That is,

$$1-f_n(0) = \alpha^{-1}(m^n).$$

Since α^{-1} is convex, $\alpha^{-1}(t)/t \uparrow$, $t \uparrow$. By (1.4), $\alpha(ms)/\alpha(s) \to m, s \to 0$. Hence, for ϵ such that $m+\epsilon < 1$ and $\lambda \in [m+\epsilon, 1]$,

$$\alpha(ms) \leq (m+\epsilon)\alpha(s) \leq \lambda\alpha(s) \leq \alpha(s)$$

for all sufficiently small s. Setting $t := \alpha(s)$, it follows that

$$1 \geq \frac{\alpha^{-1}(\lambda t)/\lambda t}{\alpha^{-1}(t)/t} \geq \frac{\alpha^{-1}(\alpha(ms))/\alpha(ms)}{\alpha^{-1}(\alpha(s))/\alpha(s)} = \frac{m\alpha(s)}{\alpha(ms)} \to 1, \; s \to 0.$$

That is, $\alpha^{-1}(t) = tL_2(t)$ with L_2 as proposed. □

From (1.4), $\alpha(1-f_n(0)) = (1-f_n(0))L(1-f_n(0))$, and we just had $\alpha(1-f_n(0)) = m^n$. Hence, by comparison

$$L_2(m^n)L(m^nL_2(m^n)) = 1.$$

1.7. THEOREM. Suppose $0 < m < 1$. Then for some constant γ

$$1-f_n(0) \sim \gamma m^n, \quad n \to \infty,$$

where $\gamma > 0$ if and only if

(X LOG X) $$\sum_{n=2}^{\infty} p_n n \log n < \infty.$$

1.8. LEMMA. Suppose $m < \infty$, $\lambda \in (0,1)$, $\epsilon \in (0,1)$. Then

(1.7) $$\sum_{\nu=1}^{\infty} r(1-\epsilon\lambda^{\nu}) < \infty$$

if and only if (X LOG X) is satisfied.

PROOF. We have

$$\int_0^{\infty} r(1-\epsilon\lambda^t)dt-m \leq \sum_{\nu=1}^{\infty} r(1-\epsilon\lambda^{\nu}) \leq \int_0^{\infty} r(1-\epsilon\lambda^t)dt.$$

Changing variables according to $s := -\log(1-\epsilon\lambda^t)$,

$$\int_0^{\infty} r(1-\epsilon\lambda^t)dt$$

$$= \frac{1}{|\log \lambda|}\int_0^{|\log(1-\epsilon)|} \{\sum_n p_n s^{-2}(e^{-ns}-1+ns)+ma(s)\}b(s)ds,$$

$$a(s) := s^{-2}(1-e^{-s}-s),$$

$$b(s) := s^2(1-e^{-s})^{-2}e^{-s}.$$

Noting that $a(s)$, $b(s)$ are bounded and bounded away from 0, rewrite

$$\int_0^{|\log(1-\epsilon)|} \sum_n p_n s^{-2}(e^{-ns}-1+ns)ds = \sum_n p_n n\int_0^{n|\log(1-\epsilon)|} \sigma^{-2}(e^{-\sigma}-1+\sigma)d\sigma,$$

and notice that

$$0 < c_1 \leq \frac{1}{\log(1+v)} \int_0^v \sigma^{-2}(e^{-\sigma}-1+\sigma)d\sigma \leq c_2 < \infty.$$

Hence (1.7) is equivalent to

$$\sum_n p_n n \log(1+n|\log(1-\epsilon)|) < \infty,$$

which is equivalent to (X LOG X). □

PROOF OF 1.7. Iterating (1.5),

$$m^{-n}(1-f_n(0)) = m^{-1}(1-f(0)) \prod_{\nu=1}^{n-1} \{1-m^{-1}r(f_\nu(0))\}.$$

That is, $\gamma > 0$ if and only if

$$\sum_{\nu=1}^{\infty} r(f_\nu(0)) < \infty.$$

If $\gamma > 0$, there exists an $\epsilon > 0$ such that $1-f_\nu(0) \geq \epsilon m^\nu$ for sufficiently large ν, so that

$$\sum_{\nu=1}^{\infty} r(1-\epsilon m^\nu) < \infty.$$

If $\gamma = 0$, then $1-f_\nu(0) \leq \epsilon m^\nu$ for sufficiently large ν, so that

$$\sum_{\nu=1}^{\infty} r(1-\epsilon m^\nu) = \infty.$$

Now apply the preceding lemma. □

1.9. COROLLARY. *Suppose* $0 < m < 1$. *Then*

$$\sum_n n\, Q(n) = 1/\gamma,$$

if (X LOG X) *is satisfied. Otherwise*

$$\sum_n n\, Q(n) = \infty.$$

PROOF. Notice that

$$\frac{1-g(1-(1-f_n(0)))}{1-f_n(0)} = \frac{m^n}{1-f_n(0)} .$$ □

1.10. REMARK. Let $m < 1$. Then the property

$$\sum_{j=n}^{\infty} jp_j \sim (\log n)^{-\alpha} L_3(\log n), \quad n \to \infty,$$

with $0 \leq \alpha < 1$ and L_3 slowly varying at infinity, is equivalent to

$$\log\left(\frac{1-f_n^{\prime}(s)}{m^n}\right) \sim \frac{|\log m|^{-\alpha}}{m(1-\alpha)} n^{1-\alpha} L_3(n), \quad n \to \infty,$$

$s \in [0,1]$, cf. Uchiyama (1976).

2. ARBITRARY INITIAL DISTRIBUTIONS AND INVARIANT MEASURES

Continuing with our study of subcritical processes, we now admit arbitrary initial distributions, characterize the domain of attraction of each limit distribution, and investigate the invariant measures and their connection to the Yaglom limits.

2.1. PROPOSITION. *Suppose* $0 < m < 1$. *If for some initial distribution* P_0 *with* $P_0(0) = 0$ *there exists a distribution* $\tilde{Q}$, *possibly defective (but not entirely concentrated at infinity), such that*

$$P(Z_n = k | Z_n \neq 0) \to \tilde{Q}(k), \; n \to \infty,$$

then $\tilde{Q}$ *is non-defective and its g.f.* $\tilde{g}$ *satisfies*

$$1-\tilde{g}(f(s)) = m^{\alpha}(1-\tilde{g}(s)), \; \tilde{g}(0) = 0, \tag{2.1}$$

with some $\alpha \in (0,1]$.

PROOF. Given (2.1), non-defectiveness follows as in III.1. Now let f_0 be the p.g.f. of P_0, and set

$$1-\tilde{g}_n(s) := \frac{1-f_0(f_n(s))}{1-f_0(f_n(0))} \, .$$

Then $\tilde{g}_n \to :\tilde{g}$, $\tilde{g}(0) = 0$, $\tilde{g} \not\equiv 0$. Moreover

$$1-\tilde{g}_n(f(s)) \to \delta(1-\tilde{g}(s)), \; n \to \infty,$$

$$\delta := \lim_n \frac{1-f_0(f_{n+1}(0))}{1-f_0(f_n(0))} = 1-\tilde{g}(f(0)).$$

Clearly $\delta < 1$. By convexity and the fact that for $s < 1$

$$\frac{1-f_{n+1}(s)}{1-f_n(s)} \to m, \; n \to \infty, \tag{2.2}$$

we have

$$1-f_0(f_{n+1}(0)) \geq 1-f_0(1-m(1-\epsilon_n)(1-f_n(0)))$$

$$\geq m(1-\epsilon_n)(1-f_0(f_n(0))$$

with $\varepsilon_n \to 0$, $n \to \infty$, so that $\delta \geq m$ and thus $\delta = m^\alpha$ with $\alpha \in (0,1]$, as proposed. □

2.2. REMARK. There exist constants $c_1 \in (0,1]$, $c_2 \in [1,\infty)$ such that for every p.g.f. solution $\tilde{g}$ of (2.1)

$$\lambda^\alpha c_1 \leq \liminf_{s\downarrow 0} \frac{1-\tilde{g}(1-\lambda s)}{1-\tilde{g}(1-s)} \leq \limsup_{s\downarrow 0} \frac{1-\tilde{g}(1-\lambda s)}{1-\tilde{g}(1-s)} \leq c_2\lambda^\alpha, \ \lambda > 0.$$

(R-O variation). Moreover, there is exactly one solution satisfying a regular variation relation,

$$1-\tilde{g}(1-s) = s^\beta L(s),$$

L slowly varying at 0, in which case $\beta = \alpha$ and

$$\tilde{g}(s) = 1-(1-g(s))^\alpha,$$

cf. Hoppe and Seneta (1978). As we have already seen, the solution is unique if $\alpha = 1$.

2.3. THEOREM. Suppose $0 < m < 1$. Then the initial distribution with p.g.f. f_0 leads to the Yaglom limit distribution with p.g.f. $\tilde{g}$, if and only if

$$\frac{1-f_0(1-s)}{1-\tilde{g}(1-s)} = L_3(s), \ 0 < s < 1, \tag{2.3}$$

where L_3 is slowly varying as $s \downarrow 0$.

PROOF. We first show that (2.3) is sufficient. If $0 < \varepsilon < 1$, then for sufficiently large n

$$(1-\varepsilon)(1-g(s))(1-f_n(0)) \leq 1-f_n(s) \leq (1+\varepsilon)(1-g(s))(1-f_n(0)). \tag{2.4}$$

This is immediate from the definition of g. Using (2.1) and (2.4),

$$\frac{1-f_0(1-(1-\varepsilon)(1-g(s))(1-f_n(0)))}{1-\tilde{g}(1-(1+\varepsilon)(1-g(s))(1-f_n(0)))} \cdot \frac{1-\tilde{g}(1-(1-\varepsilon)(1-f_n(0)))}{1-f_0(1-(1+\varepsilon)(1-f_n(0)))}$$

$$\leq \frac{1-\tilde{g}_n(s)}{1-\tilde{g}(s)}$$

$$\leq \frac{1-f_0(1-(1+\epsilon)(1-g(s))(1-f_n(0)))}{1-\tilde{g}(1-(1-\epsilon)(1-g(s))(1-f_n(0)))} \cdot \frac{1-\tilde{g}(1-(1+\epsilon)(1-f_n(0)))}{1-f_0(1-(1-\epsilon)(1-f_n(0)))} .$$

Given (2.3), it follows by convexity that

$$\left(\frac{1-\epsilon}{1+\epsilon}\right)^2 \leq \liminf_n \frac{1-\tilde{g}_n(s)}{1-\tilde{g}(s)} \leq \limsup_n \frac{1-\tilde{g}_n(s)}{1-\tilde{g}(s)} \leq \left(\frac{1+\epsilon}{1-\epsilon}\right)^2 .$$

Now let $\epsilon \downarrow 0$.

Next we prove the necessity of (2.3). For $\epsilon > 0$ fix t, t' so that $0 < t(1+\epsilon) < 1$, $0 < t'(1+\epsilon) < 1$. Then for sufficiently large n

$$1-f_n(t(1+\epsilon)) \leq (1-f_n(0))(1-g(t)) \leq 1-f_n(t(1-\epsilon))$$

and the corresponding inequalities with t' in place of t. From this

$$\frac{1-\tilde{g}_n(t(1+\epsilon))}{1-\tilde{g}_n(t'(1-\epsilon))} \leq \frac{1-f_0(1-(1-f_n(0))(1-g(t)))}{1-f_0(1-(1-f_n(0))(1-g(t')))} \leq \frac{1-\tilde{g}_n(t(1-\epsilon))}{1-\tilde{g}_n(t'(1+\epsilon))} .$$

That is,

$$\frac{1-f_0(1-(1-f_n(0))(1-g(t)))}{1-f_0(1-(1-f_n(0))(1-g(t')))} \to \frac{1-\tilde{g}(t)}{1-\tilde{g}(t')} , \quad n \to \infty.$$

Similarly,

$$\frac{1-\tilde{g}(1-(1-f_n(0))(1-g(t)))}{1-\tilde{g}(1-(1-f_n(0))(1-g(t')))} \to \frac{1-\tilde{g}(t)}{1-\tilde{g}(t')} , \quad n \to \infty.$$

Setting

$$\lambda_n := (1-f_n(0))(1-g(t')),$$

$$s := \frac{1-g(t)}{1-g(t')} ,$$

we get

$$\frac{1-f_0(1-\lambda_n s)}{1-\tilde{g}(1-\lambda_n s)} \cdot \frac{1-g(1-\lambda_n)}{1-f_0(1-\lambda_n)} \to 1, \quad n \to \infty.$$

Now apply A 13.6 . □

2.4. DEFINITION. We call $\{\pi_i\}$, $\pi_i \geq 0$, $i = 1,2,\ldots$, an invariant measure, if

$$(2.5) \qquad \pi_j = \sum_{i=1}^{\infty} \pi_i P_1(i,j), \; j \geq 1.$$

2.5. REMARK. Notice that we have excluded the state $j = 0$ in the definition 2.4. In fact, excluding the trivial case that $p_1 = 1$ and admitting $j = 0$, the only invariant measure is the trivial one, $\eta_0 = 1$, $\eta_j = 0$, $j \geq 1$: Setting $j = 0$ and recalling that $P_1(0,0) = 1$, (2.5) becomes

$$(2.6) \qquad 0 = \sum_{i=1}^{\infty} \pi_i P_1(i,0).$$

If $p_0 > 0$, then $P_i(i,0) > 0$ for $i \geq 1$ and (2.6) implies $\pi_i = 0$ for $i > 0$. If $p_0 = 0$, then $P_1(i,j) = 0$ for $i > j$, so that (2.5) becomes

$$(2.7) \qquad \pi_j = \sum_{i=1}^{j} \pi_i P_1(i,j).$$

Now let $k := \min\{j > 0 : \pi_j > 0\}$. Then (2.7) implies $\pi_k = \pi_k P_1(k,k)$. Since the exclusion of $p_1 = 1$ implies $P(k,k) < 1$ for $k > 0$, we arrive at $\pi_k = 0$ for $k > 0$.

If h is the generating function of $\{\pi_i\}$ and if $\{\pi_i\}$ is normalized such that $h(p_0) = 1$, then (2.5) is equivalent to

$$(2.8) \qquad h(f(s)) = h(s)+1.$$

2.6. PROPOSITION. There is a bijective correspondence between the g.f. solutions of (2.8) and the elements of

$$\{(\tilde{g},\alpha) : \alpha \in (0,1], \; \tilde{g} \text{ is a p.g.f. solution of } (2.1)\}$$

The proof is deferred to the end of this section.

2.7. PROPOSITION. The relation

$$(2.9) \qquad h(s) = \tilde{d}(g(s)), \; s \in [0,1),$$

defines a bijective correspondence between the g.f. solutions of (2.8) and those of

$$\tilde{d}(1-m+ms) = \tilde{d}(s)+1, \ \tilde{d}(0) = 0. \tag{2.10}$$

PROOF. Given a g.f. solution $\tilde{d}$ of (2.10), $\tilde{d}(g(s))$ obviously satisfies (2.8). Given $h(s)$, we have to construct a g.f. solution $\tilde{d}$ of (2.10) such that (2.9) is satisfied. Define

$$d_n(s) := h(1-s(1-f_n(0)))-n, \ n = 1,2,\dots .$$

By convexity and (2.8)

$$\begin{aligned} d_{n+1}(s) &= h(1-s(1-f(f_n(0))))-n-1 \\ &\geq h(f(1-s(1-f_n(0))))-n-1 \\ &= h(1-s(1-f_n(0)))-n = d_n(s). \end{aligned}$$

Moreover,

$$d_n(s) \leq h(f_\ell(0)),$$

whenever $\ell \geq \log s/\log m$. This is immediate from

$$1-f_{n+\ell}(0) \leq m^\ell(1-f_n(0)).$$

Hence, $d_n(s) \to :d(s)$, and in view of (2.2),

$$d(ms) = d(s)+1, \ d(1) = 0.$$

Since from (2.8)

$$h(s) = h(1-(1-f_n(s))-n,$$

it follows by (2.4) and the definition of d that

$$h(s) = d(1-g(s)).$$

Now take $\tilde{d}(s) := d(1-s)$. □

From (2.10)

$$\log(1-s)/\log m \leq \tilde{d}(s) \leq \log(1-s)/\log m + 1.$$

Combining this with (2.9) and (1.4) we have the following:

2.8. COROLLARY. Suppose $0 < m < 1$. Then

$$\frac{h(1-s)\log m}{\log s} \to 1, \quad s \to 0,$$

and thus (by a Tauberian theorem)

$$\sum_{k\leq\lambda} \pi_k \sim \log \lambda/|\log m|, \quad \lambda \to \infty.$$

2.9. THEOREM. For every g.f. solution $\tilde{d}$ of (2.10) there exists a probability measure ν on $[0,1]$ such that

$$\tilde{d}(s) = \int_0^1 \sum_{n=-\infty}^{\infty} (\exp\{-(1-s)m^{n-t}\}-\exp\{-m^{n-t}\})\nu(dt).$$

PROOF. We are after the invariant measures of a process $\{\tilde{Z}_n\}$ with p.g.f. $\tilde{f}(s) := 1-m+ms$. Let $\tilde{P}_n(i,j)$ be the corresponding n-step transition function, and define the Green function

$$G(i,j) := \sum_{n=1}^{\infty} \tilde{P}_n(i,j).$$

If for some sequence of integers $k_i \to \infty$ the limits

$$\pi_j := \lim_{i\to\infty} G(k_i,j), \quad j \geq 1,$$

exist and are finite, then $\{\pi_j\}$ is an invariant measure of $\{\tilde{Z}_n\}$, and the set of invariant measures is given by the set of mixtures of such sequences, cf. Kemeny, Snell, and Knapp (1966). Define

$$H(s,k) := \sum_{j=1}^{\infty} G(k,j)s^j,$$

and note that

$$H(s,k) = \sum_{n=0}^{\infty} [\tilde{f}_n(s)^k-\tilde{f}_n(0)^k]$$

$$= \sum_{n=0}^{\infty} [(1-m^n(1-s))^k-(1-m^n)^k].$$

Since for $s \in [0,1]$ and $k > 0$

$$\sum_n [(1-m^n(1-s))^k-(1-m^n)^k] < \infty,$$

$$\sum_n [\exp\{-m^n(1-s)k\}-\exp\{-m^n k\}] < \infty,$$

$$\Delta_{k,n} := |(1-m^n(1-s))^k - \exp\{-m^n(1-s)^k\}|$$

$$\leq \tfrac{1}{2} m^n(1-s)^2(k-1)\exp\{-m^n(1-s)(k-1)\} \to 0, \quad k\to\infty,$$

and

$$\sum_n \Delta_{k,n} \leq \tfrac{1}{2}\sum_n m^n(1-s)^2 \sum_k [1-m^n(1-s)]^k \leq \tfrac{1}{2}\sum_n m^n(1-s) < \infty,$$

it follows by dominated convergence that

$$\lim_{k_i\to\infty} H(s,k_i) = \lim_{k_i\to\infty} \sum_{n=0}^{\infty} [\exp\{-m^n(1-s)k_i\}-\exp\{-m^n k_i\}].$$

Defining n_i, δ_i as the integral and fractional parts of $-\log k_i/\log m$, respectively,

$$\lim_{k_i\to\infty} H(s,k_i) = \lim_{k_i\to\infty} \sum_{j=-n_i}^{\infty} [\exp\{-(1-s)m^{j-\delta_i}\}-\exp\{-m^{j-\delta_i}\}].$$

Thus $\{H(s,k_i)\}$ is a Cauchy sequence if and only if $\{k_i\}$ is a Cauchy sequence respective the metric

$$\rho(x,y) := |\delta_x-\delta_y|+|x^{-1}-y^{-1}|, \quad x,y \in \mathbb{Z},$$

$$:= |\delta_x-y|+|x^{-1}|, \quad x \in \mathbb{Z},\ y \in [0,1),$$

$$:= |x-y|, \quad x,y \in [0,1),$$

where δ_x is the fractional part of $-\log x/\log m$, i.e., if $k_i \to \infty$ such that $\delta_i \to t$,

$$\lim_{k_i\to\infty} H(s,k_i) = \sum_{j=-\infty}^{\infty} [\exp\{-(1-s)m^{j-t}\}-\exp\{-m^{j-t}\}]. \qquad \square$$

2.10. COROLLARY. <u>For every solution</u> h <u>of (2.8) there exists a probability measure</u> ν <u>on</u> $[0,1]$ <u>such that</u>

$$h(s) = \int_0^1 \sum_{n=-\infty}^{\infty} [\exp\{-(1-g(s))m^{n-t}\}-\exp\{-m^{n-t}\}]\nu(dt).$$

2.11. REMARK. If $m = 1$, there exists exactly one invariant measure, cf. Athreya and Ney (1972). If $m > 1$ and $p_0 = 0$, there are no invariant measures: If $p_0 = 0$, then $P_1(i,j) = 0$ for $i > j$ and $\pi_j \equiv 0$ is the only invariant measure, cf. 2.5. The case $m > 1$, $p_0 \neq 0$, can be reduced to the case $m < 1$. Define

$$f^*(s) := f(qs)/q,$$

which is the offspring p.g.f. of a subcritical process with transition function

$$P_1^*(i,j) := P_1(i,j)q^{j-i}.$$

Hence, $\{\pi_i\}$ is an invariant measure of the original process if and only if

$$\pi_j^* := q^j \pi_j$$

defines an invariant measure of the transformed process.

We now return to the proof of 2.6 .

2.12. LEMMA. For every $\alpha \in (0,1]$ the relation

$$\text{(2.11)} \qquad \tilde{g}(s) = \tilde{a}(g(s)), \qquad s \in 0,1 ,$$

defines a bijective correspondence between the p.g.f. solutions of (2.1) and those of

$$\text{(2.12)} \qquad 1 - \tilde{a}(1-m+ms) = m^\alpha(1-\tilde{a}(s)), \qquad \tilde{a}(0) = 0 .$$

PROOF. Given a solution of (2.12), (2.11) clearly provides a solution of (2.1). For the reverse, define

$$a_n(s) := m^{-n\alpha}(1 - \tilde{g}(1-s(1-f_n(0)))).$$

Then $a_n(s)\downarrow$, $n\uparrow$, so $a_n(s) \to a(s)$, say, and $a(ms) = m^\alpha a(s)$. Observe that

$$\tilde{a}(s) := 1 - a(1-s)$$

solves (2.12). To get (2.11), recall (2.4), which implies

$$a_n((1-\varepsilon)(1-g(s)) \leq m^{-n\alpha}(1-\tilde{g}(f_n(s)))$$

$$= 1 - \tilde{g}(s) \leq a_n((1+\varepsilon)(1-g(s)))$$

and thus in the limit $a(1-g(s)) = (1-\tilde{g}(s))$, that is, (2.11). □

PROOF OF 2.6. In view of 2.7 and 2.12, we need only to establish a bijective correspondence between the g.f. solutions of (2.10) and the elements of

$$\{(\tilde{a},\alpha) : \alpha \in (0,1],\ \tilde{a} \text{ is a p.g.f. solution of (2.12)}\}$$

Given an element of this set,

$$\tilde{d}(s) := \log(1-\tilde{a}(s)/\alpha \log m$$

is a g.f. and clearly solves (2.10). Vice versa, given a solution $\tilde{d}$ of (2.10) and $\alpha \in (0,1]$, it is easily checked that

$$\tilde{a}(s) := \int_0^s m^{(\alpha-1)\tilde{d}(u)} du \Big/ \int_0^{1-} m^{(\alpha-1)\tilde{d}(v)} dv$$

is a p.g.f. solution of (2.12). Now recall Bernstein's theorem. □

3. CRITICAL PROCESSES

Suppose $m = 1$, but $p_1 \neq 1$. Then $p_0 + p_1 < 1$, f is strictly convex and increasing, 1 is the only solution of $f(s) = s$, i.e. $q = 1$ and

$$f(s) > s, \quad s \in [0,1).$$

3.1. PROPOSITION. *If* $m = 1$, $p_1 \neq 1$, *and*

$$\mu := \frac{1}{2} f''(1-) = \frac{1}{2} \sum_{n=2}^{\infty} n(n-1)p_n < \infty,$$

then

$$\lim_{n\to\infty} \frac{1}{n}\left[\frac{1}{1-f_n(s)} - \frac{1}{1-s}\right] = \mu \tag{3.1}$$

uniformly in $s \in [0,1)$.

The proof will be based on the following lemma.

3.2. LEMMA. (Spitzer's comparison lemma.) *If* f *and* $\tilde{f}$ *are generating functions with*

$$m = \tilde{m} := \tilde{f}'(1-) = 1, \; \mu < \tilde{\mu} := \frac{1}{2}\tilde{f}''(1-) < \infty$$

then there exist integers k *and* ℓ *such that*

$$f_{n+k}(s) \leq \tilde{f}_{n+\ell}(s), \quad s \in [0,1], \quad n = 0,1,2,\dots .$$

PROOF. Note that for $0 \leq s \uparrow 1$

$$\epsilon(s) := \frac{f''(-1)}{2} - \frac{f(s)-s}{(1-s)^2} = \sum_{k=3}^{\infty} p_k \sum_{j=2}^{k-1} \sum_{i=1}^{j-1} (1-s^i) \downarrow 0.$$

Since by assumption $\mu < \tilde{\mu}$, it follows that there exists an $s_0 \in (0,1)$ such that $f(s) \leq \tilde{f}(s)$ for $s \in [s_0,1]$, hence $f_n(s) \leq \tilde{f}_n(s)$ for $s \in [s_0,1]$, $n \geq 1$, so that

$$f_{n+k}(s) \leq \tilde{f}_{n+\ell}(s), \quad s \in [s_0,1],$$

whenever $\ell > k \geq 0$. As $f_n(s) \to 1$, $\tilde{f}_n(s) \to 1$ for $s \in [0,1]$, we can pick k, ℓ so that $s_0 \leq f_k(0)$ and $f_k(s_0) \leq \tilde{f}_\ell(0)$. Consequently, for $s \in [0,s_0]$

$$s_0 \leq f_k(0) \leq f_k(s) \leq f_k(s_0) \leq \tilde{f}_\ell(0) \leq \tilde{f}_\ell(s)$$

and thus also

$$f_{n+k}(s) \leq \tilde{f}_{n+\ell}(s), \quad s \in [0,s_0]. \qquad \square$$

PROOF OF 3.1. Let $\tilde{f}$ be linear fractional with $\tilde{m} = 1$, i.e.

$$\tilde{f}(s) = \frac{p-(2p-1)s}{1-ps}$$

with some $p \in (0,1)$. Then

$$\tilde{f}_n(s) = \frac{np-(np+p-1)s}{1-p+np-nps},$$

cf. 1.1, and thus

$$(1-\tilde{f}_n(s))^{-1} - (1-s)^{-1} = np(1-p)^{-1} = \frac{1}{2} n\tilde{f}''(-1) = n\tilde{\mu}.$$

With an arbitrary $\varepsilon > 0$ take $p(1-p)^{-1} := (1+\varepsilon)\mu$. Then by 3.2 for some $k, \ell \in \mathbb{N}$

$$f_{n+k}(s) \leq \tilde{f}_{n+\ell}(s), \quad s \in [0,1], \quad n \geq 0,$$

and thus

$$(1-f_{k+n}(s))^{-1} - (1-s)^{-1} \leq (1-\tilde{f}_{n+\ell}(s))^{-1} - (1-s)^{-1} = (n+\ell)(1+\varepsilon)\mu.$$

Similarly, with $p(p-1)^{-1} := (1-\varepsilon)\mu$, for some $h, j \in \mathbb{N}$

$$(n+j)(1-\varepsilon)\mu \leq (1-f_{h+n}(s))^{-1} - (1-s)^{-1},$$

which completes the proof. $\square$

REMARK. There exists an historically older, more direct, but less elegant proof. It is contained in the subsequent section, and an extension of it will later be used for general processes.

3.3. COROLLARY. *If* $p_1 \neq 1$, $m = 1$, *and* $\mu < \infty$, *then*

$$(3.2) \qquad 1-f_n(0) = P(Z_n > 0 | Z_0 = 1) \sim (n\mu)^{-1}, \quad n \to \infty.$$

3.4. THEOREM. *If* $p_1 \neq 1$, $m = 1$, *and* $\mu < \infty$, *then*

$$(3.3) \qquad \lim_{n\to\infty} P(Z_n/n \geq \lambda | Z_n > 0,\ Z_0 = 1) = e^{-\lambda/\mu}, \quad \lambda \geq 0.$$

PROOF. Using (3.1), particularly the uniformity,

$$\begin{aligned} &E(\exp\{-Z_n s/n\} | Z_n > 0,\ Z_0 = 1) \\ &= \frac{f_n(e^{-s/n}) - f_n(0)}{1-f_n(0)} = 1 - \frac{n(1-f_n(e^{-s/n}))}{n(1-f_n(0))} \\ &\to 1 - \mu(\mu + \lim_{n\to\infty} [n(1-e^{-s/n})]^{-1})^{-1} = (1+s\mu)^{-1}. \end{aligned}$$

Now apply the continuity theorem for Laplace transforms. □

3.5. COROLLARY. *Under the assumptions of* 3.4

$$\lim_{n\to\infty} E\,(Z_n/n \,|\, Z_n > 0,\ Z_0 = 1) = \mu.$$

3.6. REMARK. The asymptotic behaviour of "near-critical" processes is "close" to that of critical processes. More precisely, fix $a, b, c > 0$, let K be the set of proper p.g.f.'s such that

$$m = f'(1-) \geq a, \quad \mu = \tfrac{1}{2} f''(1-) \geq b, \quad f'''(1-) \leq c,$$

define

$$\begin{aligned} c_n &:= \frac{m^n - 1}{m(m-1)}, \quad && m \neq 1, \\ &:= n, && m = 1, \end{aligned}$$

and denote by $o(n,f)$ any real-valued function such that

$$\forall \varepsilon > 0\ \exists \delta > 0,\ N \in \mathbb{N}: |o(n,f)| < \varepsilon\ \forall n > N,\ f \in K \cap \{|1-m| < \delta\}.$$

Then for $n \in \mathbb{N}$ and $f \in K$

$$c_n P(Z_n > 0 | Z_0 = 1) = m^n(\mu^{-1} + o(n,f)),$$

$$\sup_{\lambda \in \mathbb{R}} |P(Z_n/c_n \leq \lambda | Z_n > 0,\ Z_0 = 1) - (1-e^{-\lambda/\mu})| = o(n,f),$$

$$E(Z_n | Z_n > 0,\ Z_1 = 1) = c_n \mu (1+o(n,f))^{-1},$$

cf. Fahady, Quine, Vere-Jones (1971).

What happens if $\mu = \infty$?

3.7. THEOREM. *Suppose* $p_1 \neq 1$, $m = 1$ *and* $\mu = \infty$. *If*

(S) $$r(1-s) = s^{\alpha} L(s), \quad s \in [0,1],$$

where $\alpha \in (0,1]$ *and* $L(s)$ *is slowly varying, as* $s \downarrow 0$, *then*

$$a_n := P(Z_n > 0 | Z_0 = 1) \sim n^{-1/\alpha} L^*(n), \quad n \to \infty,$$

where L^* *is slowly varying at* ∞, *and*

$$\lim_{n\to\infty} P(a_n Z_n \leq \lambda | Z_n > 0,\ Z_0 = 1) = F(\lambda),\ \lambda \geq 0,$$

$$\int_0^{\infty} e^{-s\lambda} dF(\lambda) = 1 - s(1+s^{\alpha})^{-1/\alpha}, \quad s \geq 0.$$

Conversely, if $(a_n Z_n | Z_n > 0,\ Z_0 = 1)$ *converges in distribution to a non-degenerate limit, admitted to be defective, then* (S) *must hold.*

Do there exist normalizing constants, necessarily different from a_n, leading to a non-degenerate limit, if (S) is not satisfied?

3.8. THEOREM. *Suppose* $p_1 \neq 1$ *and* $m = 1$. *If for some normalizing sequence* $\{c_n\}$, $(c_n Z_n | Z_n > 0,\ Z_1 = 1)$ *converges in distribution to a proper limit, non-degenerate at* 0, *then* $c_n/a_n \sim C$, $n \to \infty$, *with some* $C \in (0,\infty)$.

For proofs of 3.7 and 3.8 we refer to the generalizations in VI.4. and VI. 5.

4. LOCAL LIMIT THEOREMS FOR CRITICAL PROCESSES

4.1. THEOREM. *Let* $p_1 \neq 1$, $m = 1$, *and* $\mu < \infty$. *Then*

$$\mu n^2 P_n(i,j) \underset{n\to\infty}{\to} i\Pi_j^* \geq 0, \quad i,j \geq 0,$$

where

$$(4.1) \qquad \begin{aligned} &\sum_{i=1}^{\infty} \Pi_i^* P(i,j) = \Pi_j^*, \quad j \geq 1, \\ &\sum_{j=1}^{\infty} \Pi_j^* p_0^j = 1, \end{aligned}$$

and

$$\frac{1}{n} \sum_{j=1}^{n} \Pi_j^* \underset{n\to\infty}{\to} \frac{1}{\mu} .$$

REMARK. It can be shown that with $\pi_j = j$

$$(4.2) \qquad \sum_{j=1}^{\infty} P(i,j)\pi_j = \pi_i, \quad i \geq 1,$$

and that -up to a constant factor- the solutions of (4.1) and (4.2) are unique, cf. Kesten, Ney, Spitzer (1966).

4.2. LEMMA. *Under the assumptions of 4.1*

$$n^2[f_{n+1}(s) - f_n(s)] \underset{n\to\infty}{\to} \frac{1}{\mu}, \quad s \in [0,1).$$

PROOF. By three-term Taylor expansion of f near 1,

$$(4.3) \qquad \frac{f(s)-s}{(1-s)^2} = \mu - \varepsilon(s),$$

$$\varepsilon(s) := \sum_{k=3}^{\infty} p_k \sum_{j=2}^{k-1} \sum_{\nu=1}^{j-1} (1-s^\nu), \quad s \in [0,1),$$

$$0 \leq \varepsilon(s) \downarrow 0, \quad s \uparrow 1.$$

Insert $s = f_n(t)$ into (4.3) and use 3.1. □

4.3. LEMMA. *Under the assumptions of 4.1*

$$\mu n^2[f_n(s) - f_n(0)] \underset{n\to\infty}{\to} : U(s) < \infty, \quad s \in [0,1).$$

PROOF. From (4.3)

$$(4.4) \qquad 0 \leq \epsilon(s) \leq \mu, \quad s \in [0,1].$$

Define

$$(4.5) \qquad \delta(s) := (1-s)^{-1} + \mu - (1 - f(s))^{-1}.$$

Then

$$(4.6) \qquad \delta(s) = \frac{\epsilon(s)-\mu(1-s)(\mu-\epsilon(s))}{1 - (1-s)(\mu-\epsilon(s))},$$

and by (4.4)

$$-\mu^2(1-s) \leq \epsilon(s) - \mu(1-s)(\mu-\epsilon(s)) \leq \delta(s)$$

$$\leq \frac{\epsilon(s)-\epsilon(s)(1-s)(\mu-\epsilon(s))}{1 - (1-s)(\mu-\epsilon(s))} = \epsilon(s), \quad s \in [0,1).$$

Now put $s = f_k(t)$ in (4.5) and (4.6) and sum over k from 0 to $n-1$. Keeping in mind that $f_0(t) := t$, we get

$$(4.7) \quad (1-t)^{-1}+n\mu - (1-f_n(t))^{-1} = \sum_{k=0}^{n-1} \delta(f_k(t)), \; n \geq 1, \; t \in [0,1).$$

Subtracting from (4.7) the same equation with $t = 0$ yields

$$\varphi_n(t) := (1-f_n(t))^{-1} - (1-f_n(0))^{-1} - t(1-t)^{-1} = \sum_{k=0}^{n-1} [\delta(f_k(0))-\delta(f_k(t))].$$

Using (4.5) and $\epsilon_k(s) := \epsilon(f_k(s))$,

$$(4.8) \quad \varphi_n(t) = \sum_{k=0}^{n-1} \frac{\epsilon_k(0)-\epsilon_k(t)+(\mu-\epsilon_k(0))(\mu-\epsilon_k(t))(f_k(0)-f_k(t))}{[1-(1-f_k(0))(\mu-\epsilon_k(0))][1-(1-f_k(t))(\mu-\epsilon_k(t))]}.$$

By (4.3)

$$(1-s)(\mu-\epsilon(s)) = \frac{f(s)-s}{1-s} < 1, \quad s \in [0,1),$$

so that the denominators in (4.8) are non-negative, and since $f_k(0) \leq f_k(t) \leq 1$ and $f_k(0) \uparrow 1$, as $k \to \infty$, they tend to 1, as $k \to \infty$. Moreover, for j such that $f_j(0) \geq t$,

$$0 \leq \epsilon_k(0) - \epsilon_k(t) \leq \epsilon_k(0) - \epsilon_{k+j}(0),$$

$$0 \leq f_k(t) - f_k(0) \leq f_{j+k}(0) - f_k(0),$$

so that the series

$$\sum_k(\epsilon_k(0) - \epsilon_k(t)), \quad \sum_k(f_k(t) - f_k(0))$$

converge absolutely. Hence

$$\varphi(t) := \lim_{n\to\infty} \varphi_n(t)$$

exists pointwise in $[0,1)$. Rewriting

$$n^2[f_n(t)-f_n(0)] = n[1-f_n(t)]n[1-f_n(0)][\varphi_n(t)+t(1-t)^{-1}]$$

and recalling that $n(1-f_n(t)) \to \mu^{-1}$ by 3.1, we finally arrive at 4.3. □

REMARK. Notice that (4.7) provides an alternative proof of 3.1.

PROOF OF 4.1. Step 1. We verify the first proposition of 4.1 for $i = 1$. For $s \in (0,1)$ and $k \geq 1$

$$(4.9) \qquad n^2P_n(1,k) \leq s^{-k}n^2 \sum_{j=1}^{\infty} P_n(1,j)s^j = s^{-k}n^2(f_n(s)-f_n(0)).$$

Hence 4.3 permits the choice of a subsequence $\{n'\}$ of positive integers such that

$$\mu(n')^2P_{n'}(1,k) \to: \Pi_k^*, \quad k \geq 1.$$

For fixed $t \in [0,1)$ pick $s \in (t,1)$. Then (4.9) gives bounds allowing to conclude by dominated convergence that

$$\lim_{n'\to\infty} \mu(n')^2 \sum_{k=1}^{\infty} P_{n'}(1,k)t^k = \sum_{k=1}^{\infty} \Pi_k^* t^k = U(t).$$

That is, $U(t)$ is analytic in $|t| < 1$. Hence, $n^2P_n(1,k)$ has a limit, as $n \to \infty$, for each k.

Step 2. We show that

$$P_n(i,j) \sim iP_n(1,j), \quad n \to \infty.$$

In fact,

$$(4.10)\qquad \begin{aligned} 0 &\le P_n(i,j) - iP_n(1,j)[P_n(1,0)]^{i-1} \\ &= \sum_{\nu=2}^{\min(i,j)} \binom{i}{\nu}[P_n(1,0)]^{i-\nu} \sum_{\substack{j_k>0 \\ j_1+\cdots+j_\nu=j}} P_n(1,j_1)\cdots P_n(1,j_\nu), \end{aligned}$$

which is obtained, e.g., by comparing the coefficients of s^j, $j \ge 1$, in the binomial expansion

$$\begin{aligned} &[f_n(s)]^i - [f_n(0)]^i \\ &\quad = i[f_n(0)]^{i-1}[f_n(s)-f_n(0)] + \sum_{\nu=2}^{i} \binom{i}{\nu}[f_n(0)]^{i-\nu}[f_n(s)-f_n(0)]^\nu. \end{aligned}$$

Notice now that each term in the finite sum on the right of (4.10) is of order $n^{-2\nu}$, $\nu \ge 2$.

Step 3. Next we convince ourselves of (4.1). Combining 4.2 and 4.3, we have for $t \in [0,1)$

$$(4.11)\qquad U(f(t)) = \lim_{n\to\infty} \mu n^2[f_{n+1}(t) - f_n(0)] = U(t) + 1.$$

Insert $U(t) = \Sigma_{k=1}^{\infty} \Pi_k^* t^k$ and $f(t) = \Sigma_{j=0}^{\infty} p_j t^j$, and compare coefficients of t^n.

Step 4. From (4.11)

$$U(f_n(0)) = U(0) + n = n \sim \mu^{-1}\,(1-f_n(0))^{-1},\ n \to \infty.$$

For $t \in [f_n(0), f_{n+1}(0)]$

$$(1-f_{n+1}(0))U(f_n(0)) \le (1-t)U(t) \le (1-f_n(0))U(f_{n+1}(0)).$$

Moreover,

$$\frac{1-f_{n+1}(0)}{1-f_n(0)} \underset{n\to\infty}{\to} m = 1,$$

so that

$$\sum_{k=1}^{\infty} \Pi_k^* t^k = U(t) \sim [\mu(1-t)]^{-1},\quad t \uparrow 1.$$

According to A.14.3 this proves the last proposition of 4.1. □

REMARK. A different proof of the first part of 4.1 via ratio theorems can be found in Athreya, Ney (1972).

4.4. THEOREM. *Suppose* $p_1 \neq 1$, $m = 1$, *and*

$$\sum_j p_j j^2 \log j < \infty.$$

Let $j, n \to \infty$ *such that* j/n *is bounded. Then*

$$n^2 P_n(i,j) e^{j/(n\mu)} \underset{n\to\infty}{\to} id\mu^{-2},$$

where d *is the greatest common divisor of* $\{k\colon P(1,k) > 0\}$.

A very elaborate proof of 4.4 has been given by Kesten, Ney, Spitzer (1966). The proof becomes much simpler, if one is content with higher moment assumptions, cf. Athreya, Ney (1972). A proof based on finite second moments only has apparently not yet been found.

5. SUPERCRITICAL PROCESSES

Suppose $1 < m < \infty$. Since f is convex, non-decreasing, with $0 \leq f(s) \leq f(1) = 1$, $m = f'(1-)$, this implies $q < 1$. Furthermore, f must be nontrivial, i.e. strictly convex. Thus f has exactly two fixed points, namely 1 and q, and

$$\lim_{n\to\infty} f_n(s) = q, \quad s \in [0,1).$$

Fix $s_0 \in (q,1)$, and define

$$s_n := f_n^{-1}(s_0).$$

Notice that $s_{n+k} = f_k^{-1}(s_n)$ and

$$q < s_n < 1, \quad s_n \uparrow 1, \quad n \uparrow \infty.$$

Set

$$\gamma_n = -\log s_n.$$

Expanding

$$(5.1) \qquad \begin{aligned} &1 - f(s) = m(1-s) - r(s)(1-s) \\ &r(s) \downarrow 0, \quad s \uparrow 1, \end{aligned}$$

we get

$$(5.2) \qquad \lim_{n\to\infty} \frac{\gamma_n}{\gamma_{n+1}} = \lim_{n\to\infty} \frac{1-s_n}{1-s_{n+1}} = \lim_{n\to\infty} \frac{1-f(s_{n+1})}{1-s_{n+1}} = m.$$

5.1. THEOREM. *There exists a random variable* W *such that*

$$(5.3) \qquad \lim_{n\to\infty} \gamma_n Z_n = W \quad \text{a.s.},$$

the Laplace transform $\varphi(s) = E(e^{-sW} | Z_0 = 1)$, $s \geq 0$, *satisfies*

$$(5.4) \qquad \varphi(ms) = f(\varphi(s)), \quad s \geq 0,$$

and in particular

$$(5.5) \qquad P(W = 0 | Z_0 = 1) = q, \quad P(W < \infty) = 1.$$

PROOF. Convergence in (5.3) follows by the martingale convergence theorem from the fact that $\exp\{-\gamma_n Z_n\}$ is a non-negative martingale resp. $\sigma(Z_0, \ldots, Z_n)$. Using (5.2) and dominated convergence,

$$\begin{aligned}
\varphi(sm) &= \lim_{n\to\infty} f_{n+1}(\exp\{-sm\gamma_{n+1}\}) \\
&= f(\lim_{n\to\infty} f_n(\exp\{-s(m\gamma_{n+1}/\gamma_n)\gamma_n\})) \\
&= f(\varphi(s)),
\end{aligned}$$

which not only proves (5.4) but also shows that $\varphi(0+)$ and $\varphi(\infty-)$ are fixed points of f. However, the only fixed points of f were 1 and q. Now

$$\varphi(\infty-) \leq \varphi(1) \leq \varphi(0+), \quad \varphi(1) = \lim_{n\to\infty} f_n(e^{-\gamma_n}) = s_0,$$

and, by assumption, $q < s_0 < 1$. Hence, $\varphi(\infty-) = q$ and $\varphi(0+) = 1$, which proves (5.5). □

REMARK. The general norming constants γ_n were originally introduced in an analytical proof of convergence in distribution by Seneta (1968).

5.2. THEOREM. Up to a scale factor of s there is exactly one strictly decreasing, convex solution of (5.4) satisfying $\varphi(0+) = 1$. Moreover,

$$(5.6) \qquad \lim_{s\downarrow 0} \frac{1-\varphi(\lambda s)}{1-\varphi(s)} = \lambda, \quad \lambda \geq 0.$$

PROOF. Let ψ be any strictly decreasing, convex solution of (5.4) with $\psi(0+) = 1$, and define

$$s_n(a) := \psi(am^{-n}), \quad \gamma_n(a) := -\log s_n(a), \quad a > 0.$$

Then 5.1 holds with $\gamma_n(a)$ in place of γ_n and some $W(a)$ in place of W. Hence,

$$\gamma_n(a) \sim K(a)\gamma_n, \quad 0 < K(a) < \infty.$$

We now show that $K(a) = K\cdot a$, $K \equiv$ const.. Set

$$s\chi(s) := 1-\psi(s), \quad g(s) := 1 - f(1-s),$$

then $\chi(s)\downarrow$, as $s\downarrow$, further for $s\in(0,1]$

$$\chi(ms) = g(\chi(s))/ms,$$

and thus for $\lambda\in[0,m]$

$$1 \geq \chi(\lambda s)/\chi(s) \geq \chi(ms)/\chi(s)$$

$$= g(s\chi(s))/ms\chi(s) = 1 - r(\psi(s))/m \uparrow 1, \quad s\downarrow 0,$$

which is (5.6). On the other hand

$$\chi(am^{-n}) = 1 - \psi(am^{-n}) \sim -\log \psi(am^{-n}) = \gamma_n(a).$$

Given $a, b > 0$ and $\lambda = b/a$, and letting $\{am^{-n}, n\in\mathbb{N}\}\ni s \to 0$, this yields

$$\frac{\gamma_n(b)}{\gamma_n(a)} \to \frac{b}{a}, \quad n\to\infty,$$

so that indeed $K(s) = K\cdot s$. Given any $\varepsilon > 0$,

$$(1-\varepsilon)Ks\gamma_n \leq \gamma_n(s) \leq (1+\varepsilon)Ks\gamma_n$$

for all sufficiently large n. Hence,

$$f_n(e^{-(1+\varepsilon)Ks\gamma_n}) \leq f_n(e^{-\gamma_n(s)}) = \psi(s) \leq f_n(e^{-(1-\varepsilon)Ks\gamma_n}),$$

so that $\varphi(Ks) = \psi(s)$. $\square$

By definition, (5.6) means that

$$1 - \varphi(s) = sL(s), \tag{5.7}$$

where L is slowly varying at 0.

5.3. COROLLARY. *For every* $s_0\in(q,1)$ *there exists a constant* $a > 0$ *such that*

$$s_n = \varphi(am^{-n})$$

and thus

$$\gamma_n = m^{-n}L_1(m^{-n}),$$

where L_1 *is slowly varying at* 0.

5.4. REMARK. Given $1 < m < \infty$, the relation

$$\sum_{j=n}^{\infty} jp_j \sim (\log n)^{-\alpha} L^*(\log n), \quad n \to \infty,$$

$0 \leq \alpha < 1$, L^* slowly varying at ∞, is equivalent to

$$\log(m^{-n} Z_n) \sim \frac{|\log m|^{-\alpha}}{m(1-\alpha)} n^{1-\alpha} L^*(n) \quad \text{a.s. on} \quad \{Z_n \to \infty\},$$

cf. Uchiyama (1976).

Applying A 14.1 and A 14.2 to (5.9) we get the following.

5.5. COROLLARY. *As* $x \to \infty$,

$$\int_0^x P(W > t)dt \sim L_2(x^{-1}),$$

$$P(W > x) = o(x^{-1} L_2(x^{-1})), \tag{5.8}$$

where L_2 *is slowly varying at* 0.

Notice that (5.10) implies

$$EW^{\alpha} < \infty, \quad 0 \leq \alpha < 1.$$

5.6. THEOREM. *For some* $N \in \mathbb{N}$

$$0 < m^n \gamma_n \uparrow \gamma \leq \infty, \qquad N \leq n \uparrow \infty, \tag{5.9}$$

where

$$\gamma < \infty \iff (X \text{ LOG } X). \tag{5.10}$$

PROOF. For $n, k \in \mathbb{N}$

$$0 < m^n(1-s_n) = m^n(1-f_k(s_{n+k})) \leq m^{n+k}(1-s_{n+k}),$$

which implies (5.9). Clearly, $\gamma = \infty$ if and only if $m^{-n}Z_n \to 0$ a.s.. Set

$$\psi_n(s) := E(\exp\{-m^{-n}Z_n s\} \mid Z_0 = 1).$$

Then $m^{-n}Z_n \to 0$ a.s. if and only if for any $s > 0$

$$(5.11) \qquad 1 - \psi_n(s) \to 0, \quad n \to \infty.$$

We have

$$1 - \psi_n(s) = 1 - f(\psi_{n-1}(m^{-1}s))$$

$$= m(1-\psi_{n-1}(m^{-1}s))\{1-m^{-1}r(\psi_{n-1}(m^{-1}s))\}$$

$$= m^{n-1}(1-\psi_1(m^{-n+1}s))\prod_{\nu=1}^{n-1}\{1-m^{-1}r(\psi_{n-\nu}(m^{-\nu}s))\} ,$$

where

$$m^{n-1}(1-\psi_1(m^{-n+1}s)) \to s, \quad n \to \infty.$$

Now

$$1 - \psi_{n-\nu}(m^{-\nu}) \le m^{n-\nu}(1 - e^{-m^{-n}}) \le m^{-\nu}.$$

That is, if

$$\sum_\nu r(1 - m^{-\nu}) < \infty ,$$

then $\liminf(1 - \psi_n(1)) > 0$. On the other hand, if $\gamma < \infty$, then $\psi_n(s)$ converges to $\psi(s)$, say, where $\psi(s) < 1$ for $s > 0$. Thus, by convexity,

$$(1 - \psi_n(s))/s \ge \varepsilon , \quad s \le s_0 , \quad n \ge n_0,$$

with some $\varepsilon > 0$, $s_0 > 0$, and $n_0 \in \mathbb{N}$. Hence

$$1 - \psi_{n-\nu}(m^{-\nu}) \ge \varepsilon m^{-\nu}, \qquad n-\nu \ge n_0, \quad \nu \ge \nu_0,$$

and thus

$$\sum_\nu r(1 - \varepsilon m^{-\nu}) < \infty .$$

Finally, recall 1.8. □

5.7. COROLLARY. *We have* $EW = \gamma$.

PROOF. W.l.o.g. $s_n = \varphi(m^{-n})$. With this

$$EW = -\varphi'(0+) = \lim_{n\to\infty} m^n(1-\varphi(m^{-n}))$$

$$= \lim_{n\to\infty} m^n(1-s_n) = \lim_{n\to\infty} m^n\gamma_n = \gamma. \qquad \square$$

5.8. REMARK. If $m = \infty$ there is no normalizing sequence $\{\gamma_n\}$ such that $\gamma_n Z_n$ converges (in distribution) to a proper, non-degenerate limit, cf. Seneta (1969). However, there always exist an increasing function $U: \mathbb{R}_+ \to \mathbb{R}_+$, a sequence $\{\gamma_n\} \subset \mathbb{R}_+$, and a random variable V such that

$$\lim_{n\to\infty} \gamma_n U(Z_n) = V \quad \text{a.s.},$$

$$P(V=0 \mid Z_0 = 1) = q, \quad P(V<\infty \mid Z_0 = 1) = 1,$$

cf. Schuh and Barbour (1977).

6. FURTHER PROPERTIES OF THE LIMITING DISTRIBUTION

We prove several basic properties of the limit d.f. $P(W \leq \lambda)$. It is assumed throughout that $1 < m < \infty$ and $p_n \neq 1$, $n \in \mathbb{N}$.

6.1. THEOREM. *On the positive reals the d.f. of* W *has a continuous density* w.

The proof will be conducted via the Fourier transform $\varphi(it)$. Matters are simplified by the invertible transformation

$$\overline{f}(s) := \frac{f(q+(1-q)s)-q}{1-q}, \qquad |s| \leq 1,$$

process $\{\overline{Z}_n, \overline{P}\}$ with offspring mean m. If we normalize $\overline{Z}_n$ by $\overline{\gamma}_n := -\log \overline{s}_n$, where $\overline{s}_n := \overline{f}_n^{-1}(\overline{s}_0)$, $\overline{s}_0 := (s_0 - q)/(1-q)$, i.e., $\overline{s}_n = (s_n - q)/(1-q)$, and thus $\overline{\gamma}_n \sim \gamma_n/(1-q)$, the limiting distributions are connected through

$$P(W \leq t) = q + (1-q)\overline{P}(\overline{W} \leq t), \qquad t \geq 0.$$

The advantage of $\overline{f}$ lies in the fact that $\overline{f}(0) = 0$. W.l.o.g. we may therefore assume $q = 0$ and do so for the rest of the section.

Similarly as (5.4) we have

$$\varphi(itm) = f(\varphi(it)), \qquad t \in \mathbb{R}. \tag{6.1}$$

6.2. LEMMA. *For every compactum* $K \subset \mathbb{R}$ *not containing* 0

$$\sup_{t \in K} |\varphi(it)| < 1. \tag{6.2}$$

PROOF. Since W is non-degenerate, there exists a $\delta > 0$ such that $|\varphi(it)| < 1$ for $0 < |t| < \delta$. From (6.1)

$$|\varphi(it)| \leq f_n(|\varphi(itm^{-n})|) < 1, \quad 0 < |t| < m^n\delta .$$

This holds for all $n \in \mathbb{N}$. Continuity of $\varphi(it)$ completes the proof.

□

Define

$$\varepsilon_0 := -\log_m f'(q),$$

where $f'(q) = f'(0) = p_1$ due to our assumption $q = 0$.

6.3. LEMMA. For $0 < \varepsilon < \varepsilon_0$

$$|\varphi(it)| = O(|t|^{-\varepsilon}), \quad t \neq 0. \tag{6.3}$$

PROOF. Using $f(0) = 0$, the mean-value theorem leads to

$$|f_{n+k}(\varphi(it))| \leq \prod_{j=0}^{n-1} f'(f_{k+j}(|\varphi(it)|)) f_k(|\varphi(it)|)$$

$$\leq [f'(\sup_{j \geq k} f_j(|\varphi(it)|)]^n f_k(|\varphi(it)|).$$

By 6.2 we have $|\varphi(it)| \leq c_1 < 1$ on $[1,m]$, so that $|f_k(|\varphi(it)|)| \to 0$, $k \to \infty$, uniformly in $t \in [1,m]$. Hence, there exists for every positive $\varepsilon < \varepsilon_0$ a k such that

$$\sup_{1 \leq t \leq m} |f_{n+k}(\varphi(it))| = O(m^{-n\varepsilon}) ,$$

or equivalently, by (6.1),

$$\sup_{1 \leq t \leq m} |\varphi(itm^{n+k})| = O(m^{-n\varepsilon}) .$$

That is,

$$\sup_{m^{n+k} \leq t \leq m^{n+k+1}} |\varphi(it)| \leq c_2 |t|^{-\varepsilon}$$

with c_2 independent of n. □

In the following L will denote functions slowly varying at 0, but not specified any further.

6.4. LEMMA. For $s,h \in \mathbb{R}$

$$|\varphi(i(s+h))-\varphi(is)| \leq hL(h) .$$

PROOF. Denoting $\nu(t) := P(W \leq t)$, we have

$$|\varphi(i(a+b))-\varphi(i(a-b))| \leq \int_0^\infty 2(\sin bt)d\nu(t)$$

$$\leq A \int_0^\infty (1-e^{-bt})d\nu(t) = A(1-\varphi(b)),$$

$$A := \sup_{0 < \theta \leq \pi/2} \frac{2 \sin \theta}{1 - e^{-\theta}} .$$

Set $a-b = s$ and $b = h/2$, and recall from (5.9) that $1-\varphi(s) = sL(s)$.

□

6.5. LEMMA. *For* $\varepsilon \in (0,\varepsilon_0)$ *and* $K \subset \{z \in \mathbb{C}: |z| < 1\}$ *compact there exists a* $C > 0$ *such that*

$$|f_n(z_1)-f_n(z_2)| \leq Cm^{-n\varepsilon}|z_1-z_2|$$

for all $n \in \mathbb{N}$ *and* $z_1,z_2 \in K$.

PROOF. For $z_1,z_2 \in K$

$$|f_n(z_1)-f_n(z_2)| \leq A_n|z_1-z_2| ,$$

$$A_n := \sup_{z\in K} |f_n'(z)| .$$

Let

$$B_n := \sup_{z\in K} |f'(f_{n-1}(z))| .$$

Then for $\varepsilon \in (0,\varepsilon_0)$

$$A_n \leq \prod_{k=1}^{n} B_k = O(m^{-n\varepsilon}).$$ □

6.6. LEMMA. *Let* $0 < \varepsilon < \varepsilon_0$. *Then for* $|s| \geq 1$ *and* $|h| \leq |s|/m$

$$|\varphi(i(s+h))-\varphi(is)| \leq \frac{h}{|s|^{1+\varepsilon}} L(\frac{h}{|s|}) ,$$

L *decreasing and slowly varying at* 0.

PROOF. By (6.1), 6.2, and 6.5

$$|\varphi(i(a+b))-\varphi(i(a-b))| = |f_n(\varphi(i(a+b)m^{-n})) - f_n(\varphi(i(a-b)m^{-n}))|$$

$$\leq Cm^{-n\varepsilon}|\varphi(i(a+b)m^{-n}) - \varphi(i(a-b)m^{-n})|$$

$$\leq C'm^{-n\varepsilon}|1-\varphi(bm^{-n})| .$$

Set $a-b = s$ and $b = h/2$, take $m^n \leq s \leq m^{n+1}$, and recall that $(1-\varphi(s))/s = L(s)$, L decreasing and slowly varying at 0. □

PROOF OF 6.1. For $t > 0$

$$\int_{-N}^{N} \varphi(is)e^{ist}ds = - \int_{-N-\pi/t}^{N-\pi/t} \varphi(i(s+\tfrac{\pi}{t}))e^{ist}\, ds .$$

By <u>6.3</u>

$$\int_{\pm N-\pi/t}^{\pm N} \varphi(i(s+\tfrac{\pi}{t}))e^{ist}\, ds = \int_{-\pi/2}^{0} \varphi(i(s\pm N+\tfrac{\pi}{t}))e^{i(s\pm N)}\, ds \to 0$$

$$N \to \infty .$$

Hence,

$$\lim_{N\to\infty} \int_{-N}^{N} \varphi(is)e^{ist}\, ds = \lim_{N\to\infty} \frac{1}{2} \int_{-N}^{N} (\varphi(is)-\varphi(i(s+\tfrac{\pi}{t}))e^{ist}\, ds .$$

From <u>6.4</u> and <u>6.6</u>

$$\int_{-N}^{N} |\varphi(is)-\varphi(i(s+\tfrac{\pi}{t}))|ds \leq 2\,\frac{\pi}{t}\, L(\tfrac{\pi}{t})(1+\int_{1}^{N} \frac{ds}{s^{1+\epsilon}}),\ N > 1,$$

so that by a variant of Weierstrass' M-criterion the limit

$$w(t) = \lim_{N\to\infty} \frac{1}{2\pi} \int_{-N}^{N} \varphi(is)e^{ist}\, ds$$

exists <u>uniformly</u> in $t \in [C_1,C_2]$ whenever $0 < C_1 < C_2 < \infty$. □

<u>6.7</u>. THEOREM. <u>We have</u> $w(t) > 0$ <u>for all</u> $t > 0$.

The proof relies on the following fact, which the reader will not find difficult to verify, cf. Dubuc (1971 b).

<u>6.8</u>. LEMMA. <u>If</u> $\mathbb{N} \ni n > 1$, $a \in (1,n)$, <u>and</u> $\Lambda \subset (0,\infty)$ <u>is an open set such that</u>

$$x_1,\ldots,x_n \in \Lambda \Longrightarrow \frac{1}{a} \sum_{j=1}^{n} x_j \in \Lambda ,$$

<u>then there exists a</u> $b > 0$ <u>such that</u> $(b,\infty) \subset \Lambda$.

PROOF OF <u>6.7</u>. Since W is non-degenerate and w continuous,

$$\Lambda := \{t \in (0,\infty) : w(t) > 0\}$$

is non-empty. From $\varphi(ms) = f(\varphi(s))$ we get

$$\frac{1}{m}\, w(\frac{x}{m}) = \sum_{n=1}^{\infty} p_n w_n(x) , \qquad (6.4)$$

where w_n is the n'th convolutive power of w. Since the process is non-deterministic, we can fix $n > m$ such that $p_n > 0$. If now $t_1, \ldots, t_n \in \Lambda$, then, by continuity of w, $w_n(t_1+\ldots+t_n) > 0$ and thus by (6.4) also $m^{-1}(t_1+\ldots+t_n) \in \Lambda$. Hence, according to 6.6, there exists a $b > 0$ such that $(b,\infty) \subset \Lambda$.

Recalling that we have $p_0 = 0$, there also exists a positive $k < m$ such that $p_k > 0$. If now $t \in \Lambda$, then again $w_k(kt) > 0$ and by (6.4) also $kt/m \in \Lambda$. That is, $((k/m)^n b, \infty) \subset \Lambda$ for all $n > 0$. Hence, $(0,\infty) \subset \Lambda$. □

The preceding section contained a statement on $P(W > \lambda)$ for $\lambda \to \infty$. We now look at $w(t)$ for $t \to 0$.

6.9. THEOREM. _If_ $f'(q) > 0$, _then there exist two constants_ C_1, C_2 _such that for_ $t \in (0,1)$

$$0 < C_1 t^{\epsilon_0} \le tw(t) \le C_2 t^{\epsilon_0} .$$

PROOF. Again w.l.o.g. $q = 0$. The lower bound is trivial and rather arbitrary: By assumption $p_1 = f'(0) = f'(q) > 0$. Define

$$C_1 := \inf\{w(t)/t^{\epsilon_0 - 1} : 1/m \le t \le 1\} .$$

Since w is continuous and positive on the positive reals, $C_1 > 0$. From (6.4)

$$\begin{aligned} w(t) &\ge mp_1 w(mt) \ge (mp_1)^n w(m^n t) \ge (mp_1)^n C_1 (m^n)^{\epsilon_0 - 1} t^{\epsilon_0 - 1} \\ &= C_1 t^{\epsilon_0 - 1} . \end{aligned}$$

The upper bound is much harder to get. It will emerge from the following results, which are of interest in themselves. Let

$$g_n(s) := \frac{f_n(s) - q}{f'(q)^n} .$$

6.10. PROPOSITION. _If_ $f'(q) > 0$, _then for_ $s \in [0,1)$ _the limit_ $g(s) := \lim_{n\to\infty} g_n(s)$ _exists and_ $g'(s) = \lim_{n\to\infty} g_n'(s) < \infty$.

For a proof see, e.g., Athreya and Ney (1972). Next, let

$$k := \sup\{n \in \mathbb{N} : n < \epsilon_0\} .$$

6.11. PROPOSITION. For every integer $j \in [0,k]$ the j'th derivative of w exists on $(0,\infty)$ and

$$\frac{d^j w(t)}{dt^j} = \lim_{N\to\infty} \frac{1}{2\pi} \int_{-N}^{N} (is)^j \varphi(is) e^{ist}\, ds, \quad t > 0.$$

PROOF. For $t \neq 0$ and $n \in \mathbb{Z}_+$

$$\int_{m^n}^{m^{n+1}} (is)^j \varphi(is) e^{ist} ds = m^{n(j+1)} \int_1^m (is)^j f_n(\varphi(is)) e^{im^n st} ds .$$

If h is a continuous function on $[a,b]$, then

$$(6.5) \qquad \left|\int_a^b h(t) e^{iyt} dt\right|$$

$$\leq \sup_{t\in[a,b]} |h(t)|\pi/|y| + \frac{b-a}{2} \sup_{\substack{a\leq t_1\leq t_2\leq b \\ t_2-t_1\leq \pi/|y|}} |h(t_1)-h(t_2)|.$$

Since $K := \{\varphi(is) : 1 \leq s \leq m\}$ is compact, it follows from 6.10 with $q = 0$, i.e., $f'(q) = f'(0) = p_1$, that there exists a constant A such that for all $z \in K$

$$|f_n(z)| \leq f_n(|z|) \leq Ap_1^n ,$$

$$|f_n'(z)| \leq f_n'(|z|) \leq Ap_1^n .$$

As we have seen, $|\varphi(it) - \varphi(is)| \leq |t-s| L(|t-s|)$, where L is decreasing and slowly varying near 0. Hence there exists for every $\alpha \in (0,1)$ a constant A_α such that

$$|\varphi(it)-\varphi(is)| \leq A_\alpha (t-s)^\alpha, \quad 1 \leq s \leq t \leq m.$$

cf. A 13. Using these four facts together with the mean value theorem,

$$\left|\int_1^m (is)^j f_n(\varphi(is)) e^{im^n st} ds\right|$$

$$\leq \pi A m^j p_1^n (m^n t)^{-1} + \frac{m-1}{2}[(jm^{j-1} Ap_1^n \pi m^{-n} t^{-1}) + m^j Ap_1^n A_\alpha (\pi m^{-n} t^{-1})^\alpha]$$

$$\leq B_\alpha (m^{-\epsilon_0 - \alpha})^n \max(t^{-1}, t^{-\alpha})$$

with some B_α. Taking α close enough to 1 to guarantee

$m^{k+1-\epsilon_0-\alpha} < 1$,

$$\sum_{n=0}^{\infty} \left|\int_{m^n}^{m^{n+1}} (is)^j \varphi(is) e^{ist} ds\right| \leq B_\alpha \frac{\max(t^{-1}, t^{-\alpha})}{1-m^{k+1-\epsilon_0-\alpha}},$$

that is,

$$w(t,j) := \lim_{N\to\infty} \frac{1}{2\pi} \int_{-N}^{N} (is)^j \varphi(is) e^{ist} ds$$

exists and

$$w(t,j) = \frac{1}{2\pi} \int_{-1}^{1} (is)^j \varphi(is) e^{ist} ds$$
$$+ \sum_{n=0}^{\infty} \frac{1}{2\pi} \int_{m^n}^{m^{n+1}} [(is)^j \varphi(is) e^{ist} + (-is)^j \varphi(-is) e^{-ist}] ds .$$

For $j \in \{1,2,..,k\}$ and $0 \notin (a,b)$

$$\int_a^b w(t,j)dt = \int_a^b [\lim_{n\to\infty} \frac{1}{2\pi} \int_{-m^n}^{m^n} (is)^j \varphi(is) e^{ist} ds] dt$$

$$= \lim_{n\to\infty} \frac{1}{2\pi} \int_{-m^n}^{m^n} (is)^{j-1} \varphi(is) [e^{isb} - e^{isa}] ds$$

$$= w(b,j-1) - w(a,j-1).$$

Now let ψ be any continuous function with compact support in $\mathbb{R}\setminus\{0\}$. Then it is easily verified that

$$\int_{-\infty}^{\infty} \psi(t) w(t,0) dt = \int_{-\infty}^{\infty} \psi(t) d\nu(t) ,$$

$(\nu(t) = P(W \leq t))$, i.e., $w(t,0) = w(t)$. Combining the last two statements completes the proof. □

6.12. PROPOSITION. For $0 \leq j \leq k$

$$w^{(j)}(t) = o(t^{\epsilon_0 - 1 - j}) .$$

PROOF. For $t > 0$ and $0 \leq j \leq k$

$$w^{(j)}(t) = \lim_{N\to\infty} \frac{i}{\pi} \int_{-N}^{N} (is)^j \varphi(is) \sin st \, ds .$$

Now

$$|\int_{-1}^{1} (is)^j \varphi(is)\sin st\, ds| \leq t \leq t^{\varepsilon_0 - 1 - j}, \quad 0 < t \leq 1,$$

$$\sum_{m^n \leq 1/t} |\int_{m^n}^{m^{n+1}} (is)^j \varphi(is)\sin st\, ds|$$

$$\leq A(m-1)mt \sum_{m^n \leq 1/t} (m^{j+2-\varepsilon_0})^n = O(t^{\varepsilon_0 - 1 - j}),$$

and by again using (6.5) and choosing α sufficiently close to 1,

$$\sum_{m^n > 1/t} |\int_{m^n}^{m^{n+1}} (is)^j \varphi(is)\sin st\, ds|$$

$$\leq \pi \frac{Am^j}{t} \sum_{m^n > 1/t} (m^j p_1)^n + \sum_{m^n > 1/t} (\tfrac{m-1}{2}) m^{(j+1)n} [\pi j A m^{j-1} (\tfrac{p_1}{m})^n / t$$

$$+ AA_\alpha m^j p_1^n (\tfrac{\pi}{tm^n})^\alpha] \leq D_\alpha t^{\varepsilon_0 - 1 - j}$$

with some D_α. □

REMARK. Let $f'(q) > 0$ and g as in 6.10. If

$$g(\varphi(s)) = g(\varphi(1)) s^{-\varepsilon_0}, \quad s > 0, \tag{6.6}$$

then there exists a sequence of constants c_n with $\limsup_n |c_n|^{1/n} = 0$, such that

$$w(t) = \sum_{n=1}^{\infty} c_n t^{n\varepsilon_0 - 1}$$

and thus, in particular,

$$\lim_{t \downarrow 0} t^{1-\varepsilon_0} w(t) = c_1 . \tag{6.7}$$

In fact, (6.6) and (6.7) are equivalent, cf. Dubuc (1971c). However, while (6.6) is always satisfied for processes embeddable into continuous-time processes, it is not necessarily satisfied for non-embeddable ones: For example, if f is a polynomial with $f'(q) > 0$, (6.6) fails to hold, cf. Karlin and McGregor (1968a).

7. LOCAL LIMIT THEOREM FOR SUPERCRITICAL PROCESSES

Suppose to be given a supercritical BGW-process with offspring generating function

$$f(s) = \sum_{k=0}^{\infty} p_k s^k, \text{ all } p_k \neq 1,$$

and initial generating function

$$a(s) = \sum_{k=0}^{\infty} a_k s^k, \ a_0 \neq 1 .$$

Let q, m, γ_n and $w(t)$ be as before. Then, if W_k is the sum of k independent copies of W,

$$P(W_k > t) = \int_t^{\infty} w_k(u)du ,$$

$$w_k(t) := \sum_{j=1}^{k} \binom{k}{j} q^{k-j} w^{*j}(t),$$

where w^{*j} is the j'th convolutive power of w. Let D be the _period of the process_, i.e., the greatest common divisor of $\{h-s \in \mathbb{Z}: p_h \neq 0 \wedge p_s \neq 0\}$. Then we say a _process is of type_ (D,r), if r is the residue (mod D) of any h for which $p_h \neq 0$.

7.1. THEOREM. _Suppose_ $1 < m < \infty$, _and let the process be of type_ (D,r). _If_ $\{y_n\} \subset \mathbb{N}$ _is such that_ $\gamma_n y_n \to t > 0$, _as_ $n \to \infty$, _then_

$$\lim_{n\to\infty}[P(Z_n = y_n)/\gamma_n - D \sum_{k\in I_n} a_k w_k(t)] = 0,$$

where $I_n := \{k: k > 0 \ \ kr^n \equiv y_n \pmod D\}$.

We prepare the proof by a sequence of technical lemmata. First, we shall need a stronger version of III.6.10.

7.2. LEMMA. _If_ h _is an analytic function in some neighbourhood_ $U \subset \mathbb{C}$ _of_ $z = 0$ _such that_ $h(U) \subseteq U$, $h(0) = 0$, _and_ $0 < |h'(0)| < 1$, _and if_ h_n _is the n'th iterate of_ h, _then_ $h_n(z)/h'(0)^n$ _converges, as_ $n \to \infty$, _uniformly on every compact subset of_ $\{z \in \mathbb{C}: \lim_{n\to\infty} h_n(z) = 0\}$.

For a proof see, e.g., Montel (1958).

7.3. COROLLARY. _The quotient_ $(f_n(z)-q)/f'(q)^n$ _converges, as_ $n \to \infty$, _uniformly on every compact subset of_ $\{z \in \mathbb{C}: z = q + (1-q)\zeta, |\zeta| < 1\}$.

PROOF. Simply take

$$h(z) := \frac{f(q+(1-q)z)-q}{1-q}$$

and apply 7.2. □

7.4. LEMMA. *For every* $\epsilon > 0$ *there exists an* $n_0 \in \mathbb{N}$ such that

$$\sup_{n \geq n_0} \sup_{|\theta| \in [\epsilon\gamma_n, \pi/D]} \left| \frac{f_n(e^{i\theta})-q}{1-q} \right| < 1 .$$

PROOF. We know that

$$f_n(e^{it\gamma_n}) \to \varphi(-it) := E(e^{itW}|Z_0 = 1), \quad n \to \infty,$$

uniformly on finite t-intervals, and also that

$$f_n(0) \uparrow q, \quad n \uparrow \infty .$$

Hence,

$$\frac{f_n(e^{it\gamma_n}) - f_n(0)}{1 - f_n(0)} \to \frac{\varphi(-it) - q}{1 - q}$$

uniformly on finite t-intervals. By 6.1 there exist an $r < 1$, such that

$$|\varphi(-it)| \leq r, \quad \epsilon \leq |t| \leq m\epsilon .$$

Fix $N \in \mathbb{N}$ such that

$$q - f_n(0) < \tfrac{1}{2}(1-r)(1-q), \quad n \geq N,$$

and recall that $\gamma_{n-1}/\gamma_n \leq m$. Then there exists an $r' < 1$ such that

$$\left|\frac{f_n(e^{i\theta}) - q}{1 - q}\right| \leq r' , \quad \epsilon\gamma_n \leq |\theta| \leq \epsilon\gamma_{n-1}, \quad n \geq N .$$

Set

$$\tilde{f}_k(z) := \frac{f_k(q + (1-q)z) - q}{1 - q} , \quad |z| \leq 1 .$$

Then

$$1 > \tilde{f}_k(r') \to 0, \quad k \to \infty .$$

Hence, for some $R \in [r',1)$

$$\tilde{f}_k(r') \leq R, \quad k \in \mathbb{N} .$$

Since $\tilde{f}_k(x)$ is a generating function, i.e., a power series with non-negative coefficients, and $\tilde{f}_k(0) = 0$,

$$|\tilde{f}_k(z)| \leq \tilde{f}_k(r'), \quad |z| \leq r' .$$

Hence

$$\left|\frac{f_{n+k}(e^{i\theta}) - q}{1 - q}\right| \leq R, \quad \epsilon\gamma_{n+k} \leq |\theta| \leq \epsilon\gamma_{N-1}, \quad n \geq N, \ k \geq 0.$$

Finally, $\{e^{i\theta}: \epsilon\gamma_{N-1} \leq |\theta| \leq \pi/D\}$ is a compact subset of the unit circle, not containing the D'th roots of unity. Thus for some $R' < 1$

$$\left|\frac{f_n(e^{i\theta}) - f_n(0)}{1 - f_n(0)}\right| \leq R', \quad \epsilon\gamma_{N-1} \leq |\theta| \leq \pi/D, \quad n \geq 1.$$

Now use $f_n(0) \uparrow q$ again. □

Define

$$J_k := \{\theta: \pi\gamma_k/D\gamma_0 \leq |\theta| \leq \pi\gamma_{k-1}/D\gamma_0\}, \quad k \geq 1.$$

7.5. LEMMA. *If $f'(q) > 0$, there exists a constant A such that*

$$|a(f_n(e^{i\theta})) - a(q)| \leq Af'(q)^{n-k}, \quad n \geq k, \ \theta \in J_k ,$$

and if $f'(q) = 0$, then there exists for every $\epsilon > 0$ a constant A_ϵ such that

$$|a(f_n(e^{i\theta})) - a(0)| \leq A_\epsilon \epsilon^{n-k}, \quad n \geq k, \ \theta \in J_k .$$

PROOF. Set $\epsilon := \pi/D\gamma_0$ and take $n_0 = n_0(\epsilon)$ in 7.4. Then there exists a compact subset K of $\{z = q + (1-q)\zeta: |\zeta| < 1\}$ such that

$$\{f_k(e^{i\theta}): \theta \in J_k\} \subset K, \quad k \geq n_0 .$$

By 7.3, $(f_n(z) - q)/f'(q)^n$ converges uniformly on K. Hence, there exists a B such that

$$|f_n(z) - q| \leq Bf'(q)^n, \quad z \in K, \; n \geq 0,$$

and thus a B' such that

$$|f_{k+n}(e^{i\theta}) - q| \leq B'f'(q)^n, \quad \theta \in J_k, \; n \geq 0 .$$

Since $a(s)$ is continuously differentiable on K,

$$|a(z) - a(q)| \leq C|z - q|, \quad z \in K.$$

By 7.4, $\{f_{k+n}(e^{i\theta}) : \theta \in J_k\} \subset K$, $n \geq 0$, $k \geq n_0$. Hence

$$|a(f_{k+n}(e^{i\theta})) - a(q)| \leq CB'f'(q)^n, \quad \theta \in J_k, \; k \geq 0.$$

The proof of the second part is left to the reader. □

7.6. LEMMA. *If ψ is the Laplace transform of a probability measure μ, then*

$$|\psi(-ib) - \psi(-ia)| \leq C(1 - \psi(|b - a|)), \quad a,b \in \mathbb{R},$$

with a universal constant C.

PROOF. Since

$$|\psi(-ib) - \psi(-ia)| = \left|\int_0^\infty (e^{ibt} - e^{iat})d\mu(t)\right| \leq 2 \int_0^\infty \left|\sin\frac{(b-a)t}{2}\right| d\mu(t),$$

the choice

$$C := 2 \sup_{x \geq 0} \frac{|\sin(x/2)|}{1 - e^{-x}}$$

does the job. □

For $\varphi(s) := E(e^{-sW}|Z_0 = 1)$ recall from III.5 that

$$1 - \varphi(s) = sL(s),$$

where $L(s)$ is slowly varying, as $s \downarrow 0$. Since $\{e^{-\gamma_n Z_n}, \mathfrak{F}_n\}$ is a martingale, $\{e^{-s\gamma_n Z_n}, \mathfrak{F}_n\}$ is a supermartingale for $s \in (0,1)$. That is,

$$f_n(e^{-s\gamma_n}) \geq \varphi(s) .$$

Hence,

$$1 - f_n(e^{-s\gamma_n}) \leq sL(s), \quad s \in (0,1). \tag{7.1}$$

Fix $n_0 = n_0(\varepsilon)$ according to 7.4 with $\varepsilon = \pi/D\gamma_0$.

7.7. LEMMA. *If* $f'(q) > 0$, *then there exists a function* L^* *defined on* $(0,\infty)$, *bounded on every interval not containing* 0, *and slowly varying near* 0, *such that for* $n \geq k \geq n_0$ *and* $\delta > 0$

$$\sup_{\substack{\theta_1,\theta_2 \in J_k \\ |\theta_1-\theta_2| \leq \delta}} |a(f_n(e^{i\theta_1})) - a(f_n(e^{i\theta_2}))| \leq (\delta/\gamma_k) f'(q)^{n-k} L^*(\delta/\gamma_k) .$$

PROOF. Take $\theta_1,\theta_2 \in J_k$ with $|\theta_1-\theta_2| \leq \delta$. By 7.6,

$$|f_k(e^{i\theta_1}) - f(e^{i\theta_2})| \leq C(1 - f_k(e^{-\delta})),$$

and by (7.1)

$$1 - f_k(e^{-\delta}) \leq (\delta/\gamma_k)\ L(\delta/\gamma_k), \quad \delta/\gamma_k \leq 1 .$$

Setting

$$L_1(x) := \begin{cases} CL(x), & x \leq 1, \\ C\ , & x > 1, \end{cases}$$

it follows that

$$|f_k(e^{i\theta_1}) - f_k(e^{i\theta_2})| \leq (\delta/\gamma_k) L_1(\delta/\gamma_k), \quad k \geq 1. \tag{7.2}$$

If $0 < r < 1$,

$$|f_n(z_1) - f_n(z_2)| \leq f_n'(r)|z_1-z_2|, \ |z_1|,|z_2| \leq r .$$

From 6.10 we know that $f_n'(r)/f'(q)^n$ converges. Hence,

$$|f_n(z_1) - f_n(z_2)| \leq C_r f'(q)^n |z_1-z_2|, \quad |z_1|,|z_2| \leq r, \quad n \geq 0 \tag{7.3}$$

with some C_r. According to 7.4, $\{f_k(e^{i\theta}):\theta \in J_k\}$, $k \geq n_0$, are subsets of a compact set $K \subset \{|z| < 1\}$. Thus we can combine (7.2) and (7.3) to get

$$|f_{n+k}(e^{i\theta_1}) - f_{n+k}(e^{i\theta_2})| \leq C'f'(q)^n(\delta/\gamma_k)L_1(\delta/\gamma_k), \; n \geq 0, \; k \geq n_0$$

with some C'. Now

$$\{f_n(e^{i\theta}):\theta \in J_k\} \subset K, \quad n \geq k \geq n_0$$

according to 7.4, and

$$|a(z) - a(z')| \leq C''|z - z'|, \quad z,z' \in K$$

with some C''. □

7.8. LEMMA. *Let $\psi(-is)$ be the Fourier transform of a finite measure μ on $\mathbb{R}$, with $\psi(-is) \to q$, as $|s| \to \infty$. If for some sequence of positive numbers $k_n > 0$, with $k_n \to \infty$ as $n \to \infty$, and some (necessarily continuous) function h on $\mathbb{R}$*

$$\frac{1}{2\pi}\int_{-k_n}^{k_n} (\psi(-is) - q)e^{-its}ds \to h(t), \quad n \to \infty,$$

uniformly on every compact interval not containing 0, then

$$\mu(dt) = q\delta_0(dt) + h(t)dt ,$$

where δ_0 puts unit mass on 0.

PROOF. By uniform convergence and the inversion theorem for Fourier transforms, if $0 \notin (a,b]$,

$$\int_a^b h(t)dt = \int_a^b [\lim_{n\to\infty} \frac{1}{2\pi}\int_{-k_n}^{k_n} (\psi(-is) - q)e^{-its}ds]dt$$

$$= \lim_{n\to\infty} \frac{1}{2\pi}\int_{-k_n}^{k_n} (\psi(-is) - q)\, \frac{e^{-isa} - e^{-isb}}{is}\, ds$$

$$= \mu(a,b]. \qquad □$$

7.9. LEMMA. For $0 < \alpha < \beta$

$$\lim_{n\to\infty} \frac{1}{2\pi} \int_{-\pi/D\gamma_n}^{\pi/D\gamma_n} (a(f_n(e^{is\gamma_n})) - a(q))e^{-its}ds = \Sigma_{k=1}^{\infty} a_k w_k(t)$$

uniformly on $[\alpha,\beta]$.

PROOF. Set

$$h_n(s) := \frac{1}{2\pi}[a(f_n(e^{is\gamma_n})) - a(q)],$$

$$h_n(s,t) := h_n(s)e^{-its} ,$$

$$I_{k,n}(t) := \int_{s\gamma_n \in J_{n-k}} h_n(s,t)ds .$$

Then

$$(7.4) \qquad I_{k,n}(t) = \int_{-\pi\gamma_{n-k-1}/\gamma_0 D\gamma_n}^{-\pi\gamma_{n-k}/\gamma_0 D\gamma_n} + \int_{\pi\gamma_{n-k}/\gamma_0 D\gamma_n}^{\pi\gamma_{n-k-1}/\gamma_0 D\gamma_n} h_n(s,t)ds .$$

Now $\gamma_{n-k}/\gamma_n \uparrow m^k$, $n \to \infty$, and

$$a(f_n(e^{is\gamma_n})) \to a(\varphi(-is)), \qquad n \to \infty,$$

uniformly on finite intervals. Hence, for $k \geq 0$

$$\lim_{n\to\infty} I_{k,n}(t)$$
$$= \frac{1}{2\pi} \int_{-(\pi/D\gamma_0)m^{k+1}}^{-(\pi/D\gamma_0)m^k} + \frac{1}{2\pi} \int_{(\pi/D\gamma_0)m^k}^{(\pi/D\gamma_0)m^{k+1}} [a(\varphi(-is)) - a(q)]e^{-its}ds .$$

First, suppose $0 < mf'(q) < 1$. Then it follows from (7.4),

$$\left|\int_a^b h(s)e^{-its}ds\right| < (b-a) \sup_{a\leq s\leq b} |h(s)|,$$

and 7.5 that

$$|I_{k,n}(t)| \leq (1/\gamma_0 D)(\gamma_{n-k-1}/\gamma_n)Af'(q)^k, \quad n > k \geq 1$$

and from this by $\gamma_{n-k-1}/\gamma_n \leq m^{k+1}$ that

$$\sum_{k=0}^{\infty} \sup_{n>k} \sup_{t} |I_{k,n}(t)| < \infty .$$

Dominated convergence in conjunction with Weierstrass' M-criterion then leads to

$$\int_{-\pi/D\gamma_0}^{\pi/D\gamma_0} h_n(s,t)ds + \sum_{k=0}^{n-1} I_{k,n}(t)$$

$$= \int_{-\pi/D\gamma_0}^{\pi/D\gamma_0} + \int_{-\pi/D\gamma_n}^{-\pi/D\gamma_0} + \int_{\pi/D\gamma_0}^{\pi/D\gamma_n} h_n(s,t)ds$$

$$(7.5) \qquad = \int_{-\pi/D\gamma_n}^{\pi/D\gamma_n} h_n(s,t)ds$$

$$\xrightarrow[n\to\infty]{} \lim_{k\to\infty} \frac{1}{2\pi} \int_{-\pi m^k/D\gamma_0}^{\pi m^k/D\gamma_0} [a(\varphi(-is)) - a(q)]e^{-ist}ds$$

uniformly in t.

For $f'(q) = 0$ the argument is the same, using the second half of 7.5 instead of the first.

Now suppose $mf'(q) \geq 1$. Recall from (6.5) that

$$\left|\int_a^b h(s)e^{-its}ds\right| \leq \frac{\pi}{|t|} \sup_s |h(s)| + \frac{b-a}{2} \sup_{\substack{a\leq t_1\leq t_2\leq b \\ t_2-t_1\leq\pi/|t|}} |h(t_2)-h(t_1)|$$

and restrict $|t| \geq \epsilon$, $\epsilon > 0$. Then by 7.5 and 7.7

$$|I_{k,n}(t)| \leq Af'(q)^k/|t| + \tfrac{1}{2}(\pi/\gamma_0 D|t|)mf'(q)^k L^*(\gamma_n\pi/\gamma_{n-k}|t|),$$

$$n \geq k \geq n_0 ,$$

which yields

$$\sum_{k=0}^{\infty} \sup_{n>k} \sup_{|t|\geq\epsilon} |I_{k,n}(t)| < \infty,$$

so that we again have (7.5), this time uniformly only on $\{t: |t| \geq \epsilon\}$.

What remains, is the identification of the limit. For $a(s) = s$ this is immediate from 7.8. For general $a(s)$, it also follows from

7.8, noting that $a(\varphi(-is))$ is the characteristic function of $a(q)\delta_0(t) + \Sigma_{k=1}^{\infty} a_k w_k(t)$. □

REMARK. Notice that we have derived again the existence of a continuous density $w(t)$.

PROOF OF 7.1. Suppose $a_k = 0$ for $k \not\equiv j \pmod D$, $j \in [1,D] \cap \mathbb{N}$ fixed. Then, if the process is of type (D,r) and $jr^n \not\equiv y \pmod D$, we have $P(Z_n = y) = 0$. Now let $\{y_n\} \subset \mathbb{N}$ be such that $jr^n \equiv y_n \pmod D$, $\gamma_n y_n \to t > 0$, $n \to \infty$. Then

$$P(Z_n = y_n) = \frac{1}{2\pi} \int_{-\pi}^{\pi} a(f_n(e^{is}))e^{-iy_n s}\, ds .$$

If $\omega := e^{2\pi i/D}$, then $f(\omega s) = \omega^r f(s)$, i.e.,

$$f_n(\omega^\ell s) = \omega^{\ell r^n} f_n(s), \quad n, \ell \in \mathbb{N} ,$$

and thus

$$\int_{-\pi}^{\pi} a(f_n(e^{is}))e^{-iy_n s}\, ds = \sum_{\ell=0}^{D-1} \omega^{j\ell r^n - \ell y_n} \int_{-\pi/D}^{\pi/D} a(f_n(e^{is}))e^{-iy_n s}\, ds$$

$$= D \int_{-\pi/D}^{\pi/D} a(f_n(e^{is}))e^{-isy_n} ds .$$

If $q \neq 0$, then $p_0 \neq 0$, $r = 0$, and $y_n \equiv 0 \pmod D$. If $q = 0$ and $a(0) \neq 0$, then $j = D$ and $y_n \equiv 0 \pmod D$, so that

$$\int_{-\pi/D}^{\pi/D} a(q)e^{-isy_n} ds = 0.$$

Thus

$$P(Z_n = y_n) = \frac{D}{2\pi} \int_{-\pi/D}^{\pi/D} (a(f_n(e^{is})) - a(q))e^{-iy_n s}\, ds .$$

Now apply 7.9. The result for general a follows by linearity. □

8. IMMIGRATION

We consider BGW processes with stationary immigration, restricting ourselves to the most basic and simple analytical results for subcritical and critical processes. In addition to the p.g.f.

$$f(s) = \sum_{n=0}^{\infty} p_n s^n$$

characterizing the branching, we assume to be given another p.g.f.

$$g(s) = \sum_{n=0}^{\infty} q_n s^n$$

describing the immigration into each generation. The immigration is supposed to be independent of the branching. Hence, starting with one particle and no immigration in generation zero, the p.g.f. of the population size of the first generation is given by

$$h_1(s) = g(s)f(s),$$

that of the n'th generation, $n > 1$, by

$$h_n(s) = g(s)h_{n-1}(f(s)), \tag{8.1}$$

so that

$$h_n(s) = g(s)f_n(s) \prod_{j=1}^{n-1} g(f_j(s)) . \tag{8.2}$$

8.1. THEOREM. *Given* $0 < m < 1$ *we have*

$$h(s) := \lim_{n\to\infty} h_n(s) > 0, \quad s \in (0,1]$$

if and only if

$$\sum_{n=1}^{\infty} q_n \log n < \infty , \tag{8.3}$$

and if this condition is satisfied, h *is a proper p.g.f.*

For the proof we need the following fact.

8.2. LEMMA. *For any* $\delta \in (0,1)$

$$\sum_{n=1}^{\infty} (1-g(1-\delta^n)) < \infty$$

<u>if and only if</u>

(8.4) $$\sum_{h=1}^{\infty} q_n \sum_{j=1}^{n} j^{-1} < \infty .$$

Notice that

(8.5) $$\sum_{j=1}^{n} j^{-1} \sim \log n, \; n \to \infty.$$

PROOF OF <u>8.2</u>. Without loss of generality $q_o = 0$. Then

(8.6)
$$S := \sum_{j=1}^{\infty} (1-g(1-\delta^{j}))$$
$$= \sum_{j=1}^{\infty} \sum_{n=1}^{\infty} q_n (1-(1-\delta^{j})^{n}) = \sum_{n=1}^{\infty} q_n S_n ,$$

$$S_n := \sum_{j=1}^{\infty} (1-(1-\delta^{j})^{n}) .$$

Setting

$$I_n := \int_1^{\infty} (1-(1-\delta^{x})^{n}) dx ,$$

we have

(8.7) $$0 \leq S_n - I_n \leq 1-(1-\delta)^{n} .$$

Substituting $y := 1-\delta^{x}$,

$$I_n = (-\log \delta)^{-1} \sum_{j=1}^{n} j^{-1} - L_n$$

$$L_n := (-\log \delta)^{-1} \sum_{j=1}^{n} j^{-1} (1-\delta)^{j}$$
$$\leq (-\delta \log \delta)^{-1} (1-\delta)(1-(1-\delta)^{n}) .$$

Since

$$L := \sum_{n=1}^{\infty} q_n L_n \leq (-\delta \log \delta)^{-1} (1-\delta)(1-g(1-\delta)) < \infty,$$

it follows from (8.6) and (8.7) that

$$-L \leq S-(-\log \delta)^{-1} \sum_{n=1}^{\infty} q_n \sum_{j=1}^{n} j^{-1}$$

$$< 1-g(1-\delta)-L,$$

which completes the proof. $\square$

PROOF OF 8.1. Convergence to $h(s) > 0$ for $s > 0$ is equivalent to

$$\sum_{n=1}^{\infty} [1-g(f_n(0))] < \infty . \tag{8.8}$$

Since $f_n(0) = f(f_{n-1}(0))$ and $m < 1$, the mean value theorem gives

$$1-f_n(0) = f'(\theta_{n-1})(1-f_{n-1}(0)), \quad n > 1$$

with some $\theta_{n-1} \in (f_{n-1}(0),1)$. Hence, using $f_n(0)\uparrow$ and setting $\epsilon := f'(f(0))$,

$$0 < \epsilon(1-f_{n-1}(0)) \leq 1-f_n(0)$$

$$\leq m(1-f_{n-1}(0)), \quad n > 1,$$

that is,

$$1-m^n \leq f_n(0) \leq 1-\epsilon^n, \quad n > 1$$

and thus

$$1-g(1-\epsilon^n) \leq 1-g(f_n(0)) \leq 1-g(1-m^n).$$

Recalling 8.2 and (8.5), this shows that (8.8) holds if and only if (8.3) is satisfied.

Since h is the limit of a sequence of p.g.f.'s, it remains to be shown that $h(s) \to 1$, as $s \uparrow 1$, in order to prove that h is itself a p.g.f. Given $0 < m < 1$, we have $0 < f_n(0) < 1$, $f_n(0) \uparrow 1$. Hence, it suffices to show that $h(f_n(0)) \to 1$, $n \to \infty$. From (8.1)

$$h(s) = g(s)h(f(s)), \tag{8.9}$$

which gives

$$h(f_n(0)) = h(f_{n-1}(0))[g(f_{n-1}(0))]^{-1} ,$$

so that

$$h(f_n(0)) = h(f(0))[g(f(0))\prod_{j=1}^{n-1} g(f_j(f(0)))]^{-1}$$

$$= h(f(0))f_n(f(0))/h_n(f(0))$$

$$\to 1, \quad n \to \infty. \qquad \square$$

Clearly h is the p.g.f. of an invariant measure (in the ordinary sense). Vice versa, the g.f. of any invariant measure satisfies (8.9). Is the solution of (8.9) unique (up to a constant factor)? Does (8.9) have a nontrivial solution, if (8.3) does not hold?

8.3. PROPOSITION. Let $m \leq 1$, $p_0 > 0$. Then for every $s_0 \in (0,1)$ the equation

(8.10) $$H(s) = g(s)H(f(s)), \qquad H(s_0)=1,$$

has a unique g.f. solution, given by

$$H(s) = \lim_{n\to\infty} \prod_{j=0}^{n} [g(f_j(s))/g(f_j(s_0))] .$$

PROOF. Define

$$H_n(s) := \prod_{j=0}^{n} [g(f_j(s))/g(f_j(s_0))]$$

and take any fixed $s \in [0,1)$. The case $s = s_0$ is trivial. If $s < s_0$, $H_n(s)$ is decreasing, and since it is bounded from below by 0, converging. If $s > s_0$, $H_n(s)$ increases. To see that $H_n(s)$ is bounded from above, notice we can fix a k such that

(8.11) $$0 < s \leq f_k(s_0),$$

since $f_n(s_0) \mid 1$, as $m \leq 1$, $p_0 > 0$. Thus

$$H_n(s) \leq H_n(f_k(s_0)) = \prod_{j=0}^{n} [g(f_{k+j}(s_0))/g(f_j(s_0))],$$

which for $n \geq k$ is bounded from above by $\prod_{j=0}^{k-1}[g(f_j(s_0))]^{-1}$. Hence, $H_n(s)$ converges also in this case. Passing to the limit in

$$H_{n+1}(s) = H_n(f(s))g(s)/g(f_{n+1}(s_0))$$

give (8.10).

Now let $\widetilde{H}$ be <u>any</u> g.f. solution of (8.10). Then

$$\widetilde{H}(s) = H_n(s)\widetilde{H}(f_{n+1}(s))/\widetilde{H}(f_{n+1}(s_0)). \tag{8.12}$$

First assuming $s > s_0$ and using (8.11) again,

$$1 \leq \widetilde{H}(f_{n+1}(s))/\widetilde{H}(f_{n+1}(s_0))$$

$$< \widetilde{H}(f_{n+k+1}(s_0))/\widetilde{H}(f_{n+1}(s_0))$$

$$= \prod_{j=1}^{k-1} g(f_{n+j+1}(s_0))^{-1} \to 1, \quad n \to \infty,$$

which is conjunction with (8.12) given $\widetilde{H}(s) = H(s)$.

For $s_0 < s$ interchange the roles of s_0 and s. The case $s_0 = s$ is trivial. Finally notice that as a limit of g.f's. H is itself a g.f.. □

<u>8.4</u>. REMARK. Suppose $m < 1$, and let H be a solution of (8.10). Then

$$1-f(1-s) \leq \lambda s \leq s, \quad s \in [0,1], \quad \lambda \in [m,1],$$

and thus

$$1 \leq \frac{H(1-\lambda s)}{H(1-s)} \leq \frac{H(f(1-s))}{H(1-s)}$$

$$\leq \frac{1}{g(1-s)} \to 1, \quad s \to 0,$$

i.e., $H(1-s)$ and thus, in particular, $h(1-s)$ is slowly varying, as $s \downarrow 0$.

<u>8.5</u>. THEOREM. <u>Suppose</u> $m = 1$, $0 < \mu := f''(1-)/2 < \infty$, $0 < \gamma := g'(1-) < \infty$. <u>Then</u>

$$\lim_{n\to\infty} h_n(e^{-s/\gamma n}) = (1+s/\varkappa)^{-\varkappa}, \quad \varkappa := \gamma/\varkappa, \quad s \geq 0.$$

Let $\tilde{Z}_n$ be the population size of the n'th generation in the immigration-branching process.

8.6. COROLLARY. Under the assumptions of 8.5, the d.f. of $\tilde{Z}_n/\gamma n$ converges, as $n \to \infty$, to the d.f. with density

$$\nu(x) = (\varkappa/\Gamma(\varkappa))(\varkappa x)^{\varkappa-1}e^{-\varkappa x}, \quad x \geq 0,$$

where Γ denotes the gamma function.

PROOF OF 8.5. Set

$$s_n := e^{-s/n\gamma},$$

and let $(k(n))_{n\in\mathbb{N}}$ be any sequence of integers such that

$$0 < k(n) < n, \; k(n) \to \infty, \; k(n)/n \to 0, \; n \to \infty.$$

Since $m = 1$,

$$1-g(f_j(s_n)) \leq \gamma(1-f_j(s_n)) \leq \gamma f(1-s_n) \leq s/n,$$

so that

$$\lim_{n\to\infty} \sum_{j=1}^{k(n)-1} \log g(f_j(s_n)) = 0 .$$

Writing

$$\gamma n(1-s_n) = \theta_n(s)s, \tag{8.13}$$

we have $\theta_n(s) \to 1$, $n \to \infty$. From III.3.1,

$$\frac{1}{n}[(1-f_n(s))^{-1}-(1-s)^{-1}] = \mu+\epsilon_n(s), \tag{8.14}$$

$$\epsilon_n(s) \to 0 \text{ uniformly in } s \in [0,1).$$

Expanding

$$1-g(s) = \gamma(1-s)-\rho(s)(1-s),$$

and using (8.13), (8.14), we get

$$1-g(f_j(s_n)) = \frac{(\gamma-\rho(f_j(s_n)))\theta_n(s)s}{\gamma n+\theta_n(s)s(\mu+\epsilon_j(s_n))j} .$$

Setting

$$\delta_n(s) := \rho(s_n)/\gamma ,$$

$$\epsilon_j^*(s) := \sup_{\ell \geq j} |\epsilon_\ell^*(s)|,$$

and then recalling that

$$1-g(f_j(s_n)) \leq s/n \to 0, \ n \to 0$$

uniformly in j, there exists a sequence $(\eta_n(s))_{n\in\mathbb{N}}, \eta_n(s) \to 1$, $n \to \infty$, such that for all sufficiently large n

$$\gamma(1-\delta_n(s))\theta_n(s)s \sum_{j=k(n)}^{n-1} \frac{1}{\gamma n+\theta_n(s)s(\mu+\epsilon_{k(n)}^*(s_n))j}$$

$$\leq - \sum_{j=k(n)}^{n-1} \log g(f_j(s_n))$$

$$\leq \gamma\eta_n(s)\theta_n(s)s \sum_{j=k(n)}^{n-1} \frac{1}{\gamma n+\theta_n(s)s(\mu-\epsilon_{k(n)}^*(s_n))j} .$$

Noting that $\delta_n(s) \to 0$ and $\epsilon_{k(n)}^*(s) \to 0$, as $n \to \infty$, this implies that

$$\lim_{n\to\infty} \sum_{j=k(n)}^{n-1} \log g(f_j(s_n)) = -\gamma s \lim_{n\to\infty} \sum_{j=k(n)}^{n-1} \frac{1}{\gamma n+s\mu j}$$

$$= -(\gamma/\mu) \lim_{n\to\infty} \int_{s\mu k(n)/\gamma n}^{s\mu(n+1)/\gamma n} \frac{dx}{1+x} = -(\gamma/\mu)\log(1+s\mu/\gamma),$$

which completes the proof. □

BIBLIOGRAPHICAL NOTES

Theorems 1.2 and 1.7 are standard, cf. Athreya and Ney (1972). Proposition 1.5 is due to Seneta (1971), its proof is Hoppe's (1977). Theorem 1.6 again is from Seneta (1974). In the next section, 2.1 has been taken from Seneta and Vere-Jones (1966), 2.3 and 2.6 are from Hoppe and Seneta (1978), 2.7 is from Hoppe (1976), and 2.9 from Spitzer (1967), see also Athreya and Ney (1972). More on Martin boundary theory can be found in Dubuc (1978). Spectral properties of the transition operator, which we do not discuss here, have been investigated by Karlin and McGregor (1966, 1968a).

Most of section 3 is standard, cf. Athreya and Ney (1972), theorems 3.7 and 3.8 go back to Zolotarev (1957) and Slack (1968, 1972). Section 4 is part of Kesten, Ney and Spitzer (1966).

The analytical results in Section 5, particularly the generalized norming constants, are due to Seneta (1968, 1974, 1975), the discovery of the exponential martingale is Heyde's (1970). Section 6 has been extracted from Dubuc (1970, 1971a, b, c), section 7 - except for slight modifications - is from Dubuc and Seneta (1976), see also Athreya and Ney (1970).

The limit theory for subcritical processes with immigration in Section 8 is from Heathcote (1966), except for 8.3, which is due to Seneta (1971). The proof for 8.3 given here is taken from Hoppe and Seneta (1978). The limit result in the critical case is Seneta's (1970), while our proof has been obtained by specializing a proof for processes with a general set of types from Hering (1973).

Chapter IV

Continuous Time Markov Branching Processes

1. INTRODUCTION

The Markov branching process is specified by two parameters, the offspring distribution F with point probabilities $\{p_k\}$, and the intensity β of the exponential lifetime distribution $\beta e^{-\beta x}dx$. Various constructions are possible.

Presumably the most intuitive and simple construction is the following. Given the σ-algebra $\hat{\mathfrak{F}}$ generated by a discrete time Galton-Watson process $Z_0, Z_1, \ldots$ (the embedded generation process), let the

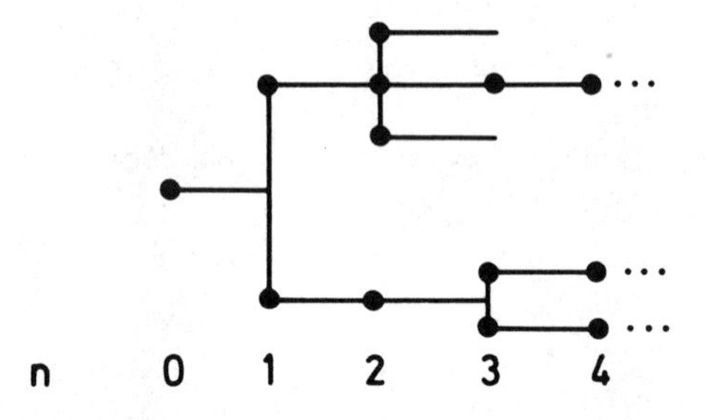

$Y_{n,i}$ $(n = 0,1,2,\ldots, i = 1,\ldots,Z_n)$ be i.i.d. exponentially distributed with intensity β (with $Y_{n,i}$ representing the lifetime of the ith individual of the nth generation). Then Z^t is the number of individuals alive at time t in the tree formed by, in the obvious manner,

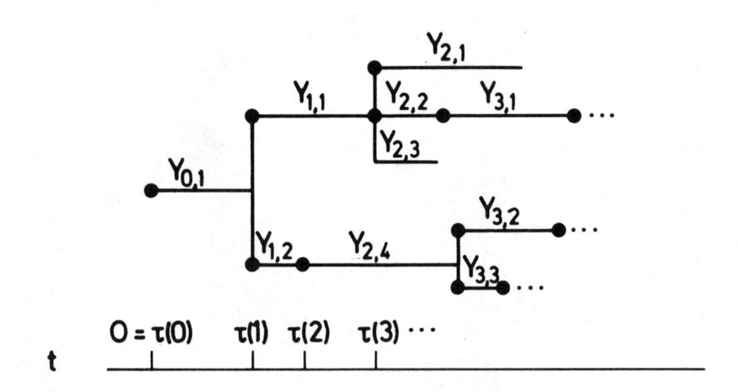

transforming the family tree of the embedded generation process with the $Y_{n,i}$. Note that $Z_t = \infty$ cannot apriori be excluded. This is the explosion problem. Obviously, we can take the paths of $\{Z_t\}_{t\geq 0}$ to be right-continuous.

The instants of reproduction (of the form $Y_{0,i(0)} + Y_{1,i(1)} + \dots + Y_{n,i(n)}$ with $i(0)$, $i(1) \dots i(n)$ representing an ancestry line) are called the split times and denoted $\tau(0), \tau(1), \dots$ in the linear ordering. Obviously, Z_t is constant when $t \in [\tau(n), \tau(n+1))$. Jumps do not necessarily occur at all τ_n if $p_1 > 0$.

Let $\mathfrak{F}_t$ be the σ-algebra containing all relevant information on the history up to time t and define $U_n := \tau(n) - \tau(n-1)$, $X_n := Z_{\tau(n)} - Z_{\tau(n-1)}$. Then $X_n + 1$ is the number of offsprings produced by the individual which died at time $\tau(n)$, i.e., distributed according to F, and clearly $X_1, X_2, \dots$ are independent. It follows that $\{Z_{\tau(n)}\}_{n \in \mathbb{N}} = \{Z_0 + X_1 + \dots + X_n\}_{n \in \mathbb{N}}$ is simply a random walk. In order to describe the U_n, note that, given $Z_{\tau_{n-1}} = k$, $U_n = A_1 \wedge \dots \wedge A_k$ where A_i is the residual lifetime of the ith individual alive at time τ_{n-1}. Since the exponential distribution is without memory, $P(A_i > a \mid \mathfrak{F}_{\tau_{n-1}}) = e^{-\beta a}$ so that $P(U_n > a \mid \mathfrak{F}_{\tau(n-1)}) = e^{-\beta a Z_{\tau(n-1)}}$. It follows similarly that the U_n are independent given $X_1, X_2, \dots$.

Conversely, starting with two sequences $\{X_n\}$ and $\{U_n\}$ with these distributional properties, we could let $\tau(n) := U_1 + \dots + U_n$, $\tilde{Z}_t := Z_0 + X_1 + \dots + X_n$ $t \in [\tau(n-1), \tau(n))$. It is then easy to see that $\{Z_t\}_{t \geq 0}$ and $\{\tilde{Z}_t\}_{t \geq 0}$ are equivalent. In fact, U_n has the same distribution as $A_1^n \wedge \dots \wedge A^n_{Z_{\tau(n-1)}}$ where the A_i^n are i.i.d. exponential and using the X_n and A_i^n, one could construct a family tree as above. Now the construction of $\tilde{Z}_t$ is the minimal construction of a Markov jump process with intensity matrix

$$Q := (q_{ij})_{i,j \in \mathbb{N}} = \beta \begin{pmatrix} 0 & 0 & 0 & 0 & \cdots \\ p_0 & p_1 - 1 & p_2 & p_3 & \cdots \\ 0 & 2p_0 & 2(p_1 - 1) & 2p_2 & \cdots \\ 0 & 0 & 3p_0 & 3(p_1 - 1) & \cdots \\ \vdots & & & & \end{pmatrix}$$

(i.e., $P(\tilde{Z}_{t+s} = j \mid \tilde{Z}_t = i) = \delta_{ij} + s q_{ij} + o(s)$). Hence $\{Z_t\}_{t \geq 0}$ is Markov with intensity matrix Q. Note that the minimal construction is slightly less detailed than the one above since quantities like the individual lifetimes are not defined.

Appealing once more to the lack of memory of the exponential distribution, we see that any of the Z_t particles at time t starts a new Markov branching process distributed as $\{Z_t\}$ (given $Z_0 = 1$), and clearly these subprocesses are independent given $\mathfrak{F}_t$. Thus we can write

$$(1.1) \qquad Z_{t+u} = \sum_{i=1}^{Z_t} Z_{t+u}^{t,i}$$

where conditionally upon $\mathfrak{F}_t$ the $Z_{t+u}^{t,i}$ are independent and distributed as Z_u. In particular, letting $t := n\delta$, $u := \delta$, we obtain

1.1. PROPOSITION. For every $\delta > 0$, $\{Z_{n\delta}\}_{n\in\mathbb{N}}$ is a Galton-Watson process.

Embedded Galton-Watson processes of the form $\{Z_{n\delta}\}$ are called discrete skeletons. An immediate question is the embedding problem: What types of Galton-Watson processes can occur as discrete skeletons of continuous-time Markov branching processes? The answer is non-trivial. We shall not go into this, but refer to Karlin and McGregor

To study the behavior of $\{Z_t\}_{t\in\mathbb{R}_+}$ we now have several possibilities. First, we may try to use the same arguments as for Galton-Watson processes, only with a continuous parameter in place of the discrete one. This works for a large part of the analytic and some of the probabilistic arguments, and some proofs even become simpler in the continuous time case, but it doesn't work everywhere. However, as additional tools we now have discrete skeletons, the embedded generation process, split times, and the Markov jump process structure. The emphasis will be on typical examples rather than completeness.

2. GENERATING FUNCTIONS IN CONTINUOUS TIME

Define

$$f(s) := \sum_{k=0}^{\infty} p_k s^k, \quad F(s,t) := E(s^{Z_t} | Z_0 = 1), \quad v(s) := \beta(f(s)-s).$$

From (1.1), we obtain then by similar arguing as when deriving the discrete time functional iteration formula

$$(2.1) \qquad F(s,t+u) = F(F(s,t),u).$$

Given $Z_0 = 1$, the population is composed according to the offspring distribution in between the two first split times $\tau(1), \tau(2)$. Hence

$$F(s,u) = sP(\tau(1) > u) + f(s)P(\tau(1) \leq u) + O(P(\tau(2) \leq u)$$

$$= s(1-\beta u) + f(s)\beta u + o(u).$$

As $u \downarrow 0$, we get $F_t(s,0) = v(s)$ and (2.1) yields

$$(2.2) \qquad F_t(s,t) = v(s)\, F_s(s,t)\ .$$

Here $\partial F/\partial t$ and $\partial F/\partial s$ are denoted by F_t and F_s resp. Replacing t by $t-u$ in (2.1) produces similarly a backward equation

$$(2.3) \qquad F_t(s,t) = v(F(s,t))\ .$$

As a first application, we consider the explosion problem. Obviously, $P(Z_t = \infty | Z_0 = 1) > 0$ if and only if $F(1,t) < 1$.

2.1. THEOREM. *If*

$$(2.4) \qquad \left| \int_{1-\epsilon}^{1} \frac{1}{v(s)}\, ds \right| = \infty \quad \text{for all small} \quad \epsilon > 0,$$

then $F(1,t) = 1$ *for all* $t > 0$. *Otherwise* $F(1,t) < 1$ *for all* $t > 0$.

2.2. COROLLARY. *If the offspring mean* $m := \Sigma_0^{\infty}\, kp_k = f'(1)$ *is finite, then*, $F(1,t) = 1$ *for all* $t > 0$.

PROOF. Clearly all $F(\cdot,t)$ are strictly increasing. Suppose $F(1,t) < 1$. Then from (5.1) $F(s,t+u) < F(1,u) \leq 1$ for all $u \geq 0$. Also $1 > F(s,t) = F(F(s,t/2),t/2)$ implies $F(s,t/2) < 1$. Iterating yields $F(1,s) < 1$ for $s = t/2, t/u, \ldots$, hence all $s > 0$. Thus either $F(1,t)=1$

for all $t > 0$ or for no $t > 0$.

If the offspring mean $f'(1)$ is finite, then $v(s) = f'(1)(1-s) + o(1-s)$ so that (2.4) fails. This shows that indeed 2.2 follows from 2.1. and yields also 2.1. in the case $f'(1) \leq 1$, since then by comparison with the (ultimately extinct) embedded generation process $Z_t \leq \hat{Z}_0 + \hat{Z}_1 + \ldots < \infty$. If $1 < f'(1) \leq \infty$, then for some $s_0 < 1$, $f(s)-s < 0$ when $s_0 < s < 1$. Chose s_1 and $t_1 > 0$ such that $s_0 < s_1 < 1$, $s_1 - \beta t_1 > s_0$. From (2.3), always $|F_t(s,t)| \leq \beta$ so that $F(s,t) \geq F(s,0) - \beta t = s - \beta t$. Hence if $s_1 \leq s < 1$, $0 \leq t \leq t_1$ then $1 > x := F(s,t) > s_0$ so that $f(x)-x \neq 0$. Thus from (2.3),

$$t_1 = \int_0^{t_1} dt = \int_0^{t_1} \frac{F_t(s,t)}{v(F(s,t))} dt = \int_s^{F(s,t_1)} \frac{1}{v(x)} dx$$

so that $\int_{F(1,t_1)}^{1} v(x)^{-1} = -t_1$. If (2.4) holds, this implies $F(1,t_1) = 1$, while otherwise we must have $F(1,t_1) < 1$. □

Some probabilistic remarks on the explosion problem are given in Section 4.

Suppose now $f'(1) < \infty$ and define $M(t) = E(Z_t | Z_0 = 1) = F_s(1,t)$. Then from (2.2) or (2.3), $M'(t) = F_{st}(1,t) = \lambda := \beta(f'(1)-1)$ so that $M(t) = e^{\lambda t}$. Similar methods apply to higher moments. We state the formula

$$\text{(2.5)} \qquad \operatorname{Var} Z_t = \begin{cases} \tau^2 e^{\lambda t}(e^{\lambda t}-1), \quad \tau^2 := \dfrac{f''(1)-f'(1)+1}{f'(1)-1}, & \lambda \neq 0 \\ f''(1)\beta t, & \lambda = 0. \end{cases}$$

The equation $M(t) = e^{\lambda t}$ could also be proved probabilistically, using the embedded generation process. In fact, let $A_i^n(t)$ be the event that the ith individual of the nth generation is alive at time t (so that $Z_t = \Sigma_0^\infty \Sigma_1^{Z_n} I(A_i^n(t)))$, i.e., that the parent of i died before t but that i died after t. Letting G be the exponential distribution, it follows that

$$P(A_i^n(t) | \hat{\mathfrak{F}}) = G^{*n}(t) - G^{*(n+1)}(t).$$

Thus

$$E(Z_t | \hat{\mathfrak{F}}) = \sum_{n=0}^{\infty} Z_n (G^{*n}(t) - G^{*(n+1)}(t)).$$

Insert $EZ_n = f'(1)^n$ and the gamma form of the G^{*n}.

2.3. EXAMPLES. Two standard examples in which the backward equation can be solved explicitly are the following:

(a) The Yule process (binary fission). Here $f(s) = s^2$, so that

$$F_t(s,t) = \beta[F^2(s,t)-F(s,t)], \quad F(s,0) = s.$$

The solution is

$$F(s,t) = se^{-\beta t}[1-(1-e^{-\beta t})s]^{-1} .$$

(b) The birth-and-death process. A birth-and-death process with birth rate $n\lambda$ and death rate $n\mu$, if the process is in the state n, is a branching process with

$$\beta = \lambda+\mu, \quad f(s) = (\mu+\lambda s^2)/\beta,$$

so that

$$F_t(s,t) = [\lambda F(s,t)^2 - (\lambda+\mu)F(s,t)+\mu], \quad F(s,0) = s.$$

The solution is

$$F(s,t) = \frac{\mu(s-1) - e^{(\mu-\lambda)t}(\lambda s-\mu)}{\lambda(s-1) - e^{(\mu-\lambda)t}(\lambda s-\mu)} .$$

We now briefly carry over some of the basic limit theorems. The reader interested in more details is referred to Chapters VI and VII, where - in a more general framework - discrete and continuous time processes are treated simultaneously.

Let us first take a look at the subcritical case with one initial particle.

2.4. THEOREM. *Suppose* $0 < m < 1$, $P(Z_0 = 1) = 1$. *Then*

$$1-F(0,t) \sim e^{\lambda t}L_1(e^{\lambda t}), \quad t \to \infty, \tag{2.6}$$

with L_1 *slowly varying at* 0, $L \equiv \gamma = \text{const} > 0$ *if and only if for some (and thus all)* $t > 0$

$$E\, Z_t \log^+ Z_t < \infty. \tag{X LOG X}$$

Further

$$\frac{F(s,t)-F(0,t)}{1-F(0,t)} \to g(s), \quad t \to \infty,$$

where g *is the unique p.g.f. solution of*

$$1-g(F(s,t)) = e^{\lambda t}(1-g(s)), \; g(0) = 0,$$

for any $t > 0$. *Finally,*

$$1-g(1-s) = sL_2(s), \; s \in [0,1],$$

with L_2 *slowly varying at* 0, *and*

$$g'(1-) = 1/L_1(0).$$

The proof is the same as in the discrete case. However, there is a shortcut in the proof of (2.6) not possible for discrete time: Set $c(s) := \lambda^{-1}\log s$ for $s \in (0,e^{\lambda})$. Then

$$\frac{1 - F(0,c(\alpha s))}{1 - F(0,c(s))} = \frac{1-F(0,c(s)+c(\alpha))}{1 - F(0,c(s))} \sim e^{\lambda c(\alpha)} = \alpha, \; s \to 0.$$

Taking $\alpha = e^{\lambda t}$ yields (2.6). We shall see in the next section that (X LOG X) is equivalent to

(x log x) $$\sum_{n=1}^{\infty} p_n \, n \log n < \infty.$$

When admitting an arbitrary initial distribution, the limit theory of subcritical processes is not as rich as in the discrete case: The non-uniqueness disappears.

2.5. THEOREM. *Suppose* $0 < m < 1$. *If for some initial* $F(s,0)$ *with* $F(0,0) = 0$ *there exists a g.f.* $\tilde{g} \not\equiv 0$ *such that*

$$\frac{F(s,t)-F(0,t)}{1 - F(0,t)} \to \tilde{g}(s), \; t \to \infty, \; s \in [0,1],$$

then there exists an $\alpha \in (0,1]$ *such that*

$$1-\tilde{g}(s) = (1-g(s))^{\alpha}, \; s \in [0,1].$$

A particular $F(s,0)$ *is in the domain of attraction of* $(1-g(s))^{\alpha}$ *if and only if*

$$L_3(s) := \frac{1-F(1-s,0)}{(1-g(1-s))^{\alpha}}$$

is slowly varying at 0. Up to a constant factor there is exactly one invariant measure, its g.f. being

$$h(s) = \lambda^{-1}\log(1-g(s)), \quad s \in [0,1).$$

For the proof one can repeat the arguments of the discrete case, the relevant functional equations now taking the form

$$1-\tilde{g}(F(s,t)) = e^{\alpha\lambda t}(1-\tilde{g}(s)), \quad \tilde{g}(0) = 0,$$

$$h(F(s,t) = h(s)+t,$$

$$1-\tilde{g}(s) = e^{\alpha\lambda h(s)},$$

$$h(s) = \tilde{d}(g(s)),$$

$$\tilde{d}(1-e^{\lambda t}+e^{\lambda t}s) = \tilde{d}(s)+t, \quad \tilde{d}(0) = 0.$$

The key observation is the following: For every solution $\tilde{d}$ of the last equation there exists a Borel measure ν on $\mathbb{R}$, such that

$$\tilde{d}(s) = \int_{-\infty}^{+\infty}(e^{-(1-s)e^{\lambda u}} - e^{-e^{\lambda u}})\nu(du),$$

$\nu([0,t)) = t$, and for every Borel set $A \subset \mathbb{R}$

$$\nu(A+t) = \nu(A), \quad t \in \mathbb{R}.$$

The only measure with this normalization and invariance property is the Lebesgue measure. This eliminates the non-uniqueness and reduces the results to 2.5.

2.6. THEOREM. Suppose $m=1, P(Z_0=1)=1$, and either $\mu = \frac{1}{2}[\partial^2 F(s,1)/\partial s^2]_{s=1-} < \infty$, or $\mu = \infty$ and

$$\text{(S)} \qquad f(s) = 1+(1-s)^{\alpha}L_4(1-s), \quad s \in [0,1],$$

$\alpha \in (0,1]$, L_4 slowly varying at 0. Then

$$a_t := 1-F(0,t) \sim (\mu t)^{-1}, \qquad \mu < \infty, \ t \to \infty$$

$$\sim t^{-1/\alpha}L^*(t), \quad \mu = \infty \quad \text{(S)}, \ t \to \infty,$$

L^* slowly varying at ∞, and

$$\frac{F(e^{-a_t s},t)-F(0,t)}{1-F(0,t)} \to (1+s)^{-1}, \qquad \mu < \infty,\ t \to \infty$$

$$\to 1-s(1+s^{\alpha})^{-1/\alpha}, \ \mu = \infty \wedge (S),\ t \to \infty,$$

where in the second case (S) is also necessary for the existence of a non-degenerate limit.

The proof is the same as in the discrete case, the only thing to check being that (S) is equivalent to

$$F(s,t) = 1-(1-s)^{\alpha} L_{4,t}(s), \quad s \in [0,1],$$

$L_{4,t}$ slowly varying at 0, cf. Chapter VI.

Next let $1 < m < \infty$. Then $F(0,t) \to q < 1$, as $t \to \infty$. Fix $s_0 \in (q,1)$, define $F^{-1}(\cdot,t)$ as the inverse of $F(\cdot,t)$ for fixed t, and set

$$\gamma_t := F^{-1}(s_0,t).$$

2.7. THEOREM. *Suppose* $1 < m < \infty$. *Then there exists a random variable* W *such that*

$$P(W = 0 | Z_0 = 1) = q, \ P(W < \infty) = 1,$$

$$\gamma_t Z_t \to W \quad \text{a.s.},\ t \to \infty.$$

For every $t > 0$ *the Laplace transform* $\varphi(s) := E(e^{-sW} | Z_0 = 1)$ *is the unique solution of*

$$\varphi(e^{\lambda t}s) = F(\varphi(s),t),\ s \geq 0,\ \varphi(0+) = 1. \tag{2.7}$$

Furthermore,

$$1-\varphi(s) = sL_5(s),\ s \geq 0,$$

$$\gamma_t = e^{-\lambda t} L_6(e^{-\lambda t}),\ t \geq 0,$$

L_5, L_6 *slowly varying at* 0,

$$0 < e^{\lambda t}\gamma_t \uparrow \gamma \leq \infty, \quad t \uparrow \infty,$$

$$EW = \gamma,$$

with $\gamma < \infty$ *if and only if* (X LOG X) *is satisfied.*

The proof is the same as in the discrete case. Again there is a shortcut, this time when proving the uniqueness of the solution of (2.7), see Chapter VII. Finally there is a more regular behaviour of the density $w(t)$ of the d.f. of W near 0. Set

$$\sigma := F_s(q,1) \quad , \epsilon_0 := -\lambda^{-1}\log \sigma .$$

2.8. LEMMA. *If* $1 < m < \infty$ *and* $\sigma > 0$ *then for* $|s| < 1$

$$Q(s) := \lim_{t\to\infty} \sigma^{-t}(F(s,t)-q) < \infty$$

exists, $Q(0) = 0$, *and in some neighbourhood of* 0 *there exists an analytic inverse function* Q^{-1} *of* Q.

A convergence proof is found, e.g., in Athreya and Ney (1972), see also Chapter VII below. For the existence and analyticity of Q^{-1} see Montel (1957).

2.9. PROPOSITION. *If* $1 < m < \infty$ *and* $\sigma > 0$, *then* $t^{1-\epsilon_0}w(t)$ *tends to a finite positive limit, as* $t \to 0$.

PROOF. Without loss of generality $q = 0$. Then from (2.7)

$$\sigma^{-t}\varphi(e^{\lambda t}) = \sigma^{-t}F(\varphi(1),t) \sim Q(\varphi(1)),\ t \to \infty,$$

that is,

$$\varphi(s) \sim s^{-\epsilon_0}Q(\varphi(1)),\ s \to \infty. \tag{2.8}$$

According to 2.8. we have

$$Q^{-1}(s) = \sum_{k=1}^{\infty} a_n s^n,\ |s| \leq R,\ R > 0.$$

Set

$$\widetilde{w}(t) := \sum_{n=1}^{\infty} (a_n/\Gamma(n\epsilon_0))t^{n\epsilon_0-1},$$

where Γ denotes the gamma function. Then there exist a constant $A(\epsilon_0)$ such that

$$\sum_{n=1}^{\infty} (1/\Gamma(n\epsilon_0))t^{n\epsilon_0-1} \leq A(\epsilon_0)te^t,\ t \geq 1,$$

and a constant B such that $|a_n| \leq B^n$, $n \in \mathbb{N}$. Hence

$$\tilde{w}(t) = O(t \exp\{B^{1/\epsilon_0} t\}), \quad t \to \infty.$$

For $s > B^{1/\epsilon_0}$ the integral $\bar{\varphi}(s) = \int_0^\infty e^{-st}\bar{w}(t)dt$ converges absolutely, so that

$$\bar{\varphi}(s) = \int_0^\infty e^{-s} \sum_{n=1}^\infty (a_n/\Gamma(n\epsilon_0))t^{n\epsilon_0 -1} dt$$

$$= \sum_{n=1}^\infty (a_n/\Gamma(n\epsilon_0)) \int_0^\infty e^{-st} t^{n\epsilon_0 -1} dt$$

$$= \sum_{n=1}^\infty a_n s^{-n\epsilon_0} = Q^{-1}(s^{-\epsilon_0}).$$

Combining this with (2.8),

$$\bar{\varphi}(s) = \varphi(Q(\varphi(1))^{1/\epsilon_0} s),$$

and the uniqueness of the Laplace transform completes the proof. □

3. THE METHOD OF DISCRETE SKELETONS

The idea is to use the results of Chapters II, III for the discrete time Galton-Watson process $\{Z_{n\delta}\}_{n\in\mathbb{N}}$, with δ small, to get information on $\{Z_t\}_{t\geq 0}$. The topic roughly splits up in two: First analytic results such as asymptotic forms of extinction probabilities, conditional probabilities, etc. Here the Croft-Kingman lemma A 9.1 is applicable. Second almost sure limit results, where the Croft-Kingman lemma does not apply to the paths and one has to bound the objects of study when $t \in [n\delta,(n+1)\delta]$.

In both cases the problem comes up of relating the properties of the offspring distribution F (specified by the p.g.f. $f(s)$ or the p_k) to those of the distribution of Z_δ. For example, one needs to know

3.1. PROPOSITION. *Either* $E(Z_\delta \log Z_\delta | Z_0 = 1) < \infty$ *for all* $\delta > 0$ *or for no* $\delta > 0$. *Furthermore,*

$$(\text{X LOG X}) \qquad (Z_\delta \log Z_\delta | Z_0 = 1) < \infty \qquad \delta > 0$$

if and only if

$$(\text{x log x}) \qquad \int_0^\infty x \log x \, dF(x) = \sum_{k=0}^\infty k \log k \, p_k < \infty.$$

PROOF. Note first that we can assume the offspring mean to be finite since otherwise already $EZ_\delta = \infty \quad \delta > 0$. It is also clear that we can replace $\log x$ by $\log^* x := x/e \quad 0 \leq x \leq e, \; := \log x \quad x \geq e$. Note that $\log^* x$ satisfies the assumptions of the moment inequality II.4.5, that $\log^* x$ is increasing and that $\varphi(x) := x \log^* x$ is convex. From (1.1), $E(Z_{t+u} | \mathfrak{F}_t) = e^{\lambda u} Z_t$ so that $\{W_t\} := \{Z_t/e^{\lambda t}\}$ is a martingale and $E(\varphi(W_\delta)|Z_0 = 1)$ is non-decreasing in δ. Furthermore, if $\alpha(\delta) := E(\varphi(Z_\delta)|Z_0 = 1) < \infty$, then by II.4.5 and (1.1) with $t = u = \delta$,

$$E(\varphi(Z_{2\delta}) | \mathfrak{F}_\delta) \leq \varphi(E(Z_{2\delta} | \mathfrak{F}_\delta)) + Z_\delta \alpha(\delta),$$

$$\alpha(2\delta) \leq E(\varphi(e^{\lambda\delta} Z_\delta)|Z_0 = 1) + e^{\lambda\delta}\alpha(\delta) < \infty.$$

Combining these facts with $E\varphi(Z_\delta) < \infty \Longleftrightarrow E\varphi(W_\delta) < \infty$ yields the first part of 3.1. For the second, note first that (X LOG X) $\Rightarrow$ (x log x) follows since by convexity

$$E\varphi(W_\delta) \geq E\varphi(W_\delta)I(\tau(1) \leq \delta)$$

$$= E\int_0^\delta \beta e^{-\beta t}E(\varphi(W_\delta)|\tau(1)=t)dt \geq \int_0^\delta \beta e^{-\beta t}E\varphi(W_{\tau(1)}|\tau(1)=t)dt$$

$$= \int_0^\delta \beta e^{-\beta t}\int_0^\infty \varphi(e^{-\beta t}x)dF(x)dt \geq \int_0^\delta \beta e^{-\beta t}dt\cdot\int_0^\infty \varphi(e^{-\beta^+\delta}x)dF(x)$$

and the finiteness of the last integral is equivalent to $(x \log x)$.

For the converse implication, define

$$\mu_{n,t}(k) := E(\varphi(Z_t)I(\tau_{n+1} > t)|Z_0 = k)$$

so that $\nu_{n,t} := \mu_{n,t}(1) \uparrow E\varphi(Z_t)$. Given $Z_0 = k$, let $A(i,n)$ be the event that there are at most n split in the ith $(i = 1,\ldots,k)$ line of descent before time t. Then

$$Z_t I(\tau_{n+1} > t) \leq \sum_{i=1}^k Z_t^{0,i} I(A(i,n)) := S$$

so that, applying II.4.5 to S,

$$\mu_{n,t}(k) \leq \varphi(ES) + kE\varphi(Z_t^{0,i}I(A_{i,n}))$$

$$\leq \varphi(ke^{\lambda t}) + k\nu_{n,t}.$$

Thus

$$\nu_{n+1,t} = \varphi(1)P(\tau(1) > t) + \int_0^t \beta e^{-\beta s}\sum_{k=0}^\infty p_k\mu_{n,t-s}(k)ds$$

and, letting $\pi_{n,t} := \sup_{0\leq s\leq t} \nu_{n,s}$, $\omega := \sup_{0\leq s\leq t} e^{\lambda s}$,

$$\pi_{n+1,t} \leq \varphi(1) + \int_0^t \beta e^{-\beta s}\sum_{k=0}^\infty p_k\{\varphi(k\omega) + k\pi_{n,t}\}ds$$

$$\leq c_1 + c_2\sum_{k=0}^\infty \varphi(k)p_k + m\beta t\pi_{n,t}.$$

Choosing t so small that $m\beta t < 1$, it follows by iteration that $\sup_n \pi_{n,t} < \infty$. Hence $E\varphi(Z_t) < \infty$. □

We give two immediate corrolaries.

<u>3.2</u>. THEOREM. <u>Suppose</u> $1 < m < \infty$. <u>Then</u> $\{W_t\}_{t\geq 0} := \{Z_t/e^{\lambda t}\}_{t\geq 0}$ <u>is a non-negative martingale with r.c. paths, hence having an a.s. limit</u>

W. <u>Furthermore</u>, $E(W|Z_0 = 1) = 1$ <u>if and only if</u> $(x \log x)$ <u>holds</u>. <u>Otherwise</u> $W = 0$ <u>a.s</u>.

In fact, the martingale property was already observed in the proof of 3.1. For the second part, write $W = \lim_{n\to\infty} W_{n\delta}$ so that by the Kesten-Stigum theorem $E(W|Z_0 = 1) = 1$ if and only if (X LOG X) holds while otherwise $W = 0$ a.s.

3.3. THEOREM. <u>Suppose</u> $0 < m < 1$. <u>Then</u> $\gamma := \lim_{t\to\infty} P(Z_t > 0)/e^{\lambda t}$ <u>exists and furthermore</u>, $\gamma > 0$ <u>if and only if</u> $(x \log x)$ <u>holds</u>. <u>Furthermore, there exists a probability measure on</u> $\{1,2,\ldots\}$ <u>with point probabilities</u> (say) $Q_1, Q_2, \ldots$ <u>such that</u> $P(Z_t = k|Z_t > 0) \to Q_k$ <u>as</u> $t \to \infty$, $k = 1,2,\ldots$ <u>Finally</u> $\Sigma_1^\infty kQ_k = \gamma^{-1}$.

In fact, since $h(t) := P(Z_t > 0)/e^{\lambda t}$ is continuous and $\gamma(\delta) := \lim_{n\to\infty} h(n\delta)$ exists for all $\delta > 0$, cf. III.1.6 + 1.7, the existence of γ is immediate from the Croft-Kingman lemma A 9.1 . The existence of limits of the $P(Z_t = k|Z_t > 0)$ follows in just the same way. Finally from the discrete time results of Chapter III, $\Sigma_1^\infty kQ_k = \gamma^{-1}$ and $\gamma > 0$ if and only if (X LOG X) holds.

The limit results for the critical case can be carried over in a similar manner. Take for example the case of finite second moments. Denote derivatives of $F(s,t)$ respective s by primes. It follows from $F(\cdot,t+s) = F(F(\cdot,t),s)$ and $F'(1-,t) \equiv 1$, that $\mu := (2t)^{-1}F''(1-,t) \equiv$ const., and, using this, from

$$(*) \quad F(s,t) = e^{-\beta t}s + \beta\int_0^t e^{-\beta u} f(F(s,t-n))du$$

and $f'(1-,t) \equiv 1$ that

$$\mu = (f''(1-)/2)\lim_{t\downarrow 0}(1-e^{-\beta t})/t + \mu\beta \lim_{t\downarrow 0}\int_0^t e^{-\beta u}(1-u/t)du = \beta f''(1-)/2.$$

By monotonicity the relation $1 - F(0,t) \sim (\mu t)^{-1}$, $t \to \infty$, for discrete skeletons implies the same in continuous time. To obtain the exponential conditional limit in continuous time from the skeleton result, notice that the limit d.f. is continuous. Hence the skeletons converge uniformly and thus also respective the Levy metric

$$d(H_1,H_2) := \inf\{\varepsilon\colon H_1(u-\varepsilon)-\varepsilon \le H_2(u) \le H_1(u+\varepsilon)+\varepsilon\}.$$

Continuity of $P(t^{-1}Z_t \leq \lambda | Z_t > 0, Z_0 = 1)$ in t respective d follows from the continuity of $P_t(Z_t = n | Z_0 = 1)$, $n \geq 0$ in t, which in turn is easily derived inductively from (*). Thus we may again apply the Croft-Kingman lemma. Similarly the local limit theorems carry over.

As a nontrivial example of extension of an a.s. limit statement from discrete to continuous time we shall now consider the LIL for $W - W_t$, cf. II.3.1 (the CLT is again just Croft-Kingman),

3.4. THEOREM. Suppose $\lambda > 0$, $f''(1-) < \infty$ and define $\tau^2 := \operatorname{Var} W = (f''(1)-f'(1)+1)/(f'(1)-1)$ (cf. (2.5)). Then a.s.

$$\overline{\lim_{t\to\infty}} \frac{e^{\lambda t}}{(2\tau^2 Z_t \log t)^{1/2}} (W-W_t) = 1, \quad \underline{\lim_{t\to\infty}} \frac{e^{\lambda t}}{(2\tau^2 Z_t \log t)^{1/2}} (W-W_t) = -1.$$

PROOF. Since $\log n\delta \cong \log n$, the skeleton version (obtained by replacing t by $n\delta$) is just II.3.1. Since always $\overline{\lim}_{t\to\infty} \geq \overline{\lim}_{n\delta\to\infty} = 1$, we only have to prove $\overline{\lim}_{t\to\infty} \leq 1$ (the proof of $\underline{\lim}_{t\to\infty} = -1$ is similar). Define for $n\delta \leq t \leq (n+1)\delta$

$$A_t := \operatorname{Var}(W_{(n+1)\delta} | \mathfrak{F}_t) = Z_t \operatorname{Var}(W_{(n+1)\delta-t} | Z_0 = 1),$$

$$B_n := \sup_{n\delta\leq t\leq (n+1)\delta} (W_{n\delta} - W_t + (2A_t)^{1/2}),$$

$$\varepsilon(n) := (1+\eta)(2\operatorname{Var}(W_\delta | Z_0 = 1) Z_{n\delta} \log n)^{1/2} = (1+\eta)(2\tau^2(1-e^{-\lambda\delta}) Z_{n\delta} \log n)^{1/2},$$

$$t^*(n) := \inf\{t \geq n\delta : W_{n\delta} - W_t + (2A_t)^{1/2} > \varepsilon(n)\}.$$

Then

$$P(W_{n\delta} - W_{(n+1)\delta} > \varepsilon(n) | \mathfrak{F}_{n\delta})$$
$$\geq P(W_{n\delta} - W_{(n+1)\delta} > \varepsilon(n), t^*(n) < (n+1)\delta | \mathfrak{F}_{n\delta})$$
$$\geq P(W_{t^*(n)} - W_{(n+1)\delta} \geq -(2A_{t^*(n)})^{1/2}, t^*(n) < (n+1)\delta | \mathfrak{F}_{n\delta})$$
$$\geq E[P(W_{t^*(n)} - W_{(n+1)\delta} \geq -(2A_{t^*(n)})^{1/2} | \mathfrak{F}_{t^*(n)}) I(t^*(n) < (n+1)\delta) | \mathfrak{F}_{n\delta}]$$
$$\geq \frac{1}{2} P(t^*(n) < (n+1)\delta | \mathfrak{F}_{n\delta}) = \frac{1}{2} P(B_n > \varepsilon(n) | \mathfrak{F}_{n\delta})$$

(using Chebycheff's inequality). The sum of the ℓ.h.s. is finite, using the converse conditional Borel-Cantelli lemma and the a.s. estimates of

$W_{(n+1)\delta}-W_{n\delta}$ provided by II.3.2. Hence the sum of the r.h.s. is finite, implying $B_n \leq \epsilon(n)$ eventually. Since $A_t = O(e^{\lambda t})$, it follows that

$$\overline{\lim_{t\to\infty}} \frac{W-W_t}{(2\tau^2 Z_t \log t)^{1/2}}$$

$$\leq e^{-\lambda\delta/2}\left[\overline{\lim_{n\to\infty}} \frac{W-W_{n\delta}}{(2\tau^2 Z_{n\delta}\log n)^{1/2}} + \overline{\lim_{n\to\infty}} \sup_{n\delta\leq t\leq(n+1)\delta} \frac{W_{n\delta}-W_t}{(2\tau^2 Z_{n\delta}\log n)^{1/2}}\right]$$

$$= e^{-\lambda\delta/2}\left[1 + \overline{\lim_{n\to\infty}} \frac{B_n}{(2\tau^2 Z_{n\delta}\log n)^{1/2}}\right] \leq e^{-\lambda\delta/2}(1+(1+\eta)(1-e^{-\lambda\delta})^{1/2}).$$

Let $\delta \downarrow 0$. □

4. THE METHOD OF SPLIT TIMES

According to Section 1, we can think of the branching process as constructed from Z_0 and two independent sequences $\{X_n\}$, $\{V_n\}$ where the X_n+1 are i.i.d. having the offspring distribution and the V_n are i.i.d. unit exponential. The nth inter split time is $U_n := V_n/(\beta Z_{\tau(n-1)})$, $U_1 := V_1/(\beta Z_0)$ and $Z_{\tau(n)} := Z_0+X_1+\ldots+X_n$. Let T be the hitting time of the random walk $\{Z_0+X_1+\ldots+X_n\}$ of 0. Then extinction is equivalent to $T < \infty$, and in that case, $\tau(n) = \infty$ when $n > T$, $Z_t = 0$ $t \geq \tau(T)$. We shall be concerned only with the case of extinction not being a.s., i.e. $P(T < \infty) < \infty$ or equivalently $1 < m \leq \infty$, i.e. $0 < EX_n \leq \infty$.

The basic observation is now that the theory of sums of i.i.d. random variables gives very precise information on $\{Z_{\tau(n)}\}$. In order to determine the behaviour of the branching process, we need thus only to describe the $\tau(n)$. This comes out from the structure $\tau(n) = \Sigma_1^n \alpha_k V_k$, with $\alpha_k := (\beta(Z_0+X_1+\ldots+X_{k-1}))^{-1}$, of $\tau(n)$ as (given $\mathfrak{G} :=$ $\sigma(Z_0, X_1, X_2, \ldots))$ a weighted sum of i.i.d. r.v. (with the weights α_k very precisely known).

A simple example is the explosion problem. Suppose for simplicity $p_0 = 0$. Then $Z_{\tau(k)} \to \infty$, $\alpha_k \downarrow 0$ and explosion is equivalent to $\sup_n \tau(n) = \Sigma_1^\infty \alpha_k V_k < \infty$, i.e. (appealing to A 4.2) to $\Sigma_1^\infty \alpha_k < \infty$. That is, the explosion problem is closely related to the problem of studying the rate of growth of α_k^{-1}, i.e. of $X_1+\ldots+X_k$. For example, if $m < \infty$, then $\alpha_k^{-1} \cong k\beta(m-1)$, $\Sigma_1^\infty \alpha_k = \infty$ and explosion does not occur, cf. 2.2. Note that by the Hewitt-Savage 0-1 law, either $\Sigma_1^\infty \alpha_k < \infty$ a.s. or $\Sigma_1^\infty \alpha_k = \infty$.

Our main example, occupying the rest of this section, will be (assuming $1 < m < \infty$ and, for simplicity, $p_0 = 0$) to show the existence of $W := \lim_{t\to\infty} Z_t/e^{\lambda t}$ without reference to martingales, reprove the necessity and sufficiency of $(x \log x)$ for non-degeneracy of W and give some estimates of the rate of growth of Z_t if $(x \log x)$ fails,

4.1. THEOREM. <u>Define</u> $\gamma_t := e^{\lambda t} e^{-\frac{1}{m-1}\int_0^\infty x(\lambda t \wedge \log x)dF(x)}$. <u>Then if</u> $\mu_{1/2} := \int_0^\infty x(\log^+ x)^{1/2} dF(x) < \infty$, Z_t/γ_t <u>has an a.s. limit in</u> $(0,\infty)$.

The proof is carried out in a series of lemmata.

4.2. LEMMA. *Without conditions beyond* $1 < m < \infty$, *there exists r.v.* A_n *having an a.s. limit* A *such that the* n^{th} *split time can be written*

$$(4.1)\quad \tau(n) = \sum_{k=1}^{n} \alpha_k + A_n = \sum_{k=1}^{n} \frac{1}{\beta(Z_0+X_1+\ldots+X_{k-1})} + A_n.$$

PROOF. Since the α_k are $\mathfrak{G}$-measurable, the $\alpha_k(V_k-1)$ are independent with mean zero conditionally upon $\mathfrak{G}$. By the LLN, $\alpha_k^{-1} \cong k\beta(m-1)$ so that $\Sigma_0^\infty \mathrm{Var}(\alpha_k(V_k-1)\,|\,\mathfrak{G}) = \Sigma_0^\infty \alpha_k^2 < \infty$. Hence $\Sigma_0^\infty \alpha_k(V_k-1)$ converges a.s. Note that $\tau(n) = \Sigma_1^n \alpha_k V_k$. □

It is clear from (4.1) that the growth rate of $\tau(n)$ does not differ much from $\Sigma_1^n(\beta k(m-1))^{-1} \cong \lambda^{-1}\log n$. In order to make this precise, define

$$Y_n := X_n - EX_n = X_n+1-m, \quad S_n := Y_1+\ldots+Y_n,$$

$$\tilde{Y}_n := (X_n+1)I(X_n+1 \le n)-m, \quad \tilde{S}_n := \tilde{Y}_1+\ldots+\tilde{Y}_n.$$

Applying the identity $1/(1+x) = 1-x+x^2/(1+x)$ yields

$$(4.2)\quad \sum_{k=2}^{n} \alpha_k = \sum_{k=2}^{n} \frac{1}{\beta(Z_0+S_{k-1}+(k-1)(m-1))}$$

$$= \sum_{k=2}^{n} \frac{1}{\beta(k-1)(m-1)(1+(Z_0+S_{k-1})/(k-1)(m-1))}$$

$$= \frac{1}{\lambda}\sum_{k=2}^{n}\left\{\frac{1}{k-1} - \frac{Z_0+S_{k-1}}{(k-1)^2(m-1)} + \frac{(Z_0+S_{k-1})^2}{(k-1)^2(m-1)(Z_0+S_{k-1}+(k-1)(m-1))}\right\}.$$

4.3. LEMMA. *Without conditions beyond* $1 < m < \infty$ *there exists r.v.* B_n *having an a.s. limit* B *such that*

$$(4.3)\quad -\sum_{k=2}^{n} \frac{S_{k-1}}{(k-1)^2} = \int_0^\infty x\log(x\wedge n)\,dF(x) + B_n.$$

PROOF. Letting $k^* := (\Sigma_k^\infty i^{-2})^{-1}$, the l.h.s. of (4.3) is

$$-\sum_{k=1}^{n-1} Y_k \sum_{i=k}^{n-1} \frac{1}{i^2} = -\sum_{k=1}^{n-1} \frac{Y_k}{k^*} + \frac{1}{n^*}\sum_{k=1}^{n-1} Y_k.$$

Since $k^*/k \to 1$, the last term tends to zero and it is a matter of routine to show that $Y_k = \tilde{Y}_k$ eventually and that $\Sigma_1^\infty(\tilde{Y}_k-E\tilde{Y}_k)/k^*$

converges a.s. Hence there exist $B_n^1 \to B^1$ such that the l.h.s. of (4.3) is $-\Sigma_1^{n-1} E\tilde{Y}_k/k^* + B_n^1$. But

$$-\sum_{k=1}^{n-1} \frac{E\tilde{Y}_k}{k^*} = \sum_{k=1}^{n-1} \frac{1}{k^*} E(X_k+1) I(X_k+1 > k) = \sum_{k=1}^{n-1} \frac{1}{k^*} \int_k^\infty x \, dF(x).$$

Now check that

$$\sum_{k=1}^{\infty} \frac{1}{k^*} I(k < x, \; k < n) - \log(x \wedge n)$$

is bounded in x (uniformly in n) and has a limit as $n \to \infty$. □

4.4. LEMMA. If $\mu_{1/2} < \infty$, then $\displaystyle\sum_{k=2}^{\infty} \frac{(Z_0+S_k)^2}{(k-1)^2(Z_0+S_{k-1}+(k-1)(m-1))} < \infty$ a.s.

PROOF. The denominator is $O(k^3)$. Since $S_k = o(k)$, $\Sigma_2^\infty (a+S_k)^2/k^3$ is easily seen to be finite for either no or all $a \in \mathbb{R}$ and since $Y_k = \tilde{Y}_k$ eventually, it follows that it suffices to show that $\Sigma_2^\infty \tilde{S}_k^2/k^3 < \infty$. We show that even the mean is finite. In fact,

$$\sum_{k=2}^{\infty} \frac{\operatorname{Var} \tilde{S}_k}{k^3} = \sum_{k=2}^{\infty} k^{-3} \sum_{n=1}^{k} \operatorname{Var}(X_n+1) I(X_n+1 \le n)$$

$$\le \sum_{k=2}^{\infty} k^{-3} \sum_{n=1}^{k} \int_0^n x^2 dF(x) = \sum_{n=1}^{\infty} O\left(\frac{1}{n^2}\right) \int_0^n x^2 dF(x) = \int_0^\infty O(x) dF(x),$$

$$\sum_{k=2}^{\infty} \frac{(E\tilde{S}_k)^2}{k^3} = \sum_{k=2}^{\infty} \frac{1}{k^3}\left(\sum_{n=1}^{k} \int_n^\infty x \, dF(x)\right)^2 \le \sum_{k=1}^{\infty} \frac{1}{k^3}\left[\int_0^\infty x(x\wedge k) dF(x)\right]^2$$

$$= \int_0^\infty \int_0^\infty xy \sum_{k=1}^{\infty} (x \wedge k)(y \wedge k)/k^3 \, dF(x) dF(y).$$

Now

$$\sum_{k=1}^{\infty} \frac{(x\wedge k)(y\wedge k)}{k^3} I(k \ge x, k \ge y) \simeq xy \sum_{x \vee y}^{\infty} \frac{1}{k^3} \simeq \frac{xy}{O((x\vee y)^2)} \simeq O(1),$$

$$\sum_{k=1}^{\infty} \frac{(x\wedge k)(y\wedge k)}{k^3} I(k < x, k \ge y) \simeq y \sum_{y}^{x} \frac{1}{k^2} \simeq y\left(\frac{1}{y} - \frac{1}{x}\right) \le 1 \quad (y < x)$$

(with a similar estimate for the $I(k < y, \; k \ge x)$ terms) so that, up to a $O(1)$ term, the above double integral is

$$\int_0^\infty \int_0^\infty xy \sum_{k=1}^{\infty} \frac{1}{k} I(k < x, k < y) dF(x) dF(y) = \int_0^\infty \int_0^\infty xy \; O(\log(x \wedge y)) dF(x) dF(y)$$

which has been shown to be finite in the proof of II.5.8. □

4.5. LEMMA. Without conditions beyond $1 < m < \infty$, $\lim_{n\to\infty}\{\tau(n+1)-\tau(n)\} = 0$ a.s.

PROOF. By the Borel-Cantelli lemma, $V_n = O(\log n)$. But $\tau(n+1)-\tau(n) = \alpha_{n+1} V_{n+1}$ and $\alpha_n = O(n^{-1})$. □

We can now easily get the existence of a limit W of $Z_t/e^{\lambda t}$ and the non-degeneracy criterion for W. Suppose first $(x \log x)$ holds. Then combining (4.2), 4.3, 4.4 shows that $\lambda\Sigma_1^n\alpha_k - \log n$ has a limit. Hence by (4.1), $\lambda\tau(n) - \log n$ has a limit, say η, and since $Z_{\tau(n)}/n \to (m-1)$, $Z_{\tau(n)}e^{-\lambda\tau(n)}$ tends to $W := e^{-\eta}$. If $\tau(n) \leq t < \tau(n+1)$, we have $Z_t = Z_{\tau(n)}$ and $e^{\lambda(t-\tau(n))} \leq e^{\lambda(\tau(n+1)-\tau(n))} \to 1$. Hence $Z_t/e^{\lambda t} \to W$. If $(x \log x)$ fails, we can disregard the last term in (4.2) to get a lower bound for $\Sigma_1^n\alpha_k$ and (1.1), (4.2), 4.3 then yield $\lambda\tau(n) - \log n \to \infty$ a.s. Thus $Z_{\tau(n)}e^{-\lambda\tau(n)} \cong ne^{-\lambda\tau(n)} \to 0$. If $\tau(n) \leq t < \tau(n+1)$, then $Z_t/e^{\lambda t} \leq Z_{\tau(n)}/e^{\lambda\tau(n)}$.

In the proof of 4.1, we need one further lemma:

4.6. LEMMA. Define $\psi(u) := \int_0^\infty x \log(x \wedge u)dF(x)$. Then if u, v tends to infinity in such a way that $\log v = \log u + O(\psi(u))$ and if $\mu_{1/2} < \infty$, we have $\psi(u) - \psi(v) \to 0$.

PROOF. Note that

$$(4.4) \quad \psi(u) = \int_0^u x \log x \, dF(x) + \log u\int_u^\infty x \, dF(x) = o((\log u)^{1/2})$$

$$\psi(v)-\psi(u) = \int_u^v x(\log x-\log u)dF(x) + (\log v-\log u)\int_v^\infty x \, dF(x)$$

$$\leq (\log v-\log u)\int_u^\infty x \, dF(x) = O(\psi(u))o((\log u)^{-1/2}) = o(1)$$

if $v \geq u$. If $v < u$, then by (4.4) $\log v = O(\log u)$ so that

$$\psi(u)-\psi(v) \leq (\log u-\log v)\int_v^\infty x \, dF(x) = O(\psi(u))o((\log v)^{-1/2}) = o(1).$$

□

PROOF OF 4.1. Let $t, n \to \infty$ in such a way that $\tau(n) \leq t < \tau(n+1)$. Then $Z_t = Z_{\tau(n)} \cong n(m-1)$ and, combining (4.1), (4,2), 4.3, 4.4, there are r.v. c_n with a.s. limit c such that

$$\lambda\tau(n) = \log n + \frac{1}{m-1}\,\psi(n) + c_n,$$

$$\lambda t = \log n + \frac{1}{m-1}\,\psi(n) + c_n + o(1)$$

(using 4.5). Hence the conditions of 4.6 hold with $u := n$, $v := e^{\lambda t}$ and we get

$$\frac{Z_t}{Y_t} = \frac{Z_t}{e^{\lambda t}}\, e^{\frac{1}{m-1}\psi(e^{\lambda t})} \cong e^{-c} e^{\frac{1}{m-1}(\psi(e^{\lambda t})-\psi(n))}(m-1) \cong (m-1)\, e^{-c} \,. \square$$

BIBLIOGRAPHICAL NOTES

Much of the material is standard and can be found in textbooks like Athreya and Ney (1972). (2.6), the slowly varying property of g, and 2.5 are (formally new) continuous-time analogues of more recent discrete-time results, cf. Chapter III and the references given there. 2.6 is a routine extension of Slack's results, see again Chapter III, and 2.9 has been taken from Dubuc (1971). Another exception is the counterpart 4.1 of formula II.5.7 for the Seneta constants. Schuh (1982) provides the connection between the analytic and the probabilistic approach to the explosion problem. For the Croft-Kingman lemma see Kingman, Proc.London Math.Soc. 13(1963),593-604.

PART C

MULTIGROUP BRANCHING DIFFUSIONS ON BOUNDED DOMAINS

CHAPTER V

FOUNDATIONS

1. EXISTENCE AND CONSTRUCTION

Let $(X,\mathfrak{A})$ be a measurable space, $X^{(n)}$ the symmetrization of the direct product X^n of n copies of X, $X^{(o)} = \{\theta\}$ with some extra point θ,

$$\hat{X} := \bigcup_{n=0}^{\infty} X^{(n)},$$

and $\hat{\mathfrak{A}}$ the σ-algebra on $\hat{X}$ induced by $\mathfrak{A}$.

An element $\langle x_1,\ldots,x_n\rangle \in X^{(n)}$ may then be regarded as representing the unordered population of n objects of type $x_1,\ldots,x_n$, respectively, while θ stands for the empty population. Accordingly, $(X,\mathfrak{A})$ is called the <u>type space</u> and $(\hat{X},\hat{\mathfrak{A}})$ the population space.

For $\hat{x} \in \hat{X}$ and $A \subset X$ define

$$\hat{x}[A] := 0; \qquad \hat{x} = \theta$$

$$:= \sum_{\nu=1}^{n} 1_A(x_\nu); \qquad \hat{x} = \langle x_1,\ldots,x_n\rangle.$$

A transition function $P_t(\hat{x},\hat{A})$, $t \in \mathbb{Z}_+$ or $t \in \mathbb{R}_+$, $\hat{x} \in \hat{X}$, $\hat{A} \in \hat{\mathfrak{A}}$, is called a <u>branching transition function</u>, if for all t

$$P_t(\theta, X^{(o)}) = 1 \tag{1.1.a}$$

and for all t, each finite measurable decomposition $\{A_1,\ldots,A_m\}$ of X, all $n_j \in \mathbb{Z}_+$, $j = 1,\ldots,m$, and every $\langle x_1,\ldots,x_k\rangle \in \hat{X}$, $k > 0$,

$$\begin{aligned} &P_t(\langle x_1,\ldots,x_k\rangle; \{\hat{x}[A_j] = n_j;\ j = 1,\ldots,m\}) \\ &= \sum_{\substack{n_{j_1}+\ldots+n_{j_k}=n_j \\ j=1,\ldots,m}} \prod_{\nu=1}^{k} P_t(\langle x_\nu\rangle, \{\hat{x}[A_j] = n_{j_\nu};\ j=1,\ldots,m\}). \end{aligned} \tag{1.1.b}$$

This is one way of expressing the <u>branching property</u> or, more precisely, the property of independent branching (1.1.b) without immigration (1.1.a). Another way is the following:

Define $\mathfrak{B}$ as the Banach algebra of all bounded, complex-valued, $\mathfrak{A}$-measurable functions ξ on X with supremum-norm

$$\|\xi\| := \sup_{x\in X} |\xi(x)|.$$

Let $\mathfrak{S}$ be the open unit ball in $\mathfrak{B}$ and $\bar{\mathfrak{S}}$ its closure. For $\eta \in \bar{\mathfrak{S}}$ and $\hat{x} \in \hat{X}$ define

$$\tilde{\eta}(\hat{x}) := 1, \qquad \hat{x} = \theta,$$
$$:= \prod_{\nu=1}^{n} \eta(x_\nu); \quad \hat{x} = \langle x_1, \ldots, x_n \rangle.$$

The <u>generating functional</u> of $P_t(\hat{x}, \cdot)$ is then given by

$$F_t(\hat{x}, \eta) := \int_{\hat{X}} \tilde{\eta}(\hat{y}) P_t(\hat{x}, d\hat{y}),$$

which is well-defined on $\bar{\mathfrak{g}}$. The branching property (1.1) now is equivalent to

$$\text{(F.1)} \qquad F_t(\hat{x}, \eta) = 1; \qquad \hat{x} = \theta,$$
$$= \prod_{\nu=1}^{n} F_t(\langle x_\nu \rangle, \eta); \quad \hat{x} = \langle x_1, \ldots, x_n \rangle,$$

which is to hold for all $t > 0$, $\hat{x} \in \hat{X}$, and $\eta \in \bar{\mathfrak{g}}$. To see the equivalence, insert $\eta = \Sigma\lambda_j 1_{A_j}$, $|\lambda_j| \leq 1$, into (F.1) and compare the coeffients of the products of λ_j- powers. This gives (1.1). For the reverse, approximate η by stepfunctions and use dominated convergence.

A useful quantity connected with $F_t(\hat{x}, \cdot)$, $\hat{x} \in \hat{X}$, is the <u>generating mapping</u> $F_t : \bar{\mathfrak{g}} \to \bar{\mathfrak{g}}$, defined by

$$F_t[\eta](x) := F_t(\langle x \rangle, \eta), \quad \eta \in \bar{\mathfrak{g}}, \; x \in X.$$

In fact, given (F.1), the Chapman-Kolmogorov relation for P_t is equivalent to the semigroup relation

$$\text{(F.2)} \qquad F_{t+s}[\eta] = F_s[F_t[\eta]], \; t,s \geq 0, \; \eta \in \bar{\mathfrak{g}},$$

whence $\{F_t\}$ is called a <u>generating semigroup</u>. To verify the equivalence, represent $F_{t+s}[\eta](x)$ as the limit of a sequence of Lebesgue-Stieltjes sums with finitely many terms each, use Chapman-Kolmogorov, then dominated convergence to interchange integration and limit operation, and finally (F.1). This gives (F.2). For the reverse, take $\eta = \Sigma\lambda_j 1_{A_j}$, which leads to the Chapman-Kolmogorov equation for a class of sets countably generating $\hat{\mathfrak{A}}$ and thus for all $\hat{A} \in \hat{\mathfrak{A}}$.

An $(\hat{X}, \hat{\mathfrak{A}})$-valued Markov process $\{\hat{x}_t, P^{\hat{x}}\}$ is called a <u>Markov branching process</u>, if it has a branching transition function. Given such a process, there is still a third possibility to express the branching property:

For $\hat{x}, \hat{y} \in \hat{X}$ define

$$\hat{x} + \hat{y} := \theta; \qquad \hat{x} = \hat{y} = \theta$$
$$:= \langle x_1, \ldots, x_n \rangle; \qquad \hat{x} = \langle x_1, \ldots, x_n \rangle, \hat{y} = \theta,$$
$$:= \langle x_1, \ldots, x_n, y_1, \ldots, y_m \rangle; \quad \hat{x} = \langle x_1, \ldots, x_n \rangle, \hat{y} = \langle y_1, \ldots, y_m \rangle.$$

If $\{\hat{x}_t, P^{\hat{x}}\}$ is a Markov process such that for all $t, s > 0$

$$(1.2.a) \qquad \hat{x}_{t+s} = \sum_{j=1}^{\hat{x}_t[X]} \hat{x}_{t+s}^{t,j} \quad \text{a.s.} \quad [P^{\hat{x}}]$$

with $\hat{x}_{t+s}^{t,j}$, $j = 1, \ldots, \hat{x}_t[X]$, conditionally independent, given $\mathfrak{F}_t := \sigma(\hat{x}_u;\ u \leq t)$, and

$$(1.2.b) \qquad P^{\hat{x}}(\hat{x}_{t+s}^{t,j} \in \hat{A} \mid \mathfrak{F}_t) = P^{\langle x_j \rangle}(\hat{x}_s \in \hat{A}) \quad \text{a.s.} \ [P^{\hat{x}}], \qquad \hat{x}_t = \langle x_1, \ldots, x_{\hat{x}_t[X]} \rangle,$$

then $\{\hat{x}_t, P^{\hat{x}}\}$ clearly is a Markov branching process. Vice versa, given a Markov branching process, there always exists an equivalent process for which (1.2) is satisfied if t and s are restricted to an arbitrary, but fixed discrete skeleton. This is a direct consequence of C. Ionescu-Tulcea's theorem. To get (1.2) for all t and s, some more structure of $(X, \mathfrak{A})$ is needed. The topological assumptions underlying the following construction are sufficient.

We now turn to the construction of a general Markov branching process in continuous time. We shall outline the basic ideas and procedures, which are, in fact, close to intuition, but shall not reproduce all the mostly straightforward and tedious technical details. Except for a slight modification in the setting, which we shall explain further down, the latter can be found in Ikeda, Nagasawa, and Watanabe (1968, 1969).

Let X be a locally compact Hausdorff space with a countable open base, $X^* := X \cup \{\partial, \Delta\}$ a two-point compactification of X, and $\mathfrak{A}^*$ the Borel algebra on X^*. Suppose to be given

(a) a right continuous strong Markov process

$$\{x_t^o, \mathfrak{F}_t^o, P_x^o, \zeta^o : x \in X\}$$

in $(X^*, \mathfrak{A}^*)$, where x_t^o is the sample path, $\{\mathfrak{F}_t^o\}$ is an adapted sequence of σ-algebras, $\{P_x^o\}$ the set of probability measures,

$$\zeta^o := \tau_\partial \wedge \tau_\Delta$$

the lifetime, τ_∂ and τ_Δ being the hitting times of ∂ and Δ, respectively, ∂ and Δ serving as traps, and x_t^o has left limits in X for $t < \tau_\partial$,

(b) a stochastic kernel

$$\pi = \pi(x,\hat{A})$$

defined on $X \otimes \hat{\mathfrak{A}}$.

The aim is, to construct from these data a right-continuous, strong Markov process

$$\{\hat{x}_t, \mathfrak{F}_t, P^{\hat{x}};\ \hat{x} \in \hat{X}\}$$

on $(\hat{X}^*, \hat{\mathfrak{A}}^*)$, $\hat{X}^* := \hat{X} \cup \{\Delta^*\}$ being the one-point compactification of $\hat{X}$ in the topology induced by the topology on $\hat{X}$ and $\hat{\mathfrak{A}}^*$ the Borel algebra on $\hat{X}^*$. The intuitive prescription is the following:

All objects ("particles", "individuals") move independently on X, each according to a copy of $\{x_t^o\}$. An object hitting ∂ disappears, i.e., is instantaneously replaced by the empty population. An object hitting Δ is instantaneously replaced by a population of new objects distributed according to $\pi(x_{\tau-},\cdot)$, where $x_{\tau-}$ is the left limit of the object being replaced at the time it hits Δ.

We first construct the process up to the time of first absorption (a particle hits ∂) or first branching (a particle hits Δ). This is done in the three steps:

(1) Define the n-fold direct product

$$\{\tilde{x}_t^n, \mathfrak{F}_t^n, P_{\tilde{x}}^n, \zeta^n;\ \tilde{x} \in X^n\}$$

of $\{x_t^o, P_x^o\}$. Here $\tilde{x}_t^n$ is a vector-valued path with n independent copies of x_t^o as components, $P_{\tilde{x}}^n$ correspondingly is first introduced as the product measure on

$$\mathfrak{G}_\infty^n := \bigcup_{t>0} \mathfrak{G}_t^n,\ \mathfrak{G}_t^n := \sigma(\tilde{x}_s^n;\ s \leq t),$$

then $\mathfrak{F}_t^n$ is taken to be $\bar{\mathfrak{G}}_{t+0}^n$, where $\bar{\mathfrak{G}}_t^n$ is the completion of $\mathfrak{G}_t^n$ respective $P_{\tilde{x}}^n$, $\tilde{x} \in X^n$, and finally the lifetime ζ^n is determined by the prescription that $\tilde{x}_t^n$ is stopped whenever one of its components terminates.

(2) Define the symmetrization

$$\{\hat{x}_t^{(n)}, \mathfrak{F}_t^{(n)}, P_{\hat{x}}^{(n)}, \zeta^{(n)};\ \hat{x} \in \hat{X}\}$$

of $\{\tilde{x}_t^{(n)}\}$: For $n = 0$ take the trivial process determined by $P_\theta^{(o)}(\hat{x}_t^{(o)} \equiv \theta) = 1$. For $n \geq 1$ set

$$\hat{x}_t^{(n)} := \gamma(\tilde{x}_t^{(n)}),\ \gamma(x_1,\ldots,x_n) := \langle x_1,\ldots,x_n\rangle,$$

then, for $\hat{x} = \gamma(\tilde{x})$, introduce $P_{\hat{x}}^n$ as $P_{\tilde{x}}^n$ restricted to

$$\mathfrak{G}_\infty^{(n)} := \bigcup_{t>0} \mathfrak{G}_t^{(n)},\ \mathfrak{G}_t^{(n)} := \sigma(\hat{x}_s^{(n)};\ s \leq t),$$

finally, $\mathfrak{F}_t^{(n)} := \overline{\mathfrak{G}}_{t+0}^{(n)}$, where $\overline{\mathfrak{G}}_t^{(n)}$ is the completion of $\mathfrak{G}_t^{(n)}$ respective $P_{\hat{x}}^{(n)}$, $\hat{x} \in X^{(n)}$, and $\zeta^{(n)} := \zeta^n$.

(3) Define the topological sum

$$\{\hat{x}_t^-, \mathfrak{F}_t^-, P_{\hat{x}}^-, \zeta^-;\ \hat{x} \in \hat{X}\}$$

of the $\{\hat{x}_t^{(n)}\}$, $n \geq 0$: Let $\Omega^{(n)}$ be the sample space of $\{\hat{x}_t^{(n)}\}$, define

$$\Omega^- := \bigcup_{n=0}^{\infty} \Omega^{(n)},$$

$$\hat{x}_t^- := \hat{x}_t^{(n)} \quad \text{on} \quad \Omega^{(n)},\ n \geq 0,$$

$$\mathfrak{G}_\infty^- := \bigcup_{t>0} \mathfrak{G}^-,\ \mathfrak{G}_t^- := \sigma(\hat{x}_s^-;\ s \leq t),$$

$$P_{\hat{x}}^-(\Gamma) = P_{\hat{x}}^{(n)}(\Gamma \cap \Omega^{(n)}),\quad \Gamma \subset \mathfrak{G}_\infty^-,$$

let $\mathfrak{F}_t^- := \overline{\mathfrak{G}}_{t+0}^-$, $\overline{\mathfrak{G}}_t^-$ being the completion of $\mathfrak{G}_t^-$ with respect to $P_{\hat{x}}^-$, $\hat{x} \in \hat{X}$, finally $\zeta^- = \zeta^{(n)}$ on $\Omega^{(n)}$.

To complete the construction we now "piece out" $\{\hat{x}_t^-\}$, taking into account the absorption or branching events at the "joints". For simplicity, let us assume that

$$P_x^o(\zeta^o = t) = 0$$

for every non-random $t \geq 0$. This implies, in particular, that almost surely only one particle at a time can be absorbed or undergo branching. Again we proceed in three steps:

(4) First define the sample space and paths: Let Ω_j^-, $\hat{X}_j^*$, $\mathfrak{F}_{\infty,j}^-$, $\hat{\mathfrak{A}}_j^*$ be copies of Ω^-, $\hat{X}^*$, $\mathfrak{F}_\infty^- := \cup_{t>0}\mathfrak{F}_t^-$, and $\hat{\mathfrak{A}}^*$, respectively, and define

$$\Omega_j^+ := \Omega_j^- \otimes \hat{X}_j^*, \quad \mathfrak{F}_j^* := \mathfrak{F}_{\infty,j}^- \otimes \hat{\mathfrak{A}}_j^*,$$

$$\Omega := \bigotimes_{j\in\mathbb{N}} \Omega_j^+, \qquad \mathfrak{F} := \bigotimes_{j\in\mathbb{N}} \mathfrak{F}_j^+ .$$

For $\omega = (\omega_1^-, \hat{x}_1;\ \omega_2^-, \hat{x}_2, \ldots) \in \Omega$ set

$$\tau_n(\omega) := \sum_{j=1}^{n} \zeta^-(\omega_j^-), \quad n \in \mathbb{N},$$

$$\begin{aligned} N(\omega) &:= \min\{j:\ \zeta^-(\omega_j^-) = 0\}; & \exists_j:\ \zeta^-(\omega_j^-) = 0, \\ &:= \infty; & \nexists_j:\ \zeta^-(\omega_j^-) = 0, \end{aligned}$$

and define

$$\begin{aligned} \hat{x}_t(\omega) &:= \hat{x}_t^-(\omega_1^-); & 0 \leq t \leq \tau_1(\omega) \\ &:= \hat{x}^-_{t-\tau_n(\omega)}(\omega_{n+1}^-); & \tau_n(\omega) < t \leq \tau_{n+1}(\omega), \\ &:= \Delta^* & \tau_{N(\omega)} \leq t. \end{aligned}$$

Consistently, $\zeta := \tau_N$.

(5) Next construct the stochastic kernel describing the transition at a "joint": The transition from a population of just one particle is determined by

$$\begin{aligned} \pi'(\omega^o, \hat{A}) &:= 1_{\hat{A}}(\theta); & \hat{A} \in \hat{\mathfrak{A}}^*,\ \zeta^o(\omega^o) = \tau_\partial(\omega^o), \\ &:= \pi(x^o_{\zeta^o(\omega^o)-}(\omega^o), \hat{A}); & \hat{A} \in \hat{\mathfrak{A}},\ \zeta^o(\omega^o) = \tau_\Delta(\omega^o), \\ &:= 0; & \hat{A} = \{\Delta^*\}, \end{aligned}$$

the transition from an ordered population of n particles by

$$\begin{aligned} &\pi^n(\omega^n, \hat{A}_1 \otimes \ldots \otimes \hat{A}_n) \\ &\quad := \sum_{j=1}^{n} 1_{\{\zeta^n(\omega^n) = \zeta^o(\omega_j^o)\}}(\omega^n)\, \pi'(\omega_j^o, \hat{A}_j) \\ &\qquad \times \prod_{i\neq j} 1_{\hat{A}_i}(\langle x^o_{\zeta^o(\omega_j^o)}(\omega_i^o)\rangle); \quad 0 < \zeta^o(\omega^n) < \infty, \\ &\quad := 1_{\hat{A}_1 \otimes \ldots \otimes \hat{A}_n}((\Delta^*, \ldots, \Delta^*)); \qquad \zeta^n(\omega^n) = 0 \quad \zeta^n(\omega^n) = \infty, \\ &\omega^n = (\omega_1^o, \ldots, \omega_n^o) \in \Omega^n, \end{aligned}$$

and the transition from an arbitrary unordered population to an unordered population by

$$\hat{\pi}(\omega^-,\hat{A}) := \pi(\omega^n,\gamma_n^{-1}(\hat{A})); \quad \omega^- = \omega^{(n)} (= \omega^n)$$

$$:= 1_{\hat{A}}(\theta); \qquad \omega^- = \omega^{(o)},$$

$$\gamma_n((\hat{x}_1,\ldots,\hat{x}_n)) := \sum_{j=1}^{n} \hat{x}_j .$$

(6) The probability measures $P_{\hat{x}}$, $\hat{x} \in \hat{X}$, are now given by

$$P^{\hat{x}}(\Gamma_1 \otimes \ldots \otimes \Gamma_n)$$

$$= \int_{\Gamma_1} P^-_{\hat{x}}(d\omega_1^-)\hat{\pi}(\omega_1^-,d\hat{x}_1^-)\int_{\Gamma_2} P^-_{\hat{x}_1}(d\omega_2^-)\hat{\pi}(d\omega_2^-,d\hat{x}_2)\int_{\Gamma_n} P^-_{\hat{x}_{n-1}}(d\omega_n^-)\hat{\pi}(\omega_n^-,d\hat{x}_n),$$

$$\Gamma_j \in \mathfrak{F}_j^+, \quad j = 1,\ldots,n, \quad n \in \mathbb{N},$$

which by C. Ionescu-Tulcea's theorem uniquely determines a probability measure P^x on $\mathfrak{F}$.

Clearly, $\{\hat{x}_t, P^{\hat{x}}\}$ is a right-continuous Markov process in the elementary sense, as defined, and its transition function, restricted to $(\hat{X},\hat{\mathfrak{A}})$, has the branching property. The technical work completing the construction is in proving the following:

1.1. PROPOSITION. There exists an increasing sequence of σ-algebras $\mathfrak{F}_t$, $\bar{\mathfrak{F}} \supset \mathfrak{F}_t = \bar{\mathfrak{F}}_{t+0} \supset \sigma(\hat{x}_s;\ s \leq t)$, where $\bar{\mathfrak{F}}$, $\bar{\mathfrak{F}}_t$ are the completions of $\mathfrak{F}$ and $\mathfrak{F}_t$ respective $P^{\hat{x}}$, $\hat{x} \in \hat{X}$, such that $\{\hat{x}_t, \mathfrak{F}_t, P_{\hat{x}}, \zeta;\ \hat{x} \in \hat{X}\}$ is strongly Markovian.

The proof, which will not be given here, is obtained by a trivial modification of the argument of Ikeda, Nagasawa, and Watanabe (1968, 1969). Their setting differs from ours insofar as they have

(i) X compact, $X^* = X \cup \{\Delta\}$,

(ii) $\pi(x,X^{(1)}) \equiv 0$.

We admit non-compact X and work with a second trap ∂, in order to incorporate in a natural way absorbing barriers, as they may occur in diffusions, and we do not require (ii), in view of multigroup branching diffusions and in order to keep a simple possibility of modelling retarded branching, cf. V.3. Formally, (ii) can always be forced by suitably redefining the respective model, but this does not always carry a practical advantage. The reason for (ii) is, of course, unicity.

What is the generating semigroup of the process just constructed?

Let $\mathcal{B}(\hat{X})$ be the Banach algebra of all bounded, $\hat{\mathfrak{U}}$-measurable, complex-valued functions on $\hat{X}$ with supremum-norm. Then the translation semigroup $\{U_t\}$ of $\{\hat{x}_t, P^{\hat{x}}\}$ is given by

$$U_t\xi(x) := E^{\hat{x}}\xi^o(\hat{x}_t),\ \xi^o|_{\hat{X}} := \xi,\ \xi^o(\Delta^*) := 0,\quad \hat{x} \in \hat{X}.$$

As before, let $\tau = \tau_1$ be the time of first branching or absorption, and write

$$(1.3)\qquad U_t\xi(\hat{x}) = E^{\hat{x}}[\xi^o(\hat{x}_t)1_{\{t \leq \tau\}}] + E^{\hat{x}}[\xi^o(\hat{x}_t)1_{\{t < \tau\}}].$$

The first term on the right can be expressed in terms of the transition semigroup $\{T_t^o\}$ of $\{x_t^o, P_x^o\}$, given by

$$T_t^o\eta(x) := E_x^o\eta^o(x_t^o),\ \eta^o|_X = \eta,\ \eta^o(\partial) := \eta^o(\Delta) := 0,\quad x \in X.$$

Simply define

$$\tilde{T}_t^o\,\tilde{\eta}(\hat{x}) := \widetilde{(T_t^o\eta)}(\hat{x}),$$

and, noticing that the linear hull of $\{\xi : \xi = \tilde{\eta},\ \eta \in \bar{\mathfrak{F}}\}$ is dense in $\mathcal{B}(\hat{X})$, extend $\tilde{T}_t^o$ to $\mathcal{B}(\hat{X})$. Then

$$E^{\hat{x}}[\xi^o(\hat{x}_t)1_{\{t \leq \tau\}}] = E^{\hat{x}}[\xi(\hat{x}_t)1_{\{t \leq \tau\}}] = \tilde{T}_t^o\xi(\hat{x}).$$

For the second term on the right of (1.3), the strong Markov property implies

$$E^{\hat{x}}[\xi^o(\hat{x}_t)1_{\{\tau \leq t\}}] = E^{\hat{x}}([E^{\hat{x}_\tau}\xi(\hat{x}_{t-s})]_{s=\tau}1_{\{\tau \leq t\}})$$
$$= \int_0^t\int_{\hat{X}} P^{\hat{x}}(\tau \in ds,\ \hat{x}_\tau \in d\hat{y})U_{t-s}\xi(\hat{y}).$$

That is, $\{U_t\xi\}$ solves

$$(1.4)\qquad \hat{u}_t(\hat{x}) = \tilde{T}_t^o\xi(\hat{x}) + \int_0^t\int_{\hat{X}} P^{\hat{x}}(\tau \in ds,\ \hat{x}_\tau \in d\hat{y})\hat{u}_t(\hat{y}).$$

<u>1.2.</u> PROPOSITION. <u>Let</u> $0 \leq \xi \in \mathcal{B}(\hat{X})$. <u>Then</u> $U_t\xi$ <u>is the limit of the iteration sequence</u> $\{\hat{u}_t^{(n)}\}$ <u>of</u> (1.4), <u>beginning with</u> $\hat{u}_t^{(o)} \equiv 0$.

PROOF. The limit of $\{\hat{u}_t^{(n)}\}$ clearly is a solution. Now

$$\hat{u}_t^{(n+1)} = \sum_{j=0}^{n} \tilde{T}_t^{(j)}\xi,\quad n \in \mathbb{N},$$

$$\tilde{T}_t^{(o)} := \tilde{T}_t^o,\ \tilde{T}_t^{(j)} := \int_0^t\int_{\hat{X}} P^{\hat{x}}(ds \in \tau, \hat{x}_t \in d\hat{y})\tilde{T}_{t-s}^{(j-1)}\xi(y),\quad j \in \mathbb{N}.$$

With τ_n as before

$$\tilde{T}_t^{(n)}\xi(\hat{x}) = E^{\hat{x}}[\xi^o(\hat{x}_t)1_{\{\tau_n \leq t < \tau_{n+1}\}}], \quad n \in \mathbb{N},$$

so that $u_t^{(n)} \uparrow U_t\xi$. □

Since any non-negative solution $\hat{u}_t$ of (1.4) satisfies $\hat{u}_t > \hat{u}_t^{(o)} \equiv 0$, hence $\hat{u}_t \geq \hat{u}_t^{(n)}$, $n \in \mathbb{N}$, and thus $\hat{u}_t \geq \lim_{t\to\infty}\hat{u}_t^{(n)} = U_t\xi$, we immediately have the following:

1.3. COROLLARY. For $0 \leq \xi \in \mathcal{B}(\hat{X})$, $U_t\xi$ is the minimal non-negative solution of (1.4).

Specializing to $\xi = \tilde{\eta}$, $\eta \in \bar{\mathfrak{S}}$, we get

$$U_t\xi(\langle x\rangle) = F_t[\eta](x), \quad t \geq 0, \quad x \in X.$$

That is, $F_t[\eta](x)$ solves

$$u_t(x) = T_t^o\eta(x) + \int_0^t\int_{\hat{X}} P^{\langle x\rangle}(\tau \in ds, \hat{x}_\tau \in d\hat{y})\tilde{u}_{t-s}(\hat{y}),$$

or more explicitly,

$$\begin{aligned} u_t(x) &= T_t^o\eta(x) + P_x^o(\zeta^o \leq t, x^o_{\zeta^o} = \partial) \\ &\quad + \int_0^t\int_X P_x^o(\zeta^o \in ds, x^o_{\zeta^o} = \Delta, x^o_{\zeta^o-} \in dy) f[u_{t-s}](y), \qquad (1.5) \\ f[\eta](x) &:= \int_{\hat{X}} \pi(x, d\hat{y})\tilde{\eta}(\hat{y}), \quad \eta \in \bar{\mathfrak{S}}, \quad x \in X. \end{aligned}$$

1.4. PROPOSITION. If u_t solves (1.5), with any $\eta \in \bar{\mathfrak{S}}$, then $\tilde{u}_t$ solves (1.4) with $\xi = \tilde{\eta}$.

If the solution of (1.5) is unique, and that is the case we shall exclusively deal with later, 1.4 is trivial. For a general verification, see Ikeda, Nagasawa, and Watanabe (1968, 1969).

1.5. COROLLARY. If $0 \leq \eta \in \bar{\mathfrak{S}}_+$, $F_t[\eta]$ is the minimal non-negative solution of (1.5) and thus equal to the limit of the iteration sequence $\{u_t^{(n)}\}$ of (1.5), beginning with $u_t^{(o)} \equiv 0$.

Let us briefly look at the important special case that $\{x_t^o, P_x^o\}$ is obtained from a process $\{x_t, P^x\}$, defined as a conservative, right-continuous, strong Markov process on $X \cup \{\partial\}$ with trap ∂, by curtailing its lifetime with a termination density $k_o \geq 0$, $k_o|_X = k \in \mathcal{B}$, $k_o(\partial) = 0$, using the second trap Δ, whenever the process is stopped before hitting ∂. In this case

$$P_x^o(\zeta^o \in ds,\ x^o_{\zeta^o} = \Delta,\ x^o_{\zeta^o -} \in dy) = T_s^o\{k1_{dy}\}(x)dx,$$

see, e.g., Dynkin (1965), further

$$\begin{aligned} P_x^o(\zeta^o \leq t,\ x^o_{\zeta^o} = \partial) &= P_x^o(\zeta^o \leq t) - P_x^o(\zeta^o \leq t,\ x^o_{\zeta^o} = \Delta) \\ &= 1 - T_t^o 1(x) + \int_0^t \int_X P_x^o(\zeta^o \in ds, x^o_o = \Delta, x^o_{\zeta^o -} \in dy)ds \\ &= 1 - T_t^o 1(x) - \int_0^t T_s^o k(x)ds, \end{aligned}$$

so that (1.5) with $u_t = F_t[\eta]$ becomes

(IF) $$1 - F_t[\eta] = T_t^o(1-\eta) + \int_0^t T_s^o k(1 - f[F_{t-s}[\eta]])ds.$$

If $\{x_t, P_x\}$ is conservative on X, i.e.

$$P_x^o(\zeta^o \leq t,\ x^o_{\zeta^o} = \partial) = 0 \quad \forall\, x \neq \partial,$$

then (IF) simplifies to

$$F_t[\eta] = T_t^o\eta + \int_0^t T_s^o kf[F_{t-s}[\eta]]ds,$$

which is the conventional form of a semigroup perturbation equation: The non-linear semigroup $\{F_t\}$ is obtained by perturbing the linear semigroup $\{T_t\}$ with the non-linear operator kf.

Define

$$(m1)(x) := \sum_{n\geq 1} n\pi(x, X^{(n)}), \quad x \in X.$$

If $km1 \in \mathcal{B}$, then

$$\| kf[\eta] - kf[\xi] \| \leq \| km1 \| \ \| \eta-\xi \|, \quad \xi,\eta \in \overline{\mathfrak{S}}.$$

<u>1.6</u>. PROPOSITION. <u>If</u> $km1 \in \mathcal{B}$, <u>then the solution of</u> (IF) <u>is unique</u>.

PROOF. Given two solutions u_t and u_t^*,

$$\| u_t - u_t^* \| \leq \| km1 \| \int_0^t \| u_s - u_s^* \| ds,$$

and by iteration $\| u_t - u_t^* \| \leq \| km1 \|^n / n! \to 0, \quad n \to \infty.$ □

The problem of uniqueness or non-uniqueness of the solution of (IF) is closely related to the question, whether or not $\{\hat{x}_t, P^{\hat{x}}\}$ is conservative on $(\hat{X},\hat{\mathfrak{A}})$. Clearly, $\{\hat{x}_t, P^{\hat{x}}\}$ is <u>not</u> conservative on $(\hat{X},\hat{\mathfrak{A}})$ if and only if there exist $t > 0$ and $x \in X$ such that $u_t 1(\langle x \rangle) < 1$, which is the same as $F_t[1](x) < 1$. In any case, $F_t[1]$ is the minimal non-negative solution of (IF) for $\eta = 1$. On the other hand, $u_t \equiv 1$ also is a non-negative solution for $\eta = 1$. Hence, we have the following:

1.7. PROPOSITION. *The process* $\{\hat{x}_t, P^{\hat{x}}\}$ *is conservative on* $(\hat{X}, \hat{\mathfrak{A}})$, *if and only if* $u_t \equiv 1$ *is the only solution of* (IF) *for* $\eta = 1$.

1.8. COROLLARY. *If the process* $\{\hat{x}_t, P^{\hat{x}}\}$ *can be constructed from* $[x_t, k, \pi]$ *such that* $km\, 1 \in \mathfrak{B}$, *then it is conservative on* $(\hat{X}, \hat{\mathfrak{A}})$.

More on "explosion", i.e., the case that $\{\hat{x}_t, P^{\hat{x}}\}$ is *not* conservative on $(\hat{X}, \hat{\mathfrak{A}})$, can be found in Savits (1969).

REMARK. How general is the (x_t^o, π)-setting? Within their framework (X compact), Ikeda, Nagasawa and Watanabe (1968, 1969), have given the following answer: Given a Markov branching process $\{\hat{x}_t, P^{\hat{x}}\}$ which is a Hunt process with reference-measure, there exists (x_t^o, π) (uniquely determined, if we require $\pi(x, X^{(1)}) \equiv 0$) such that $\{\hat{x}_t, P^{\hat{x}}\}$ is equivalent to the process constructed from (x_t^o, π).

2. GENERATING FUNCTIONALS AND MOMENTS

Let $(X,\mathfrak{A})$ and $(Y,\mathfrak{B})$ be measurable spaces, $(\hat{X},\hat{\mathfrak{A}})$ the population space corresponding to the type space $(X,\mathfrak{A})$, P a transition kernel from $(Y,\mathfrak{B})$ to $(\hat{X},\hat{\mathfrak{A}})$, and $F: \overline{\mathfrak{S}} \to \overline{\mathfrak{S}}(Y)$ the corresponding generating mapping ($\overline{\mathfrak{S}}(Y)$ defined as $\overline{\mathfrak{S}}$ with Y in place of X).

Clearly, F is isotone and convex on $\overline{\mathfrak{S}}_+ : = \{\eta \in \mathfrak{S} : = \eta \geq 0\}$, and

(a) $$F[1] = 1_Y.$$

Furthermore, it is easily verified that

(b) F <u>is analytic on</u> $\mathfrak{S}$,

that is, for $(\eta,\xi) \in \mathfrak{S} \otimes \mathfrak{B}$,

$$\delta F[\eta;\xi] : = \lim_{\lambda \to 0} \frac{F[\eta-\lambda\varepsilon]-F[\eta]}{\lambda}$$

exists respective the strong topology on $\mathfrak{B}(Y)$ ($\mathfrak{B}(Y)$ defined as $\mathfrak{B}$ with Y in place of X).

Analyticity on $\mathfrak{S}$ implies that for every $n \in \mathbb{N}$ the nth Fréchet derivative of F at η in the directions $\xi_1,\ldots,\xi_n \in \mathfrak{B}$, $\delta^n F[\eta;\xi_1,\ldots,\xi_n]$, exists, that it is symmetric, n-linear, and bounded as a function of $(\xi_1,\ldots,\xi_n) \in \mathfrak{B}^n$, and that for $||\eta|| + ||\xi|| < 1$

$$F[\eta+\xi] = \sum_{\nu=0}^{\infty} \frac{1}{\nu!} \delta^\nu F[\eta;\xi_1,\ldots,\xi_n],$$

or more generally, for $||\eta|| + ||\xi|| < 1$ and $(\zeta_1,\ldots,\zeta_2) \in \mathfrak{B}^n$,

$$\delta^n F[\eta+\xi;\zeta_1,\ldots,\zeta_n] = \sum_{\nu=0}^{\infty} \frac{1}{\nu!} \delta^{n+\nu} F[\eta;\xi,\ldots,\xi,\zeta_1,\ldots,\zeta_n],$$

both expansions converging in the strong sense, cf. Hille and Phillips (1957).

If $||\eta|| + \Sigma_{i=1}^k ||\xi|| < 1$, then

$$|\delta^{n_1+\ldots+n_k} F[\eta;\overbrace{\xi_1,\ldots,\xi_1}^{n_1},\ldots,\overbrace{\xi_k,\ldots,\xi_k}^{n_k}]| \leq \prod_{\nu=1}^{k} n_\nu!$$

by (a), (b), and Cauchy's integral formula. Using this bound in the series expansions,

$$(2.1) \qquad \| F[\eta+\xi] - F[\eta] \| \leq \frac{\|\xi\|}{1-\|\eta\|-\|\xi\|},$$

$$(2.2) \qquad \| F[\eta+\xi] - \delta F[\eta;\xi] \| \leq \frac{\|\xi\|^2}{(1-\|\eta\|)(1-\|\eta\|-\|\xi\|)},$$

if $\|\eta\| + \|\xi\| < 1$, and similarly

$$(2.3) \qquad \begin{aligned} &\| \delta^n F[\eta+\xi;\zeta_1,\ldots,\zeta_n] - \delta^n F[\eta;\zeta_1,\ldots,\zeta_n] \| \\ &\leq \frac{\|\xi\|}{a-\|\xi\|} \prod_{i=1}^{n} \frac{\|\zeta_i\|}{|a_i|}, \end{aligned}$$

$$(2.4) \qquad \begin{aligned} &\| \delta^n F[\eta+\xi;\zeta_1,\ldots,\zeta_n] - \delta^n F[\eta;\zeta_1,\ldots,\zeta_n] - \delta^{n+1} F[\eta;\zeta_1,\ldots,\zeta_n,\xi] \| \\ &\leq \frac{\|\xi\|}{a(a-\|\xi\|)} \prod_{i=1}^{n} \frac{\|\zeta_i\|}{|a_i|}, \end{aligned}$$

if $\|\eta\| + \|\xi\| < 1$, $\|\xi\| < a$, $a + \Sigma_{i=1}^{n} |a_i| \leq 1 - \|\eta\|$, $(\zeta_1,\ldots,\zeta_n) \in \mathcal{B}^n$. Notice that in all these estimates the bounds on the right are independent of F!

By the non-negativity of probabilities,

$$(c) \qquad \delta^n F[\eta;\xi_1,\ldots,\xi_n] \geq 0, \quad (\eta,\xi_1,\ldots,\xi_n) \in \mathcal{S}_+ \otimes \mathcal{B}_+^n,$$

$$\mathcal{B}_+ := \{\xi \in \mathcal{B} : \xi \geq 0\}, \quad \mathcal{S}_+ := \mathcal{S} \cap \mathcal{B}_+,$$

and in consequence of the σ-additivity,

(d) F _is sequentially continuous on_ $\bar{\mathcal{S}}$ _respective the product topologies_.

Let us call _any_ (not necessarily generating) mapping $F : \bar{\mathcal{S}} \to \bar{\mathcal{S}}(Y)$ satisfying (a), (b), (c), and (d) a _pre-generating mapping_.

When is a pre-generating mapping a generating mapping? Given (d), it is easily verified by use of (1) to (4) that for any pre-generating mapping F

(d') $\delta^n F[\eta;\xi_1,\ldots,\xi_n]$ _is sequentially continuous on bounded regions in_ $\mathcal{S} \otimes \mathcal{B}^n$ _respective the product topologies_.

That is, all we need to ensure that $\delta^n F[0;1_{A_1},\ldots,1_{A_n}]$ can be extended to a measure on $\mathfrak{A}^n$, thus making F a generating mapping, is sufficient structure of $(X,\mathfrak{A})$:

2.1. PROPOSITION. *If X is a locally compact Hausdorff space with countable open base and $\mathfrak{A}$ the Borel algebra, then any pre-generating mapping defined on $\overline{\mathfrak{S}}$ is a probability generating mapping.*

Clearly, (a), (b), and (c) together imply that

(e) *for every $y \in Y$ and any measurable decomposition $\{A_\nu\}_{1 \leq \nu \leq n}$, $n \in \mathbb{N}$, of X, $F[\Sigma_{\nu=1}^{n} 1_{A_\nu} s_\nu](y)$, $|s_\nu| \leq 1$, is an n-dimensional probability generating function.*

Given (d), the reverse is also true:

2.2. PROPOSITION. *For any mapping $F: \overline{\mathfrak{S}} \to \mathfrak{B}(Y)$ satisfying (d), the properties (a), (b), and (c) together are equivalent to (e).*

The only not completely trivial part of the proof is (e) (d) $\Rightarrow$ (b). To see this, proceed as follows: Define $\delta F[\eta;\xi]$ for stepfunctions, verify (2.1) and (2.2), again only for stepfunctions, then use these two estimates and (d) to extend the mapping $\delta F[\eta;\xi]$ continuously to $\mathfrak{S} \otimes \mathfrak{B}$, and finally verify that this extension is the first Fréchet-derivative of F.

In subsequent chapters limits of (pre-) generating mappings will occur. When is such a limit itself a (pre-) generating mapping, and what is a possible analogue of the continuity theorem for generating functions?

2.3. *Let $\{F_n\}$ be a sequence of pre-generating mappings such that $F_n \to F$ on $\mathfrak{S}_+$ and $F[\xi_n] \to 1$ for every sequence $\{\xi_n\} \subset \mathfrak{S}_+$ with $\xi_n \uparrow 1$, then F is the restriction to $\mathfrak{S}_+$ of a pre-generating mapping and $F_n \to F$ on $\mathfrak{S}$, all convergences understood respective the product topologies and as $n \to \infty$.*

PROOF. The continuity theorem for multi-dimensional probability generating functions immediately yields convergence on the set of stepfunctions in $\mathfrak{S}$ to a mapping satisfying (e). Using (2.1) and (d) for the F_n, F can be extended to a mapping, also denoted by F, which is sequentially continuous on $\mathfrak{S}$ respective the product topologies, so that F is analytic and $F_n \to F$ on $\mathfrak{S}$. Finally, use the series expansion for F to extend F to $\overline{\mathfrak{S}}$ so that (d) is satisfied. $\square$

2.4. COROLLARY. *If* X *is a locally compact Hausdorff space with countable open base and* $\mathfrak{A}$ *the Borel algebra,* $\{F_n\}$ *a sequence of probability generating mappings such that* $F_n : F$ *on* $\mathcal{S}_+$ *and* $F[\xi_n] \to 1$ *for every sequence* $\{\xi_n\} \subset \mathcal{S}_+$ *with* $\xi_n \uparrow 1$, *then* F *is the restriction of a generating mapping and* $F_n \to F$ *on* $\mathcal{S}$, *all convergences understood respective the product topologies and as* $n \to \infty$.

Notice that, if F is the generating mapping of P,

$$P(y,\{\hat{x} : \hat{x}[A_\nu] = n_\nu;\ \nu = 1,\dots,k\}) = \frac{1}{n_1!\dots n_k!}\ \delta^{n_1+\dots+n_k} F[0;\overbrace{1_{A_1},\dots,1_{A_1}}^{n_1},\dots,\overbrace{1_{A_k},\dots,1_{A_k}}^{n_k}](y)$$

for every measurable decomposition $\{A_\nu\}_{1\le\nu\le k}$, $k \in \mathbb{N}$, and that

$$\hat{\mathfrak{A}}_o := \{\hat{A} \in \hat{\mathfrak{A}} : \hat{A} = \{\hat{x} : \hat{x}[A_\nu] = n_\nu,\ \nu = 1,\dots,k\},$$
$$\{A_\nu\}_{1\le\nu\le k} \text{ measurable decomposition of } X,\ k \in \mathbb{N}\}$$

countably generates $\hat{\mathfrak{A}}$.

2.5. PROPOSITION. *Let* P_n, P *be stochastic kernels from* $(Y,\mathfrak{B})$ *to* $(\hat{X},\hat{\mathfrak{A}})$ *and* F_n, F *the corresponding generating mappings. Then* $P_n(y,\cdot) \to P(y,\cdot)$ *on* $\hat{\mathfrak{A}}_o$ *if and only if* $F_n[\cdot](y) \to F[\cdot](y)$ *on* $\mathcal{S}_+$, $y \in Y$, $n \to \infty$.

PROOF. Given the continuity theorem for multi-dimensional probability generating functions, all that remains to show is $F_n[\xi](y) \to F[\xi](y)$ for $\xi \in \mathcal{S}_+$ not necessarily a stepfunction. Choose sequences of stepfunctions $\{\xi_\nu\}$, $\{\tilde{\xi}_\nu\} \subset \mathcal{S}_+$ such that $\xi_\nu \uparrow \xi$, $\tilde{\xi}_\nu \downarrow \xi$. Then

$$F_n[\xi_\nu] \le F_n[\xi] \le F_n[\tilde{\xi}]$$

so that

$$\begin{aligned} F[\xi](y) &= \lim_{\nu\to\infty} F[\xi_\nu](y) \\ &= \liminf_{n\to\infty} F_n[\xi](y) \le \limsup_{n\to\infty} F_n[\xi](y) \\ &= \lim_{\nu\to\infty} F[\xi_\nu](y) = F[\xi](y). \end{aligned}$$

□

We now supplement the probabilistic construction of a Markov branching process $\{\hat{x}_t, P^{\hat{x}}\}$ from the data (x_t^o,π) by a corresponding, purely analytic construction of a pre-generating semigroup $\{F_t\}$ from the data (T_t^o, f), $\{T_t^o\}$ being a transition semigroup on $\mathfrak{B}$ (corresponding to $\{x_t^o\}$) and $f : \overline{\mathfrak{S}} \to \overline{\mathfrak{S}}$ a generating mapping (corresponding to π). Matters are simplified by assuming (T_t^o, π) to be determined by $[T_t, k, \pi]$ ($\{T_t\}$ corresponding to $\{x_t\}$) with k, $m1 \in \mathfrak{B}_+$, and since this is all we shall need later, we restrict ourselves to this case. The general situation is treated in Ikeda, Nagasawa, and Watanabe (1968, 1969).

Given k, $m1 \in \mathfrak{B}_+$, the solution of (IF) is unique. For every $\eta \in \overline{\mathfrak{S}}$ we now <u>define</u> $\{F_t[\eta]\}$ as the solution of (IF), $\{T_t^o\}$ taken to be the unique solution of

$$T_t^o = T_t - \int_0^t T_s \, k \, T_{t-s}^o \, ds.$$

Using the semigroup property of $\{T_t^o\}$, it then follows from (IF) that

$$\begin{aligned} 1 - F_{t+s}[\eta] &= T_t^o T_s^o (1-\eta) + \int_t^{t+s} T_u^o k(1-f[F_{t+s-u}[\eta]])du \\ &\quad + \int_0^t T_u^o k(1-f[F_{t+s-u}[\eta]])du \\ &= T_t^o \{T_s^o(1-\eta) + \int_0^s T_v^o k(1-f[F_{s-v}[\eta]])dv \\ &\quad + \int_0^t T_u^o k(1-f[F_{t+s-y}[\eta]])du, \end{aligned}$$

which by the uniqueness of the solution of (IF) implies that $\{F_t\}$ is a semigroup. Also by uniqueness, $F_t[1] = 1$. Representing $F_t[\eta]$ as the limit of the iteration sequence of (IF), beginning with 0, it is then easily verified that F_t has the properties (d) and (e). That is, for every $t > 0$, F_t is a pre-generating mapping, whence we call $\{F_t\}$ a pre-generating semigroup.

REMARK. With sufficient regularity, $F_t[\eta]$ can also be obtained as solution of a differential equation (backward equation) corresponding to (IF), cf. Ikeda, Nagasawa, and Watanabe (1968, 1969).

Next we turn to moments, beginning with first moments. Suppose to be given a pre-generating semigroup $\{F_t\}$, and <u>define</u>

$$F_t(\hat{x}, \eta) := \widetilde{F_t[\eta]}(\hat{x}), \quad \hat{x} \in \hat{X}, \ \eta \in \overline{\mathfrak{S}}.$$

For $t > 0$, $\hat{x} \in \hat{X}$, ξ non-negative and $\mathfrak{A}$-measurable, and some $\{\eta_\nu\} \subset \mathfrak{S}_+$ with $\eta_\nu(x) \uparrow 1$, $\nu \uparrow \infty$, $x \in X$, set

$$M_t(\hat{x},\xi) := \lim_{\nu\to\infty} \delta F_t(\hat{x},\eta_\nu;\xi).$$

If

(2.5) $$\sup_{x\in X} \lim_{\nu\to\infty} \delta F_t[\eta_\nu;1](x) < \infty,$$

then $M_t(\hat{x},\xi)$ exists as a linear-bounded functional of ξ on $\mathfrak{B}$, independent of the particular $\{\eta_\nu\}$. If $\{F_t\}$ is the generating semi-group of a transition function P_t, then

$$M_t(\hat{x},\xi) = \int_{\hat{X}} P_t(\hat{x},d\hat{y})\hat{y}[\xi],$$

$$\hat{x}[\xi] := 0; \qquad \hat{x} = \theta,$$

$$:= \sum_{\nu=1}^{n} \xi(x_\nu); \quad \hat{x} = \langle x_1,\ldots,x_n\rangle,$$

and (2.5) is equivalent to

(2.5') $$\sup_{x\in X} \sum_n n\, P_t(\langle x\rangle, X^{(n)}) < \infty.$$

Finally, if $\{F_t\}$ generates the transition function of a process $\{\hat{x}_t, P^{\hat{x}}\}$,

$$M_t(\hat{x},\xi) = E^{\hat{x}}\, \hat{x}_t[\xi],$$

the interpretation of which is most obvious if we take ξ to be an indicator function.

Now assume (2.5) and set

$$(M_t\xi)(x) := M_t(\langle x\rangle,\xi), \quad t > 0,\ \xi \in \mathfrak{B},\ x \in X.$$

For every $t > 0$ this defines a linear-bounded operator $M_t : \mathfrak{B} \to \mathfrak{B}$. It is non-negative and sequentially continuous respective the product topology on bounded regions. The latter is easy to see, if $\{F_t\}$ is pre-generating, and a trivial consequence of dominated convergence, if $\{F_t\}$ is generating. From (F.1)

(2.6) $$M_t(\hat{x},\xi) = \hat{x}[M_t\xi],$$

and from (F.2)

(2.7) $$M_{t+s} = M_s\, M_t.$$

That is, $\{M_t\}$ is a semigroup, accordingly called the <u>moment semigroup</u>. It is the linear semigroup obtained by linearizing the non-linear semigroup $\{F_t\}$ in its fixed point 1.

If we start with a process $\{\hat{x}_t, P^{\hat{x}}\}$ constructed from (x_t^o, π), $M_t(\hat{x},\zeta)$ must, for non-negative ζ, be the minimal non-negative solution of (1.2) with $\xi(\hat{x}) = \hat{x}[\zeta]$. In particular, $(M_t\zeta)(x)$ must solve

$$v_t(x) = T_t^o\zeta(x) + \int_0^t \int_X P_x^o(\zeta_t^o ds, x^o = \Delta,\ x^o \in dy)\ m[v_{t-s}](y), \tag{2.8}$$

$$(m\zeta)(x) := \int_{\hat{X}} \pi(x, d\hat{y})\hat{y}[\zeta] = \lim_{\nu\to\infty} \delta f[\eta_\nu;\zeta](x), \tag{2.9}$$

$\{\eta_\nu\}$ as above.

2.6. PROPOSITION. *If $\{v_t\}$ is a solution of (2.8), then $\{\hat{x}[v_t]\}$ solves (1.2) with $\xi(\hat{x}) = \hat{x}[\zeta]$.*

For a proof see Ikeda, Nagasawa, and Watanabe (1968, 1969).

2.7. COROLLARY. *For non-negative ζ, $\{(M_t\zeta)(x)\}$ is the minimal non-negative solution of (2.8) and can thus be represented as the limit of the iteration sequence of (2.8), beginning with 0.*

Returning to the simplified setting of a pre-generating semigroup $\{F_t\}$ determined by $[T_t, k, \pi]$ with $k,\ m1 \in \mathfrak{B}_+$, (2.9) defines a non-negative, linear-bounded operator $m : \mathfrak{B} \to \mathfrak{B}$, and (2.8) becomes

$$v_t = T_t^o\zeta + \int_0^t T_s^o\ km\ v_{t-s}\ ds,$$

the solution of which is unique, so that $\{M_t\}$ is the semigroup given as the unique solution of the linear semigroup perturbation equation

$$M_t = T_t^o + \int_0^t T_s^o\ km\ M_{t-s}\ ds, \tag{IM}$$

obtained by linearizing the non-linear perturbation equation (IF) in the fixed point 1. By iteration

$$\| M_t\xi \| \le e^{\|km\| t} \|\xi\|.$$

REMARK. Regularity of $(X,\mathfrak{A})$ and $[T_t, k, \pi]$ is, of course, reflected in regularity of $\{M_t\}$. For example, let $(X,\mathfrak{A})$ be a locally compact Hausdorff space with countable open base and $\mathfrak{A}$ the Borel algebra. Suppose $\{T_t\}$ has a strongly continuous restriction to

$$C_o := \{\xi \in \mathfrak{B} : \xi \text{ continuous on } X \text{ with } \xi(x) \to 0 \text{ as } X \ni x \to \bar{x} \notin X\}.$$

Then $\{T_t^o\}$ as well as $\{M_t\}$ have strongly continuous restrictions to C_o. If A is the strong generator of $\{T_t\}$ on C_o with domain $\mathcal{D}(A)$, further $kC_o \subset C_o$, then $A^o := A - k$ with domain $\mathcal{D}(A^o) = \mathcal{D}(A)$ is the strong generator of $\{T_t^o\}$ on C_o, and if also $k\,m\,C_o \subset C_o$, then $L := A^o + km$ with domain $\mathcal{D}(L) = \mathcal{D}(A^o)$ is the strong generator of $\{M_t\}$ on C_o, i.e., M_t is strongly differentiable in t on $\mathcal{D}(L)$, and for $\xi \in \mathcal{D}(L)$, $\{M_t\xi\}$ solves

$$\text{(2.10)} \qquad \begin{aligned} &\frac{dv_t}{dt} = Lv_t, \\ &\| v_t - \xi \| \to 0. \end{aligned}$$

If instead of $kC_o \subset C_o$ and $kmC_o \subset C_o$ we assume $T_t\mathbb{B} \subset C_o$ for $t > 0$, then also $T_t^o\,\mathbb{B} \subset C_o$ and $M_t\mathbb{B} \subset C_o$ for $t > 0$, and for $\xi \in$ (A), T_t^o as well as M_t are differentiable from the right in t respective the product topology, $\{M_t\}$ satisfying (2.10) with the right-hand derivative $\frac{d^+v_t}{dt}$ in place of $\frac{dv_t}{dt}$. All these statements are easily verified from the integral equations for T_t^o and M_t.

The moment semigroup and its spectral properties play a crucial role in determining the limit behaviour of Markov branching processes. The basic case will be that M_t is irreducible in the sense that every invariant subspace is dense in $\mathbb{B}$ at least respective the product topology on bounded regions. In particular, we shall work with the following condition:

(M) *The moment semigroup* $\{M_t\}$ *can be represented as*

$$M_t = \rho^t P + \Delta_t, \quad t \geq 0,$$

where $\rho \in (0,\infty)$,

$$P\xi = \Phi^*[\xi]\varphi, \quad \xi \in \mathbb{B},$$

with $\Phi^* : \mathbb{B} \to \mathbb{C}$ *linear-bounded, non-negative on* $\mathbb{B}_+$, $\varphi \in \mathbb{B}_+$, *and* $\Phi^*[\varphi]$, *further* $\Delta_t : \mathbb{B} \to \mathbb{B}$ *such that for all* $t > 0$

$$P\Delta_t\xi = \Delta_t P\xi = 0, \qquad \xi \in \mathbb{B},$$

$$-\alpha_t P\xi \leq \Delta_t\xi \leq \alpha_t P\xi, \quad \xi \in \mathbb{B}_+$$

with $\alpha_\cdot : (0,\infty) \to \mathbb{R}_+$ *satisfying*

$$\rho^{-t}\alpha_t \downarrow 0, \quad t \uparrow \infty.$$

If X is finite and M_t primitive, (M) is automatically satisfied by Perron's theorem. In general, (M) is a non-trivial restriction. It represents more structure than is implied, e.g., by the positivity theorem of Kreĭn and Rutman (1948). Notice, in particular, that we do not require merely $\rho^{-t}\|\Delta_t\| \to 0$, but $-\alpha_t P \leq \Delta_t \leq \alpha_t P$ on $\mathfrak{B}_+$, $\rho^{-t}\alpha_t \to 0$, admitting even $\inf \varphi = 0$. For a large class of branching diffusions and related models (M) has, however, been verified, cf. (Hering 1978 a,b,c). See the subsequent section for examples and counterexamples.

Let us move on to second moments. For still higher moments the theory is similar, but it will not be needed here. Again we start from a pre-generating semigroup $\{F_t\}$ and some sequence $\{\eta_\nu\} \subset \mathfrak{S}_+$ such that $\eta_\nu(x) \uparrow 1$, $\nu \uparrow \infty$, $\hat{x} \in \hat{X}$. For $\mathfrak{A}$-measurable ξ, ζ define

$$M_t^{(2)}(\hat{x};\xi,\zeta) := \lim_{\nu\to\infty} \delta^2 F_t(\hat{x},\eta_\nu;\xi,\zeta),$$

whenever the limit exists. If

$$\sup_{x\in X} \lim_{\nu\to\infty} \delta^2 F_t[\eta_\nu;1,1](x) < \infty, \tag{2.11}$$

then for every $\hat{x} \in \hat{X}$, $M_t^{(2)}(\hat{x};\xi,\zeta)$ is well-defined as a symmetric, bilinear, bounded functional on $\mathfrak{B}^2 = \mathfrak{B}\otimes\mathfrak{B}$, independent of the particular choice of $\{\eta_\nu\}$. It is called the <u>second factorial moment functional</u>. If $\{F_t\}$ is the generating semigroup of a branching transition function P_t, (2.11) is equivalent to

$$\sup_{x\in X} \sum_n n(n-1)P_t(\langle x\rangle, x^{(n)}) < \infty$$

and we have

$$M_t^{(2)}(\hat{x};\xi,\zeta) = \int_{\hat{X}} P_t(\hat{x},d\hat{y})\hat{y}^{(2)}[\xi,\zeta],$$

$$\hat{x}^{(2)}[\xi,\zeta] := 0; \qquad \hat{x}[1] \leq 1,$$

$$:= \sum_{i\neq j}\sum \xi(x_i)\zeta(x_j); \quad \hat{x} = \langle x_1,\dots,x_n\rangle,\ n \geq 2,$$

and if P_t is the transition function of a Markov branching process $\{\hat{x}_t, P^{\hat{x}}\}$,

$$M_t^{(2)}(\hat{x};\xi,\zeta) = E^{\hat{x}}\,\hat{x}_t^{(2)}[\xi,\zeta].$$

Assuming (2.11),

$$M_t^{(2)}[\xi,\zeta](x) := M_t^{(2)}(\langle x\rangle;\xi,\zeta), \quad x \in X,$$

defines a symmetric, bilinear, bounded mapping $M_t^{(2)} : \mathcal{B}^2 \to \mathcal{B}$, sequentially continuous respective the product topologies on bounded regions. If $\xi = \zeta$ we shall also write

$$M_t^{(2)}[\xi,\xi] =: M_t^{(2)}[\xi].$$

From (F.1)

$$M_t^{(2)}(\hat{x};\xi,\zeta) = \hat{x}[M_t^{(2)}[\xi,\zeta]] + \hat{x}^{(2)}[M_t\xi, M_t\zeta],$$

and from (F.2)

$$M_{t+s}^{(2)}[\xi,\zeta] = M_s M_t^{(2)}[\xi,\zeta] + M_s^{(2)}[M_t\xi, M_2\zeta].$$

If we are dealing with a process $\{\hat{x}_t, P^{\hat{x}}\}$ constructed from (x_t^o,π), $M_t^{(2)}(\hat{x};\eta,\zeta)$ must for non-negative η, ζ be the minimal non-negative solution of (1.2) with $\xi(\hat{x}) = \hat{x}^{(2)}[\eta,\zeta]$. In particular, $M_t^{(2)}[\eta,\zeta](x)$ must solve

$$v_t^{(2)}(x) = \int_0^t \int_X P_x^o(\zeta^o \in ds, x^o_{\zeta^o} = \Delta, x^o_{\zeta^o -} \in dy)[m v_{t-s}^{(2)}(y)$$

$$+ m^{(2)}[M_{t-s}\eta, M_{t-s}\zeta](y)]ds, \tag{2.12}$$

$$m^{(2)}[\eta,\zeta](x) := \lim_{\nu\to\infty} \delta^2 f[\eta_\nu;\eta,\zeta] = \int_{\hat{X}} \pi(x,d\hat{y})\hat{y}^{(2)}[\eta,\zeta].$$

Again, if v_t is a solution of (2.8) and $v_t^{(2)}$ a solution of (2.12),

$$\hat{x}[v_t^{(2)}] + \hat{x}^{(2)}[v_t, v_t]$$

is a solution of (1.2) with $\xi(\hat{x}) = \hat{x}^{(2)}[\eta,\zeta]$, i.e., $M_t^{(2)}[\eta,\zeta](x)$ is the minimal non-negative solution of (2.12). The calculations necessary to verify this are similar to those in the case of first moments, only longer and more tedious.

We shall only need the case that $\{F_t\}$ is constructed from $[T_t, k, \pi]$. Here (2.12) takes the form

$$M_t^{(2)}[\eta,\zeta] = \int_0^t T_s^o k\{m M_{t-s}^{(2)}[\eta,\zeta] + m^{(2)}[M_{t-s}\eta, M_{t-s}\zeta]\}ds, \tag{2.13}$$

the solution of which is unique if

$$\sup_{x\in X} \sum_n n(n-1)\pi(x, X^{(n)}) \ (= \sup_{x\in X} m^{(2)}[1,1](x)) < \infty.$$

The solution then is, in fact,

$$M_t^{(2)}[\eta,\zeta] = \int_0^t M_s k m^{(2)}[M_{t-s}\eta, M_{t-s}\zeta]ds. \tag{2.14}$$

With sufficient regularity assumptions, similar to those in case of first moments, (2.14) can also be obtained from the differential equation

$$\frac{dv_t^{(2)}}{dt} = A^o v_t^{(2)} + kmv_t^{(2)} + km^{(2)}[M_t\eta, M_t\zeta],$$
$$\| v_t^{(2)} \| \to 0, \quad t \to 0,$$

corresponding to (2.13).

REMARK. Similarly as in (2.14) it is true in general that the nth factorial moment functional can be expressed as a sum of integrals involving M_t, k, m, and $m^{(2)}, \ldots, m^{(n)}$, cf. Ikeda, Nagasawa, and Watanabe (1968, 1969).

Finally, let us turn to the first order Taylor expansion of F_t near its fixed point 1. As in the case of simple branching processes this will be an indispensible device in investigating the limiting behaviour. Although we could work with F_t merely pre-generating, we assume--mostly for notational comfort--that we are dealing with generating mappings. Suppose $\{M_t\}$ exists as a semigroup of bounded operators. Recalling the explicit representations of F_t and M_t in terms of P_t, we may write

(FM) $$1 - F_t[\eta] = M_t[1-\eta] - R_t(\eta)[1-\eta],$$
$$R_t(\eta)[\zeta](x) := \int_{\hat{X}} P_t(\langle x\rangle, d\hat{y})r(\hat{y};\eta,\zeta),$$
$$f(\hat{x};\eta,\zeta) := 0; \qquad \hat{x}[1] \leq 1,$$
$$:= \sum_{\nu=1}^{n} \zeta(x_\nu)\{1-\int_0^1 \prod_{\mu\neq\nu}[1-\lambda(1-\eta(x_\mu))]d\lambda\}; \quad \hat{x} = \langle x_1, \ldots, x_n\rangle,\ n > 1.$$

The mapping $R_t(\cdot)[\cdot] : \overline{\mathfrak{g}} \otimes \mathfrak{B} \to \mathfrak{B}$ is sequentially continuous respective the product topologies on bounded regions, non-increasing in the first variable, and linear-bounded in the second, such that

(RM) $$0 = R_t(1)\xi \leq R_t(\eta)\xi \leq M_t\xi, \quad (\eta,\xi) \in \overline{\mathfrak{g}}_+ \otimes \mathfrak{B}_+.$$

Note that, keeping t and η fixed, $R_t(\eta)$ is by definition a linear-bounded operator on $\mathfrak{B}$. If F_t is known merely to be pre-generating, we can define $R_t(\eta)\zeta$ only for $\zeta = 1-\eta$, using (FM) as the defining relation.

Given (M), we shall later require the following property:

(R) <u>For every</u> $t > 0$ <u>there exist a mapping</u> $g_t : \overline{\mathcal{S}}_+ \to \mathcal{B}$ <u>such that</u>

$$R_t(\xi)[1-\xi] = g_t[\xi]\rho^t\Phi^*[1-\xi]\varphi, \quad \xi \in \overline{\mathcal{S}}_+,$$

$$\lim_{\|1-\xi\|\to 0} \| g_t[\xi] \| = 0.$$

Given (M) and a finite X, (R) is trivially satisfied. For more general X, however, (R) is non-trivial:

PROPOSITION. <u>Let</u> $\{F_t\}$ <u>be a pre-generating semigroup constructed from</u> $[T_t,k,\pi]$ <u>such that</u>, $k, ml \in \mathcal{B}_+$. <u>Suppose that</u> (M) <u>is satisfied and that there exist constants</u> c, C^* <u>such that</u>

(C) $$km\, \varphi \le c\varphi,$$

(C*) $$\Phi^*[km\, \xi] \le C^*\Phi^*[\xi], \quad \xi \in \mathcal{B}_+,$$

<u>then</u> (R) <u>is satisfied with</u>

$$\| g_t[\xi] \| \underset{\|\xi-1\|\to 0}{\longrightarrow} 0 \quad \text{uniformly in } t \in [a,b],$$

<u>whenever</u> $0 < a < b < \infty$.

PROOF. Combining (IF), (IM), (FM), and the analogous expansion

$$1 - f[\xi] = m[1-\xi] - r(\xi)[1-\xi],$$

we get

$$R_t(\xi)[1-\xi] = \int_0^t T_s^0 km\, R_{t-s}(\xi)[1-\xi]ds$$

$$+ \int_0^t T_s^0 kr(F_{t-s}[\xi])[1-F_{t-s}[\xi]]ds.$$

That is, fixing $\epsilon > 0$, $R_t(\xi)[1-\xi]$, $t \ge \epsilon$, satisfies

$$w_t = A_t + B_t^\epsilon + \int_0^{t-\epsilon} T_s^0 km\, w_{t-s}ds,$$

$$A_t := \int_0^t T_s^0 kr(F_{t-s}[\xi])[1-F_{t-s}[\xi]]ds,$$

$$B_t^\epsilon := \int_0^\epsilon T_{t-s}^0 km\, R_s(\xi)[1-\xi]ds.$$

Since $k, ml \in \mathcal{B}$, the solution is unique and we can write it as

$$w_t = w_{1,t} + w_{2,t},$$

when $w_{1,t}$, $w_{2,t}$ are the unique solutions of

$$w_{1,t} = A_t + \int_0^{t-\epsilon} T_s^o km\ w_{1,t-s} ds, \quad t \geq \epsilon,$$

$$w_{2,t} = B_t^\epsilon + \int_0^{t-\epsilon} T_s^o km\ w_{2,ts-} ds, \quad t \geq \epsilon,$$

respectively. We estimate w_t by estimating $w_{1,t}$, $w_{2,t}$. The introduction of the $\epsilon > 0$ is necessitated by the impossibility to bound $\rho^{-t}\alpha_t$ for $t \to 0$.

Suppose $0 < \delta < \epsilon/2$, $\xi \in \bar{\mathfrak{S}}_+$, and let $T > 0$ be arbitrary, but fixed. By (FM), (RM),

$$F_{t-s}[\xi] \geq 1 - c_1 \| 1-\xi \|, \ \delta \leq s \leq t-s, \ t \leq \epsilon + T.$$

Furthermore,

$$T_t^o \leq M_t \quad \text{on} \quad \mathfrak{B}_+,$$

by (IM), and

$$0 = r(1)\xi \leq r(\zeta)\xi \leq m\xi, \ (\zeta,\xi) \in \bar{\mathfrak{S}}_+ \otimes \mathfrak{B},$$

similar to (RM). Using these facts together with (M), (C), and (C*),

$$\begin{aligned} A_t &\leq \left(\int_0^\delta + \int_{t-\delta}^t\right) M_s km\ M_{t-s}[1-\xi] ds \\ &\qquad + \int_\delta^{t-\delta} M_s kr(1-c_1 \| 1-\xi \|) M_{t-s}[1-\xi] ds \\ &\leq \delta(c+C^*)(1-\rho^{-\epsilon/2}\alpha_{\epsilon/2})\rho^t \Phi^*[1-\xi]\varphi \\ &\qquad + t(1+\rho^{-\delta}\alpha_\delta)(1+\rho^{-\epsilon/2}\alpha_{\epsilon/2}) \|\varphi\| \\ &\qquad\qquad \times \Phi^*[r(1-c_1 \| 1-\xi \|)\varphi]\rho^t \Phi^*[1-\xi]\varphi. \end{aligned}$$

For every $\epsilon' > 0$ we can fix δ such that the first term on the extreme right is smaller than $t'\epsilon'\Phi^*[1-\xi]\varphi$ for all $t \in [\epsilon, T+\epsilon]$. Since $\Phi^*[r(1-c_1 \| 1-\xi \|)\varphi] \to 0$, as $\| 1-\xi \| \to 0$, we can subsequently fix $\delta' > 0$ such that also the second term on the extreme right is smaller than $t\epsilon'\Phi^*[1-\xi]\varphi$, whenever $t \in [\epsilon, T+\epsilon]$ and $\| 1-\xi \| < \delta'$. That is,

$$A_t \leq t\Theta_{\epsilon,T}[\xi]\rho^t \Phi^*[1-\xi]\varphi, \quad t \in [\epsilon, \epsilon+T],$$

$$\| \Theta_{\epsilon,T}[\xi] \to 0, \quad \| 1-\xi \| \to 0.$$

Hence, by iteration,

$$w_{1,t} \leq e^{ct} t\Theta_{\epsilon,T}[\xi]\rho^t \Phi^*[1-\xi]\varphi, \quad t \in [\epsilon, \epsilon+T].$$

Now

$$B_t^\epsilon = T_{t-\epsilon}^o \int_0^\epsilon T_{\epsilon-s}^o \text{ km } R_s(\xi)[1-\xi]ds,$$

so that, by uniqueness of the solution of (IM),

$$w_{2,t} = M_{t-\epsilon}[\int_0^\epsilon T_{\epsilon-s}^o \text{ km } R_s(\xi)[1-\xi]ds]$$

$$\leq \int_0^\epsilon M_{t-s} \text{km } M_s[1-\xi]ds$$

$$\leq \epsilon(1+\rho^{-(t-\epsilon)}\alpha_{t-\epsilon})\rho^t C^*\varphi\Phi^*[1-\xi],$$

whenever $t > \epsilon$.

The estimates for $w_{1,t}$ and $w_{2,t}$ combined with the fact that $\epsilon > 0$ was arbitrary prove the proposition. □

We conclude with a little technical lemma to be used later.

LEMMA. _If_ (M) _and_ (R) _are satisfied, then_

$$\Phi^*[1-F_t[\xi]] > 0$$

for all $t > 0$ _and every_ $\xi \in \overline{\mathfrak{g}}_+ \cap \{\zeta : \Phi^*[1-\zeta] > 0\}$.

PROOF. Fix ξ as assumed. Then for $\delta \in (0,1)$

$$\Phi^*[1-F_t[\xi]] \geq \Phi^*[1-F_t[1-\delta(1-\xi)]]$$

$$\geq \rho^t \delta\Phi^*[1-\xi](1-\| g_t[1-\delta(1-\xi)] \|),$$

where we have used isotony of $F_t[\xi]$ in ξ, (M), and (R). According to the latter, we can pick δ such that

$$\| g_t[1-\delta(1-\xi)] \| < 1. \qquad \square$$

3. EXAMPLES

The simplest processes fitting into the $\{T_t, k, \pi\}$ framework are <u>p-type Markov branching processes:</u>

Let the set of types by $X = \{1,\ldots,p\}$. Then T_t is simply the $p\times p$ identity matrix for every t, k is given by a vector $(k_1,\ldots,k_p)$, T_t° is the $p\times p$ matrix with elements $-\delta_{ij}e^{-k_i t}$, δ_{ij} the Kronecker symbol, $i,j = 1,\ldots,p$, and π is given by a set of probabilities

$$\pi_i(n_1,\ldots,n_p) = \pi(i,\{x[\{j\}] = n_j,\ j = 1,\ldots,p\}),\quad n_j \in \mathbb{Z}_+,\ i,j \in X.$$

The moment semigroup $\{M_t\}$ is generated by the $p\times p$ matrix with elements $k_i(m_{ij}-\delta_{ij})$, $i,j = 1,\ldots,p$, where

$$m_{ij} = \sum_{n_1,\ldots,n_p\geq 0} n_k \pi_i(n_1,\ldots,n_p).$$

If and only if $(k_i m_{ij} - k_i\delta_{ij})_{i,j}$ is irreducible with maximal real eigenvalue λ, then (M) is satisfied with $\rho = e^{\lambda}$,

$$\Phi^*[\xi] = \sum_{i=1}^{N} \varphi^*(i)\xi(i),$$

φ^* and φ being the left and right eigenvectors of $(k_i m_{ij} - \delta_{ij}k_i)_{i,j}$ corresponding to λ, normalized such that $\Phi^*[\varphi] = 1$. This is a consequence of Perron's theorem on positive matrices.

Our predominant frame of reference will be that of <u>(multi-group) branching diffusions on bounded domains</u>, a natural generalization of multi-type Markov branching processes, which embraces the latter as a trivial special case. We formulate a setting in which the technical conditions essential for our limit theory, in particular (M) and (R), are satisfied:

Let Ω be the union of K connected open sets Ω_ν, $\nu = 1,\ldots,K$, in an N-dimensional, orientable manifold of class C^∞, let the closures $\overline{\Omega}_\nu$ be compact and pairwise disjoint, and let the boundary $\partial\Omega$ consist of a finite number of simply connected $(N-1)$-dimensional hypersurfaces of class C^3. Let X be the union of K Borel sets X_ν such that

$$\Omega_\nu \subset X_\nu \subset \overline{\Omega}_\nu,\qquad \nu = 1,\ldots,K,$$

in a way to be determined shortly, and suppose to be given a uniformly elliptic differential operator $A|\mathcal{D}(A)$, represented in local coordinates on X by

$$A := \sum_{i,j=1}^{N} \frac{1}{\sqrt{a(x)}} \frac{\partial}{\partial x^i} a^{ij}(x)\sqrt{a(x)} \frac{\partial}{\partial x^j} + \sum_{i=1}^{N} b^i(x)\frac{\partial}{\partial x^i} ,$$

$$\mathcal{D}(A) := \{u|_X : u \in C^2(\overline{\Omega}) \quad (\alpha u + \beta \frac{\partial u}{\partial n})|_{\partial\Omega} = 0\},$$

where (a^{ij}) and (b^i) are the restrictions to X of a symmetric, second-order, contravariant tensor of class $C^{2,\lambda}(\overline{\Omega})$ and a first-order, contravariant tensor of class $C^{1,\lambda}(\overline{\Omega})$,

$$a := \det(a^{ij})^{-1}$$

$$0 \leq \alpha,\beta \in C^{2,\lambda}(\partial\Omega), \quad \alpha+\beta \equiv 1,$$

$$\overline{\Omega}\setminus X := \{\beta = 0\}.$$

By $\partial/\partial n$ we denote the exterior derivative according to (a^{ij}) at $\partial\Omega$. Set

$$C^\ell := \{u|_X : u \in C^\ell(\overline{\Omega})\}, \quad C_0^\ell := \{u|_X : u \in C^\ell(\overline{\Omega}) \ u|_{\overline{\Omega}\setminus X} = 0\}.$$

The closure of $A|\{\xi \in \mathcal{D}(A) : A\xi \in C_0^0\}$ in C^0 is the C_0^0-generator of a contraction semigroup $\{T_t\}_{t\in\mathbb{R}_+}$ on $\mathcal{B}$, which determines a conservative, continuous, strong Markov process $\{\hat{x}_t, P_{\hat{x}}\}$ on $X \cup \{\partial\}$, where ∂ is a trap. This process, a diffusion, is our underlying motion process.

The semigroup $\{T_t\}$ is stochastically continuous in $t \geq 0$ on $\mathcal{B}$, and strongly continuous in $t \geq 0$ on C_0^0, with $T_t\mathcal{B} \subset C_0^2$ for $t > 0$. It can be represented in the form

$$T_t\xi(x) = \int_X p_t(x,y)\xi(y)dy,$$

where $p_t(x,y)$ is the funadmental solution of $\partial p_t/\partial t = Ap_t$. That is, $p_t(x,y)$ is given as a continuous function on $\{t > 0\} \otimes \overline{\Omega} \otimes \overline{\Omega}$, continuously differentiable in x and y for $t > 0$, such that

$$\text{(T.1)} \quad \begin{aligned} &p_t(x,y) > 0, \quad (x,y) \in X_\nu \otimes X_\nu, \quad \nu = 1,\dots,K, \\ &p_t(x,y) \equiv 0, \quad (x,y) \in X_\nu \otimes X_\mu, \quad \nu \neq \mu, \end{aligned}$$

$$(T.2) \qquad p_t(x,\cdot) = p_t(\cdot,x) \equiv 0, \quad x \in \overline{\Omega}\setminus X,$$

$$(T.3) \qquad \begin{aligned} &\frac{\partial p_t}{\partial n_x}(x,y) < 0, \quad (x,y) \in (\overline{\Omega}_\nu \setminus X_\nu) \otimes X_\nu, \\ &\frac{\partial p_t}{\partial n_y}(x,y) < 0, \quad (x,y) \in X_\nu \otimes (\overline{\Omega}_\nu \setminus X_\nu), \quad \nu = 1,\dots,K, \end{aligned}$$

and for $0 < t \leq t_0$, t_0 arbitrary but fixed,

$$(T.4) \qquad \sup_{x,y\in X} \{ |\frac{\partial p_t}{\partial x^i}(x,y)| + |\frac{\partial p_t}{\partial y^i}(x,y)| \} = O(t^{-(N+1)/2}), \quad i = 1,\dots,N,$$

$$(T.5) \qquad \sup_{x\in X} \int_X \{ |\frac{\partial p_t}{\partial x^i}(x,y)| + |\frac{\partial p_t}{\partial x^i}(y,x)| \} dy = O(t^{-1/2}), \quad i = 1,\dots,N,$$

cf. Ito (1957), Sato and Ueno (1965).

The boundary conditions, we have admitted, are not the most general probabilistically possible, but they do comprise the three most basic types of boundary behaviour, total absorption at $\{\beta = 0\}$, total instantaneous reflection at $\{\alpha = 0\}$, and elasticity (partial absorption and reflection) at $\{\alpha \neq 0\} \cap \{\beta \neq 0\}$.

Next suppose to be given a termination density $k \in \mathfrak{B}_+$, and define $k_0(x) := k(x)$ for $x \in X$ and $x_0(\partial) := 0$, and

$$\eta_t := \exp\{-\int_0^t k_0(x_s)ds\}.$$

Let $\{x_t^0, P_x^0\}$ be the η_t-subprocess of $\{x_t, P_x\}$, defined as a conservative process on $X \cup \{\partial\} \cup \{\Delta\}$, where Δ is a trap corresponding to the stopping by the multiplicative functional η_t, and let $\{T_t^0\}_{t\in\mathbb{R}_+}$ denote the associated transition semigroup.

Finally, let $\pi | X \otimes \hat{\mathfrak{U}}$ be a stochastic kernel such that

$$m\xi(x) := \int_{\hat{X}} \hat{x}[\xi]\pi(x,d\hat{x}), \quad \xi \in \mathfrak{B}, \ x \in X,$$

defines a bounded operator m on $\mathfrak{B}$.

The right-continuous, strong Markov branching process $\{\hat{x}_t, P^{\hat{x}}\}$ (up to equivalence) uniquely determined by (x_t^0, π), as described in V.1-2, is called a branching diffusion. If X or $\overline{X}$ can be decomposed into several disconnected congruent subsets, we speak of a multi-group branching diffusion.

As we recall from V.1-2, the boundedness of k and m guarantees that $\{\hat{x}_t, P^{\hat{x}}\}$ is conservative in $(\hat{X}, \hat{\mathfrak{A}})$, that $\{F_t\}$ is the unique solution of (IF), and that $\{M_t\}$ exists as a semi-group of linear-bounded operators on $\mathfrak{B}$ uniquely solving (IM).

Let us assume that the $K \times K$-matrix with elements

$$m_{\nu\mu} := \int_{X_\nu} k(x) m\, 1_{X_\mu}(x)\, dx, \quad \nu, \mu = 1, \ldots, K,$$

is irreducible. Then M_t is primitive. Suppose in addition that m has a bounded extension to $\mathfrak{L}^2$ such that

$$\sup_{\xi \in \mathfrak{B}: \|\xi\|_1 = 1} \| km^* \xi \|_\infty < \infty$$

of, if the X_ν are congruent,

$$km\xi(x) = \sum_{\nu=1}^{K} \mu_\nu(x) \xi(\varkappa_\nu x) + m_0 \xi(x),$$

$$m_0 \geq 0 \quad [\mathfrak{B}_+], \qquad \sup_{\xi \in \mathfrak{B}: \|\xi\|_1 = 1} \| m_0^* \xi \|_\infty < \infty,$$

where m^* and m_0^* are the adjoints of m and m_0, $\mu_\nu \in \mathfrak{B}_+$, $\|\cdot\|_p$ denotes the norm in $\mathfrak{L}^p$, $1 \leq p \leq \infty$, and $\varkappa_\nu x$ is the picture of x produced in X_ν by the given congruence. Then (M) is satisfied, cf. Hering (1978 b,c).

A simple example for the first kind of branching law is the following model: A branching event at x results with probability $p_{n_1, \ldots, n_k}(x)$ in $n_1 + \ldots + n_K$ new particles, n_ν of them in X_ν, $\nu = 1, \ldots, K$. The places of birth are distributed independently, a location in X_ν with distribution density $f_\nu(x, \cdot)$, $\nu = 1, \ldots, K$. That is,

$$\begin{aligned} m\xi(x) &= \int_X m(x,y) \xi(y)\, dy, \\ m(x,y) &= \sum_{\nu=1}^{K} 1_{X_\nu}(y) f_\nu(x,y) \sum_{n_1, \ldots, n_K \geq 0} n_\nu p_{n_1, \ldots, n_k}(x). \end{aligned} \tag{3.1}$$

The idea behind the second type of branching law, $m_0 = 0$, is this: There are K different kinds of particles moving on the same physical domain. To the kind ν we assign X_ν as abstract domain of diffusion, $\nu = 1, \ldots, K$. In the physical domain new particles are always born at the termination point (left limit) of their immediate ancestor. That is,

$$\pi(x,\hat{A}) = p_{0\ldots 0}(x)\, 1_{\hat{A}}(\theta) \tag{3.2}$$
$$+ \sum_{\substack{n_1 \geq 0,\ldots,n_K \geq 0 \\ n_1+\ldots+n_K > 0}} p_{n_1,\ldots,n_k}(x) 1_{\hat{A}}(\overbrace{\varkappa_1 x,\ldots,\varkappa_1 x}^{n_1},\ldots,\overbrace{\varkappa_K x,\ldots,\varkappa_K x}^{n_K}),$$

where $\{p_{n_1,\ldots,n_k}(x)\}$ is a probability distribution on $\mathbb{Z}_+^K$ for every $x \in X$. Here

$$m\xi(x) = \sum_{\nu=1}^{K} m_\nu(x)\xi(\varkappa_\nu x), \quad \xi \in \mathfrak{B}, \; x \in X,$$

$$m_\nu := \sum_{n_1 \geq 0,\ldots,n_K \geq 0} n_\nu \, p_{n_1,\ldots,n_K}, \quad \nu = 1,\ldots,K.$$

For $K = 1$, $X = X$, not necessarily simply connected, a branching law of the form (3.2) is called a <u>local branching law</u>, for $K > 1$, X_ν not necessarily simply connected, it is called <u>quasi-local</u>. A special example for a non-local law which is not quasi-local is given by (3.1).

To secure (R), given that (M) holds, it suffices to provide for (C) and (C*) to be satisfied, cf. V.2. In the present setting

$$\Phi^*[\xi] = \int_X \varphi^*(x)\xi(x)dx, \quad \xi \in \mathfrak{B},$$

$$\varphi, \varphi^* \in \overset{+}{\underset{0}{}} := \{u|_X : u \in C^1(\bar{\Omega}),\; u > 0 \text{ on } X,\; u = 0 \;\; \frac{\partial u}{\partial n} < 0 \text{ on } \bar{\Omega}\setminus X\},$$

cf. Hering (1978 b,c). Hence in case of the first kind of branching law with

$$m\xi(x) = \int_X m(x,y)\xi(y)dy, \quad \xi \in \mathfrak{B}, \; x \in X,$$

$$dy := \sqrt{a(y)}\; dy^1,\ldots,dy^N,$$

$y^1,\ldots,y^N$ local coordinates of y, it suffices for (C), (C*) to hold that

$$m(x,y) \leq \bar{m}(x,y), \quad (x,y) \in X \otimes X,$$

$$\bar{m} \in C^1(\bar{\Omega} \otimes \bar{\Omega}), \quad \bar{m}(\cdot,x) = \bar{m}(x,\cdot) \equiv 0, \quad x \in \bar{\Omega}\setminus X,$$

while in case of the second kind with $m_0 = 0$ the two conditions are always satisfied.

The simplest special case is a branching diffusion on a bounded one-dimensional interval with constant coefficients, a bounded termination density, and a homogeneous local branching law (i.e. a local branching law $\pi(x,\cdot)$ which does not depend on x). For example, take

$$X = (0,\tfrac{\pi}{2}]$$

$$A = a\frac{d^2}{dx^2} + b\frac{d}{dx}, \quad \mathcal{D}(A) = \{\xi \in C^2(0,\tfrac{\pi}{2}] : \xi(0+) = \xi'(\tfrac{\pi}{2}) = 0\},$$

$$L = A + k(m-1), \quad \mathcal{D}(L) = \mathcal{D}(A),$$

with $a > 0$, b, $k > 0$, and $m > 0$ real constants. Solving the eigenvalue problem $L\psi = \lambda\psi$ and the adjoint problem $L^*\psi^* = \lambda^*\psi^*$ yields

$$\lambda_n = \lambda_n^* = k(m-1) - \frac{b^2}{4a} - (2n+1)^2, \quad n = 0,1,2,\ldots$$

together with a complete bi-orthonormal system of eigenfunctions

$$\psi_n(x) = e^{-bx/2a}\sin[(2n+1)x],$$

$$\psi_n^*(x) = \frac{4}{\pi}e^{bx/2a}\sin[(2n+1)x], \quad x \in (0,\tfrac{\pi}{2}], \quad n = 0,1,2,\ldots .$$

That is, setting

$$\Psi_n^*[\xi] := \int_X \psi_n^*(x)\xi(x)\,dx, \quad \xi \in \mathcal{B},$$

we have $\Psi_n^*[\psi_\ell] = \delta_{n\ell}$ and, by elementary Fourier theory,

$$M_t\xi(x) = \sum_{n=0}^{\infty} e^{\lambda_n t}\Psi_n^*[\xi]\psi_n(x), \quad \xi \in \mathcal{D}(L),$$

which is immediately extendable to hold for $\xi \in \mathcal{B}$. Identifying

$$\Phi^* := \Psi_1^*, \quad \varphi := \psi_1, \quad \rho := e^{\lambda_1 t},$$

and noting that

$$|\sin(nx)| \leq n\sin x, \quad x \in (0,\tfrac{\pi}{2}],$$

$$\sum_{n=1}^{\infty}(2n+1)^2 e^{-(2n+1)^2 t} \to 0, \quad t \to \infty,$$

we see that (M) is satisfied, without having to draw on the general result quoted earlier.

REMARK. In the case of a one-dimensional bounded interval verification of (M) via complete spectral representation works well even with x-dependent a, b, k, and a general π: If π is local, one can use Sturm-Liouville techniques and asymptotic formulas for eigenvalues and eigenfunctions, cf. Hering (1974) and Asmussen and Hering (1976 a), and if π is non-local, spectral perturbation methods still apply, cf. Hering (1978 a). In fact, less regularity is needed for this, than was imposed in the general setting above. In higher dimensions, however, such methods require symmetry assumptions, which are too restrictive, particularly in the multi-group case. Thus in Hering (1978 b,c) a different approach is taken, exploiting T.1-5 in conjunction with positivity and semigroup perturbation methods.

Branching diffusions on unbounded domains do in general not satisfy (M). Two standard cases are treated in Chapter IX. More examples can be found in Watanabe (1967).

An interesting modification of the branching mechanism, motivated in Chapter 1, is retarded branching. By suitable enlargement of the type space this can formally still be fitted into the ordinary Markov branching framework. In the simplest case the modification consists in letting an exponentially distributed time elapse between the death of a particle and the birth of its offspring. A natural way to model this is the following:

Starting with a type space $(X_1,\mathfrak{A}_1)$ and a corresponding system (T_t^0,π), take the larger type space $(\tilde{X},\tilde{\mathfrak{A}})$ given by

$$\tilde{X} := X_1 \cup X_2,$$

X_2 being a copy of X_1, $\tilde{\mathfrak{A}}$ the σ-algebra generated on $\tilde{X}$ by $\mathfrak{A}_1$ and its copy on X_2, and define a corresponding system $(\tilde{T}_t^0,\tilde{\pi})$ as follows: Set

$$(\tilde{T}_t^0\xi)(x) := (T_t^0\xi_1)(x), \quad x \in X_1,$$

$$:= e^{-k_2(x)}\xi_2(x), \quad x \in X_2,$$

where ξ_1,ξ_2 are the restrictions of ξ to X_1,X_2, respectively, k_2 some bounded, non-negative, measurable function on X_2, further

$$\tilde{\pi}(x,\hat{A}) := 1_{\hat{A}}(\langle \varkappa_2 x\rangle), \quad x \in X_1,$$

$$:= \pi(\varkappa x,\hat{A}); \quad x \in X_2,$$

where $\varkappa_\nu x$ again denotes the picture of $x \in \tilde{X}$ in X_ν under the given correspondence $X_1 \to X_2$. If m is the first moment operator of π, the first moment of $\tilde{\pi}$ is now given by

$$(\tilde{m}\xi)(x) = \xi_2(\varkappa_2 x), \quad x \in X_1,$$
$$= (m\xi_1)(\varkappa_1 x), \quad x \in X_2.$$

As far as (M) is concerned, this model presents an additional difficulty. Suppose, for example, $X_1 = (0,\pi)$, let $\{T_t^0\}$ be determined by (T_t, k_1), $\{T_t\}$ generated by

$$A = a\frac{d^2}{dx^2}, \quad \mathcal{D}(A) = \{\xi \in C^2 : \xi(0+) = \xi(\pi) = 0\},$$

a and k_1 constant, and let π be local and homogeneous, i.e., m a multiplicative constant. Let k_2 also be constant. The generator $\tilde{L}$ of the moment semigroup corresponding to $(\tilde{T}_t^0, \tilde{\pi})$ is then given by

$$\tilde{L}\xi(x) = A\xi_1(x) - k_1\xi_1(x) + k_1\xi_2(\varkappa_2 x), \quad x \in X_1,$$
$$= -k_2\xi_2(x) + k_2 m\xi_1(\varkappa_1 x), \quad x \in X_2.$$

The solution of the eigenvalue problem of $\tilde{L}$ gives

$$\lambda_n^{\pm} = an^2 - k_1 + k_1\epsilon_n^{\pm},$$
$$\epsilon_n^{\pm} = \frac{1}{2k_1}(k_1 - k_2 + an^2 \pm [(k_1 - k_2 + an^2)^2 + 4k_1k_2m]^{1/2}),$$
$$\psi_n^{\pm} = \sin(nx), \quad x \in X_1,$$
$$= \epsilon_n^{\pm}\sin(nx), \quad x \in X_2.$$

Suppose (M) is satisfied. Then $\varphi = \psi_1^*$ and for $x \in (0,\pi)$ and some finite real K not depending on x or n

$$\epsilon_n^+|\sin(nx)| \leq e^{-\lambda_n^+}|M_1\psi_n^+|(\varkappa_2 x)$$
$$\leq e^{-\lambda_n^+}(M_1|\psi_n^+|)(\varkappa_2 x) \leq K\epsilon_1^+\sin x\ \Phi^*[|\psi_n^*|].$$

Since

$$\Phi^*[|\psi_n^+|] = o(\epsilon_n^+),$$

this would imply $|\sin(nx)| \leq o(1)\sin x$, $x \in (0,\pi)$, which is false.

That is, (M) cannot be satisfied in this case.

The problem can be circumvented by letting X_2 be a <u>finite</u> set of points (<u>depots</u>). This leads to an $\tilde{m}$ of the form

$$(\tilde{m}\xi)(x) = \sum_{\nu \in X_2} p_\nu(x)\xi_2(\nu), \quad x \in X_1,$$

$$= \int_{X_1} m_\nu(x)\xi_1(x)\,dx, \quad \nu \in X_2,$$

with $p_\nu(x) \geq 0$, $\Sigma_\nu p_\nu(x) = 1$. If the m_ν are bounded, then--with the appropriate irreducibility--this $\tilde{m}$ falls into the class of m admitted by the general result on (M) quoted above.

A first step towards age-dependence, which can be taken within the framework of p-type Markov branching processes, is the <u>multi-phase birth process</u>. The underlying idea is completely described by writing down the generator L of the moment semigroup,

$$L = \begin{pmatrix} -k_1 & k_1 & 0 & & & 0 \\ 0 & -k_2 & k_2 & 0 & & 0 \\ 0 & 0 & & & & \vdots \\ \vdots & \vdots & & & & 0 \\ 0 & \vdots & & & -k_{p-1} & k_{p-1} \\ k_p m_p & 0 & \cdots & & 0 & -k_p \end{pmatrix}.$$

Reinterpreting the p types as p different stages in the life of one individual, we get a branching process with a lifetime distributed according to the convolution of the p exponential distributions $1 - e^{-k_\nu t}$, $\nu = 1,\ldots,p$.

Since L is irreducible, (M) is satisfied. For example, in the simple case $k_1 = \cdots = k_p = 1$, $m = 2$, the eigenvalue problem is easily solved explicitly, yielding the spectral representation of M_t, in particular

$$\rho = e^\lambda, \quad \lambda = 2^{1/p} - 1,$$

$$\varphi(j) = (2p(1-2^{-1/p}))^{-1}2^{j/p},$$

$$\Phi^*[\xi] = \sum_{j=1}^{p} \varphi^*(j)\xi(j),$$

$$\varphi^*(j) = 2(1-2^{-1/p})2^{-j/p},$$

which we have normalized such that $\Phi^*[\varphi] = 1$ and $\Phi^*[1] = 1$.

Simple age-dependent branching processes with a general life-time distribution G can also be modelled as Markov branching processes. As we recall from Chapter I, the underlying intuitive concept is that an individual lives for a time distributed according to G and, if dying at age x, is instantaneously replaced by n particles of age 0 with probability $p_n(x)$. (The case that $\{p_n\}$ is independent of x is usually called Bellman-Harris case, while the case of x-dependent $\{p_n\}$ is named after Sevastyanov.) To fit this into the general Markovian framework, simply identify type as age and take $X = [0,\infty)$: Then

$$T_t\xi(x) = \xi(x+t), \quad x,t \geq 0$$

i.e. $\{T_t\}$ is generated by

$$A = \frac{d}{dx}, \quad (A) = \{\xi \in \mathcal{B} : \xi,\xi' \text{ uniformly continuous}\},$$

and defining

$$G^x(t) := \frac{G(x+t)-G(x)}{1-G(x)}, \quad x,t \geq 0,$$

we now have

$$(T_t^0\xi)(x) = (1-G^x(t))\xi(x+t),$$

$$\int_0^\infty P_x^0(\zeta^0 \in ds,\ x_{\zeta^0}^0 = \Delta,\ x_{\zeta^0 -}^0 \in dy)\xi(y) = \xi(x+s)dG^x(t),$$

finally

$$F[\xi](x) = \int_{\hat{X}} \tilde{\xi}(\hat{y})\pi(x,d\hat{y}) = \sum_n p_n(x)\xi^n(0).$$

Hence

$$(m\xi)(x) = m(x)\xi(0),$$

$$m(x) := \sum_n np_n(x),$$

and (IM) takes the form

$$(3.1) \qquad \begin{aligned}(M_t\xi)(x) &= (1-G^x(t))\xi(x+t) \\ &\quad + \int_0^t m(x+s)(M_{t-s}\xi)(0)dG^x(s).\end{aligned}$$

Clearly, (M) cannot be satisfied here, since it would imply--in intuitive terms--that within a sufficiently large, but fixed finite time individuals of arbitrarily large age could emerge from an ancestor of any fixed finite age, which is impossible by construction. Still, renewal theory yields a related, though weaker property:

Suppose that $H(t) := \int_0^t m(s)dG(s)$, $t \geq 0$, is non-lattice with $H(0+) = 0$ and that there exists a real constant α such that

$$\int_0^\infty e^{-\alpha t} m(s)dG(s) = 1.$$

Such an α always exists, if $H(\infty-) \geq 1$, but not necessarily otherwise. For $\xi \geq 0$ such that $e^{-\alpha x}(1-G(x))\xi(x)$ is directly Riemann integrable, standard renewal theory applied to (3.1) with $x = 0$ yields

$$e^{-\alpha t}(M_t\xi)(0) \to \frac{\int_0^\infty e^{-\alpha x}(1-G(x))\xi(x)dx}{\int_0^\infty xe^{-\lambda x} m(x)dG(x)}, \quad t \to \infty, \tag{3.2}$$

cf. Feller (1971), Karlin and Taylor (1975), or Cinlar (1975). Combining (3.2) with (3.1) for general x gives

$$e^{-\alpha t}(M_t\xi)(x) \to \Phi^*[\xi]\varphi, \quad t \to \infty,$$

$$\Phi^*[\xi] := \int_X \varphi^*(x)\xi(x)dx, \quad \varphi^*(x) := \frac{e^{-\alpha x}(1-G(x))}{\int_0^\infty e^{-\alpha y}(1-G(y)dy},$$

$$\varphi(x) := \frac{V(x)}{\Phi^*[V]}, \quad V(x) := \int_0^\infty e^{-\alpha s}m(x+s)dG^*(s) = \frac{e^{\alpha x}\int_x^\infty e^{-\alpha s}m(s)dG(s)}{1-G(x)}.$$

In common terminology $\Phi^*[1_{[0,x]}]$ is the <u>stable age distribution</u> and $V(x)$ the <u>reproductive value</u>. Notice that $\varphi > 0$ if and only if

$$\int_0^\infty ye^{-\alpha y}m(y)dG(y) < \infty. \tag{3.3}$$

As is easily verified, $e^{\alpha t}\varphi$ solves (3.1) for $\xi = \varphi$, i.e., $M_t\varphi = e^{\alpha t}\varphi$. Furthermore, if (3.3) holds and even $e^{-\alpha x}\xi(x)$ is directly Riemann integrable,

$$\Phi^*[M_t\xi] = \lim_{s\to\infty} e^{-\alpha s}M_s[M_t\xi]/\varphi = M_t[\lim_{s\to\infty} e^{-\alpha s}M_s\xi]/\varphi$$
$$= M_t[\varphi\Phi^*[\xi]]/\varphi = e^{\alpha t}\Phi^*[\xi],$$

so that for such ξ

$$M_t\xi = e^{\alpha t}\Phi^*[\xi]\varphi + \Delta_t\xi,$$
$$\Delta_t\varphi = 0, \quad \Phi^*[\Delta_t\xi] = 0, \tag{3.4}$$
$$(\Delta_t\xi)(x) = o(e^{-\alpha t}).$$

Leaving aside the restrictions on ξ, the decisive difference to (M) lies in the weaker estimate for Δ_t.

For age-dependent p-type processes, i.e., processes with p kinds of individuals with life-time distributions $G_1,\ldots,G_p$, respectively, we can proceed similarly. In the general Markovian set-up we now have

$$X = \{1,\ldots,p\} \otimes [0,\infty),$$

a single type being written as $x = (j,a)$. Equation (IM) takes the form

$$(M_t\xi)(i,a) = (1-G_i^a(t))\xi(i,a+t)$$
$$+ \sum_{j=1}^{p} \int_0^t m_{ij}(a+s)(M_{t-s}\xi)(i,0)\,dG_i^a(s),$$

where $m_{ij}(a)$ is the expected number of offspring of kind j of an individual of kind i which is dying at age a. If there exist a real α such that the matrix

$$\left(\int_0^\infty e^{-\alpha t}\, m_{ij}(s)\,dG_i(s)\right)_{i,j} \tag{3.5}$$

is irreducible with spectral radius 1, then under analogous assumptions as in the case $p = 1$ Markov renewal theory leads to (3.4) with

$$\Phi^*[\xi] = \sum_j \int \varphi^*(j,a)\xi(j,a)\,da,$$

$$\varphi^*(i,a) = \frac{v_i\, e^{-\alpha a}(1-G_i(a))}{\sum_k v_k \int_0^\infty e^{-\alpha t}(1-G_k(t))\,dt},$$

$$\varphi(i,a) = \frac{V(i,a)}{\Phi^*[V]},$$

$$V(i,a) = \sum_j u_j \int_0^\infty e^{-\alpha s} m_{ij}(a+s)\, dG_i^a(s)$$

$$= \frac{\sum_j u_j\, e^{\alpha a} \int_a^\infty e^{-\alpha s} m_{ij}(s)\, dG_i(s)}{1 - G_i(a)},$$

where $v = (v_1, \ldots, v_p)$ and $u = (u_1, \ldots, u_p)^T$ are the left and right eigenvector of (3.5) corresponding to the eigenvalue 1.

While it has been helpful to put ordinary age-dependent processes as described above into the general Markovian setting, cf. Asmussen and Hering (1976 b), it has so far not proved to be of practical value to do the same with generalized age-dependent reproduction mechanisms in the sense of Crump-Mode, Jagers, and Ryan. The latter will be treated separately in Chapter X.

As a last example consider a simple electron-photon cascade with finite cross-section:

$$X = \{1,2,3\} \otimes [0,\infty),$$

where 1 represents photons, 2 negatrons, 3 positrons, and $\varepsilon \in [0,\infty)$ energy. We now have

$$T_t^0 \xi(j,\varepsilon) = \xi(j,\varepsilon) e^{-k_j t},$$

k_j, $j = 1,2,3$, positive constants,

$$\int_X P^0_{(j,\varepsilon)}(\zeta^0 \in ds,\ x^0_{\zeta^0} = \Delta,\ x^0_{\zeta^0 -} \in dy)\xi(y)$$

$$= k_j \xi(j,\varepsilon) e^{-k_j s} ds,$$

$$\pi((1,\varepsilon),\ \{\hat{x} = \langle(2,\lambda\varepsilon),(3,(1-\lambda)\varepsilon)\rangle;\ 0 \leq \lambda \leq 1\}) = 1,$$

$$\pi((j,\varepsilon),\ \{\hat{x} = \langle(1,\lambda\varepsilon),(j,(1-\lambda)\varepsilon)\rangle;\ 0 \leq \lambda \leq 1\}) = 1, \qquad j = 2,3,$$

cf. Ikeda, Nagasawa, and Watanabe (1968, 1969). The law of conservation of energy built into this model prevents (M), and so this kind of process is not covered by the limit theory to be given in the subsequent three chapters. An extensive treatment can, however, be found in the book by Harris (1963).

4. EQUIVALENCE OF MOMENT CONDITIONS

The limit theorems we are about to prove will again involve moment conditions, in particular a generalization of (X LOG X). If a branching process $\{\hat{x}_t, P^{\hat{x}}\}$ is given by $\{T_t, k, \pi\}$, it is natural to ask for the correspondence between moment conditions in terms of $P^{\hat{x}}$ and in terms of k, π.

4.1. PROPOSITION. *Let* $\{\hat{x}_t, P^{\hat{x}}\}$ *be given by* $\{T_t, k, \pi\}$ *such that* k *and the first moment operator* m *of* π *are bounded. Suppose that* (M) *and* (C^*) *are satisfied, and let* $\chi: \mathbb{R}_+ \to \mathbb{R}_+$ *be a concave function with* $\chi(0) = 0$. *Then for every* $t > 0$

$$(4.1) \qquad \Phi^*[E^{\langle\cdot\rangle} \hat{x}_t[\varphi]\chi(\hat{x}_t[\varphi])] < \infty$$

if and only if

$$(4.2) \qquad \Phi^*[k \int_{\hat{X}} \hat{x}[\varphi]\chi(\hat{x}[\varphi])\pi(\cdot, d\hat{x})] < \infty.$$

In the proof we shall use the probabilistic equivalent of the independence property (F.1):

The independence property (F.1) can also be expressed in the following way. Let $\mathfrak{F}_t$ be the σ-algebra generated on the sample space by $\{\hat{x}_s; s \leq t\}$. For $0 \leq s \leq t$ and every non-negative, $\mathfrak{U}$-measurable η

$$(4.3) \qquad \hat{x}_t[\eta] = \sum_{i=1}^{\hat{x}_s[1]} \hat{x}_t^{s,i}[\eta] \text{ a.s. } [P^{\hat{x}}],$$

where the $\hat{x}_t^{s,i}$, $i = 1, \ldots, x_s[1]$, are $\mathfrak{F}_t$-measurable and independent conditioned on $\mathfrak{F}_s$, and for every $\hat{A} \in \hat{\mathfrak{U}}$

$$P^{\hat{x}}(\hat{x}_t^{s,i} \in \hat{A} | \mathfrak{F}_s) = P^{\langle x_i \rangle}(\hat{x}_{t-s} \in \hat{A}) \quad \text{a.s. } [P^{\hat{x}}]$$

with $\hat{x}_s^{s,i} = \langle x_i \rangle$. The sample space may not be large enough to allow (4.3) for all $s \leq t$. However, we shall need this representation only for fixed s, or for t, s restricted to sets of the form $\{n\delta : n = 0, 1, 2, \ldots\}$, $\delta > 0$. In both cases there exist processes equivalent to $\{\hat{x}_t, P^{\hat{x}}\}$ which satisfy (4.3). Hence we can use (4.3) for the process itself without loss of generality.

PROOF OF 4.1. We first assume (4.2). Let $0 < \tau_1 \leq \tau_2 \leq \ldots$ be the branching times of $\{\hat{x}_t, P^{\hat{x}}\}$, i.e., the times of discontinuities in $\hat{x}_t$

not caused by absorption via ∂. Define

$$\tilde{I}_t(\hat{x}) := E^{\hat{x}}\hat{x}_t[\varphi]\chi(\hat{x}_t[\varphi]), \quad \tilde{I}_t(x) := I_t(\langle x\rangle),$$

$$\tilde{I}_{t,n}(x) := E^{\hat{x}}\hat{x}_t[\varphi]\chi(\hat{x}_t[\varphi])1_{\{\tau_{n+1}>t\}}, \; I_{t,n}(x) := \tilde{I}_{t,n}(\langle x\rangle).$$

Then

$$I_{t,n+1}(x) = \int_0^t T_s^0\{k\int_{\hat{X}}\pi(x,d\hat{x})\tilde{I}_{t-s,n}(\hat{x})\}(x)ds + I_{t,o}(x). \tag{4.4}$$

Let τ_n^i, $n = 1,2,\ldots$, be the branching times of $\{\hat{x}_t^{0,i}, P^{\hat{x}}\}$, $i = 1,\ldots,$ $\hat{x}_0[1]$. Then

$$\hat{x}_t[\varphi]1_{\{\tau_{n+1}>t\}} \leq \sum_{i=1}^{\hat{x}_0[1]} \hat{x}_t^{0,1}[\varphi]1_{\{\tau_{n+1}^i>t\}}. \tag{4.5}$$

If S_r is the sum of r independent, non-negative random variables Z_i, then by use of Jensen's inequality

$$\begin{aligned} ES_r\chi(S_r) &\leq \sum_{i=1}^r \{EZ_i\chi(\sum_{j\neq i} EZ_j) + EZ_i\chi(Z_i)\} \\ &\leq ES_r\chi(ES_r) + \sum_{i=1}^r EZ_i\chi(Z_i). \end{aligned} \tag{4.6}$$

Applying this to (4.5), we have for $0 \leq t \leq t_0$, t_0 arbitrary but fixed,

$$\tilde{I}_{t,n}(\hat{x}) \leq \rho^t\hat{x}[\varphi]\chi(\rho^t\hat{x}[\varphi]) + \hat{x}[I_{t,n}],$$

$$\int_{\hat{X}}\pi(x,d\hat{x})\tilde{I}_{t,n}(\hat{x}) \leq c_1 + c_2\iota(x) + mI_{t,n}(x) \tag{4.7}$$

$$\iota(x) := \int_{\hat{X}}\pi(x,d\hat{x})\hat{x}[\varphi]\chi(\hat{x}[\varphi]).$$

Inserting (4.7) into (4.4) and using $T_t^0\eta \leq M_t\eta$, $\eta \in \mathcal{B}_+$, and (4.1), we get

$$\Phi^*[I_{t,n+1}] \leq c_3 + c_4\Phi^*[k\iota] + c_5 t \sup_{0\leq s\leq t}\Phi^*[I_{t,n}],$$

where $\| I_{t,0} \| = \sup_{x\in X}\varphi(x)\chi(\varphi(x))$ has been absorbed into c_3. From this, for $0 \leq t \leq t_0$ with $c_5 t < 1$,

$$\Phi^*[I_t] = \sup_n \sup_{0\leq s\leq t}\Phi^*[I_{s,n}] < \infty. \tag{4.8}$$

Applying (4.6) to $Z_i = \hat{x}^{t,i}_{t+s}[\varphi]$, $i = 1,\ldots,\hat{x}_t[1]$, $t,s \leq t_0$,

$$I_{t+s}(x) = E^{\langle x\rangle} E(\hat{x}_{t+s}[\varphi]\chi(\hat{x}_{t+s}[\varphi]) \mid {}_t)$$

$$\leq E^{\langle x\rangle}\{\rho^s\hat{x}_t[\varphi]\chi(\rho^s\hat{x}_t[\varphi])+\hat{x}_t[I_s]\} \tag{4.9}$$

$$\leq c_6+c_7I_t(x)+c_8\Phi^*[I_s]\varphi(x).$$

Thus (4.2) holds for all $t \geq 0$.

Now suppose (4.1) holds for some t. By $E^{\hat{x}}\hat{x}_t[\mathfrak{s}] = \hat{x}[M_t\mathfrak{s}]$ and (M) the process $\{\rho^{-t}\hat{x}_t[\varphi],\mathfrak{F}_t,P^{\hat{x}}\}$ is a martingale. Since $u\chi(u)$ is convex, this implies

$$\tilde{I}_s(x) \leq c_9+c_{10}\tilde{I}_t(x),\ 0 \leq s \leq t. \tag{4.10}$$

We have

$$I_s(x) \geq E^{\langle x\rangle}\hat{x}_s[\varphi]\chi(\hat{x}_s[\varphi])1_{\{\tau_1>s\}}$$

$$= \int_0^s T^0_u\{k \int_{\hat{X}}\pi(\cdot,d\hat{x})\tilde{I}_{s-u}(\hat{x})\}(x)du,\ s \leq t.$$

From (IM) and (4.1)

$$\Phi^*[T^0_s\mathfrak{s}] \geq (1-c^*s)\rho^s\Phi^*[\mathfrak{s}],\ s \geq 0, \tag{4.11}$$

for every non-negative $\mathfrak{A}$-measurable $\mathfrak{s}$. Hence, for $s \leq 1/c^*$

$$\Phi^*[I_s] \geq c_{11}s\Phi^*[k\iota]-c_{12},$$

which implies (4.3). □

<u>4.2.</u> COROLLARY. <u>Let</u> $\{\hat{x}_t,P^{\hat{x}}\}$ <u>be given as in 4.1. Then for every</u> $t > 0$

$$\Phi^*[E^{\langle\cdot\rangle}\hat{x}_t[\varphi]\log\hat{x}_t[\varphi]] < \infty \tag{X LOG X}$$

<u>if and only if</u>

$$\Phi^*[k \int_{\hat{X}}\hat{x}[\varphi]\log\hat{x}[\varphi]\pi(\cdot,d\hat{x})] < \infty. \tag{x log x}$$

PROOF. While $\log x$ does not satisfy the assumptions on χ, (4.1) with

$$\chi(x) = 1_{[0,e]}(x)x/e+1_{[e,\infty)}(x)\log x$$

is equivalent to

$$\Phi^*[E^{\langle\cdot\rangle}\hat{x}_t[\varphi]\log \hat{x}_t[\varphi]] < \infty,$$

and the same applies to (4.2). $\square$

4.3. REMARK. Although 4.1 is already more general than is needed below, the full scope of the method of proof is of interest.

(a) In order to prove that the condition

$$\Phi^*[k \int_{\hat{X}} \hat{x}[\varphi]^n \chi(\hat{x}[\varphi])\pi(\cdot, d\hat{x})] < \infty \tag{4.12}$$

is sufficient for

$$\Phi^*[E^{\langle\cdot\rangle}\hat{x}_t[\varphi]^n \chi(\hat{x}_t[\varphi])] < \infty \tag{4.13}$$

with χ as in 4.1, or 4.2, and $u = 2,3,\ldots$, the corresponding higher order analogue of the boundedness of m,

$$\int_{\hat{X}} \hat{x}[1]^n \pi(\cdot, d\hat{x}) \in \mathcal{B},$$

is needed. For finite X this is, of course, already contained in (4.12), but in general it is not, e.g., in case of a branching diffusion with a totally absorbing barrier. The necessity part of the proof goes through as before.

(b) When replacing Φ^*, or φ, the sensitive details of the proof are the following. The sufficiency part relies on (4.1) and $\Phi^*[T_t^0\eta] \leq \rho^t\Phi^*[\eta]$, $\eta \in \mathcal{B}_+$, where the factor ρ^t is insignificant. The necessity part depends on $\Phi^*[T_t^0\eta] \geq (1-c^*t)\rho^t\Phi^*[\eta]$, $\eta \in \mathcal{B}_+$, and the submartingale property of $\rho^{-t}x_t[\varphi]$, where again ρ^t could be replaced by any continuous positive function.

4.4. REMARK. In concrete relations there may be enough information about φ and Φ^* available to write down more explicit equivalents of (4.1) and (4.2), or (X LOG X) and (x log x). In particular, if $\{\hat{x}_t, P^{\hat{x}}\}$ is a branching diffusion as in V.3 , we have $\varphi \in \mathcal{D}_0^+$, and Φ^* has a density $\varphi^* \in \mathcal{D}_0^+$, so that (4.11), (4.12) with $u = 1,2,\ldots$, are simply equivalent to

$$\int_X \varphi^*(x)k(x) \int_{\hat{X}} \hat{x}[\varphi]^n \chi(\hat{x}[\varphi])\pi(x, d\hat{x})dx < \infty,$$

$$\int_X \xi^*(x) E^{\langle x\rangle} \hat{x}_t[\xi]^n \chi(\hat{x}_t[\xi])dx < \infty,$$

respectively, where ξ, ξ^* are any continuous positive functions coinciding near $\overline{\Omega}\backslash X$ with some functions in $\mathfrak{D}_0^+$. For example, in case of a branching diffusion on $X = (a,b)$ with total absorption in a and b, the choice $\xi^* = \xi = (x-a)(b-x)$ is a suitable one.

We shall use the fact that (X LOG X) is equivalent to a condition on the remainder term R_t in the first-order expansion of the generating mapping F_t. This equivalence holds in very general terms.

Let $P(\cdot,\cdot)$ be any stochastic kernel on $X \otimes \hat{\mathfrak{A}}$ such that

$$M\xi(x) := \int_{\hat{X}} \hat{x}[\xi]P(x,d\hat{x}),$$

defines a bounded operator M on $\mathfrak{B}$. Let $F[\cdot](x)$ be the generating functional of $P(x,\cdot)$, and as in (FM) expand

$$1-F[\xi] = M[1-\xi]-R(\xi)[1-\xi], \quad \xi \in \overline{\mathfrak{S}}.$$

Let Ψ^* be a non-negative, linear-bounded functional on $\mathfrak{B}$, sequentially continuous with respect to the product topology on bounded regions, and let $\psi \in \mathfrak{B}_+$ be positive on X, possibly with $\inf \psi = 0$.

4.3. PROPOSITION. _Suppose_ $\lambda \in (0,1)$. _Then_

$$\sum_{\nu=1}^{\infty} \Psi^*[R(1-\lambda^\nu\psi)\psi] < \infty \tag{4.14}$$

if and only if

$$\Psi^*[\int_{\hat{X}} \hat{x}[\psi]\log \hat{x}[\psi]P(\cdot,d\hat{x})] < \infty. \tag{4.15}$$

PROOF. We have

$$\int_0^\infty \Psi^*[R(1-\lambda^t\psi)\psi]dt - \Psi^*[M\psi] \leq \sum_{\nu=1}^{\infty} \Psi^*[R(1-\lambda^\nu\psi)\psi] \leq \int^\infty \Psi^*[R(1-\lambda^t\psi)\psi]dt.$$

Substituting $s = s(\hat{x},t) := -\hat{x}[\log(1-\lambda^t\psi)]/\hat{x}[\psi]$, we get

$$\int_0^\infty \Psi^*[R(1-\lambda^t\psi)\psi]dt$$
$$= \Psi^*[\int_{\hat{X}} \int_0^\infty (\exp\{\hat{x}[\log(1-\lambda^t\psi)]\}-1+\lambda^t\hat{x}[\psi])\lambda^{-t}dtP(\cdot,d\hat{x})$$
$$= \Psi^*[\int_{\hat{X}} \int_0^{s(x,0)} \{s^{-2}(\exp\{-\hat{x}[\psi]s\}-1+\hat{x}[\psi]s)+a(\hat{x},s)\}b(\hat{x},s)dsP(\cdot,d\hat{x})],$$

$$a(\hat{x},s(\hat{x},t)) := s^{-2}(\lambda^t - s)\hat{x}[\psi] = \frac{\hat{x}[\lambda^t\psi] - \hat{x}[|\log(1-\lambda^t\psi)|]}{(\hat{x}[\log(1-\lambda^t\psi)]/\hat{x}[\psi])^2},$$

$$b(\hat{x},s(\hat{x},t)) := -\lambda^{-t}s^2\left(\frac{\partial s}{\partial t}\right)^{-1} = \frac{1}{|\log\lambda|}\,\frac{(\hat{x}[\log(1-\lambda^t\psi)])^2}{\hat{x}[\lambda^t\psi]\hat{x}[\lambda^t\psi/(1-\lambda^t\psi)]}.$$

Since $a(\hat{x},s(\hat{x},t))$ and $b(\hat{x},s(\hat{x},t))$ are bounded as functions of $(\hat{x},t) \in \hat{X} \otimes \mathbb{R}_+$, even if $\inf \psi = 0$, the substitution $u := \hat{x}[\psi]s$ leads to the equivalence of (4.14) and

$$\Psi^*[\int_{\hat{X}} \hat{x}[\psi] \int_0^{\hat{x}[|\log(1-\psi)|]} u^{-2}(e^{-u}-1+u)du P(\cdot,d\hat{x})] < \infty. \tag{4.16}$$

For all $v > 0$

$$0 < c_{11} \leq [\log(1+v)]^{-1} \int_0^v u^{-2}(e^{-u}-1+u)du \leq c_{13} < \infty.$$

Hence (4.16) is equivalent to

$$\Psi^*[\int_{\hat{X}} \hat{x}[\psi]\log(1+\hat{x}[|\log(1-\psi)|])P(\cdot,d\hat{x})] < \infty,$$

which in turn is equivalent to (4.15). □

BIBLIOGRAPHICAL NOTES

Except for the slight modifications mentioned in the text, the probabilistic and analytic constructions are those of Ikeda, Nagasawa, and Watanabe (1968,1969). A more extensive treatment of the explosion problem has been given by Savits (1969). A detailed discussion of measures on $(\hat{X},\hat{\mathfrak{A}})$ and of generating and moment functionals was given by Moyal (1962). A readable introduction to the theory of analytic mappings and Frechet differentiation can be found in the book by Hille and Philips (1957). The proof of (R) has been taken from Hering (1077 b ,1978 c), but has been made a little more explicit in parts.

The branching diffusion setting, as formulated here, is the same as in Hering (1978 c,d), the retarded branchnig example is from Asmussen and Hering (1977). Multi-phase birth processes were introduced by Kendall (1948), the more general age-dependent models are standard, cf. Harris (1973), Athreya and Ney (1972). More on electron-photon cascades can be found in the book by Harris (1963).

Except for the admission of non-local branching laws, 4.1 and its proof are the same as in Asmussen and Hering (1976a), 4.3 and its proof the same as in Hering (1977 b ,1978 c).

Chapter VI

Limit Theory for Subcritical and Critical Processes

1. SUBCRITICAL PROCESSES WITH INITIAL DISTRIBUTIONS CONCENTRATED AT ONE POINT

Let $\{F_t\}$ be a pre-generating semigroup. For convenience let us assume $(X, \ )$ to be such that each pre-generating functional is a generating functional, cf. V.2. Since $F_t[0]$ is nondecreasing by (F.2),

$$q(x) := \lim_{t\to\infty} F_t[0](x), \quad x \in X,$$

exists and satisfies

$$q = F_t[q], \quad t > 0.$$

Note that if $\{\hat{x}_t, P^{\hat{x}}\}$ is a branching process corresponding to $\{F_t\}$,

$$P^{<X>}(\hat{x}_t = \theta) = F_t[0](x), \ t > 0, \ x \in X.$$

If the first moments exist as a semigroup of bounded operators M_t with spectral radius ρ^t, then by (FM) and (RM)

$$\begin{aligned} \| 1-F_t[\xi] \| &\leq \| 1-F_t[0] \| + \| F_t[|\xi|]-F_t[0] \| \\ &\leq 2\| 1-F_t[0] \| \leq 2\| M_t[1] \| \\ &\leq 2(1+\epsilon_t)\rho^t \end{aligned} \tag{1.1}$$

with $\epsilon_t \to 0$, as $t \to \infty$. In particular, if $\rho < 1$, then $\| 1-F_t[\xi] \| \to 0$ and a fortiori $q = 1$. On the other hand, if (M) is satisfied and $q = 1$, then

$$\begin{aligned} \| 1-F_{t+s}[0] \| &\leq \| M_s[1-F_t[0]] \| \\ &\leq (\rho^s+\alpha_s)\Phi^*[1-F_t[0]] \| \varphi \| \to 0, \ t \to \infty. \end{aligned} \tag{1.2}$$

1.1. LEMMA. *Let $\{F_t\}$ be a generating semigroup such that (M) and (R) are satisfied and $q = 1$. Then there exists for every $t > 0$ a mapping $h_t : \bar{S}_+ \to \mathbb{R}$ such that*

$$\begin{aligned} &1-F_t[\xi] = (1+h_t[\xi])\Phi^*[1-F_t[\xi]]\varphi, \ t > 0, \ \xi \in \bar{S}_+, \\ &\lim_{t\to\infty} \| h_t[\xi] \| = 0 \ \text{uniformly in} \ \xi \in \bar{S}_+. \end{aligned} \tag{1.3}$$

PROOF. If $\Phi^*[1-\xi] = 0$, then $1-F_t[\xi] \leq M_t[1-\xi] \leq (\rho^t+\alpha_t)\Phi^*[1-\xi]\varphi = 0$ $\forall \ t > 0$, and we may take $h_t[\xi] \equiv 0$. Suppose now $\Phi^*[1-\xi] > 0$. Then

$\Phi^*[1-F_t[\xi]] > 0 \ \forall \ t > 0$, according to . From (F.2) and (FM)

$$1-F_t[\xi] = M_s[1-F_{t-s}[\xi]]-R_s(F_{t-s}[\xi])[1-F_{t-s}[\xi]], \quad t > s > 0, \xi \in \bar{\mathfrak{S}}.$$

From this by (M) and (R)

$$(1-\rho^{-s}\alpha_s-\| g_s[F_{t-s}[\xi]] \|)\rho^s\Phi^*[1-F_{t-s}[\xi]]\varphi$$

$$\leq 1-F_t[\xi] \leq (1+\rho^{-s}\alpha_s)\rho^s\Phi^*[1-F_{t-s}[\xi]]\varphi.$$

Combining these inequalities with those obtained by applying Φ^* to them,

$$-\frac{2\rho^{-s}\alpha_s+\| g_s[F_{t-s}[\xi]] \|}{1+\rho^{-s}\alpha_s}\varphi \leq \frac{1-F_t[\xi]}{\Phi^*[1-F_t[\xi]]} - \varphi$$

$$\leq \frac{2\rho^{-s}\alpha_s+\| g_s[F_{t-s}[\xi]] \|}{1-\rho^{-s}\alpha_s-\| g_s[F_{t-s}[\xi]] \|}\varphi$$

for $t \geq t^*(s)$ and $s \geq s^*$ with some $t^*(s) < \infty$, $s^* < \infty$. Now use $\rho^{-s}\alpha_s \to 0$, $s \to \infty$, (R), and $\| 1-F_t[\xi] \| \leq \| 1-F_t[0] \| \to 0$, $t \to \infty$. □

<u>1.2</u>. COROLLARY. <u>Under the hypothesis of 1.1 for</u> $s > 0$

$$\Phi^*[1-F_{t+s}[\xi]] \sim \rho^s\Phi^*[1-F_t[\xi]], \quad t \to \infty,$$

<u>uniformly in</u> $\xi \in \bar{\mathfrak{S}}_+$.

PROOF. Using (F.2) and (FM),

$$\Phi^*[1-F_{t+s}[\xi]] = \rho^s\Phi^*[1-F_t[\xi]](1-\rho^{-s}\Phi^*[R_s(F_t[\xi])[(1+h_t[\xi])\varphi]]). \ \square$$

For the rest of this section we assume to be given a generating semigroup $\{F_t\}_{t\in\mathbb{R}_+}$ such that (M) and (R) are satisfied. Although all results are statements on $\{F_t\}$, or equivalently, on a branching transition probability, we express them more intuitively in terms of a corresponding process $\{\hat{X}_t, P^{\hat{X}}\}$, without actually assuming the existence of such a process.

<u>1.3</u>. PROPOSITION. <u>If</u> $\rho < 1$, <u>then</u>

$$P^{\langle x\rangle}(\hat{x}_t \neq \theta) \sim \rho^t L(\rho^t)\varphi(x),\ t \to \infty,$$

<u>uniformly in</u> $x \in X$, <u>where</u> L <u>is slowly varying at</u> 0.

PROOF. Define $c(s) := \log s/\log \rho$, $0 < s < \rho$. By <u>1.2</u>,

$$\frac{\Phi^*[1-F_{c(\lambda s)}[0]]}{\Phi^*[1-F_{c(s)}[0]]} = \frac{\Phi^*[1-F_{c(s)+c(\lambda)}[0]]}{\Phi^*[1-F_{c(s)}[0]]} \sim \rho^{c(\lambda)} = \lambda,$$

i.e.,

$$\Phi^*[1-F_t[0]] = \Phi^*[1-F_{c(\rho^t)}[0]] \sim \rho^t L(\rho^t).$$

Now apply <u>1.1</u>. □

<u>1.4</u>. REMARK. With a discrete time parameter the proof of <u>1.3</u> is more laborious, although still a straightforward extension of the proof for finite X, cf. Hoppe and Seneta (1978). All other proofs in this section are the same for discrete and continuous parameter.

<u>1.5</u>. PROPOSITION. <u>Suppose</u> $\rho < 1$. <u>Then there exists a constant</u> γ <u>with</u>

$$P^{\langle x\rangle}(\hat{x}_t \neq \theta) \sim \gamma\, \rho^t \varphi(x),\ t \to \infty, \tag{1.4}$$

<u>where</u> $\gamma > 0$ <u>if and only if</u> (X LOG X) <u>is satisfied for some</u> (<u>and thus all</u>) $t > 0$.

PROOF. Using (F.2), (FM), (M), and (RM),

$$\begin{aligned} 0 &\leq \rho^{-t-s}\Phi^*[1-F_{t+s}[\xi]] \\ &= \rho^{-t}\Phi^*[1-F_t[\xi]] - \rho^{-t-s}\Phi^*[R_s(F_t[\xi])[1-F_t[\xi]] \\ &\leq \rho^{-t}\Phi^*[1-F_t[\xi]] \leq \rho^{-t}\Phi^*[1-F_t[0]]. \end{aligned} \tag{1.5}$$

Hence, there exists a non-negative, non-increasing functional $\gamma[\cdot]$ on $\bar{\mathfrak{S}}_+$, such that

$$\rho^{-t}\Phi^*[1-F_t[\xi]] \downarrow \gamma[\xi],\ t \uparrow \infty,\ \xi \in \bar{\mathfrak{S}}_+. \tag{1.6}$$

Set $\gamma := \gamma[0]$. From (1.5) and (1.3)

$$\rho^{-n}\Phi^*[1-F_n[0]] = \rho^{-1}\Phi^*[1-F_1[0]] \prod_{\nu=1}^{n-1} \{1-\rho^{-1}\Phi^*[R_1(F_\nu[0])[(1+h_\nu[0])\varphi]]\}.$$

That is, $\gamma > 0$ if and only if

$$\sum_{\nu=1}^{\infty} \Phi^*[R_1(F_\nu[0])[(1+h_\nu[0])\varphi]] < \infty.$$

If $\gamma > 0$, there exists by (1.6), a positive real $\varepsilon < \|\varphi\|^{-1}$ such that $1-F_\nu[0] > \varepsilon\rho^\nu\varphi$ for all sufficiently large ν, so that

$$\sum_{\nu=1}^{\infty} \Phi^*[R_1(1-\varepsilon\rho^\nu\varphi)\varphi] < \infty, \tag{1.7}$$

in view of (RM). On the other hand, if $\gamma = 0$, there is for every $\varepsilon > 0$ a ν_0 such that $1-F_\nu[0] \leq \varepsilon\rho^\nu\varphi$ for all $\nu \geq \nu_0$, and (1.7) cannot hold. That is, $\gamma > 0$ if and only if (1.7) is satisfied for some $\varepsilon < \|\varphi\|^{-1}$. Now recall V.4.3. □

1.6. THEOREM. *If $\rho < 1$, there exists a probability measure P on $(\hat{X},\hat{\mathfrak{A}})$ such that*

$$\lim_{t\to\infty} P^{\hat{y}}(\hat{x}_t[1_{A_\nu}] = n_\nu,\ \nu = 1,\ldots,j \,|\, \hat{x}_t \neq \theta) \tag{1.8}$$

$$= P(\hat{x}\,[1_{A_\nu}] = n_\nu;\ \nu = 1,\ldots,j)$$

for each finite, measurable decomposition $\{A_\nu\}_{1\leq\nu\leq j}$ of X and uniformly in $\hat{y} \in X^{(n)}$ for every $n > 0$. The generating functional G of P is for any $t > 0$ the unique convex, non-decreasing solution of

$$1-G[F_t[\cdot]] = \rho^t(1-G[\cdot]),\quad G[0] = 0. \tag{1.9}$$

PROOF. The generating functional of $P^{\hat{x}}(\hat{x}_t \in \cdot\,|\hat{x}_t \neq \theta)$ is given by

$$G_t(\hat{x},\xi) = \frac{F_t(\hat{x},\xi)-F_t(\hat{x},0)}{1-F_t(\hat{x},0)} = 1 - \frac{1-F_t(\hat{x},\xi)}{1-F_t(\hat{x},0)}\,.$$

Suppose we have shown the existence of

$$\Gamma[\xi] := \lim_{t\to\infty} \frac{\Phi^*[F_t[\xi]-F_t[0]]}{\Phi^*[1-F_t[0]]} = 1 - \lim_{t\to\infty} \frac{\Phi^*[1-F_t[\xi]]}{\Phi^*[1-F_t[0]]},\quad \xi \in \bar{\mathfrak{S}}_+ \tag{1.10}$$

then 1.1 secures the existence of

$$G[\xi] := \lim_{t\to\infty} G_t(\langle x\rangle,\xi) = 1 - \lim_{t\to\infty} \frac{1-F_t[\xi](x)}{1-F_t[0](x)} = \Gamma[\xi],\quad \xi \in \bar{\mathfrak{S}}_+ \tag{1.11}$$

uniformly in $x \in X$. Rewriting (F.1) in the form

$$F_t(\langle x_1,\ldots,x_n\rangle,\xi) = \prod_{\nu=1}^{n}(1-(1-F_t[\xi](x_\nu))), \tag{1.12}$$

we then also have

$$\lim_{t\to\infty} G_t(\hat{x},\xi) = G[\xi], \quad \xi \in \bar{\mathfrak{s}}_+.$$

To prove that G is indeed the restriction to $\bar{\mathfrak{s}}_+$ of a proper generating functional, it would remain to show that for every sequence $(\xi_n)_{n\in\mathbb{N}} \subset \bar{\mathfrak{s}}_+$ with $\xi_n(x) \to 1$, $n \to \infty$, $x \in X$,

$$\lim_{n\to\infty} G[\xi_n] = 1. \tag{1.13}$$

Then setting $\xi = \Sigma_\nu 1_{A_\nu}\lambda_\nu$, $|\lambda_\nu| \leq 1$, $\nu = 1,\ldots,j$, and appealing to the continuity theorem for generating functions, we get (1.8) with the proposed uniformity. Given (1.10) and (1.11), (F.1) and 1.2 immediately yield (1.9). What we have to prove therefore is the existence of Γ, the fact that (1.9) has only one non-decreasing, convex solution, and (1.13).

Fix $t > 0$ and set

$$\Gamma_s[\xi] := \frac{\Phi^*[F_s[\xi]-F_s[0]]}{\Phi^*[1-F_s[0]]},$$

$$\Delta_s[\xi] := \frac{\Phi^*[1-F_{s+t}[\xi]]}{\Phi^*[1-F_s[\xi]]}.$$

Then

$$1-\Gamma_s[F_t[\xi]] = \Delta_s[\xi](1-\Gamma_s[\xi]) \tag{1.14}$$

Pick $t_1 < t_2 < \ldots \uparrow \infty$ and define

$$\Gamma_{\{t_n\}}[\xi] := \limsup_{n\to\infty} \Gamma_{t_n}[\xi].$$

By (1.14) and 1.2, $\Gamma_{\{t_n\}}$ satisfies (1.9) in place of G, independently of the choice of $\{t_n\}$ and, in particular, for any subsequence of a given $\{t_n\}$. Thus, in order to prove the existence of Γ and the uniqueness of the convex, non-decreasing solution of (1.9), it suffices to show that any such solution G^* coincides with $\Gamma_{\{n\}}$.

For $s \in [0,1]$ define

$$Q_n(s) := \rho^{-n}(1-G^*[1-s(1-F_n[0])]).$$

Since F_1 is convex,

$$F_1[1-s(1-F_n[0])] \leq 1-s(1-F_{n+1}[0]),$$

and since G^* is non-decreasing,

$$Q_{n+1}(s) \leq Q_n(s),$$

so that $Q_n(s)$ converges, as $n \to \infty$, to a function $Q(s)$ which is concave and non-decreasing, hence continuous in $[0,1]$, with $Q(0) = 0$, $Q(1) = 1$. By 1.1 and 1.2 there exists for every $\epsilon > 0$ an n_0 such that for all $n \geq n_0$

$$(1-\epsilon)(1-F_{n+1}[0]) \leq \rho(1-F_n[0]) \leq (1+\epsilon)(1-F_{n+1}[0]),$$

that is,

$$\rho Q_{n+1}((1-\epsilon)s) \leq Q_n(\rho s) \leq \rho Q_{n+1}((1+\epsilon)s).$$

$$Q_n(s) \to s,\ n \to \infty. \tag{1.15}$$

Take ξ such that $\Gamma_{\{n\}}[\xi] \neq 0$. (The case $\Gamma_{\{n\}}[\xi] = 0$ can be treated similarly.) Then there exists a subsequence $\{n'\}$ of integers for which $\Gamma_{n'} \to \Gamma_{\{n\}}$. Again by 1.1 and the definition of $\Gamma_{\{n\}}$, there is for every $\epsilon > 0$ an n_0 such that for $n' \geq n_0$

$$(1-\epsilon)(1-\Gamma_{\{n\}}[\xi])(1-F_{n'}[0]) \leq 1-F_{n'}[\xi]$$
$$\leq (1+\epsilon)(1-\Gamma_{\{n\}}[\xi])(1-F_{n'}[0]).$$

That is,

$$Q_{n'}((1-\epsilon)(1-\Gamma_{\{n\}}[\xi])) \leq \rho^{-n'}(1-G^*[F_{n'}[\xi]])$$
$$\leq Q_{n'}((1+\epsilon)(1-\Gamma_{\{n\}}[\xi])).$$

Using (1.15) and the fact that G^* solves (1.9) it follows that

$$1-\Gamma_{\{n\}}[\xi] = 1-G^*[\xi].$$

It remains to prove (1.13). Let $(\xi_n)_{n\in\mathbb{N}}$ be any sequence in $\bar{\mathfrak{S}}_+$ with $\xi_n(x) \to 1$, $n \to \infty$, $x \in X$. Fix $\delta > 0$, $s > 0$, $n_0 > 0$ such that

$$\rho^{-\delta}\alpha_\delta < 1,$$

$$C := \sup_{\xi\in\bar{\mathfrak{S}}_+:\, \| 1-\xi \| \le \| 1-F_s[0] \|} \| g_\delta[\xi] \| < 1-\rho^{-\delta}\alpha_\delta,$$

$$(\rho+\alpha_1)\Phi^*[1-\xi_n] \le \rho^\delta(1-\rho^{-\delta}\alpha_\delta-C)\Phi^*[1-F_s[0]],\ n \ge n_0.$$

This is clearly possible by (M), (R), and the fact that $\| 1-F_s[0] \| \to 0$, $s \to \infty$, with $\Phi^*[1-F_s[0]] > 0\ \forall\ s > 0$. Recalling the monotonicity of $F_t[0]$, (F.2), (FM), and (RM), we can then find a sequence of integers $(\ell(n))_{n\in\mathbb{N}}$ such that $\ell(n) \ge s$ if $n \ge n_0$, $\ell(n) \to \infty$ as $n \to \infty$, and

$$1-F_1[\xi_n] \le (\rho+\alpha_1)\Phi^*[1-\xi_n]\varphi \le \rho^\delta(1-\rho^{-\delta}\alpha_\delta-C)\Phi^*[1-F_{\ell(n)}[0]]\varphi$$
$$\le 1-F_{\delta+\ell(n)}[0],\ n \ge n_0.$$

Hence, by (1.9) and its special case $G[F_t[0]] = 1-\rho^t$,

$$1 \ge G[\xi_n] = 1-\rho^{-1}(1-G[F_1[\xi_n]]) \ge 1-\rho^{-1}(1-G[F_{\delta+\ell(n)}[0]])$$
$$= 1-\rho^{\delta+\ell(n)-1},\ n \ge n_0,$$

which implies (1.13). □

1.7. REMARK. If there exists a compactification $\bar{X}$ of X such that for $t > 0$, $\xi \in \bar{\mathfrak{S}}_+$ the function $(1-F_t[\xi])/\varphi$ has a continuous extension to $\bar{X}$, as is easily seen to be the case for branching diffusions as defined in V.3 , then the existence of G can also be proved by extending the argument given for finite X by Joffe and Spitzer (1978), cf. Hering (1977 a,b, 1978 c).

1.8. PROPOSITION. *Under the assumptions of 1.6*

$$1-G[1-s\varphi] \sim sL_1(s),\ s \to 0, \tag{1.16}$$

where L_1 *is slowly varying at* 0.

PROOF. Set

$$a_n := \Phi^*[1-F_n[0]].$$

Then by 1.1 for sufficiently large n

$$s(1-\epsilon)(1-F_n[0]) \leq sa_n\varphi \leq s(1+\epsilon)(1-F_n[0]).$$

Combined with (1.15), $G^* = G$, this leads to

$$\rho^{-n}(1-G[1-\lambda a_n\varphi]) \to \lambda, \ n \to \infty,$$

so that

$$\frac{1-G[1-\lambda a_n\varphi]}{1-G[1-a_n\varphi]} \to \lambda, \ n \to \infty.$$

Using A.13.6, this implies (1.16). □

1.9. COROLLARY. *Let the conditions of 1.6 be satisfied, and let* W *be a random variable with Laplace-Stieltjes transform* $\Phi(s) = G[e^{-s\varphi}]$. *Then*

$$P(W > s) = o(s^{-1}L_2(s^{-1})), \tag{1.17}$$

where $L_2(s) \sim L(s)$, $s \to 0$.

PROOF. Relation (1.16) implies that $1-\Phi(s) = sL_2(s)$ with $L_2(s) \sim L(s)$, $s \to 0$. By A 14.1, it follows that

$$\int_0^s P(W > t)dt \sim L_2(s^{-1}),$$

and by A 14.2

$$P(W > s) = o(s^{-1}L_2(s^{-1})). \quad \square$$

In particular, (1.17) implies

$$EW^\tau < \infty, \ 0 \leq \tau < 1.$$

1.10. PROPOSITION. *If* $\gamma > 0$, *then*

$$\int_{\hat{X}} \hat{x}[\xi]P(d\hat{x}) = \gamma^{-1}\Phi^*[\xi], \ \xi \in \mathcal{B}. \tag{1.18}$$

If $\gamma = 0$, then

$$\int_{\hat{X}} \hat{x}[\xi]P(d\hat{x}) = \infty \tag{1.19}$$

for every $\xi \in \mathcal{B}_+$ *with* $\Phi^*[\xi] > 0$.

PROOF. First suppose $\gamma > 0$. Then by 1.5

$$\lim_{t\to\infty} E^{\langle x\rangle}(\hat{x}_t[1]\,|\,\hat{x}_t \neq \theta) = \gamma^{-1}\Phi^*[1] < \infty.$$

Hence P has a bounded first moment functional.

Now suppose $\gamma = 0$, and define

$$\epsilon_n := \Phi^*[1-F_n[0]]/\Phi^*[1], \ n \in \mathbb{N}.$$

By $q = 1$ and monotonicity of $F_n[0]$, $0 < \epsilon_n \downarrow 0$, as $n\uparrow\infty$. Fix $t > 0$, $n_1 > 0$, $s > 0$ such that

$$\rho^{-t}\alpha_t < 1,$$

$$\rho^{-s}\alpha_s < 1, \ (\rho^t-\alpha_t-\rho^t C')/(\rho^s+\alpha_s) \geq 1,$$

$$C' := \sup_{n\geq n_1} \| g_t[1-\epsilon_n] \| .$$

Due to (M) with $\rho < 1$ and (R) this is possible. Then, using (FM), (R), and (M)

$$1-F_t[1-\epsilon_n] \geq (\rho^t-\alpha_t-\rho^t C')\Phi^*[\epsilon_n]\varphi \geq 1-F_s[F_n[0]], \ n \geq n_1.$$

Applying (1.9) and (F.2),

$$(1-G[1-\epsilon_n])/\epsilon_n = \rho^{-t}(1-G[F_t[1-\epsilon_n]])/\epsilon_n \geq \rho^{-t}(1-G[F_s[F_n[0]]])/\epsilon_n$$

$$= \rho^{s-t+n}\Phi^*[1]/\Phi^*[1-F_n[0]], \ n \geq n_1 .$$

If $\gamma = 0$, the last expression tends to ∞, as $n \to \infty$, by (1.3), (1.5). That is, in this case P cannot have a bounded first moment functional.

Let M be the first moment functional of P, bounded or not. From (1.9) for $\xi \in \mathbb{B}_+$

$$M[M_t\xi] = \rho^t M\xi.$$

By (M) therefore $M = \epsilon\Phi^*$ with some $\epsilon \in (0,\infty]$.

If $\gamma = 0$, the only possibility is $\epsilon = \infty$.

If $\gamma > 0$, then using (1.9) with $\xi = 0$ and expanding G similarly as F_t in (FM),

$$1 = \rho^{-t}(1-G[F_t[0]]) = M[\rho^{-t}(1-F_t[0])]-R(F_t[0])[\rho^{-t}(1-F_t[0])],$$

where $R(\zeta)[\xi]$ is linear-bounded in ξ and tends to 0, as $\| 1-\zeta \| \to 0$. From this, by $q = 1$, $1 = M[\gamma\varphi]$. That is, $\epsilon = \gamma^{-1}$. $\square$

2. SUBCRITICAL PROCESSES WITH ARBITRARY INITIAL DISTRIBUTIONS AND INVARIANT MEASURES

We now admit arbitrary initial distributions, characterize the domain of attraction of each possible Yaglom limit, and expose the correspondence between the limits and invariant measures. We assume throughout to be given a generating semigroup $\{F_t\}_{t\in T}$ with $T = \mathbb{N}$, or $T = \mathbb{R}_+\setminus\{0\}$, satisfying (M) and (R) with $\rho < 1$.

2.1. PROPOSITION. *Suppose for some initial distribution* $P_0|\hat{\mathfrak{A}}$ *with* $P_0(\{\theta\}) = 0$ *there exists a (possibly defective) distribution* $\tilde{P}|\hat{\mathfrak{A}}$ *with* $\tilde{P}(\hat{X}) \neq 0$, *such that for every measurable decomposition* $\{A_\nu\}_{1\leq\nu\leq j}$ *and all j-tuples of integers* $(n_1,\ldots,n_j) \neq (0,\ldots,0)$

$$P(\hat{x}_t[1_{A_\nu}] = n_\nu; \nu = 1,\ldots,j\,|\,\hat{x}_t \neq \theta) \underset{t\to\infty}{\longrightarrow} \tilde{P}(\hat{x}[1_{A_\nu}] = n_\nu; \nu = 1,\ldots,j).$$

Then $\tilde{P}$ *is non-defective* ($\tilde{P}(\hat{X}) = 1$), *and the probability generating functional* $\tilde{G}$ *of* $\tilde{P}$ *satisfies*

$$1-\tilde{G}[F_t[\xi]] = \rho^{\alpha t}(1-\tilde{G}[\xi]),\ \xi \in \bar{\mathfrak{s}}_+,\ \tilde{G}[0] = 0,\ t \in T, \tag{2.1}$$

with some $\alpha \in (0,1]$.

PROOF. Given (2.1), non-defectiveness follows as in the proof 1.6. First fix $t = 1$. Let F_0 be the p.g.f. of P_0 and

$$1-\tilde{G}_n[\xi] := \frac{1-F_0[F_n[\xi]]}{1-F_0[F_n[0]]}.$$

Then $\tilde{G}_n \to \tilde{G}$, $\tilde{G}[0] = 0$, but $\tilde{G} \not\equiv 0$. In particular,

$$1-\tilde{G}_n[F_1[\xi]] = \frac{1-F_0[F_{n+1}[\xi]]}{1-F_0[F_n[0]]} \to \gamma(1-\tilde{G}[\xi]),$$

$$\gamma := \lim_{n\to\infty} \frac{1-F_0[F_{n+1}[0]]}{1-F_0[F_n[0]]} = 1-\tilde{G}[F_1[0]].$$

Clearly, $\gamma < 1$. On the other hand, using 1.1, 1.2, the convexity of F_0, and the fact that $F_0[1] = 1$,

$$\begin{aligned} 1-F_0[F_{n+1}[0]] &\geq 1-F_0[1-\rho(1-\varepsilon_n)(1-F_n[0])] \\ &\geq \rho(1-\varepsilon_n)(1-F_0[F_n[0]]), \end{aligned}$$

$\varepsilon_n \to 0$, so that $\gamma \geq \rho$. That is, $\gamma = \rho^\alpha$, $\alpha \in (0,1]$. This proves (2.1) for $t = 1$ and thus for every fixed $t \in T$, with α possibily depending on t. Iteration of (2.1) shows that α is constant on T, if $T = \mathbb{N}$, and on the rationals, if $T = \mathbb{R}_+ \setminus \{0\}$. For arbitrary $t \in T$ in the latter case choose $\varepsilon, \delta > 0$ such that $t-\delta$ and $t+\varepsilon$ are rational. Writing α^* for the value of α on the rationals and α_t^* for its value at t, it follows from (2.1) with $\xi = 0$ and the monotony of $F_t[0]$ that

$$\rho^{\alpha^*(t-\delta)+\delta} \leq \rho^{\alpha_t^* t} \leq \rho^{\alpha^*(t+\varepsilon)-\varepsilon} .$$

Since ε and δ can be chosen arbitrarily small, $\alpha^* = \alpha_t^*$, i.e., α is constant on $\mathbb{R}_+ \setminus \{0\}$. □

Note that a functional $\tilde{G}$ solving (2.1) for fixed $t = u$ and $t = v$ satisfies (2.1) also for $t = u+v$. That is, in case $T = \mathbb{N}$ it suffices to consider (2.1) for $t = 1$. The same will be true for the two other functional equations, (2.6) and (2.8) occurring below.

2.2. REMARK. There exist constants $c_1 \in (0,1]$, $c_2 \in [1,\infty)$ such that for every p.g.f. solution $\tilde{G}$ of (2.1)

$$\lambda^\alpha c_1 \leq \liminf_{s \downarrow 0} \frac{1-\tilde{G}[1-\lambda s\varphi]}{1-\tilde{G}[1-s\varphi]} \leq \limsup_{s \downarrow 0} \frac{1-\tilde{G}[1-\lambda s\varphi]}{1-\tilde{G}[1-s\varphi]} \leq c_2\lambda^\alpha, \lambda > 0$$

(R-O variation). There is exactly one solution satisfying a regular variation relation,

$$1-\tilde{G}[1-s\varphi] = s^\beta L(s),$$

L slowly varying as $s \downarrow 0$, in which case $\beta = \alpha$ and

$$\tilde{G}[\xi] = 1-(1-G[\xi])^\alpha, \ \xi \in \bar{\mathfrak{F}}_+ .$$

These results are straightforward extensions of their specializations for finite X, cf. Hoppe and Seneta (1978). If $\alpha = 1$, we know already from 1.6 that the solution of (2.1) is unique.

2.3. THEOREM. *The initial distribution with p.g.f.* F_0 *leads to the limit distribution with p.g.f.* $\tilde{G}$ *if and only if*

$$\frac{1-F_0[1-s\varphi]}{1-\tilde{G}[1-s\varphi]} = L(s), \ s \in (0, \|\varphi\|^{-1}), \tag{2.2}$$

<u>where</u> L <u>is</u> <u>slowly</u> <u>varying</u> <u>as</u> $s \downarrow 0$.

PROOF. We first show sufficiency of (2.2). Set

$$a_t := \Phi^*[1-F_t[0]].$$

Given $\epsilon > 0$, we have for sufficiently large t

$$(2.3) \qquad (1-\epsilon)(1-G[\xi])a_t\varphi \leq 1-F_t[\xi] \leq (1+\epsilon)(1-G[\xi])a_t\varphi.$$

From this, the analogous inequalities with 0 replacing ξ, and (2.1),

$$\frac{1-F_0[1-(1-\epsilon)(1-G[\xi])a_t\varphi]}{1-\tilde{G}[1-(1+\epsilon)(1-G[\xi])a_t\varphi]} \cdot \frac{1-\tilde{G}[1-(1-\epsilon)a_t\varphi]}{1-F_0[1-(1+\epsilon)a_t\varphi]} \leq \frac{1-\tilde{G}_t[\xi]}{1-\tilde{G}[\xi]}$$

$$\leq \frac{1-F_0[1-(1+\epsilon)(1-G[\xi])a_t\varphi]}{1-\tilde{G}[1-(1-\epsilon)(1-G[\xi])a_t\varphi]} \cdot \frac{1-\tilde{G}[1-(1+\epsilon)a_t\varphi]}{1-F_0[1-(1-\epsilon)a_t\varphi]} .$$

Given (2.2), it follows by the convexity of generating functionals that

$$\left(\frac{1-\epsilon}{1+\epsilon}\right)^2 \leq \liminf_{t\to\infty} \frac{1-\tilde{G}_t[\xi]}{1-\tilde{G}[\xi]} \leq \limsup_{t\to\infty} \frac{1-\tilde{G}_t[\xi]}{1-\tilde{G}[\xi]} \leq \left(\frac{1+\epsilon}{1-\epsilon}\right)^2 .$$

Now let $\epsilon \downarrow 0$.

Next we show the necessity of (2.2). For convenience $\|\varphi\| < 1$. Then for $0 < \epsilon < 1-\|\varphi\|$, $\xi = c\varphi$, $c \in (0,1]$, and sufficiently large t

$$1-F_t[\xi(1+\epsilon)] \leq a_t(1-G[\xi])\varphi \leq 1-F_t[\xi(1-\epsilon)].$$

Using this,

$$\frac{1-\tilde{G}_t[\xi(1+\epsilon)]}{1-\tilde{G}_t[\varphi(1-\epsilon)]} \leq \frac{1-F_0[1-a_t(1-G[\xi])\varphi]}{1-F_0[1-a_t(1-G[\varphi])\varphi]} \leq \frac{1-\tilde{G}_t[\xi(1-\epsilon)]}{1-\tilde{G}_t[\varphi(1+\epsilon)]} ,$$

so that

$$(2.4) \qquad \frac{1-F_0[1-a_t(1-G[\xi])\varphi]}{1-F_0[1-a_t(1-G[\varphi])\varphi]} \to \frac{1-\tilde{G}[\xi]}{1-\tilde{G}[\varphi]} .$$

Similarly

$$(2.5) \qquad \frac{1-\tilde{G}[1-a_t(1-G[\xi])\varphi]}{1-\tilde{G}[1-a_t(1-G[\varphi])\varphi]} \to \frac{1-\tilde{G}[\xi]}{1-\tilde{G}[\varphi]} .$$

Setting

$$\lambda_t := a_t(1-G[\xi]),$$

$$s := \frac{1-G[\xi]}{1-G[\varphi]},$$

we see from (2.4) and (2.5) that

$$\frac{1-F_0[1-\lambda_t s\varphi]}{1-\tilde{G}[1-\lambda_t s\varphi]} \cdot \frac{1-\tilde{G}[1-\lambda_t\varphi]}{1-F_0[1-\lambda_t\varphi]} \to 1$$

as $t \to \infty$. If $T = \mathbb{R}_+ \setminus \{0\}$, this already concludes the proof. If $T = \mathbb{N}$, we invoke A.13.6. □

2.4. DEFINITION. *A generating functional* H *generates an invariant measure if and only if*

$$H[F_t[\xi]] = H[\xi]+t, \quad \xi \in S_+, \quad H[0] = 0, \quad t \in T, \tag{2.6}$$

up to a constant multiplying H.

2.5. PROPOSITION. *There exists a bijective correspondence between the generating functional solutions of* (2.6) *and the elements of*

$$\{(\tilde{G},\alpha) : \alpha \in (0,1],\ \tilde{G} \text{ is a probability generating functional solution of (2.1)}\}$$

A proof will be given at the end of this section.

2.6. PROPOSITION. *The relation*

$$H[\xi] = \tilde{d}(G[\xi]), \quad \xi \in S_+, \tag{2.7}$$

defines a bijective correspondence between the solutions of (2.6) *and those of*

$$\tilde{d}(1-\rho^t+\rho^t s) = \tilde{d}(s)+t, \quad s \in [0,1), \quad \tilde{d}(0) = 0, \quad t \in T. \tag{2.8}$$

PROOF. The non-trivial direction is to construct a $\tilde{d}$ solving (2.7), given H. First consider $t = 1$, and define

$$d_n(s) = H[1-s(1-F_n[0])]-n, \quad n \in \mathbb{N}.$$

By convexity and (2.6),

$$\begin{aligned} d_{n+1}(s) &= H[1-s(1-F_1[F_n[0]])]-n-1 \\ &\geq H[F_1[1-s(1-F_n[0])]]-n-1 \\ &= H[1-s(1-F_n[0])]-n = d_n(s). \end{aligned}$$

Fix $s \in (0,1]$ and choose $\ell \in \mathbb{N} : 2(\rho^{\ell}+\alpha_{\ell}) \leq s$, then recall (1.3) and choose $n_0 \in \mathbb{N} : \| h_n[0] \| < \frac{1}{2} \ \forall\, n \geq n_0$. Then

$$\begin{aligned} 1-F_{n+\ell}[0] &\leq M^{\ell}[1-F_n[0]] \leq (\rho^{\ell}+\alpha_{\ell})\Phi^*[1-F_n[0]]\varphi \\ &\leq (\rho^{\ell}+\alpha_{\ell})(1+h_n[0])^{-1}(1-F_n[0]) \leq s(1-F_n[0]) \ \forall\, n \geq n_0 \end{aligned}$$

and thus

$$d_n(s) \leq H[F_{\ell}[0]].$$

That is, for fixed s the sequence $d_n(s)$ is non-decreasing and bounded from above. Hence

$$d_n(s) \to : d(s).$$

From the concluding lemma of V.3, 1.1, and 1.2

$$\frac{1-F_{n+1}[0]}{1-F_n[0]} \to \rho ,$$

and using this,

$$d(\rho s) = d(s)+1, \ d(1) = 0.$$

Since by (2.6)

$$H[\xi] = H[1-(1-F_n[\xi])-n,$$

it follows by (2.3) and the definition of d that

$$H[\xi] = d(1-G[\xi]).$$

Now set $\tilde{d}(s) := d(1-s)$.

We have constructed a solution $\tilde{d}$ for $t = 1$ and with that for every fixed $t \in T$. However, since G takes all values in $[0,1)$ on $\mathcal{S}_+$, solutions $\tilde{d}$ for two different fixed t must coincide. □

2.8. COROLLARY. *If H is the generating functional of the invariant measure Π, then*

$$H[1-s\varphi] \sim \frac{\log s}{\log \rho}, \ s \to 0,$$

and equivalently

$$\int_{\{\hat{x}[\varphi]\leq\lambda\}} \Pi(d\hat{x}) \sim \frac{\log \lambda}{|\log \rho|}, \quad \lambda \to \infty .$$

PROOF. Using (2.7), (2.8), and (1.16), the argument is the same as for simple branching processes, III.2. □

2.9. THEOREM. *The solutions of* (2.8) *are given by*

$$\tilde{d}(s) = \int_{-\infty}^{+\infty} (e^{-(1-s)\rho^u} - e^{-\rho^u})\nu(du),$$

where ν *varies over all measures such that for every Borel set* $A \subset \mathbb{R}$ *and each* $t \in T$

$$\nu(A+t) = \nu(A), \quad \nu([0,t)) = t.$$

This result follows immediately from III.2.9 . In view of the uniqueness theorem for the Haar measure, it implies, in particular, that for $T = \mathbb{R}_+ \setminus \{0\}$ there is exactly one solution, ν having to be the Lebesgue measure. The integral representing $\tilde{d}$ can then be evaluated,

$$\tilde{d}(s) = \frac{\log(1-s)}{\log \rho},$$

which leads to

$$1-\tilde{G}[\xi] = (1-G[\xi])^{\alpha},$$

cf. 2.2.

We now come to the proof of 2.5.

2.10. LEMMA. *For every* $\alpha \in (0,1]$ *the relation*

$$\tilde{G}[\xi] = \tilde{a}(G[\xi]), \quad \xi \in \tag{2.9}$$

defines a bijective correspondence between the probability generating functional solutions of (2.1) *and the p.g.f. solutions of*

$$1 - \tilde{a}(1 - \rho^t + \rho^t s) = \tilde{a}(s) + t, \quad s \in [0,1], \; \tilde{a}(0) = 0, \; t \in T. \tag{2.10}$$

PROOF. The proof is analogous to the proof in the BGW case. Given a solution of (2.10), (2.9) clearly supplies a solution of (2.1). For the reverse it suffices to consider $t=1$ fixed, cf. the proof of 2.6. Define

$$a_n(s) := \rho^{-n\alpha}(1 - G[1 - s(1 - F_n[0])]).$$

By convexity of F and (2.1), $a_n(s)$ is non-increasing in n, so that $a_n(s) \to a(s)$, say, and $a(\rho s) = \rho^{\alpha} a(s)$. Notice that

$$\tilde{a}(s) := 1 - a(1 - s)$$

solves (2.10). To get (2.9), observe that if ξ and $\varepsilon > 0$ are fixed, then for all sufficiently large n

$$(1-\varepsilon)(1 - G[\xi])(1 - F_n[0]) \leq 1 - F_n[\xi] \leq (1+\varepsilon)(1 - G[\xi])(1 - F_n[0])$$

and thus

$$a_n((1-\varepsilon)(1 - G[\xi])) \leq \rho^{-n\alpha}(1 - \tilde{G}[F_n[\xi]])$$

$$= 1 - \tilde{G}[\xi] \leq a_n((1+\varepsilon)(1 - G[\xi])).$$

Hence

$$a(1 - G[\xi]) = 1 - \tilde{G}[\xi],$$

which is equivalent to (2.9). □

PROOF OF 2.5. In view of 2.6 and 2.10 we have to establish only the bijectice correspondence between the g.f. solutions of (2,8) and the elements of $\{(\tilde{a},\alpha) : \alpha \in (0,1], \tilde{a}$ is a p.g.f. solution of (2.10)$\}$. But this is achieved by a trivial adaptation of the proof of III,2,6. □

3. CRITICAL PROCESSES WITH FINITE SECOND MOMENT PARAMETER

Suppose to be given a generating semigroup $\{F_t\}_{t\in T}$, $T = \mathbb{N}$, or $T = \mathbb{R}_+ \setminus \{0\}$, satisfying (M) with $\rho = 1$. Extend Φ^* in the obvious way to the set of not necessarily bounded, non-negative, $\mathfrak{A}$-measurable functions.

3.1. PROPOSITION. *Given* $\rho = 1$, *the quantity*

$$\mu: = (2t)^{-1}\Phi^*[M_t^{(2)}[\varphi]]$$

is constant as a function of $t \in T$, $0 \leq \mu \leq \infty$.

PROOF. From the recursion relation

$$M_{t+s}^{(2)}[\xi] = M_t^{(2)}[M_s\xi] + M_tM_s^{(2)}[\xi], \quad s, t \in T, \xi \in \mathfrak{B}_+$$

and (M) with $\rho = 1$,

$$(*) \qquad \Phi^*[M_t^{(2)}[\varphi]] = t\Phi^*[M_1^{(2)}[\varphi]]$$

for all t, if $T = \mathbb{N}$, and for all rationals, if $T = \mathbb{R}_+ \setminus \{0\}$. Also

$$\Phi^*[M_t^{(2)}[\varphi]] \geq \Phi^*[M_s^{(2)}[\varphi]], \quad t \geq s.$$

That is, (*) holds for all $t \in T$ in any case. □

3.2. PROPOSITION. *Given a process constructable from* $\{T_t, k, \pi\}$ *with bounded* k *and* m, *satisfying* (M) *with* $\rho = 1$ *and* (C^*),

$$\mu = \frac{1}{2}\Phi^*[k\int_{\hat{X}}\pi(\cdot, d\hat{x})\{\hat{x}[\varphi]^2 - \hat{x}[\varphi^2]\}]$$

PROOF. Extend T_t^0, m and M_t to all non-negative, $\mathfrak{A}$-measurable functions. By (IF) the function $M_t^{(2)}[\xi](x)$, $t > 0$, $x \in X$, finite or not, solves

$$z_t(x) = \int_0^t T_s^0\{kmz_{t-s} + km^{(2)}[M_{t-s}\xi]\}(x)ds,$$

$$m^{(2)}[\xi](x): = \int_{\hat{X}}\pi(x, d\hat{x})\{\hat{x}[\xi]^2 - \hat{x}[\xi^2]\}, \quad \xi \in \mathfrak{B}_+, \ x \in X.$$

Using (M) with $\rho = 1$, the fact that $T_t^0 \leq M_t$, and (C^*),

$$0 \leq \Phi^*[\int_0^t T_s^0\{kmM_{t-s}^{(2)}[\varphi]\}ds \leq tc^* \sup_{0\leq s\leq t} \Phi^*[M_s^{(2)}[\varphi]] = 2c^*t^2\mu,$$

$$t(1-c^*t)\Phi^*[km^{(2)}[\varphi]] \leq \Phi^*[\int_0^t T_s^0\{km^{(2)}[\varphi]\}ds] \leq t\Phi^*[km^{(2)}[\varphi]], \ t > 0.$$

Divide by t and let $t \downarrow 0$. □

Quite generally either $\mu = 0$ and $\delta F_t[0;1] \equiv 1$, or $\mu > 0$ and $q \equiv 1$, where we recall that $\delta F_t[0;1](x) = P^{\langle x\rangle}(\hat{x}_t[1] = 1)$, if $\{\hat{x}_t, P^{\hat{x}}\}$ is a process corresponding to $\{F_t\}$. The proof is simplified by additional regularity assumptions. For example, if X is a locally compact Hausdorff space with countable open base, $\mathfrak{A}$ the Borel algebra, if $\varphi > 0$ and Φ^* has a density $\varphi^* > 0$, and if $\delta F_t[0;1] \in C^0$, then the equivalence of $\mu = 0$ and $\delta F_t[0;1] = 1$ is trivial, and it remains to show that $\mu > 0$ implies $q = 1$.

3.3. PROPOSITION. <u>Let</u> $(X,\mathfrak{A})$ <u>be a topological measurable space, let</u> q <u>be continuous</u>, $M_t C^0 \subset C^0$ <u>for some</u> $t > 0$, <u>and</u> Φ^* <u>such that</u> $\xi \in C_+^0 \wedge \Phi^*[\xi] = 0$ <u>implies</u> $\xi = 0$. <u>Then</u> $\mu > 0$ <u>implies</u> $q = 1$.

Given $[T_t, k, \pi]$ with k and m bounded and $T_t\mathfrak{B} \subset C^0$ for $t > 0$, the assumptions on q, M_t, and Φ^* are easily seen to be satisfied by use of (IF) and (IM). This applies, e.g., to the diffusion setting in V.3.

PROOF OF 3.3. Fix t such that $M_t C^0 \subset C^0$. Given (M) with $\rho = 1$,

$$M_t(1 - q) - (1 - q) = 0 .$$

On the other hand

$$1 - q = 1 - F_t[q] \leq M_t(1 - q) .$$

Hence

$$M_t(1 - q) = 1 - q$$

and thus, by (M) with $\rho = 1$,

$$1 - q = \Phi^*[1 - q]\,\varphi .$$

If $\Phi^*[1 - q] = 0$, we are done. Suppose $\Phi^*[1 - q] > 0$. Without loss of generality $\varphi > 0$. Then

$$0 \leq q < 1 .$$

But

$$M_t(1 - q) = 1 - q = 1 - F_t[q] .$$

That is,

$$R_t(q)(1 - q) = 0.$$

Having $0 \le q < 1$, this is incompatible with $\mu > 0$. Hence $\Phi^*[1-q] = 0$ and thus $q = 1$. □

For the rest of this section we assume to be given a generating semigroup satisfying (M) and (R) with $\rho = 1$. Again we formulate statements on the transition function in terms of a corresponding process $\{\hat{X}_t, P^{\hat{x}}\}$ without actually assuming the existence of a process.

3.4. THEOREM. *If* $0 < \mu < \infty$ *and* $q = 1$, *then*

$$\lim_{t\to\infty} tP^{\hat{x}}(\hat{X}_t \neq \theta) = \mu^{-1}\hat{x}[\varphi] \tag{3.1}$$

uniformly in $x \in X^{(n)}$ *for each* $n \in \mathbb{N}$.

The proofs of 3.4 and the subsequent theorem rest on the following lemma.

3.5. LEMMA. *If* $\mu < \infty$ *and* $q = 1$, *then for every* $\delta \in T$

$$\lim_{\mathbb{N}\ni n\to\infty} \frac{1}{n\delta}\{\Phi^*[1-F_{n\delta}[\xi]]^{-1} - \Phi^*[1-\xi]^{-1}\} = \mu \tag{3.2}$$

uniformly in $\xi \in \bar{\mathfrak{S}}_+ \cap \{\Phi^*[1-\xi] > 0\}$.

PROOF. Fix $\xi \in \bar{\mathfrak{S}}_+ \cap \{\Phi^*[1-\xi] > 0\}$. Then $\Phi^*[1-F_t[\xi]] > 0$ for all $t \in T$. Using (F.2)

$$\frac{1}{n\delta}\{\Phi^*[1-F_{n\delta}[\xi]]^{-1}-\Phi^*[1-\xi]^{-1}\}$$

$$= \frac{1}{n}\sum_{\nu=0}^{n-1} \frac{1}{\delta}\{\Phi^*[1-F_\delta[F_{\nu\delta}[\xi]]]^{-1}-\Phi^*[1-F_{\nu\delta}[\xi]]^{-1}\}$$

$$= \frac{1}{n}\sum_{\nu=0}^{n-1} \frac{1}{\delta}(1-\Phi^*[1-F_{\nu\delta}[\xi]]\Lambda_\delta[F_{\nu\delta}[\xi]])^{-1}\Lambda_\delta[F_{\nu\delta}[\xi]],$$

$$\Lambda_\delta[\zeta] := \Phi^*[1-\zeta]^{-2}\{\Phi^*[1-\zeta] - \Phi^*[1-F_\delta[\zeta]]\}.$$

If $\mu < \infty$, then for $\zeta \in \bar{\mathfrak{S}}_+$

$$\Phi^*[1-F_t[\zeta]] = \Phi^*[M_t[1-\zeta]] - \tfrac{1}{2}\Phi^*[M_t^{(2)}[1-\zeta]] + \tfrac{1}{2}\Phi^*[R_t^{(2)}(\zeta)[1-\zeta]],$$

and for $t \in T$ and $(\xi, \eta) \in \bar{\mathfrak{S}}_+ \oplus \mathfrak{B}_+$

$$0 = \Phi^*[R_t^{(2)}(1)[\eta\varphi]] \leq \Phi^*[R_t^{(2)}(\xi)[\eta\varphi]]$$
$$\leq \Phi^*[M_t^{(2)}[\eta\varphi]] \leq 2t\mu\|\eta\|^2.$$

Using $\rho = 1$ and 1.1

$$\Lambda_\delta[F_t[\xi]] = \frac{1}{2}\Phi^*[M_\delta^{(2)}[(1+h_t[\xi])\varphi]]$$
$$- \frac{1}{2}\Phi^*[R_\delta^{(2)}(F_t[\xi])[(1+h_t[\xi])\varphi]].$$

Since $1 \geq F_t[\xi] \geq F_t[0] \uparrow 1$, as $t \uparrow \infty$,

$$\lim_{t\to\infty} \Lambda_\delta[F_t[\xi]] = \delta\mu$$

uniformly in ξ. □

PROOF OF 3.4. Combining (1.3), (3.2), and (F.1) yields (3.1) with t restricted to sets of the form $\{n\delta: n \in \mathbb{N}\}$, $\delta > 0$. Since $F_t[0]$ is monotone in t this implies (3.1) with $t \in T$. □

3.6. CONDITION. *If* $T = \mathbb{R}_+ \setminus \{0\}$, *then* $P^{\langle x\rangle}[\hat{X}_t[1_{A_\nu}] = n_\nu,\ \nu = 1,\dots,j)$ *is continuous in* $t > 0$ *for every* $x \in X$, *any decomposition* $\{A_\nu\}_{1\leq\nu\leq j}$ *of* X *with* $A_\nu \in \mathfrak{A}$, $\nu = 1,\dots,j$, *and every* j-*tupel* $(n_1,\dots,n_j)$, $j \in \mathbb{N}$.

3.7. PROPOSITION. *Let* $\{\hat{X}_t, P^{\hat{x}}\}$ *be a branching diffusion as defined in* V.3 *and* $\{Y_\nu\}_{1\leq\nu\leq j}$ *any finite collection of sets in* $\mathfrak{A}$. *Then* $P^{\langle x\rangle}(\hat{x}_t[1_{Y_\nu}] = n_\nu;\ \nu = 1,\dots,j)$ *is continuous in* $x \in X$ *for every* $t > 0$ *and continuous in* $t > 0$ *for every* $x \in X$.

PROOF. It suffices to prove 3.7 for finite decompositions of X. For any such decomposition

$$P^{\langle x\rangle}(\hat{x}_t[1_{Y_\nu}] = n_\nu;\ \nu = 1,\dots,j) = \begin{cases} H_t(x) + I_t(x), & \sum_\nu n_\nu = 0 \\ \sum_\nu 1_{n_\nu=1} T_t^0 1_{Y_\nu}(x) + I_t(x), & \sum_\nu n_\nu = 1 \\ I_t(x), & \sum_\nu n_\nu > 1 \end{cases}$$

$$H_t(x) := 1 - T_t^0 1(x) - \int_0^t T_s^0 k(x)\,ds,$$

$$I_t(x) := \int_0^t T_s^0\{k\int_{\hat{X}} \pi(\cdot, d\hat{x}) P^{\hat{x}}(\hat{x}_{t-s}[1_{Y_\nu}] = n_\nu;\ \nu = 1,\ldots,j)\}(x)ds.$$

This follows from (IF). The continuity of $H_t(x)$ and $T_t^0 1_{Y_\nu}(x)$ in x and t and that of $I_t(x)$ in x follows immediately from $\| T_t^0 \| \leq 1$, $T_t^0 \mathcal{B} \subseteq C_0^0$, $t > 0$, and the continuity of T_t^0 in t. As for the continuity of I_t in t, note that

$$\| I_{t+\delta} - I_t \| \leq \| T_\delta^0(T_\epsilon^0 I_{t-\epsilon}) - T_\epsilon^0 I_{t-\epsilon} \| + 3\| k \| \epsilon,$$

$$\| I_{t-\delta} - I_t \| \leq \| T_{\epsilon-\delta}^0(T_\epsilon^0 I_{t-2\epsilon}) - T_\epsilon^0(T_\epsilon^0 I_{t-2\epsilon}) \| + 4k\| \epsilon \|,$$

whenever $0 < 2\delta < 2\epsilon < t$. □

3.8. THEOREM. *Suppose* $0 < \mu < \infty$ *and* $q = 1$, *and let* 3.6 *be satisfied. Then for every finite, measurable decomposition* $\{A_\nu\}_{1\leq\nu\leq j}$ *of* X *and any* $\hat{x} \neq \theta$

$$\lim_{t\to\infty} P^{\hat{x}}(t^{-1}\hat{x}_t[1_{A_\nu}] \leq \lambda_\nu;\ \nu = 1,\ldots,j \mid \hat{x}_t \neq \theta) \tag{3.3}$$

$$= \begin{cases} 0, & \min_\nu \lambda_\nu \leq 0 \\ 1-\exp\{-\min_\nu[(\mu\Phi^*[1_{A_\nu}])^{-1}\lambda_\nu]\}, & \min_\nu \lambda_\nu > 0 \end{cases}$$

uniformly in $(\lambda_1,\ldots,\lambda_j) \in \mathbb{R}^j$.

In more intuitive terms (3.3) says that the conditional distribution function of the vector

$$t^{-1}(\hat{x}_t[1_{A_1}],\ldots,\hat{x}_t[1_{A_j}]), \text{ given } \hat{x}_t \neq \theta,$$

converges to the d.f. of a vector of the form

$$(\Phi^*[1_{A_1}],\ldots,\Phi^*[1_{A_j}])W,$$

$$P(W > \lambda) = e^{-\lambda/\mu},\ \lambda > 0. \tag{3.4}$$

PROOF. The Laplace transform $L_t^{\hat{x}}(s_1,\ldots,s_j)$ of $Q_t^{\hat{x}}(\lambda_1,\ldots,\lambda_j) := P^{\hat{x}}(t^{-1}\hat{x}_t[1_{A_\nu}] \leq \lambda_\nu;\ \nu = 1,\ldots,j)$ is given by

$$L_t^{\hat{x}} = \frac{F_t(\hat{x},\xi_t) - F_t(\hat{x},0)}{1 - F_t(x,0)} = 1 - \frac{1-F_t(\hat{x},\xi_t)}{1-F_t(\hat{x},0)},$$

$$\xi_t := e^{-\xi/t}, \quad \xi := \sum_{\nu=1}^{j} s_\nu 1_{A_\nu}.$$

Note that

$$t\Phi^*[1-\xi_t] \to \Phi^*[\xi], \quad t \to \infty.$$

Using this, it follows again from (1.3), (3.2), and (F.1) that

$$\nu\delta(1-F_{\nu\delta}(\hat{x},\xi_{\nu\delta})) \to (1+\mu\Phi^*[\xi])^{-1}\Phi^*[\xi]\hat{x}[\varphi], \quad \mathbb{N} \ni \nu \to \infty.$$

From this by (3.1)

$$\lim_{\mathbb{N}\ni\nu\to\infty} L^{\hat{x}}_{\nu\delta} = (1+\mu\Phi^*[\xi])^{-1}, \quad \delta > 0.$$

The expression on the right is the Laplace transform of the limit d.f. proposed in (3.3). Denote this d.f. by Q_∞. Now suppose $T = \mathbb{R}_+ \setminus \{0\}$. By the continuity theorem $Q^{\hat{x}}_{\nu\delta} \to Q_\infty$, $\nu \to \infty$, and since Q_∞ is continuous, we have uniform convergence. Hence, we have convergence respective the metric

$$d(Q_1,Q_2) := \inf\{\varepsilon: Q_1(\lambda_1-\varepsilon,\ldots,\lambda_j-\varepsilon)-\varepsilon \le Q_2(\lambda_1,\ldots,\lambda_j)$$

$$\le Q_1(\lambda_1+\varepsilon,\ldots,\lambda_j+\varepsilon)+\varepsilon, \ \lambda_\nu \in [0,\infty), \ \nu = 1,\ldots,j\},$$

defined for all pairs of j-dimensional distribution functions Q_1, Q_2 with $Q_1(0,\ldots,0) = Q_2(0,\ldots,0) = 0$. Writing

$$Q^{\hat{x}}_t(\lambda_1,\ldots,\lambda_j) = \sum_{\substack{n_\nu \le t\lambda_\nu;\ \nu=1,\ldots,j \\ n_1+\ldots+n_j > 0}} \frac{P^{\hat{x}}(\hat{x}_t[1_{A_\nu}]=n_\nu;\ \nu=1,\ldots,j)}{P^{\hat{x}}(\hat{x}_t \neq \theta)},$$

it follows from 3.6 and (F.1) that $Q^{\hat{x}}_t$ is continuous in $t > 0$ respective d. By the Croft-Kingman lemma A 9.1 therefore

$$\lim_{\mathbb{R}_+\ni t\to\infty} d(Q^{\hat{x}}_t, Q_\infty) = 0$$

which implies (3.3). □

3.9 . REMARK. Condition 3.6 and the application of the Croft-Kingman lemma can be avoided, if a continuous-time version of (3.2) is at hand. The latter is easily proved if

$$(3.5) \qquad \delta^{-2}\Phi^*[R_t^{(2)}[1-\delta\varphi]] \to 0$$

uniformly in t on some bounded, open interval. With finite third moments (3.5) is trivial. But in the branching diffusion setting of V.3 (3.5) can be verified without such additional assumptions. We shall return to this in Section 6.

The proof of 3.8 also implies that under the conditions of the theorem

$$t^{-1}\hat{x}_t[\xi]\,|\,\hat{x}_t \neq \theta \overset{D}{\to} \Phi^*[\xi]W, \quad \xi \in \mathfrak{B},$$

with W distributed as in (3.4).

3.10. COROLLARY. Given a process $\{\hat{x}_t, P^{\hat{x}}\}$ satisfying the assumptions of 3.8

$$\left.\frac{\hat{x}_t[\xi]}{\hat{x}_t[\varphi]}\right| \hat{x}_t \neq \theta \overset{P}{\to} \Phi^*[\xi], \; \xi \in \mathfrak{B}.$$

Noting that

$$\hat{x}[M_t\xi] = E^{\hat{x}}\hat{x}_t[\xi] = P^{\hat{x}}(\hat{x}_t \neq \theta)E^{\hat{x}}(\hat{x}_t[\xi]\,|\,\hat{x}_t \neq \theta)$$

and recalling (3.2), we have

$$\lim_{t\to\infty} t^{-1}E^{\langle x\rangle}(\hat{x}_t[\xi]\,|\,\hat{x}_t \neq \theta) = \mu\Phi^*[\xi], \quad \xi \in \mathfrak{B}.$$

Finally, if $\hat{x}_t[1] = \hat{x}_0[1]$ a.s. for all t, then by (FM) and (M) with $\rho = 1$,

$$\lim_{t\to\infty} P^{\langle x\rangle}(\hat{x}_t[1_A] = 1) = \frac{\Phi^*[1_A]}{\Phi^*[1]}$$

for all $x \in X$ and $A \in \mathfrak{A}$.

4. CRITICAL PROCESSES WITH INFINITE SECOND MOMENT PARAMETER

The Setting is the same as in the preceding section except that we now assume $\mu = \infty$. A crucial role will be played by the following condition:

(S) *For some* $\varepsilon > 0$

$$\Phi^*[R_t(1-s\varphi)\varphi] = s^\alpha L_t(s), \quad 0 < t \leq \varepsilon,$$

where $0 < \alpha \leq 1$, *independent of* t, *and* $L_t(s)$ *is slowly varying, as* $s \downarrow 0$.

Using (FM), (R), and (RM),

$$\Phi^*[R_{2t}(1-s\varphi)\varphi] = \Phi^*[R_t(1-s\varphi)\varphi]$$

$$+ \Phi^*[R_t(F_t[1-s\varphi])\frac{1-F_t[1-s\varphi]}{s}, \tag{4.1}$$

$$(1-g_t[1-s\varphi])s\varphi \leq 1 - F_t[1-s\varphi] \leq s\varphi.$$

Hence, if (S) holds for some $\varepsilon > 0$, it is satisfied for all $\varepsilon > 0$, and by the uniform convergence property for slowly varying functions

$$\lim_{u\downarrow 0} \frac{\Phi^*[R_t(1-u\varphi)\varphi]}{\Phi^*[R_s(1-u\varphi)\varphi]} = \frac{t}{s}, \quad t, s > 0.$$

4.1. PROPOSITION. *Suppose to be given a process determined by* $\{T_t, k, \pi\}$ *with bounded* k *and* m, *satisfying* (M), (R), *and* (C^*). *Then* (S) *is equivalent to*

$$\Phi^*[kr(1-s\varphi)\varphi] = s^\alpha L(s), \tag{4.2}$$

where $L(s)$ *is slowly varying as* $s \downarrow 0$. *In fact,*

$$\lim_{s\downarrow 0} \frac{L_t(s)}{L(s)} = t. \tag{4.3}$$

PROOF. Recalling (IF) and (IM), $R_t(1-u\varphi)\varphi$ solves

$$w_t = w_t^1 + \int_0^t T_s^0 km\, w_{t-s}\, ds,$$

$$w_t^1 := \int_0^t T_s^0 kr(F_{t-s}[1-u\varphi])\frac{1-F_{t-s}[1-u\varphi]}{u}\, ds.$$

By (IM), $T_s^0 \leq M_s$, so that $\Phi^* T_t^0 \leq \rho^t \Phi^*$. Hence using $r(\cdot) \leq m$ and (C^*),

$$(4.4) \qquad \Phi^*[w_t^1] \leq \Phi^*[w_t] \leq e^{c^* t} \sup_{0<s\leq t} \Phi^*[w_s^1],$$

and using the last inequality of (4.1),

$$(4.5) \qquad \sup_{0<s\leq t} \Phi^*[w_s^1] \leq t\Phi^*[kr(1-u\varphi)\varphi].$$

From (IM) and (C^*), $\Phi^* T_s^0 \geq (1-c^* s)\rho^s \Phi^*$. Set

$$\gamma_{\epsilon,t}(u) := \sup_{\epsilon\leq s\leq t} \| g_s[1-u\,\varphi] \|.$$

Then, using the first inequality of (4.1),

$$(4.6) \qquad \begin{aligned} \Phi^*[w_t^1] &\geq (1-\gamma_{\epsilon,t}(u))(1-c^* t)(t-\epsilon) \\ &\quad \times \Phi^*[kr(1-(1-\gamma_{\epsilon,t}(u))u\varphi)\varphi]. \end{aligned}$$

Assuming (S), set $\gamma_{\epsilon,t}^v := \sup_{u\leq v} \gamma_{\epsilon,t}(u)$. From (4.4-6)

$$\frac{\Phi^*[kr(1-\lambda u\varphi)\varphi]}{\Phi^*[kr(1-u\varphi)\varphi]} \geq \frac{t-\epsilon}{t}\cdot\frac{1-c^* t}{e^{c^* t}}(1-\gamma_{\epsilon,t}(u))\frac{\Phi^*[R_t(1-\lambda u\varphi)\varphi]}{\Phi^*[R_t(1-u(1-\gamma_{\epsilon,t}^v)^{-1}\varphi)\varphi]}$$

$$= \left(\frac{\lambda}{u}(1-\gamma_{\epsilon,t}^v)\right)^{\alpha} \frac{L_t(\lambda u)}{L_t((1-\gamma_{\epsilon,t}^v)^{-1}u)}.$$

Choose t, ϵ, and v in this order. It follows that $\liminf_{u\downarrow 0}(\ell.h.s) \geq \lambda^{\alpha}$. Similarly $\limsup_{u\downarrow 0}(\ell.h.s) \leq \lambda^{\alpha}$. That is, (S) implies (4.2).

Now assume (4.2) and define $\gamma_t^0(s) := 1$,

$$\gamma_t^n(s) := \gamma_t^{n-1}(s)(1-\| g_{2^{-n}t}[1-\gamma_t^{n-1}(s)s\varphi] \|), \quad n \in \mathbb{N}.$$

Then by induction, using (4.1) and (4.4-6),

$$\frac{\Phi^*[R_t(1-\lambda s\varphi)\varphi]}{\Phi^*[R_t(1-s\varphi)\varphi]} \leq \frac{2^{-n_t}e^{c^* 2^{-n_t}}}{(1-\gamma_{\varepsilon,2^{-n_t}}(s))(1-c^* 2^{-n_t})(2^{-n_t}-\varepsilon)\gamma_t^n(s)}$$

$$\times \frac{\Phi^*[kr(1-\lambda s\varphi)\varphi]}{\Phi^*[kr(1-(1-\gamma_{\varepsilon,2^{-n_t}}(\gamma^n(s)s))\gamma_t^n(s)s\varphi)\varphi]} .$$

Fixing t, n, ε in this order leads to $\limsup_{s\downarrow 0}(\ell.h.s) \leq \lambda^\alpha$. Similarly, $\liminf_{s\downarrow 0}(\ell.h.s) \geq \lambda^\alpha$. That is, (4.2) implies (S).

The same estimates also yield (4.3). □

4.2. THEOREM. *Suppose* (S) *is satisfied. Then*

$$(4.7) \qquad P^{\langle x\rangle}(\hat{x}_t \neq \theta) \sim t^{-1/\alpha}L^*(t)\,\varphi(x), \quad t \to \infty ,$$

where L^* *is slowly varying at infinity. Moreover, for*

$$a_t := \Phi^*[P^{\langle\cdot\rangle}(\hat{x}_t \neq \theta)]$$

and any measurable decomposition $\{A_\nu\}_{1\leq\nu\leq j}$ *of* X

$$(4.8) \qquad a_t(\hat{x}_t[1_{A_1}],\ldots,\hat{x}_t[1_{A_j}])\,|\,\hat{x}_t \neq \theta \xrightarrow{D} (\Phi^*[1_{A_1}],\ldots,\Phi^*[1_{A_j}])W,$$

where the d.f. *of* W *has the Laplace-Stieltjes transform*

$$(4.9) \qquad \Phi(t) = 1 - t(1+t^\alpha)^{-1/\alpha}.$$

The proof will rely on the following lemma.

4.3. LEMMA. *Given* (S),

$$t\Phi^*[R_\varepsilon(1-a_t\varphi)\varphi] \to \frac{\varepsilon}{\alpha}, \quad t \to \infty.$$

PROOF. It suffices to prove that for every $\varepsilon \in T$

$$\lim_{\mathbb{N}\ni n\to\infty} n\Phi^*[R_\varepsilon(1-a_{n\varepsilon}\varphi)\varphi] = \frac{1}{\alpha} .$$

The continuous-parameter result then follows by the monotony of a_t in t and $R_\varepsilon(\xi)$ in ξ.

W.ℓ.o.g. $\varepsilon = 1$. Let

$$\Lambda(s) := \Phi^*[R_1(1-s\varphi)\varphi],$$

$$\Delta_n := \Phi^*[R_1(F_n[0])(1-F_n[0])] .$$

Then, using (FM), $a_{n+1} = a_n - a_n\Delta_n$, and by the mean-value theorem

$$\Lambda(a_n) - \Lambda(a_{n+1}) = a_n\Delta_n\Lambda'(a_n - \theta_n a_n \Delta_n)$$

with some θ_n, $0 < \theta_n < 1$. Thus

$$\frac{1}{\Lambda(a_{n+1})} - \frac{1}{\Lambda(a_n)} = \frac{a_n\Delta_n\Lambda'(a_n-\theta_n a_n\Delta_n)}{\Lambda(a_n)\Lambda(a_n-a_n\Delta_n)} = A_nB_nC_nD_n,$$

$$A_n := \frac{a_n}{a_n-\theta_n a_n\Delta_n}, \qquad B_n := \frac{(a_n-\theta_n a_n\Delta_n)\Lambda'(a_n-\theta_n a_n\Delta_n)}{\Lambda(a_n-\theta_n a_n\Delta_n)},$$

$$C_n := \frac{\Lambda(a_n - \theta_n a_n\Delta_n)}{\Lambda(a_n - a_n\Delta_n)}, \qquad D_n := \frac{\Delta_n}{\Lambda(a_n)} .$$

Clearly $A_n \to 1$. Using (S) and A 13.5 , $B_n \to \infty$ By the uniform convergence properties of slowly varying functions, $C_n \to 1$, and by (1.3), $D_n \to 1$. Hence

$$\Lambda(a_{n+1})^{-1} - \Lambda(a_n)^{-1} \to \alpha.$$

Cesaro summation completes the proof. □

PROOF OF 4.2. From 4.3 and (S)

$$a_{\lambda t}/a_t \sim \lambda^{-1/\alpha}[L(a_{\lambda t})/L(a_t)]^{-1/\alpha}, \quad t \to \infty.$$

It suffices to consider $\lambda > 1$. By monotonicity of a_t

$$1 \geq a_{\lambda t}/a_t \geq a_{[\lambda t]+1}/a_{[t]} = \prod_{\nu=[t]}^{[\lambda t]} a_{\nu+1}/a_\nu = \prod_{\nu=[t]}^{[\lambda t]} (1-\Delta_\nu).$$

In the notation of the proof of 4.3, $\Delta_\nu \sim \Lambda(a_\nu) \sim (\alpha\nu)^{-1}$, $\nu \to \infty$, so that for sufficiently large t

$$a_{\lambda t}/a_t \geq [1-(\alpha[t])^{-1}]^{[\lambda t]+1} \geq c > 0$$

with c independent of t. Hence, by the uniform convergence property, $L(a_{\lambda t})/L(a_t) \to 1$, thus $a_{\lambda t}/a_t \sim \lambda^{-1/\alpha}$, i.e., $a_t = t^{-1/\alpha}L^*(t)$, L^* slowly varying at infinity, where in case $T = \mathbb{N}$ we have used A.13.6 . Recalling (1.3) this implies (4.7).

We now turn to (4.8). Setting

$$\xi := \sum_{\nu=1}^{j} \lambda_\nu 1_{A_\nu}, \quad \xi_t := e^{-a_t \xi}$$

and again referring to (1.3), it suffices to show

$$a_t^{-1}\Phi^*[1-F_t[\xi_t]] \to 1 - \Phi(\Phi^*[\xi]). \tag{4.10}$$

The $\lambda_\nu > 0$ are arbitrary but fixed. Pick $\varepsilon_1, \varepsilon_2 \in (0,1)$. Since $0 < a_t \downarrow 0$, we can choose $u(t) \to \infty$, as $t \to \infty$, such that

$$a_{u(t)} \leq \Phi^*[\xi]a_t(1+\varepsilon_1)/(1-\varepsilon_2) \leq a_{u(t)-1}. \tag{4.11}$$

By (1.3) for sufficiently large t

$$(1-\varepsilon_2)a_t\varphi \leq 1 - F_t[0]. \tag{4.12}$$

Using (FM), (RM), and $1 - e^{-x} \sim x(1+0(x))$, $x \downarrow 0$,

$$1 - \Phi^*[\xi](1-\varepsilon_1)a_t\varphi \leq F_1[\xi_t]. \tag{4.13}$$

From the second inequality in (4.11) and (4.12)

$$\Phi^*[\xi](1+\varepsilon_1)a_t\varphi \leq (1-\varepsilon_2)a_{u(t)-1}\varphi \leq 1 - F_{u(t)-1}[0].$$

From this and (4.13)

$$1 - F_1[\xi_t] \leq \Phi^*[\xi](1+\varepsilon_1)a_t\varphi \leq 1 - F_{u(t)-1}[0]. \tag{4.14}$$

Similarly, there is some $v(t) \to \infty$, $t \to \infty$, such that for sufficently large t

$$a_{v(t)+1} \leq \Phi^*[\xi]a_t(1-\varepsilon_1)/(1+\varepsilon_2) \leq a_{v(t)}$$

$$1 - F_{v(t)}[0] \leq (1+\varepsilon_2)a_v\varphi,$$

$$F_1[\xi_t] \leq 1 - \Phi^*[\xi](1-\epsilon_1)a_t\varphi,$$

and from this

(4.15) $$1 - F_1[\xi_t] \geq 1 - F_{v(t)+1}[0].$$

Combining (4.14) and (4.15), we arrive at

(4.16) $$1 - F_{v(t)+t+1}[0] \leq 1 - F_t[\xi_t] \leq 1 - F_{u(t)+t-1}[0].$$

Using (FM) and (1.3),

$$a_u/a_{u-1} = 1 - \Phi^*[R_1(F_{u-1}[0])(1-F_{u-1}[0])/a_{u-1}] \to 1, \quad u \to \infty.$$

Hence, by (4.11),

$$a_{u(t)}/a_t \to \Phi^*[\xi](1+\epsilon_1)/(1-\epsilon_2), \quad t \to \infty.$$

From this, by 4.3 and the uniform convergence property of slowly varying functions,

(4.17) $$t/u(t) \to \Phi^*[\xi]^\alpha[(1+\epsilon_1)/(1-\epsilon_2)]^\alpha, \quad t \to \infty.$$

Also by 4.3

(4.18) $$a_{u(t)+t-1}/a_t \sim [(u(t)+t-1)/t]^{-1/\alpha}[L(a_{u(t)+t-1})/L(a_t)]^{-1/\alpha}, \quad t \to \infty.$$

Proceeding as in the first part of the proof, $L(a_{u(t)+t-1})/L(a_t) \to 1$, so that by (4.18), (4.17),

$$a_{u(t)+t-1}/a_t \to (1+\Phi^*[\xi]^{-\alpha}[(1-\epsilon_2)/(1+\epsilon_1)]^\alpha)^{-1/\alpha}.$$

Similarly,

$$a_{v(t)+t+1}/a_t \to (1+\Phi^*[\xi]^{-\alpha}[(1+\epsilon_2)/(1-\epsilon_1)]^\alpha)^{-1/\alpha}.$$

Applying Φ^* to (4.16), then using the last two relations, and finally letting $\epsilon_1, \epsilon_2 \to 0$ yields (4.10), thus completing the proof. □

The above proof also implies that under the assumptions of the theorem

$$a_t x_t[\xi] | \hat{x}_t \neq \theta \overset{D}{\rightarrow} \Phi^*[\xi] W, \quad \xi \in \mathcal{B},$$

with W as in the theorem.

4.4. COROLLARY. *Under the conditions of 4.2*

$$\left. \frac{\hat{x}_t[\xi]}{\hat{x}_t[\varphi]} \right| \hat{x}_t \neq \theta \overset{P}{\rightarrow} \Phi^*[\xi], \quad \xi \in \mathcal{B}.$$

4.5. THEOREM. *If for some* $\xi \in \mathcal{B}_+$ *with* $\Phi^*[\xi] > 0$

$$a_t \hat{x}_t[\xi] | \hat{x}_t \neq \theta \overset{D}{\rightarrow} W_\xi,$$

where W_ξ *has a non-degenerate, admittedly defective* d.f., *then* (S) *is satisfied*.

The proof will be attacked via the following two propositions.

4.6. PROPOSITION. *If the hypothesis of 4.5 is satisfied for one* $\xi \in \mathcal{B}_+$ *with* $\Phi^*[\xi] > 0$, *then it is satisfied for all such* ξ.

PROOF. Fix $\lambda > 0$. For each $\epsilon > 0$ we can choose $\delta > 0$ such that for all $t > \delta$

$$a_t^{-1} \Phi^*[1 - F_t[e^{-\lambda a_t \xi}]]$$

$$\leq (a_{t-\delta}/a_t) a_{t-\delta}^{-1} \Phi^*[1 - F_{t-\delta}[1 - (1+\epsilon)\Phi^*[\xi]\lambda a_t \varphi]].$$

Here we have used (F.2), (FM), and (RM). Similarly, drawing also on (R), we can choose for every $\epsilon > 0$ a $\delta > 0$ and $t_0 > \delta$ such that for all $t > t_0$

$$a_t^{-1} \Phi^*[1 - F_t[e^{-\lambda a_t \xi}]]$$

$$\geq (a_{t-\delta}/a_t) a_{t-\delta}^{-1} \Phi^*[1 - F_{t-\delta}[1 - (1-\epsilon)\Phi^*[\xi]\lambda a_t \varphi]].$$

Since, by assumption, $a_t^{-1}\Phi^*[1 - F_t[\exp\{-\lambda a_t \xi\}]]$ converges to a limit continuous in $\lambda > 0$, further $a_{t-\delta}/a_t \rightarrow 1$, as $t \rightarrow \infty$, it follows that $a_t^{-1}\Phi^*[1 - F_t[1 - \lambda \Phi^*[\xi] a_t \varphi]]$ converges to the same limit. Now reverse the argument, using again the continuity in λ and the fact that ξ enters only through the numerical factor $\Phi^*[\xi]$ of λ. $\square$

4.7. PROPOSITION. The hypothesis of 4.5 implies

(4.19) $$a_t = t^{-1/\alpha} L_1(t)$$

where $0 < \alpha \leq 1$ and L_1 is slowly varying at infinity.

PROOF. Step 1. We show that

(4.20) $$\liminf_{n\to\infty} n\,(1-a_{n+1}/a_n) \geq 1.$$

Let $\lambda \in [0,1]$ and for convenience $\|\varphi\| = 1$. Define

$$A(\lambda) := \Phi^*[1] - \Phi^*[F_1[1-\varphi+\lambda\varphi]] - 1 + \lambda,$$

$$B(\lambda) := -(1-\lambda)/A(\lambda).$$

Since the derivative $A'(\lambda)$ is concave,

$$\tfrac{1}{2}(1-\lambda)A'(\lambda) \leq A(\lambda) \leq (1-\lambda)A'(\lambda),$$

and from this

$$0 \leq (1-\lambda)B'(\lambda)/B(\lambda) \leq 1.$$

That is, $B(\lambda)$ is non-decreasing and $(1-\lambda)B(\lambda)$ is non-increasing. Hence, for $0 \leq \lambda_1 < \lambda_2 < 1$,

$$0 \leq B(\lambda_2) - B(\lambda_1) \leq B(\lambda_1)(\lambda_2-\lambda_1)/(1-\lambda_2).$$

With $\lambda_1 := 1 - a_n$ and $\lambda_2 := 1 - a_{n+1}$

$$0 \leq B(1-a_{n+1}) - B(1-a_n)$$

$$= \frac{a_n}{a_{n+1}} \frac{a_n - a_{n-1}}{a_n - \Phi^*[1-F_1[1-a_n\varphi]]} .$$

By (FM) and (1.3), $a_n/a_{n+1} \to 1$. Furthermore, for $0 \leq \eta \leq \xi \leq 1$,

$$0 \leq 1 - \frac{\Phi^*[F_1[\xi]-F_1[\eta]]}{\Phi^*[\xi-\eta]} \leq \frac{\Phi^*[R_1(\eta)(\xi-\eta)]}{\Phi^*[\xi-\eta]} ,$$

so that

$$\frac{a_n - a_{n+1}}{a_n-\Phi^*[1-F_1[1-a_n\varphi]]}$$

(4.21)
$$= \frac{\Phi^*[F_1[1-a_n\varphi]-F_n[0]]}{\Phi^*[1-a_n\varphi-F_{n-1}[0]]} \cdot \frac{\Phi^*[F_n[0]-F_{n-1}[0]]}{\Phi^*[F_{n+1}[0]-F_n[0]]} \to 1.$$

Hence, $\limsup_{n\to\infty} n^{-1}B(1-a_n) \leq 1$. Recalling the definition of $B(\lambda)$ and applying (4.21) once more yields (4.20).

Step 2. We show (4.19) with some α not necessarily in $(0,1]$. According to A.13.7 it suffices to show that $a_{nk}/a_n \to c_k > 0$ for every integer $k \geq 1$. We proceed by induction respective k. That is, we assume $a_{nj}/a_n \to c_j > 0$ and proceed to $a_{n(j+1)}/a_n \to c_{j+1} > 0$. Define

$$\psi_n(t) := a_n^{-1}\Phi^*[1-F_n[e^{-ta_n\varphi}]].$$

By (1.3) there is for every $\varepsilon > 0$ and n_0 such that for $n \geq n_0$

$$e^{-r_n a_n\varphi} \leq F_{nj}[0] \leq e^{-t_n a_n\varphi},$$

$$r_n := (1+\varepsilon)a_{nj}/a_n, \qquad t_n := (1-\varepsilon)a_{nj}/a_n.$$

That is,

(4.22) $\qquad \psi_n(t_n) \leq a_{n(j+1)}/a_n \leq \psi_n(r_n).$

By hypothesis and 4.6 $\psi_n(t)$ converges, as $n \to \infty$, to a function $\psi(t)$ continuous in $t > 0$. Since

$$0 \leq \psi_n(r) - \psi_n(t) \leq a_n^{-1}\Phi^*[e^{-ra_n\varphi}-e^{-ta_n\varphi}] \leq t - r,$$

the family $\{\psi_n\}$ is equicontinuous, so that the convergence is uniform on compact t-intervals not containing 0. Hence, letting $n \to \infty$ in (4.22) yields

$$\psi((1-\varepsilon)c_j) \leq \liminf_{n\to\infty} \frac{a_{n(j+1)}}{a_n} \leq \psi((1+\varepsilon)c_j).$$

Now let $\varepsilon \to \infty$, and recall the non-degeneracy assumption.

Step 3. It remains to show $\alpha \in (0,1]$. It suffices to verify

$$(4.23) \qquad n(1-a_{n+1}/a_n) \to 1/\alpha, \quad n \to \infty.$$

Using (FM),

$$0 \leq \Phi^*[F_{n+1}[0]-F_n[0]] = \Phi^*[R_1(F_n[0])(1-F_n[0])] \downarrow 0.$$

Hence, if $\lambda > 1$,

$$a_n - a_{[\lambda n]+1} = \sum_{\nu=n}^{[\lambda n]} (a_\nu - a_{\nu+1}) \leq (\lambda-1+1/n)n(a_n-a_{n+1}),$$

so that, by Step 2,

$$\liminf_{n\to\infty} n(1-a_{n+1}/a_n) \geq \frac{1-\lambda^{-1/\alpha}}{\lambda-1} \to \frac{1}{\alpha}, \quad \lambda \downarrow 1.$$

Similarly, with $\lambda \uparrow 1$,

$$\limsup_{n\to\infty} n(1-a_{n+1}/a_n) \leq 1/\alpha. \quad \square$$

PROOF OF 4.5. It suffices to consider $t = 1$. The proof is the same for any other $t > 0$. For $0 < s < a_1$ we can choose $\lambda = \lambda(s)$ such that

$$a_{\lambda+1} < s \leq a_\lambda.$$

Then

$$a_\lambda^{-1}\Phi^*[F_1[1-a_{\lambda+1}\varphi] - (1-a_{\lambda+1}\varphi)]$$

$$\leq \Phi^*[R_1(1-s\varphi)\varphi]$$

$$\leq a_{\lambda+1}^{-1}\Phi^*[F_1[1-a_\lambda\varphi] - (1-a_\lambda\varphi)].$$

Multiplying through by λ, applying (4.21) and (4.23), then letting $s \downarrow 0$ leads to

$$(4.24) \qquad \lambda(s)\Phi^*[R(1-s\varphi)\varphi] \to 1/\alpha, \quad s \to 0.$$

From the preceding proposition

$$1/t \sim a_t^\alpha L_1(t)^{-\alpha}, \quad t \to \infty.$$

From this by

$$1/t \sim a_t^{\alpha} L_2(1/a_t), \quad t \to \infty,$$

where L_2 is slowly varying at infinity. By definition of λ, $a_{\lambda(s)}/s \to 1$, $s \to 0$. Thus by the uniform convergence property $L_2(1/a_\lambda) \sim L_2(1/s)$, $s \to 0$. Substituting $t = \lambda(s)$ yields

$$1/\lambda(s) \sim s^{\alpha} L_2(1/s), \quad s \to \infty.$$

Combined with (4.24) this completes the proof. □

5. CRITICAL PROCESSES WITHOUT PROPER CONDITIONAL LIMIT

Let there be given a generating semigroup $\{F_t\}_{t\in T}$, $T = \mathbb{N}$, or $T = \mathbb{R}_+ \setminus \{0\}$ satisfying (M) and (R) with $\rho = 1$. Suppose $\mu = \infty$ and (S) does not hold. Does there exist a normalization $(c_t)_{t\in T}$, necessarily essentially different from $(a_t)_{t\in T}$, leading to a non-trivial proper conditional limit?

5.1. THEOREM. *Suppose for some* $(c_t)_{t\in T}$, *monotone or continuous, and for all* $x \in X$

$$\frac{P_t(\langle x\rangle, \{c_t\hat{X}[\varphi]\leq\lambda\}\cap\{\hat{X}\neq\theta\})}{P_t(\langle x\rangle, \{\hat{X}\neq\theta\})} \xrightarrow[t\to\infty]{D} D_\varphi(\lambda), \quad \lambda > 0,$$

where D_φ *is proper and non-degenerate at zero, then*

$$c_t \sim Ca_t, \quad t \to \infty,$$

with $0 < C < \infty$.

By 1.1 D_φ is necessarily independent of $x \in X$. The hypothesis of 5.1 is that

$$(5.1) \qquad a_t^{-1}\Phi^*[1-F_t[e^{-c_t\varphi u}]] \xrightarrow[t\to\infty]{} 1 - \Psi^*(u), \quad u > 0,$$

where Ψ^* is the Laplace-Stieltjes transform of a proper d.f., non-degenerate at 0. We shall repeatedly use the fact that the convergence in (5.1) is automatically uniform on bounded u-intervals.

It suffices to prove 5.1 for $T = \mathbb{N}$. In fact, if $T = \mathbb{R}\setminus\{0\}$ and c_t monotone,

$$\frac{c_{[t]}}{a_{[t]+1}} \cdot \frac{a_{[t]+1}}{a_{[t]}} \leq \frac{c_t}{a_t} \leq \frac{c_{[t]}}{a_{[t]}} \cdot \frac{a_{[t]}}{a_{[t]+1}},$$

so that, if $c_{[t]}/a_{[t]} \to C$, also $c_t/a_t \to C$. If we know instead that c_t is continuous,

$$c_t^{-1}(a_{[t]}-(t-[t])(a_{[t]}-a_{[t]+1})) \sim \frac{a_{[t]}}{c_t} \geq \frac{a_t}{c_t}$$

$$\geq \frac{a_{[t]+1}}{c_t} \sim c_t^{-1}(a_{[t]}-(t-[t])(a_{[t]}-a_{[t]+1})).$$

The quotient on the extreme right and left is continuous, so that the continuous parameter result follows from the discrete parameter one by the Croft-Kingman lemma.

5.2. LEMMA. *Suppose for some* $m, n: \mathbb{N} \to \mathbb{N}$ *with* $m(k), n(k) \to \infty$, *as* $k \to \infty$,

$$a_{m+n}/a_n \to K, \quad k \to \infty.$$

Then

$$\Phi^*[1-F_n[e^{-a_m\varphi}]]/a_n \to K, \quad k \to \infty,$$

and vice versa.

PROOF. It suffices to prove the explicitly stated direction. Clearly $K \in [0,1]$. First let $K \in (0,1)$. We are done, if we can show $c_n^{-1}a_m \to \gamma$, where γ is the (unique) solution of $1-\Psi^*(\gamma) = K$. Suppose for some subsequence $\{k'\} \subset \{k\}$, $c_{n(k')}^{-1}a_{m(k')} \to \lambda \neq \gamma$. If $\gamma < \lambda \leq \infty$, then, using 1.1, for some $\varepsilon > 0$ and sufficiently large k',

$$-c_{n(k')}\log F_{m(k')}[0] \geq (\gamma+\varepsilon)\varphi,$$

so that

$$a_{n(k')}^{-1}a_{n(k')+m(k')} \geq a_{n(k')}\Phi^*[1-F_{n(k')}[e^{-c_{n(k')}(\gamma+\varepsilon)\varphi}]]$$

$$\to 1 - \Psi^*(\gamma+\varepsilon)$$

and thus $K \geq 1 - \Psi^*(\gamma+\varepsilon)$. On the other hand, $K = 1 - \Psi^*(\gamma)$, so that $\Psi^*(\gamma+\varepsilon) \geq \Psi^*(\gamma)$, which is impossible. Similarly the assumption that $0 \leq \lambda < \gamma$ leads to $\Psi^*(\gamma-\varepsilon) \leq \Psi^*(\gamma)$, which again is impossible. Hence, $\lambda = \gamma$.

Next let $K = 0$. Then it suffices to show that $c_n^{-1}a_m \to 0$. Suppose the latter is not the case. Then, for some $\varepsilon > 0$ and some subsequence $\{k'\} \subset \{k\}$, $c_{n(k')}^{-1}a_{m(k')} \geq \varepsilon > 0$, and thus, by 1.1, for some $\delta > 0$ and sufficiently large k',

$$-c_{n(k')}^{-1}\log F_{m(k')}[0] \geq \delta\varphi > 0,$$

that is,

$$a^{-1}_{n(k')} a_{n(k')+m(k')} \geq a^{-1}_{n(k')} \Phi^{*}[1-F_{n(k')}[e^{-c_{n(k')}\delta\varphi}]]$$

$$\to 1 - \Psi^{*}(\delta) > 0.$$

Since the quotient on the left tends to $K = 0$, we have a contradiction. Hence, $c_n^{-1} a_m \to 0$.

The case $K = 1$ is handled similarly. □

Fix $k\colon \mathbb{N} \to \mathbb{N}$ such that

$$\Phi^{*}[F_{k(n)}[0]] \leq e^{-c_n} \leq \Phi^{*}[F_{k(n)+1}[0]].$$

Then $a_{k(n)} \sim c_n$, $n \to \infty$, and in view of (5.1),

$$a_n^{-1} a_{n+k(n)} \to 1 - \Psi^{*}(1) \equiv: c, \quad n \to \infty. \tag{5.2}$$

5.3. LEMMA. *There exists a sequence* $n'\colon \mathbb{N} \to \mathbb{N}$ *with* $n'(n) \to \infty$, *as* $n \to \infty$, *such that* $a^{-1}_{n'(n)} c_{n'(n)}$ *tends to a finite positive number, as* $n \to \infty$.

PROOF. Define

$$1 - \delta_n : = \frac{1 - \| h_{n+k}[0] \|}{1 + \| h_n[0] \|}, \quad n \in \mathbb{N}.$$

Then, using 1.1 and convexity,

$$a_{n+2k} \geq \Phi^{*}[1-F_k[1-a_n^{-1} a_{n+k}(1-\delta_n)(1-F_n[0])]]$$

$$\geq a_n^{-1} a_{n+k}(1-\delta_n) a_{n+k},$$

so that

$$\frac{a_{n+k}}{a_n} \geq \frac{a_{n+2k}}{a_n} = \frac{a_{n+2k}}{a_{n+k}} \cdot \frac{a_{n+k}}{a_n} \geq (1-\delta_n)\left(\frac{a_{n+k}}{a_n}\right)^2 \to c^2, \quad n \to \infty.$$

Hence, there exist $d \in [c^2, c]$ and $n'\colon \mathbb{N} \to \mathbb{N}$, $n'(n) \to \infty$, $n \to \infty$, such that

$$a^{-1}_{n'} a_{n'+k(n')} \to d, \quad n \to \infty.$$

Define $k' := k(n')$. Then, by 5.2, as $n \to \infty$,

$$a_{n'}^{-1}\Phi^*[1-F_{n'}[e^{-a_{2k(n')}\varphi}]] \to d,$$

from this, by (5.1),

$$a_{2k(n')}^{-1} \sim c_{n'}\Psi^{*-1}(1-d) \sim a_{k(n')}\Psi^{*-1}(1-d) := a,$$

i.e., $a_{k(n')}^{-1}a_{2k(n')} \to a$. From this, again by 5.2,

$$a_{k(n')}^{-1}\Phi^*[1-F_{k(n')}[e^{-a_{k(n')}\varphi}]] \to a,$$

which, by (5.1), implies

$$a_{k(n')} \sim c_{k(n')}\Psi^{*-1}(1-a), \; n \to \infty. \quad \square$$

Lemma 5.3 implies that c_n/a_n can tend neither to 0, nor to ∞. More precisely,

$$a_n^{-1}\Phi^*[1-e^{-c_n u\varphi}] = a_n^{-1}\Phi^*[M_t[1-e^{-c_n u\varphi}]]$$

$$\geq a_n^{-1}\Phi^*[1-F_n[e^{-c_n u\varphi}]] \to 1 - \Psi^*(u),$$

so that

$$\liminf_{n\to\infty} \frac{c_n}{a_n} \geq \frac{1-\Psi^*(u)}{u} \geq 0 \quad \forall\, u.$$

Assume the c_n be normalized in such a way that

$$\liminf_{n\to\infty} \frac{c_n}{a_n} = 1,$$

so that, in particular, $m := -\Psi^{*\prime}(0+) \leq 1$. If $c_n/a_n \not\to 1$, there must exist a $K > 1$ and $i: \mathbb{N} \to \mathbb{N}$, $i(n) \to \infty$, $n \to \infty$, such that $c_i/a_i \to K$, $n \to \infty$.

5.4. LEMMA. *If there exist* $K > 1$, *and* $i: \mathbb{N} \to \mathbb{N}$ *with* $i(n) \to \infty$, $n \to \infty$, *such that* $c_i/a_i \to K$, $n \to \infty$, *then*

$$\limsup_{n\to\infty} \frac{c_n}{a_n} = \infty.$$

PROOF. Under the hypothesis of the lemma

$$a_{2i} \sim a_i^{-1}\Phi^*[1-F_i[e^{-a_i\varphi}]]$$

$$\sim a_i^{-1}\Phi^*[1-F_i[e^{-c_i\varphi/K}]] \to 1 - \Psi^*(K^{-1}).$$

Defining

$$g_1(u) \equiv g(u) := 1 - \Psi^*(u/K),$$

$$g_{N+1}(u) := g(g_N(u)), \quad N \in \mathbb{N},$$

it follows by induction that, for every N,

$$a_{(N+1)i}/a_i \to g_N(1) := g_N, \quad n \to \infty.$$

Since g is concave, non-decreasing with $g(0) = 0$, and $g'(0+) = m/K < 1$, we have

$$g_N \downarrow 0, \quad N \uparrow \infty,$$

$$1 > m/K \geq g_{N+1}/g_N = \lim_{n\to\infty}(a_{(N+1)i}/a_{Ni})$$

$$= \lim_{n\to\infty} a_{Ni}^{-1} \Phi^*[1-F_{Ni}[e^{-a_i\varphi}]]$$

$$= \lim_{N\to\infty} a_{Ni}^{-1} \Phi^*[1-F_{Ni}[e^{-c_i\varphi/K}]].$$

Using (5.1), this implies,

$$c_i/c_{Ni} \to K\Psi^{*-1}(1-g_{N+1}/g_N),$$

so that

$$\frac{c_{Ni}}{a_{Ni}} = \frac{c_{Ni}}{c_i} \cdot \frac{a_i}{a_{Ni}} \cdot \frac{c_i}{a_i}$$

$$\underset{n\to\infty}{\sim} [K\Psi^{*-1}(1-g_{N+1}/g_N)]^{-1} g_{N-1}^{-1} K$$

$$\geq [\Psi^{*-1}(1-m/K)]^{-1} g_{N-1}^{-1} \xrightarrow[N\to\infty]{} \infty. \qquad \square$$

This means that, if $c_n/a_n \not\to 1$, there exists for <u>every</u> $K > 1$ a subsequence $(i(n))_{n\in\mathbb{N}}$ of $(n)_{n\in\mathbb{N}}$ on which $c_i/a_i \to K$, as $n \to \infty$. We now lead this to a contradiction to the non-degeneracy assumption on the limiting d.f.

5.5. LEMMA. Under the hypothesis of 5.1

$$c_{n+1}/c_n \to 1, \quad n \to \infty.$$

PROOF. Using (5.2), 5.2, and the fact that $a_{n+1}/a_n \to 1$, as $n \to \infty$,

$$\begin{aligned} c &= \lim_{n\to\infty}(a_{n+1+k(n+1)}/a_{n+1}) = \lim_{n\to\infty}(a_{n+k(n+1)}/a_n) \\ &= \lim_{n\to\infty} a_n^{-1}\Phi^*[1-F_n[e^{-a_{k(n+1)}\varphi}]] \\ &= \lim_{n\to\infty} a_n^{-1}\Phi^*[1-F_n[e^{-c_{n+1}\varphi}]], \end{aligned}$$

and from this, by (5.1), $c_{n+1}/c_n \to \Psi^{*-1}(1-c)=1$. □

5.6. LEMMA. Let $(i(n))_{n\in\mathbb{N}}$ be a subsequence of $(n)_{n\in\mathbb{N}}$ such that $c_i/a_i \to K$, as $n \to \infty$. If K is sufficiently large, then

$$\liminf_{n\to\infty} \frac{c_{2i}}{a_{2i}} > K.$$

PROOF. In the proof of 5.4 we had

$$\lim_{n\to\infty} \frac{a_{(N+1)i}}{a_{Ni}} \leq \frac{m}{K} \leq \frac{1}{K} .$$

Setting $N = 2,3$,

$$\begin{aligned} K^{-2} \geq g_3/g_1 &= \lim_{n\to\infty} (a_{4i}/a_{2i}) \\ &= \lim_{n\to\infty} a_{2i}^{-1}\ \Phi^*[1-F_{2i}[e^{-a_{2i}\varphi}]]. \end{aligned}$$

Hence, by (5.1),

$$(5.3) \qquad a_{2i}/c_{2i} \to : K^* \geq \Psi^{*-1}(1-K^{-2}), \quad n \to \infty.$$

Suppose $K^* > (2K)^{-1}$. Then, by monotonicity and concavity of Ψ^*,

$$\frac{1-\Psi^*(K^*)}{(2K)^{-1}} > \frac{1-\Psi^*((2K)^{-1})}{(2K)^{-1}} > 1 - \Psi^*(1) = c,$$

so that, by (5.3), $c < 2/K$. For sufficiently large K this is impossible. Hence $K^* \leq (2K)^{-1}$ and thus

$$\lim_{n\to\infty} (c_{2i}/a_{2i}) \geq 2K. \qquad \square$$

5.7. LEMMA. *Suppose* $c_n/a_n \not\to 1$, *as* $n \to \infty$, *and* $K > 2/c$. *Then there exist subsequences* $(j(n))_{n\in\mathbb{N}}$, $(\ell(n))_{n\in\mathbb{N}}$, *and* $(m(n))_{n\in\mathbb{N}}$ *of* $(n)_{n\in\mathbb{N}}$ *such that*

$$j(n) \leq \ell(n) \leq m(n) \leq 2j(n), \quad n \in \mathbb{N},$$

$$c_\ell/a_\ell \to 1, \quad c_m/a_m \to K, \quad n \to \infty.$$

PROOF. Set $\gamma_n := c_n/a_n$, $n \in \mathbb{N}$. Since $a_{n+1}/a_n \to 1$ and, by 5.5, also $c_{n+1}/c_n \to 1$, as $n \to \infty$, there exists an increasing sequence of integers $M(n)$ such that for $N \geq M(n)$

$$|\gamma_{N+1}/\gamma_N - 1| \leq 2^{-n}/K.$$

Since $\limsup \gamma_n = \infty$, there exist $N(n) \geq M(n)$ such that $\gamma_{N(n)} > K$, and since $\liminf \gamma_n = 1$, there exist $\ell(n) > M(n)$ such that

$$|\gamma_{\ell(n)} - 1| \leq 2^{-n}/K < K, \quad n \in \mathbb{N}.$$

Let $j(n) - 1$ be the largest integer smaller $\ell(n)$ such that $\gamma_{j(n)-1} \geq K$. Then

$$K > \gamma_{j(n)} = (\gamma_{j(n)}/\gamma_{j(n)-1})\gamma_{j(n)-1} \geq K(1-2^{-n}/K) = K - 2^{-n},$$

that is,

$$\gamma_{j(n)} = c_{j(n)}/a_{j(n)} \to K, \quad n \to \infty.$$

By 5.6 this implies $\gamma_{2j(n)} > K$ for sufficiently large n. By definition of $j(n)$, any integer $m \in [j(n), \ell(n)]$ satisfies $\gamma_m < K$, so that $2j(n) > \ell(n)$. Now let $m(n) - 1$ be the smallest integer larger $\ell(n)$ such that $\gamma_{m(n)-1} < K$. Then $\ell(n) < m(n) \leq 2j(n)$ and

$$K \leq \gamma_{m(n)} = (\gamma_{m(n)}/\gamma_{m(n)-1})\gamma_{m(n)-1} \leq K(1+2^{-n}/K) = K + 2^{-n},$$

that is,

$$\gamma_{m(n)} = c_{m(n)}/a_{m(n)} \to K, \quad n \to \infty. \qquad \square$$

PROOF OF 5.1. Suppose $c_n/a_n \not\to 1$ and let K, $\ell(n)$, and $m(n)$ be as in 5.7. Then, since $a_{n+1}/a_n \to 1$, as $n \to \infty$, and $m(n) \leq 2\ell(n) \leq 2m(n)$ $\forall\, n \in \mathbb{N}$,

$$(a_{2\ell(n)}/a_{\ell(n)})^4 = \prod_{\nu=0}^{\ell(n)} (a_{\ell(n)+\nu+1}/a_{\ell(n)+\nu})^4$$

$$\leq (a_{2\ell(n)+1}/a_{2\ell(n)})^{4\ell(n)} \leq (a_{2\ell(n)+1}/a_{2\ell(n)})^{2m(n)}$$

$$\leq (a_{2m(n)+1}/a_{2m(n)})^{2m(n)} \leq \prod_{\nu=0}^{2m(n)-1} a_{2m(n)+\nu+1}/a_{2m(n)+\nu}$$

$$= a_{4m(n)}/a_{2m(n)}.$$

From $c_{\ell(n)}/a_{\ell(n)} \to 1$ as in the proof of 5.4 $a_{2\ell(n)}/a_{\ell(n)} \to 1 - \Psi^*(1) \equiv c$. From $c_{m(n)}/a_{m(n)} \to K$ on the other hand $a_{4m(n)}/a_{2m(n)} \to g_3/g_2 \leq K^{-2}$. Hence, $c^4 \leq K^{-2}$. Since we can choose K arbitrarily large, this contradicts $c > 0$. Thus $c_n/a_n \to 1$, as $n \to \infty$. $\square$

6. SUBCRITICAL AND CRITICAL PROCESSES WITH IMMIGRATION

We consider immigration-branching processes in which the immigration does - in a sense to be made precise - not overpower the effects of branching. In addition to a generating semigroup $\{F_t\}_{t\in T}$, or a branching process $\{\hat{x}_t, P^{\hat{x}}\}$, $T = \mathbb{N}$, or $T = (0,\infty)$, we assume to be given an immigration process $\{\tau_\nu, \hat{y}_\nu, P\}$, where $(\tau_\nu)_{\nu\in\mathbb{N}} \subset T$, $\tau_\nu \uparrow \infty$, is a sequence of random times and $(\hat{y}_\nu)_{\nu\in\mathbb{N}}$ a random sequence in $(\hat{X},\hat{\mathfrak{A}})$, both defined on the same probability space with measure P. Denote

$$N_t := \max\{\nu : \tau_\nu \le t\},$$

and define

$$\hat{x}+\hat{y} := \hat{x}; \qquad \hat{y} = \theta,$$

$$:= \langle x_1,\ldots,x_n,y_1,\ldots,y_\ell\rangle; \; \hat{x} = \langle x_1,\ldots,x_n\rangle, \hat{y} = \langle y_1,\ldots,y_\ell\rangle.$$

Given a process $\{\hat{x}_t, P^{\hat{x}}\}$, let $\{\hat{x}_{\nu,t}; t \ge \tau_\nu\}$ be the branching process initiated at τ_ν by $\hat{y}_\nu$. The immigration-branching process $\{\hat{z}_t, \tilde{P}\}$ is then given by

$$\hat{z}_t = \sum_{\nu\le N_t} \hat{x}_{\nu,t}$$

as a superposition of independent processes defined on the appropriate product space with probability measure $\tilde{P}$. The generating functional of the distribution of $\hat{z}_t$ is

$$\tilde{F}_t[\xi] = E\exp\{\sum_{\nu=1}^{N_t} \hat{y}_\nu[\log F_{t-\tau_\nu}[\xi]]\}, \xi \in \bar{\mathfrak{g}} \quad .$$

Given merely a generating semigroup $\{F_t\}$, this expression serves, of course, as definition of the immigration-branching generating functional.

First suppose (M) is satisfied with $\rho < 1$. If we also have (R), then we know that uniformly in $x \in X$

$$P^{\langle x\rangle}(\hat{x}_t \ne \theta) \sim \gamma_t \varphi(x), \; t \to \infty, \tag{6.1}$$
$$\gamma_t = L(\rho^t)\rho^t,$$

where L is slowly varying at 0 and $L \sim \gamma = \text{const.} > 0$ if and only if (X LOG X) is satisfied.

6.1. THEOREM. *Let the* τ_ν *the epochs of a renewal process and the* $\hat{y}_\nu$ *i.i.d. and independent of* $\{\tau_\nu\}$. *If*

(6.2) $$\sum_{\nu=1}^{\infty} \rho^{\tau_\nu} \hat{y}_\nu[\varphi] < \infty \quad \text{a.s.,}$$

then

(6.3) $$\hat{z}_t[\xi] \xrightarrow{D} \ , \ t \to \infty, \ \xi \in \ .$$

Conversely, if (R) *and* (6.3) *hold, then*

(6.4) $$\sum_{\nu=1}^{\infty} \gamma_{\tau_\nu} \hat{y}_\nu[\varphi] < \infty \quad \text{a.s.}$$

6.2. PROPOSITION. *Let* $\{\tau_\nu\}$, $\{\hat{y}_\nu\}$ *be as in* 6.1. *Then the condition*

(6.5) $$E \log^+ \hat{y}_1[\varphi] < \infty$$

is sufficient for (6.2), *as well as* (6.4), *and if the mean interarrival time is finite, it is also necessary*.

PROOF OF 6.1. From the assumptions on $\{\tau_\nu, \hat{y}_\nu\}$

$$\hat{z}_t = \sum_{\nu \leq N_t} \hat{x}_{\nu, 2\tau_\nu} \quad \text{in distribution.}$$

That is, $\hat{z}_t[\xi]$ converges in distribution for all $\xi \in \mathbb{B}$ if and only if

(6.6) $$\sum_{\nu=1}^{\infty} \hat{x}_{\nu, 2\tau_\nu}[1] < \infty \quad \text{a.s.}$$

The sufficiency of (6.2) now follows by

$$E(\sum_\nu \hat{x}_{\nu, 2\tau_\nu}[1] \,|\, \{\tau_\nu, \hat{y}_\nu\}) \leq (1+\rho^{-\tau_1}\alpha_{\tau_1})\Phi^*[1] \sum_\nu \rho^{\tau_\nu} \hat{y}_\nu[\varphi].$$

Since the summands in (6.6) are integer-valued, at most finitely many of them may be greater than 0, if (6.6) is to hold. Conditioning again on $\{\tau_\nu, \hat{y}_\nu\}$, the Borel-Cantelli lemma implies that the condition

$$\sum_\nu \hat{y}_\nu[P^{\langle\cdot\rangle}(\hat{x}_{\tau_\nu} \neq \theta)] < \infty \quad \text{a.s.}$$

is necessary for (6.6). Now recall (6.1). □

PROOF OF 6.2. Let

$$\beta < 1,\ K(y) := P(\hat{y}_\nu[\varphi] \leq y),\ A_\nu := \{\beta^\nu \hat{y}_\nu[\varphi] \leq 1\}.$$

Then

$$\sum_{\nu=1}^{\infty} P(A_\nu) = \sum_{\nu=1}^{\infty} \int_{\beta^{-\nu}}^{\infty} dK(y) = \int_0^\infty \sum_{\nu=1}^{\infty} 1_{\{y \geq \beta^{-\nu}\}} dK(y) = \int_0^\infty O(\log^+ y) dK(y)$$

and similarly

$$\sum_{\nu=1}^{\infty} E\, \beta^\nu \hat{y}_\nu[\varphi] 1_{A_\nu} = \int_0^\infty O(1) dK(y),$$

$$\sum_{\nu=1}^{\infty} E(\beta^\nu \hat{y}_\nu[\varphi] 1_{A_\nu})^2 = \int_0^\infty O(1) dK(y).$$

Thus it follows by Kolmogorov's three series criterion that (6.5) is equivalent to

$$\sum_{\nu=1}^{\infty} \beta^\nu \hat{y}_\nu[\varphi] < \infty \text{ a.s.}$$

Now condition on $\{\tau_\nu\}$ and note that τ_ν/ν tends a.s. to the mean interarrival time λ, that is, for some $\beta_1, \beta_2 < 1$

$$0 < \gamma_{\tau_\nu} \cdot \text{const.} \leq \rho^{\tau_\nu} \leq \beta_1^\nu,\ \nu \geq \nu_0,\ 0 < \lambda \leq \infty,$$

$$\text{const.}\ \rho^{\tau_\nu} \geq \gamma_{\tau_\nu} \geq \beta_2^\nu,\ 0 < \lambda < \infty. \quad \square$$

From now on suppose that (M) and (R) are satisfied with $\rho = 1$, and let $q = 1$, $0 < \mu < \infty$.

6.3. THEOREM. Suppose

$$\delta^{-2} \Phi^*[R_t^{(2)}(1-\delta\varphi)\varphi] \to 0,\ \delta \to 0 \tag{6.7}$$

uniformly on closed bounded intervals not containing 0. Let the τ_ν be the epochs of a renewal process, $E\tau_1 < \infty$, and the $\hat{y}_\nu$ i.i.d., independent of $\{\tau_\nu\}$, $E\hat{y}_1[1] < \infty$. Then for every measurable decomposition $\{A_\nu\}_{1 \leq \nu \leq j}$ of X

$$t^{-1}(\hat{z}_t[1_{A_1}], \ldots, \hat{z}_t[1_{A_j}]) \xrightarrow[t\to\infty]{D} (\Phi^*[1_{A_1}], \ldots, \Phi^*[1_{A_1}])\tilde{W},$$

$$\widetilde{P}(\widetilde{W} \leq z) = \int_0^{z/\sigma} (\varkappa/\Gamma(\varkappa))(\varkappa y)^{\varkappa-1} e^{-\varkappa y} dy,$$

$$\sigma := E\hat{y}_1[\varphi]/E\tau_1, \quad \varkappa := \gamma/\mu,$$

Γ denoting the Gamma function.

6.4. PROPOSITION. Suppose to be given a generating semigroup determined by $\{T_t, k, \pi\}$ with bounded k and m, satisfying (M), (R), and (C^*) with $\rho = 1$ and $\mu < \infty$. Then (6.7) is satisfied uniformly on closed bounded intervals not containing 0.

PROOF OF 6.4. Expanding $F_t[\xi]$ and $f[\xi]$ up to second order in ξ near 1 and using the integral equations for F_t, M_t, and $M_t^{(2)}$,

$$\begin{aligned} R_t^{(2)}(1-\delta\varphi)[\delta\varphi](x) &= \int_0^t T_s^0 km^{(2)}[R_{t-s}(1-\delta\varphi)[\delta\varphi]](x)ds \\ &+ \int_0^t T_s^0 kr^{(2)}[1-F_{t-s}[1-\delta\varphi]](x)ds \\ &+ \int_0^t T_s^0 km \; R_{t-s}^{(2)}(1-\delta\varphi)[\delta\varphi](x)ds. \end{aligned}$$

Always $T_s^0 \leq M_s$. Furthermore, if $\rho = 1$, $1-F_s[1-\delta\varphi] \leq \delta\varphi$. Then, also using ($C^*$), for $0 < \epsilon \leq t$

$$\Phi^*[R_t^{(2)}(1-\delta\varphi)[\delta\varphi]] \leq A_t(\delta)+B_t(\delta)+C^\epsilon(\delta)+c^*\int_0^{t-\epsilon} \Phi^*[R_{t-s}^{(2)}(1-\delta\varphi)[\delta\varphi]ds,$$

$$A_t(\delta) := \int_0^t \Phi^*[km^{(2)}[R_s(1-\delta\varphi)[\delta\varphi]]]ds,$$

$$B_t(\delta) := \int_0^t \Phi^*[kr^{(2)}(1-\delta\varphi)[\delta\varphi]]ds,$$

$$C^\epsilon(\delta) := \int_0^\epsilon \Phi^*[R_s^{(2)}(1-\delta\varphi)[\delta\varphi]]ds.$$

Using (RM) and (R),

$$A_t(\delta) \leq c_1\{\epsilon + \sup_{[\epsilon,t]} \| g_s[1-\delta\varphi] \| (t-\epsilon)\}$$

$$B_t(\delta) \leq \delta^2 o(\delta)t$$

$$C^\epsilon(\delta) \leq c_2\delta^2\epsilon,$$

where c_1, c_2 are constant if $t \leq T$, T fixed. By iteration

$$\Phi^*[R_t^{(2)}(1-\delta\varphi)[\delta\varphi]] \leq \delta^2 O(1)(\epsilon + O_\epsilon(\delta))e^{c^* T}$$

for $\epsilon \leq t \leq T$. □

6.5. LEMMA. Given $\rho = 1$, $q = 1$, $0 < \mu < \infty$, and (6.7) uniformly on bounded intervals not containing 0, we have

$$t^{-1}\{\Phi^*[1-F_t[\xi]]^{-1} - \Phi^*[1-\xi]^{-1}\} = \mu + k_t[\xi],$$

$$|k_t[\xi]| \to 0, \; t \to \infty,$$

uniformly in $\xi \in \bar{S}_t \cap \{\Phi^*[1-\xi] > 0\}$.

The proof of 6.5 requires merely a slight modification of the proof of 3.5: Replace n by $[t]$ and δ by $\delta(t) := t/[t]$. To handle the dependence of δ on t use the uniformity of (6.7).

PROOF OF 6.3. From (1.3) and 6.5

$$1-F_s[\eta] = (1+s\Phi^*[1-\eta](\mu+k_s[\eta]))^{-1} \times \Phi^*[1-\eta](1+h_s[\eta])\varphi.$$

Define

$$\xi := \sum_{\nu=1}^{j} \lambda_\nu 1_{A_\nu}, \quad \xi_t := e^{-\xi/t},$$

$$\vartheta_t[\xi] := t\Phi^*[1-\xi_t]/\Phi^*[\xi].$$

Under our assumptions on $\{\tau_\nu, \hat{y}_\nu\}$,

$$E \exp\{\sum_{\nu=1}^{N_t} \hat{y}_\nu[\log F_{t-\tau_\nu}[\xi]]\} = E \exp\{\sum_{\nu=1}^{N_t} \hat{y}_\nu[\log F_{\tau_\nu}[\xi]]\}.$$

By the law of large numbers

$$\limsup_{t\to\infty} |\sum_{\nu=1}^{N_{\sqrt{t}}} \hat{y}_\nu[\log F_{\tau_\nu}[\xi_t]]|$$

$$\leq \limsup_{t\to\infty} |\sum_{\nu=1}^{N_{\sqrt{t}}} \hat{y}_\nu[\log(1-M_{\tau_\nu}[1-e^{-\xi/t}])]|$$

$$\leq O(\|\xi\|) \lim_{t\to\infty} t^{-1} \sum_{\nu=1}^{N_{\sqrt{t}}} \hat{y}_\nu[1] = 0 \quad \text{a.s.} \; .$$

Moreover,

$$E \exp\{ \sum_{N_{\sqrt{t}}+1}^{N_t} \hat{y}_\nu[\log F_{\tau_\nu}[e^{-\xi/t}]]\}$$

$$= E \exp\{ \sum_{N_{\sqrt{t}}+1}^{N_t} \hat{y}_\nu[\log\{1-(1+h_{\tau_\nu}[\xi_t])\vartheta_t[\xi]\Phi^*[\xi]\varphi$$

$$\times (t+\vartheta_t[\xi]\Phi^*[\xi](\mu+k_{\tau_\nu}[\xi_t])\tau_\nu)^{-1}\}]\}$$

$$\sim E \exp\{ \sum_{N_{\sqrt{t}}+1}^{N_t} \hat{y}_\nu[\varphi]\Phi^*[\xi](t+\Phi^*[\xi]\mu\tau_\nu)^{-1}\}$$

$$\sim \sigma\Phi^*[\xi]EN_t^{-1} \sum_{N_{\sqrt{t}}+1}^{N_t} (1+\Phi^*[\xi]\mu\nu/N_t)^{-1}$$

$$\sim \sigma\Phi^*[\xi]t^{-1}\int_{\sqrt{t}}^{t} (1+\Phi^*[\xi]\mu s/t)^{-1}ds$$

$$\sim \varkappa \log(1+\Phi^*[\xi]\mu), \; t \to \infty. \qquad \square$$

Next we consider critical processes with (not necessarily homogeneous) Poisson immigration. Continuing with our general assumptions on the branching part, we assume $\{\tau_\nu\}$ to be a Poisson process and the $\hat{y}_\nu$ to be independent conditioned on $\{\tau_\nu\}$, with the distribution of $\hat{y}_\nu$ depending only on τ_ν. That is,

$$\tilde{F}_t[\xi] := \prod_{s=0}^{t} \{1-p_s(1-F_s^I[F_{t-s}[\xi]])\}, \; T = \mathbb{N},$$

$$:= \exp\{-\int_0^t p_s(1-F_s^I[F_{t-s}[\xi]])\}, \; T = (0,\infty),$$

where p_s, or $\int_s^{s+\Delta} p_t dt + o_s(\Delta)$ is the probability that an immigration event occurs at t, respectively in $(t,t+\Delta)$, while F_s^I is the generating functional of $P(\hat{y} \in \;|\tau_\nu = s)$. We assume that the latter has a bounded first moment functional M_s^I and define for $\tau < t$

$$\beta_{\tau,t}[\xi] := \sum_{s=0}^{\tau} p_{t-s}M_{t-s}^I[\xi], \; T = \mathbb{N},$$

$$:= \int_0^\tau p_{t-s}M_{t-s}^I[\xi], \quad T = (0,\infty),$$

$$\beta_t := \beta_{t,t}[\varphi]$$

$$\beta^*_{\tau,t}[\xi] := \sum_{j=0}^{\tau} p_{t-s} M^I_{t-s}[\xi](1+s)^{-1}, \; T = \mathbb{N},$$

$$:= \int_0^{\tau} p_{t-s} M^I_{t-s}[\xi](1+s)^{-1}, \; T = (0,\infty),$$

$$\beta^*_t := \beta^*_{t,t}[\varphi].$$

The remainder term in the first order expansion of F^I_s near 1 is denoted by R^I_s.

<u>6.6.</u> THEOREM. <u>Let</u> $M^I_t[1]$ <u>be bounded on bounded</u> t-<u>intervals</u>,

$$0 < \beta^*_t \to 0, \; t \to \infty,$$

<u>let there exist a</u> $\tau: T \to T$, $\tau(t) < t$, $\tau(t) \uparrow \infty$ $(t \uparrow \infty)$, <u>such that</u>

$$\beta^*_{\tau(t)}[1]/\beta^*_t \to 0, \; t \to \infty,$$

<u>and suppose</u>

$$\sup_{s>0} R^I_s((1-\epsilon)1)[\varphi]/M^I_s[\varphi] \to 0, \; \epsilon \to 0. \tag{6.8}$$

<u>Then</u>

$$\tilde{P}(\hat{Z}_t \neq \theta) \sim \beta^*_t/\mu, \; t \to \infty. \tag{6.9}$$

<u>Suppose in addition that</u>

$$L(\lambda) := 1-\lambda\mu \lim_{t\to\infty} \sum_{s=\tau(t)+1}^{t} p_{t-s} M^I_{t-s}[\varphi](\beta_t+\lambda\mu\beta^*_t s)^{-1}, \; T = \mathbb{N},$$

$$:= 1-\lambda\mu \lim_{t\to\infty} \int_{\tau(t)}^{t} p_{t-s} M^I_{t-s}[\varphi](\beta_t+\lambda\mu\beta^*_t s)^{-1} ds, \; T = (0,\infty),$$

<u>exists for</u> $\lambda > 0$ <u>and that, if</u> $T = (0,\infty)$, <u>at least either</u> <u>3.6</u> <u>and</u>

$$\sup_{t\geq 0} M^I_t[1]/M^I_t[\varphi] < \infty,$$

<u>or</u> (6.7) <u>with the indicated uniformity are satisfied</u>. <u>Then for every measurable decomposition</u> $\{A_\nu\}_{1\leq\nu\leq j}$ <u>of</u> X

$$(\beta_t^*/\beta_t)(\hat{Z}_t[1_{A_1}],\dots,\hat{Z}_t[1_{A_j}]|\hat{Z}_t \neq \theta$$

(6.10)

$$\xrightarrow{D} (\Phi^*[1_{A_1}],\dots,\Phi^*[1_{A_j}])W^*, \quad t \to \infty,$$

where the d.f. of W^* has the Laplace-Stieltjes transform $L(\lambda)$.

6.7. EXAMPLES. Assume $M_t^I[1] < \infty$ $(t \in T)$, $M_t^I[1] \sim \text{const.}\ M_t^I[\varphi]$ $(t \to \infty)$, and (6.8). The latter holds at least if

$$\sup_{s>0} M_s^{I,(2)}[1]/M_s^I[\varphi] < \infty,$$

and it is vacuous if F_s^I does not depend on s.

(a) If $p_t M_t^I[\varphi] = o(t^{-1-\epsilon})$ with some $\epsilon > 0$, but $\beta_t > 0$, then $\beta_t^* \sim t^{-1}\lim_{t\to\infty}\beta_t = \text{const.}\ t^{-1}$, $\beta_t^*/\beta_t \sim t^{-1}$, and $L(\lambda) = (1+\lambda\mu)^{-1}$, $\lambda \geq 0$, i.e., $\tilde{P}(W^* \leq z) = 1-e^{-z/\mu}$, $z \geq 0$.

(b) Given $p_t M_t^I[\varphi] \sim ct^{-1}$, $c > 0$, we have $\beta_t^* \sim 2ct^{-1}\log t$, $\beta_t^*/\beta_t \sim 2t^{-1}$, $L(\lambda) = (1+2\lambda\mu)^{-1}(1+\lambda\mu)$, $\lambda \geq 0$, i.e., $\tilde{P}(W^* \leq z) = 1_{[0,\infty)}(z)(1-\frac{1}{2}e^{-z/\mu})$, $-\infty < z < \infty$.

(c) If $p_t M_t^I[\varphi] \sim ct^{-\epsilon}$, $c > 0$, $0 < \epsilon < 1$, then $\beta_t^* \sim ct^{-\epsilon}\log t$, $\beta_t^*/\beta_t \sim t^{-1}\log t$, but $L(\lambda) = 1$, $\lambda \geq 0$, i.e., $\tilde{P}(W^* = 0) = 1$.

There is arbitrariness in the selection of τ. The choice $\tau(t) = \log(1+t)$ serves (a)-(c).

6.8. THEOREM. Let $M_t^I[1]$ be bounded on bounded t-intervals,

$$\beta_t^* \to \infty, \quad t \to \infty,$$

let there exist a $\tau: T \to T$, $\tau(t) < t$, $\tau(t) \uparrow \infty$ $(t \uparrow \infty)$, such that

$$\beta_{\tau(t),t}[1]/\beta_t \to 0, \quad t \to \infty,$$

and let

(6.11) $$\sup_{s \leq t} R_s^I(1-c1/\beta_t)[\varphi]/M_s^I[\varphi] \to 0, \quad t \to \infty, \quad c > 0.$$

Suppose

$$-\log L^*(\lambda) := \lambda \lim_{t\to\infty} \sum_{s=\tau(t)+1}^{t} p_{t-s} M_{t-s}^I[\varphi](\beta_t+\lambda\mu s)^{-1}, \quad T = \mathbb{N},$$

$$:= \lambda \lim_{t\to\infty} \int_{\tau(t)}^{t} p_{t-s} M_{t-s}^I[\varphi](\beta_t+\lambda\mu s)^{-1} ds, \quad T = (0,\infty),$$

exists for $\lambda > 0$ and that, if $T = (0,\infty)$, at least either 3.10 and

$$\sup_{t>0} M_t^I[1]/M_t^I[\varphi] < \infty,$$

or (6.7) with the indicated uniformity are satisfied. Then for every measurable decomposition $\{A_\nu\}_{1\le\nu\le l}$ of X

$$(6.12)\qquad (1/\beta_t)(\hat{z}_t[1_{A_1}],\ldots,\hat{z}_t[1_{A_j}]) \xrightarrow{d} (\Phi^*[1_{A_1}],\ldots,\Phi^*[1_{A_j}])\widetilde{W}^*,$$

$$t \to \infty,$$

where the d.f. of $\widetilde{W}^*$ has the Laplace-Stieltjes transform $\widetilde{L}(\lambda)$.

6.9. EXAMPLES. Assume $M_t^I[1] \sim$ const. $M_t^I[\varphi]$, $t \to \infty$, and (6.11), which is vacuous if F_s^I does not depend on s.

(a) If $p_t M_t^I \sim c > 0$, then $\beta_t \sim ct$, $\widetilde{L}(\lambda) = (1+\lambda/\varkappa)^{-\varkappa}$, $\varkappa = c/\mu$, i.e.,

$$\widetilde{P}(\widetilde{W}^* \le z) = \int_0^t (\varkappa/\Gamma(\varkappa))(\varkappa y)^{\varkappa-1}e^{-\varkappa y}dy,\ z \ge 0.$$

(b) Suppose $t^{-1}\beta_t \to \infty$, $t \to \infty$, such that $\beta_{\tau(t),t}[1]/\alpha_t \to 0$, e.g., $p_t M_t^I[\varphi] \sim ct^{\epsilon}$, $c > 0$, $\epsilon > 0$, then $\widetilde{L}(\lambda) = e^{-\lambda}$, i.e., $\widetilde{P}(\widetilde{W}^* = 1) = 1$.

The proofs of 6.6 and 6.8 are very similar. Since 6.8 is the more important of the two, we prove (6.9) and 6.8.

PROOF OF (6.9). First consider the discrete parameter case. We have to show

$$-\lim_{t\to\infty}(1/\beta_t^*)\sum_{s=0}^{t} \log\{1-p_{t-s}(1-F_{t-s}^I[F_s[0]])\} = 1/\mu.$$

We have

$$(6.13)\qquad 0 \le 1-F_{t-s}^I[F_s[\eta]] \le M_{t-s}^I[(1+s)(1-F_s[0])](1+s)^{-1},\ \eta \in \bar{\mathfrak{s}}_+,$$

and from 1.1 and 3.5

$$(6.14)\qquad \begin{aligned} &(s+1)(1-F_s[0]) = (1+\epsilon_s)\varphi/\mu,\\ &\lim_{s\to\infty}\|\epsilon_s\| = 0. \end{aligned}$$

Since $\beta_t^* \to 0$ and $\beta_{\tau(t)}^*[1]/\beta_t^* \to 0$ by assumption, it follows that

$$(6.15)\qquad -\lim_{t\to\infty}(1/\beta_t^*)\sum_{s=0}^{\tau(t)} \log\{1-p_{t-s}(1-F_{t-s}^I[F_s[\eta]])\} = 0 \text{ uniformly in } \eta \in \bar{\mathfrak{s}}_+.$$

Using (6.14),

$$(6.16)\quad \begin{aligned} &1-F^I_{t-s}[F_s[0]] \\ &= (1/\mu)\{M^I_{t-s}[(1+\epsilon_s)\varphi]-R^I_{t-s}(F_s[0])[(1+\epsilon_s)\varphi]\}(1+s)^{-1}. \end{aligned}$$

Setting

$$\epsilon^*_\tau := \sup_{s>\tau} \| \epsilon_s \|,$$

$$\delta^*_\tau[\eta] := \sup_{s>\tau} \sup_{t\geq 0} R^I_t(F_s[\eta])\varphi/M^I_t\varphi, \quad \eta \in \bar{\mathfrak{s}}_+ ,$$

it follows from (6.16) by $\beta^*_t \to 0$ that

$$\begin{aligned} &\{1-\epsilon^*_\tau-(1+\epsilon^*_\tau)\delta^*_\tau[0]\}(\beta^*_t-\beta^*_{\tau,t})/\mu \\ &\leq - \sum_{s=\tau+1}^{t} \log\{1-p_{t-s}(1-F^I_{t-s}[F_s[0]])\} \\ &\leq g^*_t\cdot\{1-\epsilon^*_\tau\}(\beta^*_t-\beta^*_{\tau,t})/\mu, \\ &\lim_{t\to\infty} g^*_t = 1. \end{aligned}$$

By (6.8) and $\| F_t[0]-1 \| \to 0,\ t \to \infty,$

$$\lim_{\tau\to\infty} \delta^*_\tau[\eta] = 0 \quad \text{uniformly in} \quad \eta \in \bar{\mathfrak{s}}_+ ,$$

so that

$$-\lim_{t\to\infty}(1/\beta^*_t) \sum_{s=\tau(t)+1}^{t} \log\{1-p_{t-s}(1-F^I_{t-s}[F_s[0]])\} = 1/\mu.$$

Since $F_t[0]$ is monotone, (6.13), (6.14), and thus (6.16) carry over to the continuous parameter case. Hence,

$$\begin{aligned} 0 &\leq \lim_{t\to\infty}(1/\beta^*_t)\int_0^{\tau(t)} p_{t-s}(1-F^I_{t-s}[F_s[\eta]])ds \\ &\leq \lim_{t\to\infty}(1/\beta^*_t)\int_0^{\tau(t)} p_{t-s}M^I_{t-s}[1-F_s[0]]\, ds = 0, \quad \eta \in \bar{\mathfrak{s}}_+, \end{aligned}$$

$$\begin{aligned} &\{1-\epsilon^*_\tau-(1+\epsilon^*_\tau)\delta^*_\tau[0]\}(\beta^*_t-\beta^*_{\tau,t})/\mu \leq \int_\tau^t p_{t-s}(1-F^I_{t-s}[F_s[0]])ds \\ &\leq \{1+\epsilon^*_\tau\}(\beta^*_t-\beta^*_{\tau,t})/\mu , \end{aligned}$$

so that

$$\lim_{t\to\infty}(1/\beta_t^*)\int_0^t p_{t-s}(1-F_{t-s}^I[F_s[0]])ds = 1/\mu,$$

which proves (6.9). □

PROOF OF 6.8. Part I: Discrete parameter. Set

$$\xi = \sum_{\nu=1}^{j}\lambda_\nu 1_{A_\nu}, \quad \xi_t := e^{-\xi/\beta_t}.$$

Then

$$0 \leq 1-F_t^I[F_s[\xi_t]] \leq \| M^s\xi \| \beta_t^{-1}M_{t-s}^I[1].$$

Since, by assumption, $\beta_{\tau(t),t}[1]/\beta_t \to 0$, as $t \to \infty$, it follows that

$$\lim_{t\to\infty}\sum_{s=0}^{\tau(t)}\log\{1-p_{t-s}(1-F_{t-s}^I[F_s[\xi_t]])\} = 0.$$

As $\beta_t \to 0$,

$$\beta_t\Phi^*[1-\xi_t] = \tilde{\vartheta}_t[\xi]\Phi^*[\xi],$$

$$\lim_{t\to\infty}\tilde{\vartheta}_t[\xi] = 1.$$

$$
\begin{aligned}
&1-F_{t-s}^I[F_s[\xi_t]]\\
(6.17)\quad &= \tilde{\vartheta}_t[\xi]\Phi^*[\xi]\{M_{t-s}^I[(1+h_s[\tilde{\xi}_t])\varphi]\\
&\quad -R_{t-s}^I(F_s[\xi_t])[(1+h_s[\xi_t])\varphi]\}(\beta_t+\tilde{\vartheta}_t[\xi]\Phi^*[\xi](\mu+k_s[\xi_t])s)^{-1}.
\end{aligned}
$$

Setting

$$\tilde{\delta}_t[\xi] := \sup_{s\leq t} R_s^I(\tilde{\xi}_t)\varphi/M_s^I\varphi$$

(6.11) implies

$$\lim_{t\to\infty}\tilde{\delta}_t[\xi] = 0.$$

From (6.17), using $\beta_t^* \to \infty$, $\beta_{\tau(t),t}[1]/\beta_t \to 0$, $t \to \infty$,

$$
\begin{aligned}
&\tilde{\vartheta}_t[\xi]\Phi^*[\xi]\{1-h_\tau^*[\tilde{\xi}_t]+(1+h_\tau^*[\tilde{\xi}_t])\tilde{\delta}_t[\xi]\}\\
&\quad\times\sum_{s=\tau+1}^{t}p_{t-s}M_{t-s}^I\varphi(\beta_t+\tilde{\vartheta}_t[\xi]\Phi^*[\xi](\mu+k_\tau^*[\tilde{\xi}_t])s)^{-1}
\end{aligned}
$$

$$\leq -\sum_{s=\tau+1}^{t} \log\{1-p_{t-s}(1-F_{t-s}^{I}[F_s[\tilde{\xi}_t]])\}$$

$$\leq \tilde{g}_t[\xi]\tilde{\vartheta}_t[\xi]\Phi^*[\xi](1+h_\tau^*[\tilde{\xi}_t])$$

$$\times \sum_{s=\tau}^{t} p_{t-s}M_{t-s}^{I}\varphi(\beta_t+\tilde{\vartheta}_t[\xi]\Phi^*[\xi](\mu-k_\tau^*[\tilde{\xi}_t])s)^{-1}$$

$$h_\tau^*[\eta] := \sup_{s>\tau}\| h_s[\eta] \|,$$

$$k_\tau^*[\eta] := \sup_{s>\tau}| k_s[\eta] |,$$

$$\lim_{t\to\infty} \tilde{g}_t[\xi] = 1.$$

Hence,

$$\lim_{t\to\infty} \sum_{s=\tau(t)+1}^{t} \log\{1-p_{t-s}(1-F_{t-s}^{I}[F_s[\tilde{\xi}_t]])\} = \log \tilde{L}(\Phi^*[\xi]).$$

Part II: Continuous parameter, first assuming (6.7). In any case

$$0 \leq \lim_{t\to\infty} \int_0^{\tau(t)} p_{t-s}(1-F_{t-s}^{I}[F_s[\tilde{\xi}_t]])ds$$

(6.18)

$$\leq \sup_{s\geq 0}\| M^s\xi \|\cdot\lim_{t\to\infty} \beta_{\tau(t),t}[1]/\beta_t = 0.$$

From (6.17), using (6.7),

$$\tilde{\vartheta}_t[\xi]\Phi^*[\xi]\{1-h_\tau^*[\tilde{\xi}_t]-(1+h_\tau^*[\tilde{\xi}_t])\tilde{\sigma}_t[\xi]\}$$

$$\times\int_\tau^t p_{t-s}M_{t-s}^{I}\varphi(\beta_t+\tilde{\vartheta}_t[\xi]\Phi^*[\xi](\mu+k_\tau^*[\tilde{\xi}_t])s)^{-1}ds$$

$$\leq \int_\tau^t p_{t-s}(1-F_{t-s}^{I}[F_s[\tilde{\xi}_t]])ds$$

$$\leq \tilde{\vartheta}_t[\xi]\Phi^*[\xi](1+h_\tau^*[\tilde{\xi}_t]])ds$$

$$\times\int_\tau^t p_{t-s}M_{t-s}^{I}\varphi(\beta_t+\tilde{\vartheta}_t[\xi]\Phi^*[\xi](\mu-k_\tau^*[\tilde{\xi}_t])s)^{-1}ds,$$

so that

$$\lim_{t\to\infty}\int_{\tau(t)}^{t} p_{t-s}(1-F_{t-s}^{I}[F_s[\tilde{\xi}_t]])ds = -\log\tilde{L}(\Phi^*[\xi]).$$

Dropping (6.7), we still have (6.18) and

$$\vartheta_{n\Delta}[\xi]\Phi^*[\xi]\{1-h^*_{\nu\Delta}[\tilde{\xi}_{\nu\Delta}]-(1+h^*_{\nu\Delta}[\tilde{\xi}_{n\Delta}])(D_{\nu,\Delta}[\xi_{n\Delta}]+\delta^*_{\nu\Delta}[\tilde{\xi}_{n\Delta}])\}$$

$$\times\int_{\nu\Delta}^{n\Delta} p_{n\Delta-s}M^I_{n\Delta-s}\varphi(\beta_{n\Delta}+\tilde{\vartheta}_{n\Delta}[\xi]\Phi^*[\xi](\mu+\tilde{k}_{\nu,\Delta}[\tilde{\xi}_{n\Delta}])s)^{-1}ds$$

$$\leq \int_{\nu\Delta}^{n\Delta} p_{n\Delta-s}(1-F^I_{n\Delta-s}[F_s[\tilde{\xi}_{n\Delta}]])ds$$

$$\leq \tilde{\vartheta}_{n\Delta}[\xi]\Phi^*[\xi](1+h^*_{\nu\Delta}[\tilde{\xi}_{n\Delta}])$$

$$\times\int_{\nu\Delta}^{n\Delta} p_{n\Delta-s}M^I_{u\Delta-s}\varphi(\beta_{n\Delta}+\tilde{\vartheta}_{n\Delta}[\xi]\Phi^*[\xi](\mu-\tilde{k}_{\nu,\Delta}[\tilde{\xi}_{n\Delta}])s)^{-1}ds,$$

$$\tilde{k}_{\nu,\Delta}[\eta] := \sup_{\varkappa\geq\nu}|k_{\varkappa\Delta}[\eta]|,$$

$$D_{\nu,\Delta}[\eta] := \sup_{t\geq 0}(M^I 1/M^I\varphi) \sup_{0\leq s\leq\Delta} \| M^{(2)}_s[\varphi,1] \| \sup_{\varkappa\geq\nu} \|1-F_{\varkappa\Delta}[\eta]\|.$$

Since

$$\lim_{\nu\to\infty} D_{\nu,\Delta}[\eta] = 0 \quad \text{uniformly in} \quad \eta \in \bar{\mathfrak{S}}_+ ,$$

it follows for $\nu = \nu(n) := [\tau(n)]$ that

$$\lim_{\mathbb{N}\ni n\to\infty} \int_{\nu(n)\Delta}^{n\Delta} p_{t-s}(1-F^I_{n\Delta-s}[F_s[\xi_{n\Delta}]])ds = -\tilde{L}(\Phi^*[\xi]).$$

In order to complete the proof by an application of the Croft-Kingman lemma, it suffices to show that

$$\tilde{H}_t(u_1,\ldots,u_j) :=$$

$$= \sum_{\substack{i_\nu\leq\beta_t u_\nu,\ \nu=1,\ldots,j \\ i_1+\ldots+i_j>0}} P(\mathfrak{Z}_t[1_{A_\nu}] = i_\nu,\ \nu = 1,\ldots,j)$$

is continuous with respect to the metric

$$d(H,H') := \inf\{\epsilon: H(u_1-\epsilon,\ldots,u_j-\epsilon)-\epsilon \leq H'(u_1,\ldots,u_j)$$

$$\leq H(u_1+\epsilon,\ldots,u_j+\epsilon)+\epsilon,\ -\infty < u_\nu < \infty,\ \nu = 1,\ldots,j\}.$$

In view of V.(2.2-4) the derivative

$$(*) \qquad \frac{\partial^{i_1+\dots+i_j}}{\partial\lambda_1^{i_1}\dots\partial\lambda_j^{i_j}} F_s^I[F_{t-s}[\sum_{i=1}^{j}\lambda_i 1_{A_i}]]\Big|_{\lambda_1=\dots=\lambda_j=0}$$

exists uniformly in $s > 0$ and $t-s > 0$, so we may interchange differentiation and integration in

$$\tilde{P}(\hat{Z}_t[1_{A_\nu}] = i_\nu;\ \nu = 1,\dots,j)$$

$$= \frac{1}{i_1!\dots i_j!}\frac{\partial^{i_1+\dots+i_j}}{\partial\lambda_1^{i_1}\dots\partial\lambda_j^{i_j}}\int_0^t p_s F_s^I[F_{t-s}[\sum_{i=1}^{j}\lambda_i 1_{A_i}]]dt\Big|_{\lambda_1=\dots=\lambda_j=0},$$

and it remains to see that (*) is continuous in $t \geq s$ for every $s > 0$. In fact, our continuity assumption implies that $F_t[\Sigma_{i=1}^j \lambda_i 1_{A_i}](x)$ is continuous in t uniformly in $|\lambda_i| \leq \epsilon$, $i = 1,\dots,j$, for every $\epsilon < 1$, $x \in X$. As F_s^I is sequentially continuous with respect to the product topology on $\bar{\mathfrak{S}}$, this ensures continuity of $F_s^I[F_{t-s}[\Sigma_{i=1}^j \lambda_i 1_{A_i}]]$ in t with the same uniformity. Again using V.(2.2-4), we finally obtain from this the continuity of (*). □

BIBLIOGRAPHICAL NOTES

Except for 1.8 and 1.9, which are straightforward extensions of p-type results by Hoppe and Seneta (1978), Section 1 is essentially extracted from Hering (1977 a,b). However, part of the proof of 1.6 has been changed to put it in line with Hoppe's p-type argument, cf. III.1, rather than Joffe and Spitzer's (1967). Section 2 extends the p-type, discrete-time theory of Hoppe and Seneta (1978).

Section 3 is again essentially extracted from Hering (1977 a,b) and is in the spirit of the p-type theory of Joffe and Spitzer (1967). To be mentioned in this context, however, is the paper by Mullikin (1963), which has been a milestone in the development of the limit theory of branching processes with a general set of types.

Sections 4 and 5 have been taken from Hering and Hoppe (1981), while Section 6 is a mixture of new and published results, 6.2 being from Asmussen and Hering (1976 b , 6.6, 6.7, and 6.8 from Hering (1973).

Chapter VII

Basic Limit Theory for Supercritical Processes

1. EXTINCTION PROBABILITY AND TRANSIENCE

We assume to be given a sufficiently regular set of types and a system $[T_t,k,\pi]$ such that T_t has a continuously differentiable kernel for $t > 0$, satisfying T.1-5, k is bounded, and π has a bounded first moment m. The exact form of the r.h.s. of T.4-5 is immaterial so long as the expressions are integrable on bounded t-intervals. Suppose that (M),(C), and (C*) are satisfied with $\rho > 1$ and $\varphi,\varphi^* \in \mathcal{D}_0^+$. This implies that (R) is satisfied also, cf.V.2. The case of a finite set of types is is a trivial specialization.

Recall that

$$q(x) := \lim_{t\to\infty} P^{\langle x\rangle}(\hat{x}_t = \theta), \quad x \in X,$$

always exists.

1.1. PROPOSITION. We have $1 - q \in \mathcal{D}_0^+$.

PROOF. For the moment fix $t > 0$. By (FM), (M) with $\rho > 1$, and (R) we can find an $\varepsilon > 0$ such that $\Phi^*[1-F_t[1-\xi]] > \Phi^*[\xi]$ whenever $\|\xi\| < \varepsilon$. Suppose $\Phi^*[1-q] = 0$. Then $\Phi^*[1-F_s[0]] \to 0$, as $s \to \infty$. By (F.2), (FM), (RM), and (M) there must then exist an $s > 0$ such that $\|1-F_s[0]\| < \varepsilon$ and consequently $\Phi^*[1-F_{t+s}[0]] > \Phi^*[1-F_s[0]]$. But this contradicts the fact that $F_s[0]$ is non-decreasing. Hence, $q < 1$ on a set of positive measure. From (IF) and $q = F_t[q]$, $t > 0$,

$$(1.1) \qquad 1-q = T_t^0(1-q) + \int_0^t T_s^0\{k(1-f[q])\}ds.$$

By (T.1), the boundedness of k, and the irreducibility of $(m_{\nu\mu})$, iteration of this equation yields $q < 1$ on X. Using $T_s^0 \mathcal{B} \subseteq C_0^1$, $s > 0$, and (1.1) with (T.3-5) we get

$$1 - q \in \mathcal{D}_0^+. \qquad \square$$

A prerequisite of the convergence theorem we are aiming at is a sufficiently strong transience result. We shall need that $F_t[\xi] \to q$, as $t \to \infty$, for a rather large class of $\xi \in \bar{\mathbf{S}}_+$. Particularly if $\sup q = 1$, as is the case for branching diffusions with $\{\beta = 0\} \neq \emptyset$, the following transformation helps:

The functional $\bar{F}_t(x,\cdot)|\bar{S}$ given by

$$\overline{F}_t[\xi]: = \frac{F_t[q+(1-q)\xi]-q}{1-q},$$

$$\overline{F}_t(\hat{x},\xi): = 1; \qquad \hat{x} = \theta,$$

$$: = \prod_{\nu=1}^{n} \overline{F}_t[\xi](x_\nu); \quad \hat{x} = \langle x_1,\ldots,x_n\rangle,$$

generates the transition function of a Markov branching process on $(\hat{X},\hat{\mathfrak{U}})$. In fact, from (IF)

$$(\overline{\mathrm{IF}}) \qquad 1 - \overline{F}_t[\xi] = \overline{T}_t^0(1-\xi) + \int_0^t \overline{T}_s^0\, \overline{k}(1-\overline{f}[\overline{F}_{t-s}[\xi]])ds,$$

$$\overline{T}_t^0\xi: = \frac{T_t^0[(1-q)\xi]}{1-q}, \qquad \xi \in \mathfrak{B},$$

$$\overline{k}: = \frac{1-f[q]}{1-q}\, k,$$

$$\overline{f}[\xi]: = \frac{f[q+(1-q)\xi]-f[q]}{1-q}, \quad \xi \in \overline{\mathfrak{S}}.$$

Clearly, $\overline{T}_t^0$ is a non-negative contraction semigroup on B. It is stochastically continuous in $t \geq 0$ on B, and using the continuous differentiability of $p_t(x,y)$ with (T.5) and $1-q \in \mathfrak{D}_0^+$, we have $\overline{T}_t^0\mathfrak{B} \subset \mathbf{C}^0$ for $t > 0$. Hence $\overline{T}_t^0$ has a restriction to $\mathbf{C}^0$ which is strongly continuous in $t \geq 0$, cf. Dynkin (1965). Expanding

$$1 - f[q] = m(1-q) - r(q)(1-q)$$

in analogy to (FM), it follows from (C) that $\overline{k}$ is bounded. From (1.1)

$$1 = \overline{T}_t^0 1 + \int_0^t \overline{T}_s^0\overline{k}\, ds.$$

That is, the process determined (up to equivalence) by $\overline{T}_t^0$ is the sub-process, corresponding to stopping with density $\overline{k}$, of a conservative process, whose transition semigroup $\{T_t\}$ is simply the (unique) solution of

$$\overline{T}_t = \overline{T}_t^0 + \int_0^t \overline{T}_s^0\overline{k}\, \overline{T}_{t-s}\, ds.$$

REMARK. Assuming $1 - q \in \mathbf{C}_0^2$, we can formally calculate the differential generator of $\overline{T}_t$ as

$$\underset{\sim}{L}\xi = \frac{\underset{\sim}{A}[(1-q)\xi]}{1-q} - k\xi + k\,\frac{1-f[q]}{1-q}\,\xi .$$

Using

$$0 = \frac{\partial q}{\partial t} = \frac{\partial F_t[q]}{\partial t} = \underset{\sim}{A}F_t[q] + k(f(F_t[q]] - F_t[q])$$

$$= \underset{\sim}{A}q + k(f[q] - q)$$

this becomes

$$\underset{\sim}{L}\xi = \frac{\underset{\sim}{A}[(1-q)\xi]}{1-q} - \frac{\underset{\sim}{A}(1-q)}{1-q}\cdot\xi,$$

or explicitly for branching diffusions,

$$\underset{\sim}{L} = \underset{\sim}{A} - 2\sum_{i,j}\frac{a_{ij}}{1-q}\left(\frac{\partial}{\partial x_i}q\right)\frac{\partial}{\partial x_j},$$

$$\mathcal{D}(\underset{\sim}{L}) = \{\xi|_X : \xi \in C^2(\overline{\Omega}) \quad (1_{\{\beta=0\}}\xi + 1_{\{\beta>0\}}\frac{\partial \xi}{\partial n})|_{\partial\Omega} = 0\}.$$

That is, recalling $1 - q \in \mathcal{D}_0^+$, the transformation preserves reflecting barriers, turns elastic barriers into reflecting ones, and makes absorbing barriers inaccessible.

Let $\overline{\pi}$ be the stochastic kernel generated by $\overline{f}$ and $(\hat{x}_t, \hat{P}^x)$ the Markov branching process determined by $(\overline{T}_t^0, \overline{\pi})$. By definition, the extinction probability of $(\hat{x}_t, \hat{\overline{P}}^x)$ is zero,

$$\overline{F}_t[0] \equiv 0.$$

Its moment semigroup is given by

$$\overline{M}_t\xi(x) = \overline{E}^{\langle x\rangle}\hat{x}_t[\xi] = \lim_{\epsilon\to 1}\frac{\partial}{\partial\lambda}\overline{F}[\epsilon 1 + \lambda\xi](x)|_{\lambda=0}$$

$$= (1-q)^{-1}M_t[(1-q)\xi](x).$$

Defining

$$\overline{\varphi} := (1-q)^{-1}\varphi, \qquad \overline{\varphi}^* := (1-q)\varphi^*,$$

$$\overline{\Delta}_t\xi := (1-q)^{-1}\Delta_t[(1-q)\xi],$$

the following statement is an immediate consequence of (M) and $1-q \in \mathcal{D}_0^+$:

The semigroup $\{\overline{M}_t\}$ is stochastically continuous on $\mathcal{B}$ and strongly continuous on C^0 in $t \geq 0$, and it can be represented in the form

$$(\overline{M}) \qquad \overline{M}_t = \rho^t \overline{\varphi}\,\overline{\Phi}^* + \overline{\Delta}_t, \qquad t > 0,$$

$$\overline{\Phi}^*[\xi] := \int_X \overline{\varphi}^*(x)\xi(x)dx, \qquad \xi \in \mathcal{B},$$

with $\overline{\varphi}, \overline{\varphi}^* \in C^0$, $\inf \overline{\varphi} > 0$, $\inf \overline{\varphi}^* > 0$, and $\overline{\Delta}_t$: $\mathcal{B} \to \mathcal{B}$ such that

$$\overline{\varphi}\,\overline{\Phi}^* \Delta_t = \overline{\Delta}_t \overline{\varphi}\,\overline{\Phi}^* = 0,$$

$$-\alpha_t \overline{\varphi}\,\overline{\Phi}^* \leq \overline{\Delta}_t \leq \alpha_t \overline{\varphi}\,\overline{\Phi}^*, \qquad t > 0,$$

where ρ and α_t are the same as in (M). Recall that $\rho^{-t}\alpha_t \downarrow 0$, as $t \uparrow \infty$.

Similarly, we can expand

$$1 - \overline{F}_t[\xi] = \overline{M}_t[1-\xi] - \overline{R}_t(\xi)[1-\xi],$$

$$\overline{R}_t(\xi)[1-\xi] := (1-q)^{-1}\overline{R}_t(q+(1-q)\xi)[(1-q)(1-\xi)],$$

and obtain the following analog of (R):

For every $t > 0$ there exists a map $\overline{g}_t$: $\overline{S}_+ \to \mathcal{B}_+$, namely

$$\overline{g}_t[\xi] := g_t[q+(1-q)\xi]$$

such that

$$(\overline{R}) \qquad \overline{R}_t(\xi)[1-\xi] = \overline{g}_t[\xi]\overline{\Phi}^*(1-\xi]\overline{\varphi},$$

$$\lim_{\|1-\xi\|\to 0} \|\overline{g}_t[\xi]\| = 0,$$

uniformly in $t \in [a,b]$, $0 < a < b < \infty$.

Thus, we can switch freely between $\{\hat{x}_t, P^{\hat{x}}\}$ and $\{\hat{x}_t, \overline{P}^{\hat{x}}\}$, according to convenience. The advantage of the second process is its monotonicity, which follows from the fact that $\overline{T}_t$ is conservative and $\overline{f}[0] = 0$, i.e., $\overline{\pi}(x,\{\theta\}) \equiv 0$.

1.2. PROPOSITION. For every $n > 0$ and $\hat{x} \neq \theta$

$$\lim_{t\to\infty} \overline{P}^{\hat{x}}(\hat{x}_t[1] \leq n) = 0$$

and even

$$P^{\hat{x}}(\hat{x}_t[1] \to \infty, \quad \text{as} \quad t \to \infty) = 1.$$

PROOF. Irreducibility of $(m_{\nu\mu})$ implies irreducibility of $(\overline{m}_{\nu\mu})$,

$$\overline{m}_{\nu\mu} := \int_{X_\mu} \overline{k}(x)\overline{m}\, 1_{X_\nu}(x)dx.$$

Hence, $\{\overline{k} > 0\} \cap X_\nu$ has positive measure for every ν. Moreover, since $\rho > 1$, there exist a μ and a $\delta > 0$ such that even $\{\overline{k} > 0\} \cap \{x\colon \overline{\pi}(x,\hat{x}[1] > 1) \geq \delta\} \cap X_\mu$ has positive measure. Define

$$A_\nu := \{\overline{k} > 0\} \cap \{\overline{\pi}(\cdot,\hat{x}[1] > 1) \geq \delta\} \cap X_\nu, \quad \nu = \mu,$$

$$:= \{\overline{k} > 0\} \cap X_\nu, \qquad \nu \neq \mu,$$

$$\overline{\tau} := \inf\{t > 0\colon \hat{x}_{t-}[1] \neq \hat{x}_t[1]\}.$$

Then, using (T.1-5) and $1 - q \in \mathcal{D}_0^+$,

$$\overline{P}^{\langle x\rangle}\{\overline{\tau} \leq 1, \hat{x}_{\overline{\tau}-}[1_{A_\nu}] = 1\} = \int_0^1 \overline{T}_s^0\, \overline{k}1_{A_\nu}(x)ds$$

$$\geq e^{-\|\overline{k}\|} \int_0^1 (1-q)^{-1} T_s[(1-q)\overline{k}1_{A_\nu}](x)ds \geq \epsilon_\nu > 0, \qquad x \in X_\nu,$$

ϵ_ν independent of x. Hence, by irreducibility and monotonicity, there exists an integer n such that

$$\overline{P}^{\langle x\rangle}(\hat{x}_n[1] > 1) \geq \delta\epsilon^n, \qquad x \in X,$$

$$\epsilon := \min_\nu \epsilon_\nu.$$

From this, by homogeneity,

$$\overline{P}^{\langle x\rangle}(\hat{x}_{nm}[1] = 1) \leq (1-\delta\epsilon^n)^m, \qquad m = 1,2,\ldots\ .$$

The proposition now follows by the branching independence, the monotonicity, and in case of the second statement also the Borel-Cantelli lemma. □

1.3. COROLLARY. For all $\xi \in \; := \{\eta \in \mathcal{B} : \|\eta\| < 1\}$

$$\lim_{t\to\infty} \| \overline{F}_t[\xi] \| = 0.$$

PROOF. Pointwise convergence follows from (2.2) by

$$|\overline{F}_t[\xi](x)| \leq \sum_{n=1}^{\ell} \overline{P}^{\langle x\rangle}(\hat{x}_t[1]=n) \|\xi\|^n + (1-\|\xi\|)^{-1}\|\xi\|^{\ell+1},$$

convergence in norm from pointwise convergence by

$$\|\overline{F}_t[\xi]\| = \|\overline{F}_s[\overline{F}_{t-s}[\xi]] - \overline{F}_s[\overline{F}_{t-s}[0]]\|$$

$$\leq \|\overline{M}_s[|\overline{F}_{t-s}[\xi] - \overline{F}_{t-s}[0]|]\| = \|\overline{M}_s[|\overline{F}_{t-s}[\xi]|]\|$$

and $(\overline{M})$. □

1.4. COROLLARY. If $\xi = q + (1-q)\zeta$, $\zeta \in \mathcal{S}$, then

$$\lim_{t\to\infty} \| F_t[\xi] - q \| = 0.$$

1.5. PROPOSITION. For $\xi \in \overline{\mathcal{S}}_+$ with $\xi < 1$ on a set of positive Lebesgue measure

$$\lim_{t\to\infty} \| F_t[\xi] - q \| = 0.$$

PROOF. We have $F_t[0](x) \to q(x)$ for every x. As in the proof of 1.3, $\| q-F_t[0] \| \to 0$. Now fix ξ as assumed. Clearly,

$$F_t[0] \leq F_t[\xi] \leq F_t[\xi \vee q],$$

so that we may consider $\xi \vee q$ instead of ξ. By (F.2) and (2.4) it suffices to show for some $t > 0$ that $(1-q)^{-1}(1-F_t[\xi\vee q])$ is bounded from below by a positive constant. For $\eta \in \overline{\mathcal{S}}_+$ define

$$T_t^{(0)}\eta := T_t^0\eta, \qquad T_t^{(n+1)}\eta := \int_0^t T_s^0 k(1-f[1-T_{t-s}^{(n)}(1-\eta)])ds.$$

The irreducibility of $(m_{\nu\mu})$ and (T.1) imply the existence of an n such that $\{k(1-f[1-T_s^{(n)}(1-\xi\vee q)]) > 0\} \cap X_\nu$ has positive measure for all $s > 0$ and ν. Hence, using (IF) and (T.1-5),

$$\frac{1-F_t[\xi\ q]}{1-q} \geq \frac{e^{-\|k\|t}}{1-q} \int_0^t T_s(1-f[1-T_{t-s}^{(n)}(1-\xi\vee q)])ds \geq c_1 > 0,$$

which completes the proof. $\square$

1.6. COROLLARY. *For $\xi \in \bar{\mathcal{S}}_+$ with $\xi < 1$ on a set of positive measure*

$$\lim_{t\to\infty} \| \bar{F}_t[\xi] \| = 0.$$

2. NORMALIZING CONSTANTS

We continue within the setting of the previous section. What we shall actually use, are (M), (IF), continuity properties of $\{T_t^0\}$, the fact that $\{T_t^0\}$ has a representation similar to (M), further $q < 1$, and 1.5 for the first two propositions. Thereafter the results of the latter together with (M), (R), and 1.5 will suffice.

The aim is to find constants γ_t, $t \in T$, such that $\gamma_t \hat{x}_t[\varphi]$ converges at least in distribution to a proper, non-degenerate limit without any additional moment assumptions. Such normalizing constants can be obtained via solving the backward iterate problem for F_t. We call $(\xi_t)_{t \in \mathbb{R}_+} \subset \bar{\mathfrak{s}}_+$ a <u>sequence of backward iterates</u>, if

$$\xi_t = F_s[\xi_{t+s}], \qquad t,s \geq 0.$$

Such a sequence is <u>non-trivial</u>, if for some $t \geq 0$ neither $\xi_t = 1$ a.e., nor $\xi_t = q$ a.e.

2.1. PROPOSITION. <u>There exists a non-trivial sequence of backward iterates.</u>

PROOF. Let $\mathcal{E} := \{\xi \in \bar{\mathfrak{s}}_+ : \xi \geq q\}$. Since q and 1 are fixed points, $F_t[\mathcal{E}]$ is decreasing in t, by (F.2) and the monotonicity of $F_t[\xi]$ in ξ. The continuity of $F_t[\xi]$ in ξ implies connectedness and compactness of $F_t[\mathcal{E}]$ in the topology of pointwise convergence. Hence, $\mathcal{E}_\infty := \bigcap_{n \in \mathbb{N}} F_n[\mathcal{E}]$ is connected, and as $q, 1 \in \mathcal{E}_\infty$ and $q < 1$, there exists a $\xi_0 \in \mathcal{E}_\infty$ such that $q < \xi_0 < 1$ on a set of positive measure. By definition of $\mathcal{E}_\infty$, there exists for every $n \in \mathbb{N}$ a finite sequence $(\xi_{n,j})_{j=0,1,\ldots,n-1} \subset \mathcal{E}$ such that $\xi_{n,j} = F_1[\xi_{n,j+1}]$.

It follows from (IF) and the continuity properties of T_t^0 that the family $\{\xi_{n,j} : j < n,\ n \in \mathbb{N}\}$ is equicontinuous and thus, by Arzela's theorem, relatively compact in the topology of uniform convergence on $\mathcal{E} \cap C_0^0$. Hence, there exists for every $j \in \mathbb{N}$ a sequence $(\xi_{n_\ell, j})_{\ell \in \mathbb{N}}$, $n_\ell \to \infty$, as $\ell \to \infty$, converging in norm to some $\xi_j \in \mathcal{E}$, and by continuity of $F_1[\xi]$ in ξ, $\xi_j = F_1[\xi_{j+1}]$. Finally, define $\xi_t := F_{[t+1]-t}[\xi_{[t+1]}]$ for $t \in \mathbb{R}_+$ and recall (F.2). □

If $\xi < 1$ on a set of positive measure, it follows from (T.1) and the irreducibility of $(m_{\nu\mu})$, by iterating (IF), that $F_t[\xi] < 1$ on X for all $t > 0$. Similarly, if $\xi > q$ on a set of positive measure, it follows by iterating the equation obtained by subtracting (IF) from 1.1 that $F_t[\xi] > q$ on X for all $t > 0$. In particular, let (ξ_t) be a non-trivial sequence of backward iterates. By definition, $\xi_s < 1$ on a set of positive measure for some $s > 0$. Consequently, $\xi_t < 1$ on X for all $t < s$, but also $\xi_t < 1$ on a set of positive measure, depending on t, for $t > s$. Hence, $\xi_t < 1$ on X for all t. On the other hand, $\xi_t \geq F_n[0] \to q$, $n \to \infty$. That is, $\xi_t \geq q$ on X for all t. By definition $\xi_s > q$ on a set of positive measure. Hence, by the same argument as before, $\xi_t > q$ on X for all t.

2.2. PROPOSITION. *If (ξ_t) is a non-trivial sequence of backward iterates, then*

$$\lim_{t\to\infty} \| 1-\xi_t \| = 0.$$

PROOF. In view of

$$1-\xi_t = F_s[1] - F_s[\xi_{t+s}] \leq M_s[1-\xi_{t+s}]$$

and (M), it suffices to show $\Phi^*[1-\xi_t] \to 0$. Suppose $\Phi^*[1-\xi_t] \not\to 0$. Then there exist a ν, a sequence $t_j \uparrow \infty$, and a constant $c_2 > 0$ such that

$$\Phi^*[1_{X_\nu}(1-\xi_{t_j})] \geq c_2, \qquad j \in \mathbb{N}.$$

Restricted to functions on X_ν, the semigroup $\{T_t^0\}$ has a representation analogous to (M): The role of $k(m-1)$ is simply taken by $-k$. Thus for s fixed sufficiently large, there exists a $c_3 > 0$ such that

$$T_s^0\xi \geq c_3\Phi^*[1_{X_\nu}\xi]1_{X_\nu}\varphi, \qquad \xi \in \mathcal{B}_+.$$

Recalling (F.2) and (IF),

$$\xi_0 = F_{t_j-s}[F_s[\xi_{t_j}]] \leq F_{t_j-s}[1-T_s^0(1-\xi_{t_j})]$$

$$\leq F_{t_j-s}[1-c_2c_3 1_{X_\nu}\varphi], \qquad j \in \mathbb{N}.$$

By 1.5 the righthand side converges to q, as $j \to \infty$. This is a contradiction, and 2.2 is proved. □

<u>2.3</u>. PROPOSITION. *Let* (ξ_t) *be a non-trivial sequence of backward iterates, and define*

$$\zeta_t := -\log \xi_t, \qquad \gamma_t := \Phi^*[\zeta_t].$$

Then there exists, for every $t > 0$, *a sequence* (ε_t) *in* $\mathbb{B}$ *such that*

$$\zeta_t = (1+\varepsilon_t)\gamma_t\varphi, \qquad t > 0$$

$$\lim_{t\to\infty} \| \varepsilon_t \| = 0.$$

Furthermore,

$$\gamma_t = \rho^{-t}L(\rho^{-t}), \qquad t > 0,$$

where $L(s)$ *is slowly varying, as* $s \to 0$.

The statement follows from the next two lemmata, which will be needed again later.

<u>2.4</u>. LEMMA. *For every non-trivial sequence of backward iterates* (ξ_t) *there exists a sequence* $(\tilde{h}_t)$ *in* $\mathbb{B}$ *such that*

$$1-\xi_t = (1+\tilde{h}_t)\Phi^*[1-\xi_t]\varphi, \qquad t > 0,$$

$$\lim_{t\to\infty} \| \tilde{h}_t \| = 0.$$

PROOF. The proof is similar to that of VI.<u>1</u>.<u>1</u>. Using (F.2), (FM), (M), and (R),

$$\rho^s(1-\rho^{-s}\alpha_s-\| g_s[\xi_{t+s}] \|)\Phi^*[1-\xi_{t+s}]\varphi$$

$$\leq 1-F_s[\xi_{t+s}] \leq \rho^s(1+\rho^{-s}\alpha_s)\Phi^*[1-\xi_{t+s}]\varphi.$$

Combining these inequalities with those obtained by applying Φ^* to them yields

$$-\frac{2\rho^{-s}\alpha_s+\| g_s[\xi_{t+s}] \|}{1+\rho^{-s}\alpha_s}\varphi \leq \frac{1-F_s[\xi_{t+s}]}{\Phi^*[1-F_s[\xi_{t+s}]]} - \varphi$$

$$\leq \frac{2\rho^{-s}\alpha_s+\| g_s[\xi_{t+s}] \|}{1-\rho^{-s}\alpha_s-\| g_s[\xi_{t+s}] \|}\varphi.$$

Replace $F_s[\xi_{t+s}]$ by ξ_t, and let $t \to \infty$, using 2.2 and (R). Then let $s \to \infty$. □

2.5. LEMMA. *If* (ξ_t) *is a non-trivial sequence of backward iterates, then*

$$\lim_{t\to\infty} \frac{\Phi^*[1-\xi_t]}{\Phi^*[1-\xi_{t+s}]} = \rho^s, \qquad s > 0.$$

PROOF. Note that, by the definition of (ξ_t), (FM), (M), and (R),

$$\Phi^*[1-\xi_t] = \rho^s \Phi^*[1-\xi_{t+s}]\Phi^*[(1-g_s[\xi_{t+s}])\varphi]. \qquad □$$

PROOF OF 2.3. Note that $1-\xi_t = \zeta_t(1+\underset{\sim}{o}(\zeta_t))$, where $\| \underset{\sim}{o}(\xi) \| = \underset{\sim}{o}(\| \xi \|)$, recall 2.2, and apply 2.4-5. □

2.6. PROPOSITION. *Let* (ξ_t) *be a non-trivial sequence of backward iterates. Then there exists a random variable* W *such that*

$$\lim_{t\to\infty} \hat{\mathfrak{X}}_t[\zeta_t] = \lim_{t\to\infty} \gamma_t \hat{\mathfrak{X}}_t[\varphi] = W \quad \text{a.s.} \quad [P^{\hat{x}}],$$

$$P^{\hat{x}}(W = 0) = \hat{q}(\hat{x}), \qquad P^{\hat{x}}(W < \infty) = 1, \qquad \hat{x} \in \hat{X},$$

with $\Phi(s)(x) := E^{\langle x \rangle}\exp\{-sW\}$, $s \in \mathbb{R}_+$, $x \in X$, *satisfying*

$$(\Phi) \qquad \Phi(s\rho^t) = F_t[\Phi(s)], \qquad s,t \geq 0.$$

PROOF. Let $\mathfrak{F}_t = \sigma\{\hat{\mathfrak{X}}_s;\ s \leq t\}$. Then for $t,s \geq 0$

$$E^{\hat{x}}(\hat{\xi}_{t+s}(\hat{\mathfrak{X}}_{t+s})|\mathfrak{F}_t) = F_{t+s}(\hat{\mathfrak{X}}_t, \xi_{t+s}) = \hat{\xi}_t(\hat{\mathfrak{X}}_t) \quad \text{a.s.} \quad [P^{\hat{x}}].$$

By the martingale convergence theorem and 2.5, this implies the convergence statements. Using (F.2), 2.5, and dominated convergence,

$$\Phi(s\rho^t) = \lim_{u\to\infty} F_{u+t}[\exp\{-s\rho^t\gamma_{t+u}\varphi\}]$$
$$= F_t[\lim_{u\to\infty} F_u[\exp\{-s(\rho^t\gamma_u^{-1}\gamma_{t+u})\gamma_u\varphi\}]] = F_t[\Phi(s)].$$

It follows that $\Phi(0+)$ and $\Phi(\infty-)$ are fixed points of F_t in $C^0 \cap \bar{\mathfrak{S}}_+$. By 1.5 the only fixed points in $C^0 \cap \bar{\mathfrak{S}}_+$ are q and 1. Clearly, $\Phi(\infty-) \leq \Phi(1) \leq \Phi(0+)$. On the other hand,

$$\Phi(1) = \lim_{t\to\infty} F_t[e^{-\zeta_t}] = \xi_0 \in C^0 \cap \bar{S}_+,$$

and $q < \xi_0 < 1$ on X. Hence, $\Phi(\infty-) = q$ and $\Phi(0+) = 1$. $\square$

2.7. PROPOSITION. For every non-trivial sequence of backward iterates

$$0 < \lim_{t\to\infty} \rho^t \gamma_t = \gamma \leq \infty,$$

$$E^{\hat{x}} W = \gamma \hat{x}[\varphi], \qquad \hat{x} \neq \theta.$$

Here, $\gamma < \infty$ if and only if for some and thus all $t > 0$

(X LOG X) $\qquad \Phi^*[E^{\langle\cdot\rangle} \hat{x}_t[\varphi] \log \hat{x}_t[\varphi]] < \infty.$

PROOF. Using (F.2) and (FM),

$$\rho^t \Phi^*[1-\xi_t] = \rho^t \Phi^*[1-F_s[\xi_{t+s}]] \leq \rho^{t+s} \Phi^*[1-\xi_{t+s}].$$

Hence,

$$\lim_{t\to\infty} \rho^t \gamma_t = \lim_{t\to\infty} \rho^t \Phi^*[1-\xi_t] = \gamma$$

exists and is positive, possibly infinite.

Since $\{\rho^{-t} \hat{x}_t[\varphi]\}$ is a non-negative martingale respective $\{\sigma(\hat{x}_s;\ s \leq t)\}$, it converges a.s. to a limit W'. Define

$$\psi_t(\lambda)(x) := E^{\langle x\rangle} \exp\{-\rho^{-t} \hat{x}_t[\varphi]\lambda\} = F_t[\exp\{-\rho^{-t}\varphi\lambda\}](x),$$

$$\psi(\lambda)(x) := E^{\langle x\rangle} e^{-W'\lambda}, \qquad \lambda \geq 0,\ x \in X.$$

Then

$$\psi_{t+s}(\lambda) = F_t[\psi_s(\rho^{-t}\lambda)], \qquad t,s > 0,\ \lambda \geq 0, \tag{2.1}$$

$$\psi(\lambda) = F_t[\psi(\rho^{-t}\lambda)], \qquad \lambda > 0,\ t > 0.$$

The last equation implies

$$E^{\langle x\rangle} W' = M_t[\rho^{-t} E^{\langle\cdot\rangle} W'](x), \qquad x \in X,\ t > 0.$$

By (M) we therefore have either $E^{\langle x\rangle}W' = \varphi(x)$, $x \in X$, or $E^{\langle x\rangle}W' = 0$, $x \in X$. Given this alternative, the first occurs if and only if

$$\lim_{n\to\infty} \Phi^*[1-\psi_{nt}(s)] > 0. \tag{2.2}$$

for any $t\in T, s>0$. We show that (2.2) is equivalent to (X LOG X). W.l.o.g. $t = 1$. By (FM) and (2.1)

$$\begin{aligned}\Phi^*[1-\psi_n(s)] &= \Phi^*[1-F_1[\psi_{n-1}(\rho^{-1}s)]] \\ &= \rho\Phi^*[1-\psi_{n-1}(\rho^{-1}s)]\left\{1-\Phi^*\left[R_1(\psi_{n-1}(\rho^{-1}s)\frac{1-\psi_{n-1}(\rho^{-1}s)}{\Phi^*[1-\psi_{n-1}(\rho^{-1}s)]}\right]\rho^{-1}\right\} \\ &= \rho^{n-1}\Phi^*[1-\psi_1(\rho^{-n+1}s)]\prod_{\nu=1}^{n-1}\left\{1-\Phi^*\left[R_1(\psi_{n-\nu}(\rho^{-\nu}s))\frac{1-\psi_{n-\nu}(\rho^{-\nu}s)}{\Phi^*[1-\psi_{n-\nu}(\rho^{-\nu}s)]}\right]\rho^{-1}\right\}.\end{aligned}$$

Using (FM), (M), and (R),

$$\lim_{n\to\infty} \rho^{n-1}\Phi^*[1-\psi_1(\rho^{-n+1}s)] = s.$$

Now

$$1 - \psi_{n-\nu}(\rho^{-\nu}) \le M_{n-\nu}[1 - e^{-\varphi/\rho^n}] \le \rho^{-\nu}\varphi .$$

That is, if

$$\sum_{\nu} \Phi^*[R_1(1 - \rho^{-\nu}\varphi)\varphi] < \infty,$$

then $\lim \Phi^*[1 - \psi_n(1)] > 0$. If on the other hand (2.2) is satisfied, then by convexity

$$\rho^{-1}\Phi^*[1 - \psi_n(s)] \ge \varepsilon, \quad 0 < s \le s_0,\ n \ge n_0,$$

for some $\varepsilon > 0$, $s_0 > 0$, and $n_0 \in \mathbb{N}$. Hence

$$\Phi^*[1 - \psi_{n-\nu}(\rho^{-\nu})] \ge \varepsilon\rho^{-\nu}, \quad n-\nu \ge n_0,\ \nu \ge \nu_0,$$

for some ν_0. Using (FM), (M), and (R) again,

$$1 - \psi_{n-\nu}(\rho^{-\nu}) \ge \varepsilon'\Phi^*[1 - \psi_{n-m-\nu}(\rho^{-\nu-m})]\varphi, \quad n > \nu \ge \nu_1,$$

for some $\nu_1, m \in \mathbb{N}$ and $\varepsilon' > 0$. That is,

$$1 - \psi_{n-\nu}(\rho^{-\nu}) \ge \varepsilon'\varepsilon\rho^{-\nu-m}\varphi, \quad \nu \ge \nu_0 \vee \nu_1,\ n-\nu \ge n_0+m,$$

and thus

$$\sum_{\nu} \Phi^*[R_1(1 - \varepsilon'\varepsilon \rho^{-\nu} \varphi] < \infty .$$

Finally, recall V.4.3. □

REMARK. The second part of 3.7 can be proved also by a probabilistic argument using (M), but not (R). We shall return to this in section VII.5.

2.8. PROPOSITION. For any $t > 0$ the solutions of

$$\psi(\rho^t s) = F_t[\Psi(s)], \qquad s \geq 0,$$

in $\bar{S}_+$ are

$$\Psi_0(s) \equiv 1, \qquad \Psi_\alpha(s) = \Phi(\alpha s), \quad 0 < \alpha < \infty, \qquad \Psi_\infty(s) \equiv q.$$

PROOF. We have to show that, given a solution $\Psi(s)$ not identical 1 or q, there exists a positive real α such that $\Psi(s) = \Phi(\alpha s)$, $s > 0$. Consider the family of backward iterate sequences

$$\xi_t := \Phi(\rho^{-t}), \qquad \xi_t^{(s)} := \Psi(s\rho^{-t}), \quad s > 0.$$

Since by 2.6 the corresponding normalizing functions γ_t, $\gamma_t^{(s)}$, lead to non-degenerate, finite, strong limits W, $W^{(s)}$, we must have

$$\lim_{t\to\infty} \gamma_t^{-1} \gamma_t^{(s)} = \alpha(s) > 0, \qquad s > 0.$$

Using 2.5,

$$\lim_{t\to\infty} \frac{\gamma_t^{(a)}}{\gamma_t^{(b)}} = \lim_{t\to\infty} \frac{\Phi^*[1-\xi_t^{(a)}]}{\Phi^*[1-\xi_t^{(b)}]} = \lim_{t\to\infty} \frac{\Phi^*[1-\xi_t^{(a)}]}{\Phi^*[1-\xi^{(a)}_{t+\log_\rho(a/b)}]} = \frac{a}{b} .$$

That is, $\alpha(s) = \alpha s$, α a positive real constant. Accordingly,

$$(1-\delta_t)\alpha s\zeta_t \leq \zeta_t^{(s)} \leq (1+\delta_t)\alpha s\zeta_t,$$

$$\lim_{t\to\infty} \delta_t = 0.$$

Taking the exponential function and applying F_t,

$$\Phi_t((1-\delta_t)\alpha s) \leq F_t[\Psi(s\rho^{-t})] \leq \Phi_t((1+\delta_t)\alpha s).$$

Note that the middle term is equal to $\Psi(s)$, and let $t \to \infty$. □

2.9. COROLLARY. *For every non-trivial sequence of backward iterates* (ξ_t) *there exists an* $a \in \mathbb{R}$ *such that*

$$\xi_t = \Phi(\rho^{a-t}), \qquad t > 0.$$

3. EXTINCTION PROBABILITY AND TRANSIENCE CONTINUED

We continue with the considerations of Section VII.1 and with an additional indecomposibility assumption prove a rate of convergence result corresponding to 1.3. It will be used to obtain the behaviour of the distribution function of W near zero.

Let $\delta F_t[\eta;\xi]$ be the Fréchet derivative of F_t at $\eta \in \mathcal{S}$ in the direction of ξ. Since M_t is bounded, the pointwise limit

$$\delta F_t(q)\xi := \lim_{\epsilon \uparrow 1} \delta F_t[\epsilon q;\xi], \qquad \xi \in \mathcal{B}$$

defines a linear-bounded operator on $\mathcal{B}$. Since q is a fixed point of F_t for all t, $\{\delta F_t(q)\}$ is a semigroup. The same is true for $\{\delta \overline{F}_t(0)\}$. The two are connected through

$$(3.1) \qquad \delta \overline{F}_t(0)\xi = (1-q)^{-1}\delta F_t(q)[(1-q)\xi], \qquad \xi \in \mathcal{B},\ t > 0.$$

3.1. LEMMA. *Suppose the matrix* $(\overline{f}_{\nu\mu})$,

$$\overline{f}_{\nu,\mu} := \int_{X_\mu} \overline{k}(x)\delta\overline{f}(0)1_{X_\nu}(x)dx; \qquad \nu,\mu = 1,\ldots,K,$$

is irreducible. Then $\{\delta\overline{F}_t(0)\}$ *can be represented in the form*

$$\delta\overline{F}_t(0) = \sigma^t \overline{\psi}\,\overline{\Psi}^* + \overline{\Gamma}_t, \qquad t > 0,$$

$$\overline{\Psi}^*[\xi] = \int_X \overline{\psi}^*(x)\xi(x)dx, \qquad \xi \in \ ,$$

where $\sigma \in (0,1)$, $\overline{\psi},\overline{\psi}^* \in C^0$, $\inf \overline{\psi} > 0$, $\inf \overline{\psi}^* > 0$, *and* $\overline{\Gamma}_t: \mathcal{B} \to \mathcal{B}$ *such that*

$$\overline{\psi}\,\overline{\Psi}^*\overline{\Gamma}_t = \overline{\Gamma}_t\overline{\psi}\,\overline{\Psi}^* = 0,$$

$$-\beta_t\overline{\psi}\,\overline{\Psi}^* \leq \overline{\Gamma}_t \leq \beta_t\overline{\psi}\,\overline{\Psi}^*, \qquad t > 0,$$

$$\sigma^{-t}\beta_t \downarrow 0, \qquad t \uparrow \infty.$$

PROOF. It follows from $\overline{k}\delta\overline{f}(0)\xi = k(1-q)^{-1}\delta f(q)[(1-q)\xi]$ that irreducibility of $k\delta f(q)$ and $\overline{k}\delta f(0)$ are equivalent. From (IF)

$$\delta F_t(q)\xi = T_t^0\xi + \int_0^t T_s^0 k\delta f(q)\delta F_{t-s}(q)\xi\, ds.$$

Also,

$$\delta f(q) \leq m \qquad [\mathcal{B}_+].$$

Hence, the proof of (M) also applies to $\{\delta F_t(q)\}$. Application of (3.1) to the resulting representation leads to the proposed representation for $\{\delta\overline{F}_t(0)\}$. To see that $\sigma < 1$, note that

$$\delta\overline{F}_t(0)1_X(x) = \overline{P}^{\langle x\rangle}\{\hat{X}_t[1] = 1\} \to 0, \qquad x \in X,$$

by <u>1.2</u>, further

$$\delta\overline{F}_{t+s}(0)1_X(x) \leq (\sigma^s+\beta_s)\overline{\Psi}\,\overline{\Psi}^* [\delta\overline{F}_t(0)1_X],$$

so that $\| \delta\overline{F}_t(0)1_X \| \to 0, \quad t \to \infty.$ □

3.2. PROPOSITION. <u>There exists a functional</u> $\overline{Q}$ <u>on</u> $\overline{\mathcal{S}}_+$ <u>such that for all</u> $\xi \in \overline{\mathcal{S}}_+$

$$\sigma^{-t}\overline{F}_t[\xi] = (1+\delta_t[\xi])\overline{Q}[\xi]\overline{\Psi},$$

$$\lim_{t\to\infty} \| \delta_t[\xi] \| = 0.$$

<u>We have</u> $\overline{Q}[\xi] = 0$ <u>if and only if</u> $\xi = 0$ a.e., <u>while</u> $\overline{Q}[\xi] = \infty$ <u>if and only if</u> $\xi = 1$ a.e.

The proof will be based on three lemmata. Note that

$$(3.2)\qquad \begin{aligned} \overline{F}_t[\xi] &= \delta\overline{F}_t(0)\xi + \overline{G}_t[\xi], \\ \overline{G}_t[\xi] &:= \sum_{\nu=2}^{\infty} \frac{1}{\nu!}\, \delta^\nu\overline{F}_t[0;\xi,\ldots,\xi]. \end{aligned}$$

<u>3.3</u>. LEMMA. <u>For every</u> $t > 0$ <u>there exists a mapping</u> $a_t: \overline{\mathcal{S}}_+ \to \mathcal{B}$ <u>such that</u>

$$\overline{G}_t[\xi] = a_t[\xi]\sigma^t\overline{\Psi}\,\overline{\Psi}^*,$$

$$\lim_{\|\xi\|\to 0} \| a_t[\xi] \| = 0$$

<u>uniformly</u> <u>in</u> $t \in [a,b]$, $0 < a < b < \infty$.

PROOF. The proof resembles the proof of (R). We have

$$\overline{f}[\xi] = \delta\overline{f}(0)\xi + \overline{g}[\xi],$$

$$\overline{F}_t[\xi] = \overline{T}^0_t\xi + \int_0^t \overline{T}^0_s \overline{k}\,\overline{f}[\overline{F}_{t-s}[\xi]]ds,$$

$$\delta\overline{F}_t(0)\xi = \overline{T}^0_t\xi + \int_0^t \overline{T}^0_s \overline{k}\delta\overline{f}(0)\delta\overline{F}_{t-s}(0)\xi\, ds.$$

For every $\varepsilon > 0$ and $\xi \in \overline{\mathfrak{S}}_+$, $\overline{G}_t[\xi]$ is the only bounded solution in $[\varepsilon, \varepsilon+\lambda]$, $\lambda > 0$, of

$$v_t = A_t + B^\varepsilon_t + \int_0^{t-\varepsilon} \overline{T}^0_s \overline{k}\delta\overline{f}(0)v_{t-s}ds,$$

$$A_t := \int_0^t \overline{T}^0_s \overline{k}\,\overline{g}[\overline{F}_{t-s}[\xi]]ds,$$

$$B^\varepsilon_t := \int_0^\varepsilon \overline{T}^0_{t-s}\overline{k}\delta\overline{f}(0)\overline{G}_t[\xi]ds.$$

By the inherent positivity, this solution equals the limit of the iteration sequence $(v^{(\nu)}_t(x))_{\nu\in\mathbb{N}}$, $v^{(0)}_t \equiv 0$, which we now estimate.

By the mean-value theorem we have for every $\xi \in \overline{\mathfrak{S}}_+$

$$\overline{g}[\xi] \leq \overline{f}[\xi] \leq \delta\overline{f}(\xi)\xi \leq \overline{m}\xi,$$

$$\overline{G}_t[\xi] \leq \overline{F}_t[\xi] \leq \delta\overline{F}_t(\xi)\xi \leq \overline{M}_t\xi.$$

Fixing λ, let $0 < \delta < \varepsilon/2$ and $\varepsilon \leq t \leq \varepsilon+\lambda$. Then using $(\overline{M})$,(C),(C*),

$$A_t \leq \int_0^\delta + \int_{t-\delta}^\delta \overline{M}_s\overline{k}\,\overline{m}\,\overline{M}_{t-s}\xi\, ds + \int_\delta^{t-\delta} \overline{M}_s\overline{k}\,\overline{g}[\overline{M}_{t-s}\xi]ds$$

$$\leq \delta C_1(\varepsilon)\rho^t\overline{\varphi}\,\overline{\Phi}^*[\xi]$$

$$+ C_2(\delta,\varepsilon)\overline{\Phi}^*[\overline{g}[C_3(\delta)\overline{\Phi}^*[\xi]\varphi]]C_3(\delta)\rho^t\overline{\varphi}\,\overline{\Phi}^*[\xi],$$

where C_1, C_2, C_3 are constants depending on the choice of ε and δ, as indicated. Hence, there exists for every ε a functional Θ_ε on $\overline{\mathfrak{S}}_+$ such that

$$(3.3)\qquad \begin{aligned} &A_t \leq \Theta_\varepsilon[\xi]\rho^t\overline{\Phi}^*[\xi]\overline{\varphi},\\ &\lim_{\|\xi\|\to 0}\Theta_\varepsilon[\xi] = 0.\end{aligned}$$

Secondly,

$$(3.4)\qquad \begin{aligned} &B_t^\varepsilon \leq \overline{T}^0_{t-\varepsilon}\int_0^\varepsilon \overline{M}_{\varepsilon-s}\overline{k}\,\overline{m}\,\overline{M}_s\xi\,ds\\ &\overline{M}_{t-s}\int_0^\varepsilon \overline{M}_{\varepsilon-s}\overline{k}\,\overline{m}\,\overline{M}_s[\xi]ds \leq \varepsilon c_4(1+\rho^{-t+\varepsilon}\alpha_{t-\varepsilon})\rho^t\overline{\Phi}^*[\xi]\overline{\varphi}\ .\end{aligned}$$

Using (3.3), (3.4), $(\overline{M})$, and the fact that

$$\overline{\varphi} \leq c\overline{\Psi}, \qquad \overline{\Phi}^* \leq c^*\overline{\Psi}^* \qquad [\mathcal{B}_+],$$

we get

$$\lim_{t\to\infty} v_t^{(\nu)} \leq CC^*\left(\frac{\rho}{\sigma}\right)^t(e^{ct}\Theta_\varepsilon[\xi] + \varepsilon c_4(1+\rho^{-t+\varepsilon}\alpha_{t+\varepsilon}))\sigma^t\overline{\Psi}^*[\xi]\overline{\Psi}.$$

Since ε was arbitrary, this proves 3.3. □

3.4. LEMMA. *If* $(\overline{F}_{\nu\mu})$ *is irreducible, there exists for every* $t > 0$ *a mapping* $b_t\colon \overline{\mathcal{S}}_+ \to \mathcal{B}$ *such that*

$$\overline{F}_t[\xi] = (1 + b_t[\xi])\overline{\Psi}^*[\overline{F}_t[\xi]]\overline{\Psi},$$

$$\lim_{t\to\infty}\|b_t[\xi]\| = 0$$

for every $\xi \in \overline{\mathcal{S}}_+$ *with* $\xi < 1$ *on a set of positive measure.*

PROOF. The proof is similar to the proof of 2.4. From (F.2), (3.2),

$$\overline{F}_s[\overline{F}_{t-s}[\xi]] = \delta\overline{F}_s(0)\overline{F}_{t-s}[\xi] + \overline{G}_s[\overline{F}_{t-s}[\xi]],$$

and from this, by 3.1 and 3.3,

$$\begin{aligned}(1-\sigma^{-s}\beta_s)\sigma^s\overline{\Psi}^*[\overline{F}_{t-s}[\xi]]\overline{\Psi} &\leq \overline{F}_t[\xi]\\ &\leq (1+\sigma^{-s}\beta_s + \|a_s[\overline{F}_{t-s}[\xi]]\|)\sigma^s\overline{\Psi}^*[\overline{F}_{t-s}[\xi]]\overline{\Psi}.\end{aligned}$$

To estimate $(\overline{\Psi}^*[\overline{F}_t[\xi]])^{-1}\overline{F}_t[\xi] - \overline{\Psi}$, combine these two inequalities with those obtained by applying $\overline{\Psi}^*$ to them. First let $t \to \infty$, then

$s \to \infty$, recalling that $\| \overline{F}_t[0] \| \to 0$, as $t \to \infty$, by 1.5. □

3.5. LEMMA. *For* $t > 0$ *and* $\eta \in \mathbb{S}_+$ *let* $\sigma_t(\eta)$ *be the spectral radius of* $\delta\overline{F}_t(\eta)$. *Then*

$$\lim_{\|\eta\|\to 0} \sigma_t(\eta) = \sigma^t.$$

PROOF. For $t > 0$ and $\eta \in \mathbb{S}_+$

$$\delta\overline{F}_t(\eta)1 = \delta\overline{F}_t(0)1 + \sum_{\nu=1}^{\infty} \frac{1}{\nu!}\, \delta^{\nu+1}\overline{F}_t[0,\eta,\dots,\eta,1]$$
$$\leq \delta\overline{F}_t(0)1 + 2\| \eta \|(1-\| \eta \|)^{-2}1.$$

Hence, for $\| \eta \| \leq \frac{1}{2}$,

$$\| \delta\overline{F}_t(\eta)^n 1 \| \leq \| \sum_{\nu=0}^{n} \binom{n}{\nu} \| 8\eta \|^{n-\nu} \delta\overline{F}_t(0)^{\nu} 1 \|$$
$$\leq c_5 \sum_{\nu=0}^{n} \binom{n}{\nu} \| 8\eta \|^{n-\nu} \sigma^{t\nu}$$
$$= c_5(8\| \eta \| + \sigma^t)^n.$$

Since we are dealing with positive operators, this proves 3.5. □

PROOF. For $\xi \in \overline{\mathbb{S}}_+$, using (3.2),

$$\sigma^{-t-s}\overline{\Psi}^*[\overline{F}_{t-s}[\xi]]$$
$$= \sigma^{-t-s}\overline{\Psi}^*[\delta\overline{F}_s(0)\overline{F}_t[\xi]] + \sigma^{-t-s}\overline{\Psi}^*[\overline{G}_s[\overline{F}_t[\xi]]]$$
$$\geq \sigma^{-t}\overline{\Psi}^*[\overline{F}_t[\xi]].$$

That is, $\sigma^{-t}\overline{\Psi}^*[\overline{F}_t[\xi]]$ converges to some functional $\overline{Q}[\xi]$. Combined with 3.4, this implies $\sigma^{-t}\overline{F}_t[\xi] \to \overline{Q}[\xi]\overline{\Psi}$ in the way proposed.

If $\xi > 0$ on a set of positive measure, then by ($\overline{\text{IF}}$) and (T.1)

$$\overline{F}_t[\xi] \geq \overline{T}_t^0\xi \geq e^{-\|k\|t}(1-q)^{-1}T_t[(1-q)\xi] > 0$$

on a set of positive measure, thus $\overline{Q}[\xi] > 0$.

Finally, suppose $\xi < 1$ on a set of positive measure, $\xi \not\equiv 0$. Using (3.2),

$$\sigma^{-n}\Psi^*[F_n[\xi]] = \sigma^{-1}\Psi^*[\overline{F}_1[\xi]]\prod_{\nu=1}^{n-1}\{1+\sigma^{-1}\frac{\Psi^*[\overline{G}_1[\overline{F}_\nu[\xi]]]}{\Psi^*[\overline{F}_\nu[\xi]]}\},$$

$$\sum_{\nu=1}^{n}\frac{\Psi^*[\overline{G}_1[\overline{F}_\nu[\xi]]]}{\Psi^*[\overline{F}_\nu[\xi]]} \leq \sum_{\nu=1}^{n}(1-\|\overline{F}_\nu[\xi]\|)^{-1}\frac{\|\overline{F}_\nu[\xi]\|}{\Psi^*[\overline{F}_\nu[\xi]]}\|\overline{F}_\nu[\xi]\|.$$

In view of 3.4, it suffices to show that $\|\overline{F}_n[\xi]\| = O(\lambda^n)$ with some $\lambda < 1$, in order to secure that the limit of $\sigma^{-n}\Psi^*[\overline{F}_n[\xi]]$ is finite. Since $\overline{F}_t[0] \equiv 0$, (F.2) and the mean-value theorem yield

$$\overline{F}_{t+s}[\xi] \leq \delta\overline{F}_t(\overline{F}_s[\xi])\overline{F}_s[\xi].$$

Iterating this inequality, we get

$$\overline{F}_{n+j}[\xi] \leq \delta\overline{F}_1(\overline{F}_{j+n-1}[\xi])\delta\overline{F}_1(\overline{F}_{j+n-2}[\xi])\ldots\delta\overline{F}_1(\overline{F}_j[\xi])\overline{F}_j[\xi]$$

$$\leq \delta\overline{F}_1(\sup_{\nu\geq j}\|\overline{F}_\nu[\xi]\|)^n 1, \qquad n,j \in \mathbb{N}.$$

Recall that $\|\overline{F}_\nu[\xi]\| \to 0$, as $\nu \to \infty$, and apply 3.5. $\square$

4. PROPERTIES OF THE LIMIT DISTRIBUTION

We prove within the setting of VII.1-3 that the distribution function of W has a positive density and look at its behaviour near zero and infinity. As before let σ be the spectral radius of $\delta F_1(q)$, and define

$$\epsilon_0 := -\log_\rho \sigma .$$

4.1. PROPOSITION. *For every $\hat{x} \neq \theta$ there exists a measurable function $\hat{w}_x | \mathbb{R}_+$ such that*

$$P^{\hat{x}}(W \leq \lambda) = \hat{q}(\hat{x}) + \int_0^\lambda w_{\hat{x}}(u)du, \quad \lambda \geq 0.$$

If we have $\hat{x}[1]\epsilon_0 > 1$, or if (XLOGX) is satisfied, then $w_{\hat{x}}(u)$ is bounded and continuous in u.

It suffices to prove that the distribution function of the limit $\overline{W}$ for the transformed process $\{\hat{x}_t, \overline{P}^{\hat{x}}\}$ has a Lebesgue density $\overline{w}_{\hat{x}}$ and that in case $\hat{x}[1]\epsilon_0 > 1$, or (XLOGX) is satisfied, this density is bounded and continuous. Recall that $\delta F_1(q)$ and $\delta \overline{F}_1(0)$ have the same spectral radius, and note that (XLOGX) is equivalent to its analog for $\{\hat{x}_t, \overline{P}^{\hat{x}}\}$. Let $\overline{\Phi}(s)(x)$ denote the Laplace transform of $\overline{W}$ respectively $\overline{P}^{\langle x \rangle}$.

4.2. LEMMA. *For every bounded, closed interval $I \subset \mathbb{R}$ not containing zero,*

$$\sup_{t \in I} \| \overline{\Phi}(it) \| < 1.$$

PROOF. Suppose $|\overline{\Phi}(it)| \equiv 1$ for all t. Then

$$\overline{\Phi}(it) = e^{it\xi} ,$$

ξ measurable, finite. Inserting this into

$$(4.1) \qquad \overline{\Phi}(i\rho^s t) = \overline{F}_s[\overline{\Phi}(it)]$$

yields

$$\overline{E}^{\hat{x}} \exp\{it(\hat{x}_s[\xi] - \rho^s \hat{x}[\xi])\} = 0, \qquad t \geq 0.$$

From this

$$\hat{x}_s[\xi] = \rho^s \hat{x}_0[\xi] \quad \text{a.s.,}$$

$$\overline{M}_s \xi(x) = \overline{E}^{\langle x \rangle} \hat{x}_s[\xi] = \rho^s \xi(x), \qquad x \in X.$$

That is, $\xi = \alpha\overline{\varphi}$, where α is a constant. Since $\overline{W}$ is non-degenerate, $\alpha \neq 0$. However, since our process is super-critical and $\overline{\varphi}$ continuous, we cannot have $\hat{X}_s[\overline{\varphi}] = \rho^s \hat{X}_0[\overline{\varphi}]$ a.s. for all s. Hence, for some $t \neq 0$, $x \in X$,

$$\overline{\Phi}(it)(x) < 1.$$

Since $\overline{F}_s[\xi](x)$ is continuous in x, by $(\overline{IF})$ and the continuity and boundedness of $\overline{T}^0_s$, so is $\overline{\Phi}(it)(x)$, by (4.1). We can therefore find for every $x \in X$ a $\delta(x) > 0$, and for every t with $0 < |t| \leq \delta(x)$ a $U(x,t) \subset X$ of positive measure such that

$$|\overline{\Phi}(it)(y)| < 1, \qquad 0 \leq |t| \leq \delta(x), \qquad y \in U(x,t).$$

If $\eta \in \overline{\mathcal{S}}_+$ is positive on a set of positive measure in X_ν, then $\overline{T}^0_s\eta \geq e^{-\overline{k}s}\overline{T}_s\eta$, which is uniformly positive in X_ν, by (T.1-5) and $1-q \in \mathcal{D}^+_0$. Iterating

$$1-|\overline{\Phi}(\rho^s it)| \geq 1-\overline{F}_s[|\overline{\Phi}(it)|]$$

$$= \overline{T}^0_s(1-|\overline{\Phi}(it)|) + \int_0^s \overline{T}^0_u\overline{k}(1-\overline{f}[\overline{F}_{s-u}[|\overline{\Phi}(it)|]])du$$

and recalling the irreducibility of $(\overline{m}_{\nu\mu})$, we get

$$\| \overline{\Phi}(it) \| < 1, \qquad t \neq 0.$$

Since $\overline{\Phi}(it)$ is pointwise continuous in t and bounded, it follows by

$$\| \overline{\Phi}(it)-\overline{\Phi}(i(t+\epsilon)) \| \leq \| \overline{M}_s[|\overline{\Phi}(it\rho^{-s})-\overline{\Phi}(i(t+\epsilon)\rho^{-s})|] \|$$

and $(\overline{M})$ that $\overline{\Phi}(it)$ is strongly continuous in t. Hence $\|\overline{\Phi}(it)\|$ is continuous, and the proof is complete. □

<u>4.3</u>. LEMMA. *For every positive* $\epsilon < \epsilon_0$

$$\| \overline{\Phi}(it) \| = o(|t|^{-\epsilon}), \qquad t \neq 0.$$

PROOF. As in the proof of <u>3.2</u>

$$|\overline{F}_{n+j}[\overline{\Phi}(it)]| \leq \overline{F}_n[\overline{F}_j[|\overline{\Phi}(it)|]]$$

$$\leq \delta\overline{F}_1(\sup_{\nu\geq j}\| \overline{F}_\nu[|\overline{\Phi}(it)|] \|)^n\overline{F}_j[|\overline{\Phi}(it)|].$$

By 4.2, $\| \overline{\Phi}(it) \| \leq c_6 < 1$ on $[1,\rho]$, so that $\| \overline{F}_\nu[|\overline{\Phi}(it)|] \| \to 0$, as $\nu \to \infty$, uniformly in $t \in [1,\rho]$. According to 3.5, there then exists for every positive $\epsilon < \epsilon_0$ a j such that

$$\sup_{1 \leq t \leq \rho} \| \overline{F}_{j+n}[\overline{\Phi}(it)] \| = O(\rho^{-n\epsilon}),$$

or equivalently,

$$\sup_{1 \leq t \leq \rho} \| \overline{\Phi}(it\rho^{j+n}) \| = O(\rho^{-n\epsilon}).$$

Hence,

$$\sup_{\rho^{n+j} \leq t \leq \rho^{n+j+1}} \| \overline{\Phi}(it) \| \leq c_7 |t|^{-\epsilon}$$

and c_7 independent of n. □

4.4. COROLLARY. If (XLOGX) is satisfied, then

$$\| \tfrac{\partial}{\partial t}\overline{\Phi}(it) \| = O(|t|^{-1-\epsilon}), \qquad t \neq 0.$$

PROOF. Denoting $\partial\overline{\Phi}(s)/\partial s$ by $\overline{\Phi}'(s)$

$$\rho^{n+j}|\overline{\Phi}'(it\,\rho^{n+j})| \leq \delta\overline{F}_n(\overline{F}_j[|\overline{\Phi}(it)|])\delta\overline{F}_j(|\overline{\Phi}(it)|)|\overline{\Phi}'(it)| .$$

Also,

$$\delta\overline{F}_j(|\overline{\Phi}(it)|) \leq \overline{M}_j, \qquad [\mathcal{B}_+],$$

and by assumption of (XLOGX),

$$\overline{E}^{\langle x \rangle}\overline{W} = \overline{\varphi}(x), \qquad x \in X.$$

That is, $|\overline{\Phi}'(it)| < \overline{\varphi}$. Continue as in the proof of 4.3. □

PROOF OF 4.1. Given $\hat{x}_0[1]\epsilon_0 > 1$, or (XLOGX), the characteristic function of $\overline{W}$ is absolutely integrable, by 4.3-4, and this implies that the distribution function of $\overline{W}$ has a bounded continuous density.

Now let $\hat{x}_0[1]\epsilon_0 < 1$. For every $\hat{x} \neq \theta$ the probability space carrying $\{\hat{x}_t\}$ can be enlarged in such a way that it also carries a set of independent random variables $\overline{W}_x$, $x \in X$, which are independent of $\hat{x}_t$ and satisfy

$$\overline{P}^{\hat{x}}(\overline{W}_x \in I) = \overline{P}^{\langle x \rangle}(\overline{W} \in I), \qquad x \in X,$$

for every Borel set $I \subset \mathbb{R}_+$. Since the characteristic functions of $\rho^{-n}\hat{x}_n[\overline{W}_.]$ and $\overline{W}$ coincide, so do their distributions. That is, for every I of Lebesgue measure zero,

$$\overline{P}^{\hat{x}}(\overline{W} \in I) = \int_{\hat{X}} \overline{P}^{\hat{x}}(\hat{x}_n \in d\hat{y})\overline{P}^{\hat{x}}(\hat{y}[\overline{W}_.] \in \rho^n I)$$

$$\leq \overline{P}^{\hat{x}}(\hat{x}_n[1]\varepsilon_0 \leq 1) + \int_{\{\hat{y}[1]\varepsilon_0 > 1\}} \overline{P}^{\hat{x}}(\hat{x}_n \in d\hat{y})\overline{P}^{\hat{x}}(\hat{y}[\overline{W}_.] \in \rho^n I),$$

the first term on the right vanishes, as $n \to \infty$, by 1.2, and the second term is identical zero for all n. □

4.5. PROPOSITION. *For every* $\hat{x} \neq \theta$, *the density* $\overline{w}_{\hat{x}}$ *is positive on the positive reals.*

PROOF. Let $\lambda^{(n)}$ be the measure induced on $\mathfrak{A}^{(n)} := \{\hat{A} \cap X^{(n)} : \hat{A} \in \hat{\mathfrak{A}}\}$ by the Lebesgue measure on $\mathfrak{A}$, and let $\mathfrak{P}^{(n)}$ be the class of elements of $\mathfrak{A}^{(n)}$ which have positive $\lambda^{(n)}$-measure. Defining

$$\langle x_1,\ldots,x_n\rangle + \langle y_1,\ldots,y_j\rangle := \langle x_1,\ldots,x_n,y_1,\ldots,y_j\rangle ,$$

we have

$$\overline{w}_{\hat{x}+\hat{y}}(s) = \int_{0+}^{s} \overline{w}_{\hat{x}}(s-u)\overline{w}_{\hat{y}}(u)du .$$

Let $\overline{P}_t(\hat{x},\hat{A})$ be the transition function of $\{\hat{x}_t,\overline{P}^{\hat{x}}\}$. From (4.1)

$$\overline{w}_{\hat{x}}(s) = \int_{\hat{X}} \overline{P}_t(\hat{x},d\hat{y})\rho^t\overline{w}_{\hat{y}}(\rho^t s). \tag{4.2}$$

Step 1. For $\hat{x} \in X^{(n)}$, $n\varepsilon_0 > 1$, $\overline{w}_{\hat{x}}(s)$ is continuous, and since $\overline{W}$ is non-degenerate, it is also positive somewhere. That is, $\overline{w}_{\hat{x}} > 0$ on some open interval $I_{\hat{x}} \in \mathbb{R}_+$. It follows by (4.2) that for each $u \in \rho^t I_{\hat{x}}$ there exist a $j = j(\hat{x},u) \in \mathbb{N}$ and an $\hat{A}_{\hat{x},k} \in \mathfrak{P}^{(j)}$ such that $\overline{P}_t(\hat{x},\hat{A}_{\hat{x},u}) > 0$ and $\overline{w}_{\hat{y}}(u) > 0$ for $\hat{y} \in A_{\hat{x},u}$. Because of (T.1) and $k \in \mathfrak{B}$, there then exists a neighbourhood $\hat{U}_{s,\hat{x}} \in \mathfrak{P}^{(n)}$ of $\hat{x}$ on which $\overline{P}_t(\cdot,\hat{A}_{\hat{x},u}) > 0$, and thus

$$\overline{w}_{\hat{y}}(s) > 0, \qquad s \in I_{\hat{x}}, \qquad \hat{y} \in \hat{U}_{s,\hat{x}} . \tag{4.3}$$

Step 2. In view of (T.1), the irreducibility of $(m_{\nu\mu})$, and $\rho > 1$, there exists an integer d such that for $\hat{x} \neq \theta$

$$(4.4) \qquad \overline{P}_t(\hat{x},\hat{A}) > 0, \quad t > 0, \quad \hat{A} \in \mathfrak{B}^{(\varkappa d)}, \ \varkappa \in \mathbb{N}.$$

If $\overline{w}_{\hat{x}} > 0$ on I and $\overline{w}_{\hat{y}} > 0$ on J, then $\overline{w}_{\hat{x}+\hat{y}} > 0$ on $I+J := \{z = x+y : x \in I,\ y \in J\}$. That is, given any $s_0 > 0$, we can choose $\hat{y} \in X^{(\varkappa d)}$, $\varkappa d \epsilon_0 > 1$, such that $\overline{w}_{\hat{y}} > 0$ on some interval (a,b), $b > s_0$, and thus, by (4.2-4), $\overline{w}_{\hat{x}} > 0$ on $(0,b)$. Now let $s_0 \to \infty$. □

4.6. PROPOSITION. *There exists a function* $L(s)$ *on* $\mathbb{R}_+$, *slowly varying as* $s \to 0$, *such that for* $x \in X$

$$P^{\langle x \rangle}(W > \lambda) = o(\lambda^{-1}L(\lambda^{-1}))\varphi(x), \qquad \lambda \to \infty.$$

PROOF. By 2.4-5 with $\xi_t = \Phi(\rho^{-t})$,

$$1-\Phi(s) = (1+h_s^*)sL(s)\varphi,$$

$$\lim_{x\to 0} \| h_s^* \| = 0.$$

From this, by Karamata's Tauberian theorem, A 14.1,

$$\int_0^\lambda P^{\langle x \rangle}(W > u)du \sim L(\lambda^{-1})\varphi(x), \qquad \lambda \to \infty.$$

Now apply Seneta's version of the Tauberian theorem of Landau and Feller, A 14.2. □

4.7. PROPOSITION. *If* $(f_{\nu\mu})$ *is irreducible, then for* $x \in X$

$$P^{\langle x \rangle}(W \leq \lambda) - q(x) \sim \lambda^{\epsilon_0} \frac{\overline{Q}[\overline{\Phi}(1)]}{\Gamma(\epsilon_0+1)}(1-q(x))^{\epsilon_0}\psi(x), \quad \lambda \to 0,$$

where Γ *is the Gamma function and* $\psi := (1-q)\overline{\Psi}$. *If in addition* $\epsilon_0 > 1$, *then*

$$\overline{w}_{\langle x \rangle}(\lambda) \sim \epsilon_0 \lambda^{\epsilon_0 - 1} \frac{\overline{Q}[\overline{\Phi}(1)]}{\Gamma(\epsilon_0+1)} \overline{\Psi}(x), \qquad \lambda \to 0.$$

PROOF. From (Φ) and 3.2

$$\sigma^{-t}\overline{\Phi}(\rho^t)(x) = \sigma^{-t}\overline{F}_t[\overline{\Phi}(1)](x) \sim \overline{Q}[\overline{\Phi}(1)]\overline{\Psi}(x), \qquad t \to \infty.$$

That is,

$$\overline{\Phi}(s)(x) \sim s^{-\epsilon_0}\,\overline{Q}[\overline{\Phi}(1)]\overline{\Psi}(x), \qquad s \to \infty.$$

Now apply Karamaṭa's Tauberian theorem, to obtain the first statement. If $\varepsilon_0 > 0$, notice that $\overline{w}_{\langle x \rangle}(\cdot)$ is differentiable and by partial integration

$$\overline{w}'_{(x)}(\lambda) = \frac{1}{2\pi}\int_{-\infty}^{\infty} e^{iu\lambda}\Phi(iu)(x)du$$

$$= -\frac{1}{2\pi\lambda^3}\int_{-\infty}^{\infty} e^{iv}(w+i)\Phi'(i\,\tfrac{v}{\lambda})(x)dv = O(\lambda^{\varepsilon_0 - 2}) .$$

Now apply A 14.2. □

5. ALMOST SURE CONVERGENCE WITH GENERAL TEST FUNCTIONS

We want to prove without assumptions on moments of higher than the first order that $\gamma_t \hat{X}_t[\xi] \to \Phi^*[\xi] W$ a.s., as $t \to \infty$, for a large class of ξ. For this we assume a metric X, a process $\{\hat{X}_t, P^{\hat{X}}\}$ constructable from $\{T_t, k, \pi\}$ such that T_t has a bounded kernel w.r. to a finite measure on $[\alpha, s] \otimes X \otimes X$ whenever $s > \alpha > 0$, k and m are bounded, and (14) is satisfied with $\rho > 1$.

For weaker results such as P-convergence with normalization γ_t and a.s. or P-convergence with normalization ρ^{-t} a less explicit structure will do. For P-convergence the existence of the process $\{\hat{X}_t, P^{\hat{X}}\}$, (M) with $\rho > 1$, and the normalizing constants suffice. The same is true for a.s. skeleton convergence with normalization ρ^{-t}. For the transition from a.s. skeleton convergence to a.s. convergence in continuous time metrizability of X and right continuity of the process are needed.

5.1. THEOREM. *For every a.e. continuous* $\eta \in \mathbb{B}$

$$\lim_{t\to\infty} \gamma_t \hat{X}_t[\eta] = \Phi^*[\eta] W \quad \text{a.s. } [P^{\hat{X}}].$$

To prove 5.1 we first show

$$(5.1) \qquad \lim_{\mathbb{N} \ni n \to \infty} \frac{\hat{X}_{n\delta}[\xi]}{\hat{X}_{n\delta}[\varphi]} = \Phi^*[\xi] \quad \text{a.s. on } \{W > 0\}$$

for every $\delta > 0$ and then make the transition to the continuous time limit. Both steps rely on the following lemma.

5.2. LEMMA. *For* $0 < \delta \in \mathbb{R}_+$ *let* $Y_n^{\delta,j}, Z_n^{\delta,j}, \beta_n^{\delta}$, $j = 1, \ldots, \hat{X}_{n\delta}[1]$, $n = 0, 1, 2, \ldots$, *be random variables such that*

$$0 \leq Z_n^{\delta,j} \leq Y_n^{\delta,j}, \qquad \beta_n^{\delta} \geq 0 \quad \text{a.e. } [P^{\hat{X}}].$$

Suppose the $Z_n^{\delta,j}$ *are independent conditioned on* $\mathfrak{F}_{n\delta}$, *the same is true of the*

$$\tilde{Z}_n^{\delta,j} := Z_n^{\delta,j} 1_{\{Y_n^{\delta,j} \leq \beta_{n-1}^{\delta}\}}, \qquad i = 1, \ldots, \hat{X}_{n\delta}[1],$$

and the distribution $G^{\delta}_{\langle x \rangle}$ *of* $Y_n^{\delta,j}$ *depends only on* $\langle x_j \rangle := \hat{x}_{n\delta}^{n\delta,j}$,

$$\Phi^*[\int_0^\infty \lambda dG^\delta_{\langle\cdot\rangle}(\lambda)] < \infty.$$

Suppose further β_n^δ is $\mathfrak{F}_{n\delta}$-measurable, $\{\beta_n^\delta > 0\} \supset \{\beta_{n+1}^\delta > 0\}$,

(5.2) $$\liminf_{n\to\infty} (\beta_n^\delta)^{-1}\beta_{n+1}^\delta > 1 \quad \text{a.e. on} \quad \Gamma_\delta := \bigcap_{n\in\mathbb{N}} \{\beta_n^\delta > 0\},$$

and $(\beta_n^\delta)^{-1}\hat{\mathfrak{X}}_{n\delta}[\varphi]1_{\{\beta_n^\delta > 0\}}$ is bounded a.e. $[P^{\hat{x}}]$. Define

$$S_n^\delta := 1_{\{\beta_{n-1}^\delta > 0\}} (\beta_{n-1}^\delta)^{-1} \sum_{j=1}^{\hat{x}_{n\delta}[1]} Z_n^{\delta,j},$$

$$\tilde{S}_n^\delta := 1_{\{\beta_{n-1}^\delta > 0\}} (\beta_{n-1}^\delta)^{-1} \sum_{j=1}^{\hat{x}_{n\delta}[1]} \tilde{Z}_n^{\delta,j} .$$

Then

$$\lim_{n\to\infty} \{S_n^\delta - E^{\hat{x}}(\tilde{S}_n^\delta | \mathfrak{F}_{n\delta})\} = 0 \quad \text{a.s.} \quad [P^{\hat{x}}].$$

PROOF. Omitting the superscripts $\hat{x}$ and δ, setting $\delta = 1$ elsewhere, and using (M), (5.2), and $E^{\hat{x}}\hat{\mathfrak{X}}_t[\xi] = \hat{\mathfrak{X}}[M_t\xi]$,

(5.3) $$\sum_{n=1}^\infty E\{[\tilde{S}_n - E(\tilde{S}_n|\mathfrak{F}_n)]^2|\mathfrak{F}_{n-1}\}$$

$$\leq \sum_{n=1}^\infty E\{(\beta_{n-1})^{-2} \sum_{j=1}^{\hat{x}_n[1]} E((\tilde{Z}_n^j)^2|\mathfrak{F}_n)|\mathfrak{F}_{n-1}\}1_{\{\beta_{n-1} > 0\}}$$

$$= \sum_{n=1}^\infty (\beta_{n-1})^{-2}\hat{\mathfrak{X}}_{n-1}[M_1[\int_0^{\beta_{n-1}} \lambda^2 dG_{\langle\cdot\rangle}(\lambda)]]1_{\{\beta_{n-1} > 0\}}$$

$$\leq c_1 \sum_{n=1}^\infty \beta_{n-1}^{-1} \int_0^{\beta_{n-1}} \lambda^2 d\Phi^*[G_{\langle\cdot\rangle}(\lambda)]1_{\{\beta_{n-1} > 0\}}$$

$$\leq c_2 \int_0^\infty \lambda d\Phi^*[G_{\langle\cdot\rangle}(\lambda)] + c_3 ,$$

$$(5.4) \quad \sum_{n=1}^{\infty} P\{S_n \neq \tilde{S}_n | \mathfrak{F}_{n-1}\}$$

$$= \sum_{n=1}^{\infty} E\{ \sum_{i=1}^{\hat{x}_n[1]} P(Z_n^j > \beta_{n-1} | \mathfrak{F}_n) | \mathfrak{F}_{n-1}\} 1_{\{\beta_{n-1} > 0\}}$$

$$\leq \sum_{n=1}^{\infty} \hat{x}_{n-1}[M_1[\int_{\beta_{n-1}}^{\infty} dG_{\langle\cdot\rangle}(\lambda)]] 1_{\{\beta_{n-1} > 0\}}$$

$$\leq C_4 \sum_{n=1}^{\infty} \beta_{n-1} \int_{\beta_{n-1}}^{\infty} d\Phi^*[G_{\langle\cdot\rangle}(\lambda)] 1_{\{\beta_{n-1} > 0\}}$$

$$\leq C_5 \int_0^{\infty} \lambda d\Phi^*[G_{\langle\cdot\rangle}(\lambda)] + C_6 \ .$$

The $C_1, \ldots, C_6$ are finite, but in general random. Chebychev's inequality and the conditional Borel-Cantelli lemma complete the proof. □

PROOF OF (5.1). W.l.o.g. we again omit the superscript δ and set $\delta = 1$ elsewhere. Recall the representation

$$\hat{x}_{n+\ell}[\xi] = \sum_{j=1}^{\hat{x}_n[1]} \hat{x}_{n+\ell}^{n,j}[\xi] \quad \text{a.s. } [P^{\hat{x}}]$$

and set

$$\beta_n := \hat{x}_n[\varphi], \qquad 1_k := 1_{\{\beta_n > 0\}}, \qquad Z_{n,\ell}^j := \hat{x}_{n+\ell}^{n,j}[\xi].$$

<u>Step 1</u>. In the notation of <u>5.2</u> with $Z_n^j = Z_{n,\ell}^j$, $S_n = S_{n,\ell}$, $\tilde{S}_n = \tilde{S}_{n,\ell}$ we have

$$1_{n+\ell}\beta_{n+\ell}^{-1}\hat{x}_{n+\ell}[1]$$

$$= 1_{n+\ell}\beta_{n+\ell}^{-1}\beta_{n-1}(S_{n,\ell} - E^{\hat{x}}(\tilde{S}_{n,\ell}|\mathfrak{F}_n) - \epsilon_{n,\ell} + E^{\hat{x}}(S_{n,\ell}|\mathfrak{F}_n)),$$

$$\epsilon_{n,\ell} := E^{\hat{x}}(S_{n,\ell} - \tilde{S}_{n,\ell}|\mathfrak{F}_n).$$

Since

$$\lim_{n\to\infty} \beta_n^{-1}\beta_{n+\ell} = \lim_{n\to\infty} \gamma_{n+\ell}^{-1}\gamma_n = \rho^{\ell} > 1 \quad \text{a.s. on } \Gamma,$$

it follows by <u>5.2</u> that

$$\lim_{n\to\infty}(S_{n,\ell}-E^{\hat{x}}(\tilde{S}_{n,\ell}|\mathfrak{F}_n)) = 0 \quad \text{as} \quad [P^{\hat{x}}].$$

Step 2. Using (M),

$$\epsilon_{n,\ell} \leq 1_{n-1}\beta_{n-1}^{-1}\hat{x}_n[M_\ell[\xi]] \leq (\rho^\ell+\alpha_\ell)\beta_{n-1}^{-1}\beta_n\Phi^*[\xi].$$

Hence, in particular,

$$\limsup_{n\to\infty} \beta_n^{-1}\hat{x}_n[1] < \infty \quad \text{a.s. on} \quad \{W > 0\}.$$

Now

$$\epsilon_{n,\ell} \leq (1+\|\xi\|)1_n\beta_{n-1}^{-1}\hat{x}_n[1]\sup_{x\in X}\int_{\beta_{n-1}/\|\xi\|}^\infty \lambda \, dP^{\langle x\rangle}(\hat{x}_\ell[1]\leq\lambda).$$

From (IF)

$$\int_y^\infty \lambda \, dP^{\langle x\rangle}(\hat{x}_\ell[1]\leq\lambda) = \int_0^\ell T_s^0 kN_{\ell-s}^y(x)ds, \; y > 1,$$

$$N_t^y(x) := \int_{\hat{X}} \sum_{n\geq y} n\pi(x,d\hat{x})P^{\hat{x}}(\hat{x}_t[1] = n).$$

Notice that

$$\lim_{y\to\infty} N_s^y(x) = 0, \qquad s \geq 0, \; x \in X,$$

$$N_s^y \leq m[M_s[1]] \leq e^{\|kms\|}m[1], \qquad s \geq 0, \; y \geq 0.$$

Hence, by dominated convergence

$$\lim_{y\to\infty}\sup_{x\in X}\int_y^\infty \lambda \, dP^{\langle x\rangle}(\hat{x}_\ell[1]\leq\lambda) = 0,$$

using $T_s^0 \leq T_s$, $\|T_s\| \leq 1$, and the boundedness of $p_s(x,y)$ on $[\alpha,t] \otimes X \otimes X$ for $\alpha > 0$. Thus, since $\beta_n \to \infty$ on $\{W > 0\}$,

$$\lim_{n\to\infty} \epsilon_{n,\ell} = 0 \quad \text{a.s.} \; [P^{\hat{x}}] \quad \text{on} \quad \{W > 0\}, \; \ell > 0.$$

Step 3. We have

$$1_{n+\ell}\beta_{n+\ell}^{-1}\beta_{n-1}E^{\hat{x}}(S_{n,\ell}|\mathfrak{F}_n) = 1_{n+\ell}\beta_{n+\ell}^{-1}x_n[M^\ell[\xi]].$$

Again by (M) and (5.3),

$$\limsup_{n\to\infty} 1_{n+\ell}\beta_{n+\ell}^{-1}\hat{x}_n[M^\ell[\xi]] \leq (1+\rho^{-\ell}\alpha_\ell)\Phi^*[\xi],$$

$$\liminf_{n\to\infty} 1_{n+\ell}\beta_{n+\ell}^{-1}\hat{x}_n[M^\ell[\xi]] \geq (1-\rho^{-\ell}\alpha_\ell)\Phi^*[\xi] .$$

Now let $\ell \to \infty$. □

5.3. REMARK. With normalization $\beta_t := \rho^t$ (and thus a totally degenerate limit W' if (XLOGX) is not satisfied) step 2 can be taken without explicit reference to (IF) and properties of $\{T_t^0\}$. In fact, all that is needed in this case is (M) with $\rho > 1$:

Let η_n, $\eta \in \mathcal{B}$ such that $0 \leq \eta_n \leq \eta$. If

$$\epsilon'_{n,\ell} := \rho^{-n} \sum_{j=1}^{\hat{x}_n[1]} E(x_{n+\ell}^{n,j}[\eta_{n+\ell}]1_{\{\hat{x}_{n+\ell}^{n,j}[\eta] > \rho^n\}}|\mathfrak{F}_n) \xrightarrow{\text{a.s.}} 0,$$

then as before

$$\rho^{-n}\hat{x}_n[\eta_n] - \Phi^*[\eta_n]W' \xrightarrow{\text{a.s.}} 0 .$$

That is, if we can show that the choice

$$\eta_n(x) := E^{\langle x\rangle}\hat{x}_\ell[\xi]1_{\{\hat{x}_\ell[\xi] > \rho^n\}}$$

results in $\epsilon'_{n,\ell} \to 0$, then

$$\limsup_{n\to\infty} \epsilon'_{n,\ell} = \limsup_{n\to\infty} \Phi^*[\eta_n]W' = 0.$$

In fact,

$$\eta_n \leq \rho^\ell(1+\rho^{-\ell}\alpha_\ell)\Phi^*[\xi]\varphi ,$$

i.e., $\eta \leq c_1\varphi$. That is,

$$0 \leq \epsilon'_{n,\ell} \leq c_2\rho^{-n}\hat{x}_n[\xi_{n,\ell}] \leq c_3\rho^\ell\rho^{-n}\hat{x}_n[\varphi],$$

$$\xi_{n,\ell}(x) := E^{\langle x\rangle}\hat{x}_\ell[\varphi]1_{\{\hat{x}_\ell[\varphi] > \rho^n\}} .$$

Hence, if (XLOGX) is not satisfied, $\epsilon'_{n,\ell} \to 0$ a.s. If (XLOGX) is satisfied,

$$\sum_{n=0}^{\infty} E\epsilon'_{n,\ell} \leq c_4 \sum_{n=0}^{\infty} \Phi^*[\xi_{n,\ell}]$$

$$= c_4 \int_0^{\infty} y \sum_{n=0}^{\infty} 1_{(\rho^n,\infty)}(y) d\Phi^*[\hat{x}_\ell[\varphi] \leq y] < \infty,$$

where we have used the fact that

$$\sum_{n=0}^{\infty} 1_{(\rho^n,\infty)}(y) \sim \frac{\log y}{\log \rho}, \qquad y \to \infty.$$

So again $\epsilon'_{n,\ell} \to 0$ a.s. Notice that the argument goes through for $\eta, \eta_n, \xi \in L^1_{\Phi^*}$.

5.4. LEMMA. *Let* X *be metric. If* $\{\beta_t, t \in \mathbb{R}_+\}$ *is a rightcontinuous process such that the* $\beta_n^\delta := \beta_{n\delta}$, $n \in \mathbb{N}$, $\delta > 0$, *satisfy the assumptions of* 5.2 *with*

$$\lim_{t\to\infty} \beta_t^{-1}\beta_{t+s} = \beta^s \quad \text{a.s. on} \quad \Gamma = \bigcap_{t\geq 0} \{\beta_t > 0\}, \qquad s > 0,$$

and $\widetilde{W}$ *a random variable such that*

$$\lim_{\mathbb{N}\ni n\to\infty} \beta_{n\delta}^{-1}\hat{x}_{n\delta}[\xi] = \Phi^*[\xi]\widetilde{W} \quad \text{a.s. on} \quad \Gamma \tag{5.5}$$

for every $\delta > 0$ *and* $\xi \in \mathbb{B}_+$, *then*

$$\lim_{t\to\infty} \beta_t^{-1} x_t[\eta] = \Phi^*[\eta]\widetilde{W} \quad \text{a.s. on} \quad \Gamma$$

for any almost everywhere continuous $\eta \in \mathbb{B}$.

PROOF. For every $U \in \mathfrak{U}$ define

$$\xi_U^\delta(x) := P^{\langle x\rangle}(\hat{x}_t[1_U] = \hat{x}_t[1] \,\forall\, t \in [0,\delta]).$$

Clearly, $\xi_U^\delta(x) \uparrow 1_U(x)$, as $\delta \downarrow 0$, for every $x \in X$. Set

$$Z_n^{\delta,j} := 1_{\{\hat{x}_t^{n\delta,j}[1_U] = \hat{x}_t^{n\delta,j}[1] \,\forall\, t\in[n\delta,(n+1)\delta]\}}, \qquad Y_n^{\delta,j} = 1.$$

Then

$$\hat{x}_t[1_U] \geq S_n^\delta, \qquad t \in [n\delta,(n+1)\delta],$$

and by 5.2 and (5.5)

$$(5.6)\qquad \liminf_t \beta_t^{-1}\hat{x}_t[1_U] \geq \beta^{-\delta} \liminf_n S_n^\delta$$

$$= \beta^{-\delta} \liminf_n E^{\hat{x}}(\tilde{S}_n^\delta | \mathfrak{F}_{n\delta}) = \beta^{-\delta} \liminf_n E^{\hat{x}}(S_n^\delta | \mathfrak{F}_{n\delta})$$

$$= \beta^{-\delta} \liminf_n \beta_{n\delta}^{-1}\hat{x}_{n\delta}[\xi_U^\delta] = \beta^{-\delta}\Phi^*[\xi_U^\delta]\tilde{W} \uparrow \Phi^*[1_U]\tilde{W},\ \delta \downarrow 0$$

a.e. on Γ.

Next, set

$$Y_n^{\delta,j} = Z_n^{\delta,j} = \hat{x}_{(n+1)\delta}^{n\delta,j}[1] + \#\{t: \hat{x}_{t-}^{n\delta,j}[1] > \hat{x}_t^{n\delta,j}[1], n\delta < t \leq (n+1)\delta\}.$$

Then

$$E^{\langle x\rangle} Y_{o,1}^\delta \leq e^{\alpha\delta}, \qquad \alpha = \|k\|\cdot(\|m\|+1),$$

and again by 5.2 and (5.5)

$$(5.7)\qquad \limsup_t \beta_t^{-1}\hat{x}_t[1] \leq \beta^{-\delta} \limsup_n S_n^\delta$$

$$= \beta^{-\delta} \limsup_n E^{\hat{x}}(\tilde{S}_n^\delta | \mathfrak{F}_{n\delta}) \leq \beta^{-\delta} \limsup_n E^{\hat{x}}(S_n^\delta | \mathfrak{F}_{n\delta})$$

$$\leq e^{\alpha\delta} \limsup_n \beta_{n\delta}^{-1}\hat{x}_{n\delta}[1] = e^{\alpha\delta}\Phi^*[1]\tilde{W} \downarrow \Phi^*[1]\tilde{W} \quad \text{a.e.}\ \Gamma,\ \delta \downarrow 0.$$

From (5.6) and (5.7) with $U = X$

$$\lim_{t\to\infty} \beta_t^{-1}\hat{x}_t[1] = \Phi^*[1]\tilde{W} \quad \text{a.e. on } \Gamma,$$

and from this and (5.6) for any U with a boundary of measure zero

$$\limsup_t \beta_t^{-1}\hat{x}_t[1_U] = \Phi^*[1]\tilde{W} - \liminf_t \beta_t^{-1}\hat{x}_t[1_{U'}]$$

$$\leq \Phi^*[1_U]\tilde{W} \quad \text{a.e. on } \Gamma.$$

Now take an appropriate denumerable class of such U's and apply a standard result on weak convergence. □

5.5. REMARK. All we have used for the transition from discrete skeletons to continuous parameter was a separable metric X, right-continuity, and the existence of non-negative random variables Γ_t, $t \geq 0$, such that $\hat{x}_s[1] \leq \Gamma_t\ \forall\, s \in [0,t], t \geq 0$, and $\|E^{\langle\cdot\rangle}\Gamma_t\| \downarrow 1$, as $t \downarrow 0$.

5.6. REMARK. Matters simplify considerably if we restrict overselves to proving P-convergence. Proceeding as in the proof of 5.2, but omitting the summation over n and replacing n with $t \in T$, we get

$$S_t^\delta - E^{\hat{x}}(\tilde{S}_t^\delta | \mathfrak{F}_t) \xrightarrow{p} 0, \quad t \to \infty,$$

so that

$$S_{t,\ell} - E^{\hat{x}}(\tilde{S}_{t,\ell} | \mathfrak{F}_t) \xrightarrow{p} 0, \quad t \to \infty.$$

Moreover,

$$E^{\hat{x}}(S_{t,\ell} - \tilde{S}_{t,\ell} | \mathfrak{F}_t) = 1_{t-1} \beta_{t-1}^{-t} \hat{x}_t [\int_{\beta_{t-1}}^{\infty} \lambda \, dP^{\langle\cdot\rangle}(\hat{x}_\ell[\xi] \leq \lambda)],$$

$$E 1_{\{W>0\}} \epsilon_{t,\ell} = E\{1_{t-1} \beta_{t-1}^{-1} E(1_{\{W>0\}} \hat{x}_t [\int_{\beta_{t-1}}^{\infty} \lambda \, dP^{\langle\cdot\rangle}(\hat{x}_\ell[\xi] \leq \lambda)] | \mathfrak{F}_{t-1})\}$$

$$\leq (\rho + \alpha_1) E \, 1_{\{W>0\}} \int_{\beta_{t-1}}^{\infty} \lambda \, d\Phi^*[P^{\langle\cdot\rangle}(\hat{x}_\ell[\xi] \leq \lambda)] \to 0, t \to \infty.$$

Since $\epsilon_{t,\ell} \geq 0$, this implies $\epsilon_{t,\ell} 1_{\{W>0\}} \xrightarrow{p} 0$, $t \to \infty$, for any T. That is we only need (M) with $\rho > 1$ and the normalizing constants. Again we can admit $\xi \in L^1_{\Phi^*}$.

We conclude with a probabilistic proof of 2.7. It is more laborious than the analytic proof, but it does not require (R).

5.7. PROBABILISTIC PROOF OF 2.7. Assume only (M) with $\rho > 1$. First, a direct proof that (XLOGX) is satisfied with a fixed $t > 0$ if and only if it holds for all $t > 0$: Set

$$\log^* := x/e; \quad x \in [0,e],$$
$$:= \log x; \quad x \in (e,\infty),$$
$$\iota_t^*(x) := E^{\langle x\rangle} \hat{x}_t[\varphi] \log^* \hat{x}_t[\varphi],$$
$$I_t^*(x) := \Phi^*[\iota_t^*].$$

Since $x \log^* x$ is $\hat{x}_t[\varphi]$ is a submartingale, I_t^* is non-decreasing in t. Thus it suffices to show that (XLOGX) for some $t > 0$ implies (XLOGX) for 2t. Using V.4.6,

$$\iota^*_{2t}(x) = E^{\langle x\rangle} E^{\langle x\rangle}(\hat{x}_{2t}[\varphi]\log^*\hat{x}_{2t}[\varphi]\,|\,\mathfrak{F}_t)$$
$$\leq E^{\langle x\rangle}(\rho^t\hat{x}_t[\varphi]\log^*\rho^t\hat{x}_t[\varphi]+\hat{x}_t[\iota^*_t])$$
$$\leq c_1+c_2\iota^*_t(x)+c_3 I^*_t$$

with some constants c_1, c_2, c_3. Now apply Φ^*.

Let $\xi_{n,\ell}$ be as in 5.3, and set

$$\tilde{\varepsilon}_{n,\ell} := \rho^{-(n+1)\ell}\hat{x}_{n\ell}[\xi_{n\ell,\ell}],$$

$$W_n := \rho^{-n}\hat{x}_n[\varphi],$$

$$\tilde{W}_{n+1} := \rho^{-n-1}\sum_{j=1}^{\hat{x}_n[1]} \hat{x}^{n,j}_{n+1}[\varphi]1_{\{\hat{x}^{n,j}_{n+1}[\varphi]\leq\rho^n\}}.$$

Then

$$E(\tilde{W}_{n+1}|\mathfrak{F}_n) = W_n-\tilde{\varepsilon}_{n,1},$$

so that $\tilde{W}_{n+1}-W_n-\tilde{\varepsilon}_{n,1}$ is a martingale increment and

$$E(\tilde{W}_{n+1}-W_n-\tilde{\varepsilon}_{n,1})^2 \leq E\tilde{W}^2_{n+1} \leq \operatorname{Var}(\tilde{W}_{n+1}-E(\tilde{W}_{n+1}|\mathfrak{F}_n)).$$

It follows by (5.8) and the convergence theorem for $\mathcal{L}^2$-bounded martingales, A 2.3, that

(5.8) $$\sum_{n=0}^{\infty}(\tilde{W}_{n+1}-W_n+\tilde{\varepsilon}_{n,1})$$ <u>converges</u> <u>a.s.</u> <u>and</u> <u>in</u> $\mathcal{L}^1$.

Suppose (XLOGX) is satisfied. Then, as in 5.3 for $\varepsilon'_{n,1}$,

$$\sum_{n=0}^{\infty} E\,\tilde{\varepsilon}_{n,1} < \infty.$$

Since $\tilde{\varepsilon}_{n\,1} \geq 0$, it follows that $\Sigma_n\varepsilon_{n,1}$ and thus by (5.8) also $\Sigma_n(\tilde{W}_{n+1}-W_n)$ converges in $\mathcal{L}^1$. As $\tilde{W}_{n+1} \leq W_{n+1}$, we get

$$E^{\langle x\rangle}W' = E^{\langle x\rangle}(W_0+\sum_{n=0}^{\infty}\{W_{n+1}-W_n\})$$
$$\geq E^{\langle x\rangle}(W_0+\sum_{n=0}^{N}\{W_{n+1}-W_n\}+\sum_{n=N+1}^{\infty}\{\tilde{W}_{n+1}-W_n\})$$

$$= \varphi(x) + E^{\langle x\rangle}\Big(\sum_{n=N+1}^{\infty}\{\widetilde{W}_{n+1} - W_n\}\Big) \xrightarrow[N\to\infty]{} \varphi(x),$$

so that $E^{\langle x\rangle}W' \geq \varphi(x)$. From $E^{\langle x\rangle}W_n = \varphi(x)$ and Fatou's lemma $E^{\langle x\rangle}W' \leq \varphi(x)$.

The same argument that led to (5.8) also yields a.s. convergence of $\Sigma_n\{\widetilde{W}_{(n+1)\ell} + W_{n\ell} - \widetilde{\varepsilon}_{n,\ell}\}$. By (5.4)

$$\sum_{n=0}^{\infty} P(\widetilde{W}_{(n+1)\ell} \neq W_{(n+1)\ell}) < \infty,$$

so that also $\Sigma_n\{W_{(n+1)\ell} - W_{n\ell} - \widetilde{\varepsilon}_{n,\ell}\}$ converges a.s. Since $W_{n\ell} \to W'$ a.s., this implies that

(5.9) $$\Sigma_n \widetilde{\varepsilon}_{n,\ell} \ \underline{\text{converges}}\ \underline{\text{a.s.}}$$

Furthermore,

$$\xi_{n\ell,\ell}(x) = E^{\langle x\rangle} E(\hat{X}_\ell[\varphi] 1_{\{\hat{X}_\ell[\varphi] > \rho^{n\ell}\}} | \mathfrak{F}_{\ell-1})$$

$$\geq E^{\langle x\rangle} E\Big(\sum_{j=1}^{\hat{X}_{\ell-1}[1]} \hat{X}_\ell^{\ell-1,j}[\varphi] 1_{\{\hat{X}_\ell^{\ell-1,j}[\varphi] > \rho^{n\ell}\}} | \mathfrak{F}_{\ell-1}\Big)$$

$$= E^{\langle x\rangle} \hat{X}_{\ell-1}[\xi_{n\ell,1}] \geq \rho^{\ell}(1-\rho^{-\ell}\alpha_\ell)\Phi^*[\Phi_{n\ell,1}]\varphi(x).$$

That is, for sufficiently large ℓ,

(5.10) $$\widetilde{\varepsilon}_{n,\ell} \geq c\ \Phi^*[\xi_{n\ell,1}]W_{n\ell}$$

with some $c > 0$.

Now suppose (XLOGX) does not hold. Then

$$\sum_{n=0}^{\infty} \Phi^*[\xi_{n\ell,1}] = \infty \quad \text{a.s.},$$

c.f. <u>5</u>.<u>3</u>. Combining (5.9) and (5.10),

$$\infty > \sum_{n=0}^{\infty} \widetilde{\varepsilon}_{n,\ell} \geq c' \inf_n W_{n\ell} \sum_{n=0}^{\infty} \Phi^*[\xi_{n\ell,1}] \geq 0,$$

where ℓ can be chosen so large that $c' > 0$. Since $W_{n\ell} \to W$ a.s., this implies $P(W > 0) = 0$. □

6. SUPERCRITICAL PROCESSES WITH IMMIGRATION

As in VI.6 let $\{\hat{x}_t, P^{\hat{x}}\}$ be a Markov branching process, $\{\tau_\nu, \hat{y}_\nu, P\}$ an immigration process, and $\{z_t, \tilde{P}\}$ the corresponding immigration-branching process. We admit $T = \mathbb{N}$, or $T = (0,\infty)$, and assume that (M) is satisfied with $\rho > 1$.

Let there exist a sequence of backward iterates $(\xi_t)_{t \in T}$, set

$$\gamma_t := \Phi^*[-\log \xi_t],$$

and suppose that

$$\gamma_t^{-1} \gamma_{t-s} \xrightarrow[t\to\infty]{} \rho^s, \quad s \geq 0,$$

$$\gamma_t \hat{x}_t[\xi] \xrightarrow[t\to\infty]{} \Phi^*[\xi] W \quad \text{a.s. } [P^{\hat{x}}],$$

$$P^{\hat{x}}(W < \infty) = 1,$$

for $\xi \in \mathfrak{B}$, if $T = \mathbb{N}$, and for a.e. continuous $\xi \in \mathfrak{B}$, if $T = (0,\infty)$.

If $T = (0,\infty)$ and $\inf \varphi = 0$, we assume in addition that the regularity conditions formulated in 5.5 are satisfied.

Let $\tilde{\mathfrak{F}}_t$ be the σ-algebra generated by $\{\hat{z}_s;\ s \leq t\}$, and define

$$W_\nu := \lim_{t\to\infty} \gamma_{t-\tau_\nu} \hat{x}_{\nu,t}[\varphi],$$

$$\tilde{W}^* := \sum_{\nu=1}^{\infty} \rho^{-\tau_\nu} W_\nu .$$

6.1. PROPOSITION. *If*

$$\sum_{\nu=1}^{\infty} \gamma_{\tau_\nu} \hat{y}_\nu[\varphi] < \infty ,$$

then

$$\tilde{W} = \lim_{t\to\infty} \gamma_t \hat{z}_t[\varphi]$$

exists almost surely, and

$$\tilde{W} = \tilde{W}^* < \infty \quad \text{a.s.} \quad [\tilde{P}] .$$

PROOF. We condition throughout on

$$\mathfrak{Z} := \sigma(\tau_\nu, \hat{y}_\nu; \nu \in \mathbb{N}) .$$

Step 1. The limit $\tilde{W}^*$ always exists, but may be infinite. Since

$$\Big(\prod_{\nu \le N_t} \hat{\xi}_t(\hat{x}_{\nu,t}), \tilde{\mathfrak{Z}}_t \Big)$$

is a positive supermartingale, $\tilde{W}$ also exists, but again may be infinite.

Step 2. We first show that $\tilde{W}$ and $\tilde{W}^*$ are finite almost everywhere. The Laplace transform of the distribution function of $z_t[-\log \xi_t]$ is

$$\Psi_t(s) = \prod_{\nu \le N_t} F_{t-\tau_\nu}(\hat{y}_\nu, \xi_t^s).$$

Let $\hat{y}_\nu = \langle y_\nu^1, \ldots, y_\nu^n \rangle$. For $0 < u < t < \infty$ and $0 < s \le 1$

$$\sum_{\nu=N_u+1}^{N_t} \sum_{j=1}^{\hat{y}_\nu[1]} (1-F_{t-\tau_\nu}[\xi_t^s](y_\nu^j)) \le \sum_{\nu=N_u+1}^{N_t} \sum_{j=1}^{\hat{y}_\nu[1]} (1-\xi_{\tau_\nu}(y_\nu^j))$$

$$\le (1+\varepsilon_u) \sum_{\nu=N_u+1}^{\infty} \gamma_{\tau_\nu} \hat{y}_\nu[\varphi] ,$$

where (ε_u) is a numerical null sequence. Hence, still for $0 < s \le 1$,

$$\Psi(s) := \lim_{t\to\infty} \Psi_t(s) = \prod_{\nu=1}^{\infty} \lim_{t\to\infty} F_{t-\tau_\nu}(\hat{y}_\nu, \xi_t^s)$$

$$= \prod_{\nu=1}^{\infty} \hat{\Phi}(\rho^{-\tau_\nu} s)(\hat{y}_\nu) = \tilde{E} \exp\{-\tilde{W}^* s\} =: \Psi^*(s)$$

with

$$\tilde{\Psi}(0+) = \tilde{\Psi}^*(0+) = 1.$$

That is,

$$\tilde{P}(\tilde{W} < \infty) = \tilde{P}(\tilde{W}^* < \infty) = 1.$$

Step 3. We now show that $\tilde{W}$ and $\tilde{W}^*$ are equal almost surely. Let

$$W_{\nu,t} := \gamma_{t-\tau_\nu} \hat{x}_{\nu,t}[\varphi], \qquad \tilde{W}_t := \gamma_t \hat{z}_t[\varphi].$$

Then

$$\tilde{W}-\tilde{W}^* = (\tilde{W}-\tilde{W}_t) + \sum_{\tau_\nu \leq s} (\gamma_{t-\tau_\nu}^{-1}\gamma_t W_{\nu,t} - \rho^{-\tau_\nu} W_\nu)$$

$$+ \sum_{s<\tau_\nu\leq t} \gamma_{t-\tau_\nu}^{-1}\gamma_t W_{\nu,t} - \sum_{\tau_\nu > s} \rho^{-\tau_\nu} W_\nu \ .$$

As $t \to \infty$, for fixed s, the first term on the right tends to zero and the third term to a finite limit $U_s \geq 0$, non-increasing in s. Next let $s \to \infty$. Then U_s tends to a finite limit $U \geq 0$, and the last term to zero. Thus, we have $\tilde{W} = \tilde{W}^* + U$, all three variables being finite and non-negative. Since the Laplace transforms $\tilde{\Psi}(s)$ and $\tilde{\Psi}^*(s)$ of $\tilde{W}$ and $\tilde{W}^*$ coincide for $s \in (0,1]$, this can only be true if $U = 0$ almost surely. □

6.2. REMARK. Again matters simplify if we normalize by ρ^{-t} instead of γ_t, thus having to be content with a degenerate limit, if (XLOGX) is not satisfied. Since $\tilde{W}_t' := \rho^{-t}\hat{z}_t[\varphi]$ is a submartingale conditioned upon $\mathfrak{Y} = \sigma(\tau_1,\tau_2,\ldots,\hat{y}_1,\hat{y}_2,\ldots)$ with

$$\sup_t \tilde{E}(\tilde{W}_t'|\mathfrak{Y}) = \sum_{\nu=1}^{\infty} \rho^{-\tau_\nu}\hat{y}_\nu[\varphi] \text{ a.s.},$$

Doob's condition is satisfied for a.a. realizations of the immigration process if and only if

$$\sum_{\nu=1}^{\infty} \rho^{-\tau_\nu}\hat{y}_\nu[\varphi] < \infty \quad \text{a.s.} \tag{6.2}$$

Since (6.2) implies that

$$\tilde{W}' := \lim_{t\to\infty} \tilde{W}_t'$$

exists and is finite a.s. With

$$W_\nu' := \lim_{t\to\infty} \rho^{-t+\tau_\nu}\hat{x}_{\nu,t}[\varphi]$$

clearly

$$\sum_{\nu=1}^{\infty} \rho^{-\tau_\nu} W_\nu' \leq \tilde{W}' \quad \text{a.s.}$$

On the other hand, by Fatou's lemma,

$$\tilde{E}(\tilde{W}'|\mathfrak{Y}) \leq \sum_{\nu=1}^{\infty} \rho^{-\tau_\nu}\hat{y}_\nu[\varphi] \quad \text{a.s.}$$

Just for the next proposition let us assume in addition that (R) is satisfied. Then there exists $(\theta_t)_{t\in T} \subset \bar{\mathfrak{S}}_+$ such that

$$tP^{\langle x\rangle}(W>t) = \theta_t(x)\int_0^t P^{\langle x\rangle}(W>u)du, \qquad t>0,$$

$$\lim_{t\to\infty} \theta_t(x) = 0, \qquad x \in X,$$

cf. the proof of 4.6. Note further that always

$$\liminf_{t\to\infty} \tilde{W}_t \geq \tilde{W}^* \quad \text{a.s. } [\tilde{P}].$$

(6.3) PROPOSITION. *If*

$$\limsup_{t\to\infty} \| \theta_t \| < 1,$$

then

$$\tilde{W}^* = \infty \text{ a.s. } \underline{\text{on}} \ \{\sum_{\nu=1}^{\infty} \gamma_{\tau_\nu} \hat{y}_\nu[\varphi] = \infty\}.$$

PROOF. Condition on $\mathfrak{Y}$. Then either $\tilde{W}^* < \infty$ a.s., or $\tilde{W}^* = \infty$ a.s., by Kolmogorov's zero-one law. Moreover, $\tilde{W}^* < \infty$ a.s. only if

$$\tilde{S} := \sum_{\nu=1}^{\infty} \rho^{-\tau_\nu} \hat{y}_\nu[E^{\langle\cdot\rangle} W \, 1_{\{\rho^{-\tau_\nu} W \leq 1\}}] < \infty,$$

by Kolmogorov's three series criterion. Since $\gamma_t \sim \rho^{-t} L(\rho^{-t})$, cf. 2.3, and

$$\Phi^*[\int_0^t P^{\langle\cdot\rangle}(W>u)du] \sim L(t^{-1}), \qquad t\to\infty,$$

cf. the proof of 4.6, there exists for every $s>0$ and $\mu \geq \mu_s$, μ_s sufficiently large, a $C_{s,\mu}$ such that

$$\tilde{S} \geq C_{s,\mu} \sum_{\nu\geq\mu} \gamma_{\tau_\nu} \frac{\hat{y}_\nu[\int_0^{\rho^{\tau_\nu}} uP^{\langle\cdot\rangle}(W\leq u)du]}{\Phi^*[\int_0^{\rho^{\tau_\nu}} P^{\langle\cdot\rangle}(W>\rho^s u)du]} .$$

Observing that

$$\int_0^t uP^{\langle x\rangle}(W\leq u)du = (1-\theta_t(x))\int_0^t P^{\langle x\rangle}(W>u)du$$

and

$$P^{\langle x\rangle}(W>u) = E^{\langle x\rangle} P^{\langle x\rangle}(W>u\,|\,\mathfrak{F}_s) \geq E^{\langle x\rangle} x_s[P^{\langle\cdot\rangle}(W>\rho^s u)]$$

$$\geq (\rho^s-\alpha_s)\varphi(x)\Phi^*[P^{\langle\cdot\rangle}(W>\rho^s u)],$$

we get

$$\tilde{S} \geq c_{s,\mu}(\rho^s - \alpha_s) \inf_{t \geq \rho^\mu} (1 - \| \theta_t \|) \sum_{\nu \geq \mu} \gamma_{\tau_\nu} \hat{y}_\nu[\varphi] \ .$$

By (M), we can choose s such that $\alpha_s < \rho^s$. □

6.4. REMARK. If we assume that $\{\hat{x}_t, P^{\hat{x}}\}$ is determined by $\{T_t, k, \pi\}$, it follows from (IF) that

$$P^{\langle x \rangle}(W > t) = \int_0^t T_s^0 k \int_{\hat{X}} \pi(\cdot, d\hat{x}) P^{\hat{x}}(W > \rho^s t) ds.$$

Using this, it is easily verified that the existence of a $c_8 > 0$ such that

$$k(x) \int_{\hat{X}} \pi(x, d\hat{x}) h(\hat{x}) \leq c_8 \Phi^*[k \int_{\hat{X}} \pi(\cdot, d\hat{x}) h(x)], \qquad x \in X,$$

for all bounded measurable h, is a sufficient condition for $\| \theta_t \| \to 0$.

6.5. PROPOSITION. *If*

$$\sum_{\nu=1}^{\infty} \gamma_{\tau_\nu} \hat{y}_\nu[1] < \infty \quad \text{a.s.} \quad [P],$$

then

$$\lim_{t \to \infty} \gamma_t \hat{z}_t[\eta] = \Phi^*[\eta] \tilde{W} \quad \text{a.s.} \quad [\tilde{P}],$$

for all $\eta \in \mathcal{B}$, *if* $T = \mathbb{N}$, *and for all a.e. continuous* $\eta \in \mathcal{B}$, *if* $T = (0, \infty)$.

We break the proposition up into two lemmata. Define

$$\mathcal{U}_+ := \{\xi \in \mathcal{B}_+ : \lim_{t \to \infty} \gamma_t \hat{x}_t[\xi] = \Phi^*[\xi] W\}.$$

6.6. LEMMA. *Given a* $\quad \in \bar{S}_+$ *such that*

$$\lim_{t \to \infty} \gamma_t \hat{z}_t[\vartheta] = \Phi^*[\vartheta] \tilde{W} \quad \text{a.s.} \quad [\tilde{P}],$$

we have for every $\xi \in \mathcal{U}_+$

$$\lim_{t \to \infty} \gamma_t \hat{z}_t[\vartheta \xi] = \Phi^*[\vartheta \xi] \tilde{W} \quad \text{a.s.} \ [\tilde{P}].$$

PROOF. From the definition of $\hat{z}_t$ as a superposition

$$\liminf_{t \to \infty} \gamma_t \hat{z}_t[\eta] \geq \Phi^*[\eta] \tilde{W}$$

for $\eta \in \mathcal{B}_+$, in particular, for $\eta = \vartheta\zeta$. Clearly $\zeta \in \mathcal{U}_+$ and $\zeta(1-\vartheta) \in \mathcal{U}_+$. Thus

$$\limsup_{t\to\infty} \gamma_t \hat{z}_t[\vartheta\zeta] = \lim_{t\to\infty} \gamma_t \hat{z}_t[\vartheta] - \liminf_{t\to\infty} \gamma_t \hat{z}_t[\vartheta(1-\zeta)]$$

$$\leq \Phi^*[\vartheta]\widetilde{W} - \Phi^*[\vartheta(1-\zeta)]\widetilde{W} = \Phi^*[\vartheta\zeta]\widetilde{W}. \qquad \square$$

This already proves 6.5 if $\inf \varphi > 0$. In that case we simply take $\vartheta := \varphi$.

6.7. LEMMA. *If*

$$\sum_{\nu=1}^{\infty} \gamma_{\tau_\nu} \hat{y}_\nu[1] < \infty \quad \text{a.s. } [P],$$

then

$$\limsup_{t\to\infty} \gamma_t \hat{z}_t[1] \leq \Phi^*[1]\widetilde{W} \quad \text{a.s. } [\widetilde{P}].$$

PROOF. We first consider discrete time processes, respectively discrete skeletons of continuous time processes assuming τ_ν to take its values on the time skeleton, and then reduce the continuous time case to the treatment of skeletons.

Part I. First let τ_ν be integer valued. Define

$$\widetilde{\beta}_n := \hat{z}_n[\varphi], \qquad \widetilde{I}_n := \{\widetilde{\beta}_n > 0\},$$

$$z_{n,\ell} := \sum_{\nu=1}^{N_n-1} \hat{x}_{\nu,n+\ell}[1] = \sum_{j=1}^{z_{n,0}} z^j_{n,\ell},$$

$$\widetilde{S}_{n,\ell} := \widetilde{I}_{n-1}\widetilde{\beta}^{-1}_{n-1} z_{n,\ell}, \qquad \widetilde{S}^*_{n,\ell} := \widetilde{I}_{n-1}\widetilde{\beta}^{-1}_{n-1} \sum_{j=1}^{z_{n,0}} z^j_{n,\ell} 1_{\{z^j_{n,\ell} \leq \widetilde{\beta}_{n-1}\}},$$

$$\Delta_{n,\ell} := \widetilde{I}_{n+\ell}\widetilde{\beta}^{-1}_{n+\ell} \sum_{n-1<\tau_\nu\leq n+\ell} \hat{x}_{\nu,n+\ell}[1].$$

Then

$$\widetilde{I}_{n+\ell}\widetilde{\beta}^{-1}_{n+\ell}\hat{z}_{n+\ell}[1] \leq \widetilde{I}_{n+\ell}\widetilde{\beta}^{-1}_{n+\ell}[(\widetilde{S}_{n,\ell} - \widetilde{E}(\widetilde{S}^*_{n,\ell}|\widetilde{\mathfrak{F}}_n)) + \widetilde{E}(\widetilde{S}_{n,\ell}|\widetilde{\mathfrak{F}}_n)] + \Delta_{n,\ell}\,.$$

Step 1. As in (5.3)

$$\lim_{n\to\infty}(\widetilde{S}_{n,\ell} - \widetilde{E}(\widetilde{S}^*_{n,\ell}|\widetilde{\mathfrak{F}}_n)) = 0 \quad \text{a.s. } [\widetilde{P}], \quad \ell > 0.$$

Step 2. Define

$$\Delta^*_{n,\ell} := \gamma_{n+\ell}\tilde{\beta}_{n+\ell}\Delta_{n,\ell} \, .$$

Then

$$\sum_{n=1}^{\infty} \tilde{E}(\Delta^*_{n,\ell}|\mathfrak{Z}) \leq \sum_{n=1}^{\infty} \gamma_{n+\ell} \sum_{n-1<\tau_\nu\leq n+\ell} \hat{y}_\nu[M_{n+\ell-\tau_\nu}[1]]$$

$$\leq c_9(\ell+1)\sum_{\nu=1}^{\infty} \gamma_{\tau_\nu}\hat{y}_\nu[1].$$

That is, $\Delta^*_{n,\ell} \to 0$ a.s., and thus

$$\lim_{n\to\infty} \Delta_{n,\ell} = 0 \quad \text{a.s. } [\tilde{P}], \quad \ell > 0.$$

Step 3. Using (M), we have

$$\limsup_{n\to\infty} \tilde{I}_{n+\ell}\tilde{\beta}^{-1}_{n+\ell}\tilde{\beta}_{n-1}\tilde{E}(\tilde{S}_{n,\ell}|\tilde{\mathfrak{Z}}_n) = \limsup_{n\to\infty} \tilde{I}_{n+\ell}\tilde{\beta}^{-1}_{n+\ell}z_n[M_\ell[1]]$$

$$\leq (1+\rho^{-\ell}\alpha_\ell)\Phi^*[1], \quad \ell > 0.$$

Combining Steps 1 to 3 yields

$$\limsup_{n\to\infty} \tilde{\beta}^{-1}_n\hat{z}_n[1] \leq \Phi^*[1] \quad \text{a.s. } [\tilde{P}].$$

Part II. We now come to the continuous time case with general τ_ν. For every $\epsilon > 0$ the process $\{\hat{z}_{n\epsilon}\}$ can be considered as discrete-time process with immigration

$$\tau^*_\nu := ([\tau_\nu/\epsilon]+1)\epsilon, \qquad \hat{y}^*_\nu := \hat{x}_{\nu,\tau^*_\nu} \, .$$

Let us do this from now on. For convenience let us further work with an explicit choice of Γ_t, assuming the existence of left-hand limits of $\hat{x}_t$. There is no loss of generality in this. Define $\hat{z}^{s,j}_t$ in analogy to $\hat{x}^{s,j}_t$,

$$\Gamma_\epsilon := \hat{x}_\epsilon[1] + \#\{t:\hat{x}_{t-}[1] > \hat{x}_t[1]: \; 0 < t \leq \epsilon\},$$

$$\tilde{\Gamma}^{\epsilon,j}_n := \hat{z}^{n\epsilon,j}_{(n+1)\epsilon}[1] + \#\{t:\hat{z}^{n\epsilon,j}_{t-}[1] > \hat{z}^{n\epsilon,j}_t[1]; \; n\epsilon < t \leq (n+1)\epsilon\}.$$

Step 1. Proceeding as in part I, with $z^j_{n,\ell}$ replaced by $\tilde{\Gamma}^{\epsilon,j}_n$ and $\ell = 1$,

$$\limsup_{t\to\infty} \tilde{\beta}_t^{-1}\hat{z}_t[1] \leq \limsup_{t\to\infty} \tilde{\beta}_t^{-1} \sum_{j=1}^{\hat{z}_{n\epsilon}[1]} \tilde{\Gamma}^{\epsilon,j}_n$$

$$\leq \limsup_{n\to\infty} \tilde{\beta}_{n\epsilon}^{-1}\hat{z}_{n\epsilon}[E^{\langle\cdot\rangle}\Gamma_\epsilon] \text{ a.s. } [\tilde{P}] \text{ on } \{\tilde{W} > 0\}.$$

Step 2. We have

$$\tilde{E}(\hat{y}^*_\nu[1]\,|\,\_) \leq \| E^{\langle\cdot\rangle}\Gamma_\epsilon \| \hat{y}_\nu[1],$$

$$\sum_{\nu=1}^{\infty} \gamma_{\tau^*_\nu} \hat{y}^*_\nu[1] < \infty.$$

Hence, according to Part I,

$$\limsup_{n\to\infty} \tilde{\beta}_{n\epsilon}^{-1}\hat{z}_{n\epsilon}[1] \leq \Phi^*[1] \quad \text{a.s. } [\tilde{P}] \quad \text{on } \{\tilde{W} > 0\}.$$

Step 3. Recall that by assumption

$$\| E^{\langle\cdot\rangle}\Gamma_\epsilon \| \downarrow 1, \qquad \epsilon \downarrow 0.$$

Combine steps 1 to 3 to complete the proof. □

6.8. EXAMPLES. With say $\eta \in \mathcal{B}_+$ the relation

$$\sum_{\nu=1}^{\infty} \gamma_{\tau_\nu} \hat{y}_\nu[\eta] < \infty \text{ a.s.} \tag{6.3}$$

holds at least if

$$E \sum_{\nu=1}^{\infty} \gamma_{\tau_\nu} \hat{y}_\nu[\eta] < \infty \text{ a.s.} \tag{6.4}$$

(a) If $\{\tau_\nu\}$ is a (possibly inhomogeneous) Poisson process with density $p(t)$, if the $\hat{y}_\nu$ are independent conditioned on $\{\tau_\nu\}$, and if the distribution of $\hat{y}_\nu$ depends only on τ_ν, then (6.4) reduces to

$$\int_0^\infty \gamma_t p(t) E(\hat{y}_\nu[\eta]\,|\,\tau_\nu = t)dt < \infty.$$

(b) In a two-component decomposable branching process we have immigration from the first into the second component. If the corresponding spectral radii ρ_1, ρ_2 satisfy $\rho_1 < \rho_2 < \infty$, then (6.4) is satisfied.

(c) If the τ_ν are the epochs of a renewal process and the $\hat{y}_\nu$ are i.i.d. and independent of $\{\tau_\nu\}$, then with $\gamma_t = \rho^{-t}L(\rho^{-t})$, $L(s)$ slowly varying at 0, the condition

$$E \log^+ \hat{y}_1[\eta] < \infty$$

is sufficient, and in case the mean interarrival time is finite, also necessary for (6.3), cf. VI.6.2.

REMARK. The case of a finite X is almost a triviality: Finiteness of X implies $\inf \varphi > 0$ and $\| \theta_t \| \to 0$, $t \to \infty$.

BIBLIOGRAPHICAL NOTES

Most of this chapter has been taken from Hering (1978d). Exceptions are the remark 5.3 and the proof 5.7, which are based on Asmussen and Hering (1976a), furthermore 5.4, which has been taken from Hering (1978c), with a preceding version in Asmussen and Hering (1976a), and finally the remark 6.2, which is from Asmussen and Hering (1976a). Of the preceding papers on p-type processes we mention only Kesten und Stigum (1966a) and Hoppe (1976). A more detailed historical account can be found in Hering (1978d).

Chapter VIII

More on the Limiting Behaviour of Linear Functionals

1. INTRODUCTION. I

The limit theorems given so far for the critical case $(\rho = 1)$ and the supercritical case $(\rho > 1)$ determine the growth rate of the total population size and state that the type distribution in a population which has grown large is asymptotically the stable type distribution Φ^*.

The motivation for the investigations of the present chapter arises from the observation that a description along these lines is by no means complete, even with regularity assumptions like (X LOG X) or finite variance. For example in the supercritical case, the a.s. convergence of $\rho^{-t}\hat{x}_t[\xi]$ to $W\Phi^*[\xi]$ gives only the upper bound $|\hat{x}_t[\xi]| = o(\rho^t)$ if ξ belongs to the hyperplane determined by the equation $\Phi^*[\xi] = 0$, and leaves the possibility open of obtaining a non-degenerate limit under a weaker norming. More generally one might ask for the rate of convergence of $\rho^{-t}\hat{x}_t[\eta]$ to $W\Phi^*[\eta]$ also if $\Phi^*[\eta] \neq 0$. Writing $\eta := \Phi^*[\eta]\varphi + \xi$ (so that $\Phi^*[\xi] = 0$) this reduces, however, as we shall see, essentially to the problem of in addition to study the rate of convergence of $W_t := \rho^{-t}\hat{x}_t[\varphi]$ to W. Similar remarks apply to the critical case.

We now introduce the set-up used in the main part of the chapter. We are concerned with a discrete time process $\{\hat{x}_N\}_{N\in\mathbb{N}}$ with a general set X of types and satisfying (M). The offspring mean operator (matrix in case of a finite set of types) M is given by $M\eta(x) = E^{\langle x\rangle}\hat{x}_1[\eta]$ and we denote its Nth iterate by M^N. The relevant second moments are assumed to be finite, which amounts to assumptions of the form

$$(1.1) \qquad \Phi^*[\mathrm{Var}^{\langle\cdot\rangle}\hat{x}_1[\eta]] < \infty.$$

If (1.1) holds, we say for brevity that the process has finite variance w.r.t. η. Though complete results and proofs are given both for the critical case and the supercritical case, the latter is the main one in the sense that in the critical case the results are less complex in form and the main ideas in their proofs are similar to some of the easier parts of the arguments used in the supercritical case. We remark for completeness that, due to the lack of an asymptotic type distribution, in the subcritical case $\rho < 1$ there could be no immediate analogous theory.

Extensions to continuous time and infinite variances are given in separate sections.

In the results, no specific assumptions are usually made concerning the initial P-distribution of $\hat{x}_0$ except that $P(\hat{x}_0 = 0) < 1$. We let P_N, resp. P_∞ be the probability law obtained by conditioning upon no extinction before time N, resp. within finite time,

$$P_N: = P(\cdot|\hat{x}_N \neq 0) = P(\cdot|\hat{x}_n \neq 0 \quad n = 0,1,\ldots,N),$$

$$P_\infty: = P(\cdot|\hat{x}_n \neq 0 \quad n = 0,1,\ldots).$$

The regularity conditions imposed will always ensure that $P_\infty = P(\cdot|W>0)$ if $\rho > 1$. 'A.s. $[P_\infty]$' is then the same as 'a.s. on $\{W > 0\}$'.

The problem of studying the rate of convergence of W_N to W turns out to be somewhat more elementary than to deal with $\hat{x}_N[\xi]$, $\Phi^*[\xi] = 0$. We have here

1.1. THEOREM. *Suppose that $\rho > 1$ and that the process has finite variance w.r.t. φ, i.e. $\sigma^2: = \Phi^*[\mathrm{Var}^{\langle\cdot\rangle}\hat{x}_1[\varphi]] < \infty$. Then $\tau^2: = \Phi^*[\mathrm{Var}^{\langle\cdot\rangle}W] = \sigma^2/(\rho^2-\rho) < \infty$. Suppose that $\sigma^2 > 0$, and define*

$$U_N: = \left(\frac{\rho^N}{\tau^2 W_N}\right)^{1/2}(W-W_N).$$

Then: (i) *For all* $y \in \mathbb{R}$,

$$\lim_{N\to\infty} P_N(U_N \leq y|\mathfrak{F}_N) = \Phi(y) \quad \text{a.s. on } \{W > 0\}. \tag{1.2}$$

In particular, the limiting distribution w.r.t. P_∞ of U_N exists and is standard normal. (ii) *On $\{W > 0\}$, it holds a.s. that*

$$\overline{\lim_{N\to\infty}} \frac{U_N}{(2\log N)^{1/2}} = 1, \quad \underline{\lim_{N\to\infty}} \frac{U_N}{(2\log N)^{1/2}} = -1. \tag{1.3}$$

The results are analogous in form to II.3.1 and also the generalizations needed in the proof do not go beyond ideas already met in Chapter VII. The rather obvious analogue of II.3.2 becomes

1.2. PROPOSITION. *Let $Y: = Y(\hat{x}_0, \hat{x}_1, \ldots)$ be some functional of the process and $Y_{N,i}$ the same functional evaluated in the process initiated by the ith particle alive at time N. Then the $Y_{N,i}$ are independent conditionally upon $\mathfrak{F}_N$, with $P(Y_{N,i} \leq y|\mathfrak{F}_N) = P^{\langle x_i\rangle}(Y \leq y)$* (where x_i is the type of the ith particle alive at time N). *Define $S_N: = \Sigma_1^{|\hat{x}_N|} Y_{N,i}$ and suppose that $\rho > 1$, $E^{\langle x\rangle}Y = 0$ $\forall x \in X$, $0 < \tau^2: = \Phi^*[\mathrm{Var}^{\langle\cdot\rangle}Y] < \infty$. Then:* (i) *For all* $y \in \mathbb{R}$

$$(1.4)\quad \lim_{N\to\infty} P_N\left(\frac{S_N}{(\tau^2 W\rho^N)^{1/2}} \leq y \mid \mathfrak{F}_N\right) = \Phi(y) \quad \text{a.s. on} \quad \{W > 0\}.$$

In particular, the limiting distribution w.r.t. P_∞ _of_ $S_N/(\tau^2 W\rho^N)^{1/2}$ _exists and is standard normal_. (ii) _On_ $\{W > 0\}$, _it holds a.s._ that

$$(1.5)\quad \overline{\lim_{N\to\infty}} \frac{S_N}{(2\tau^2 W\rho^N \log N)^{1/2}} \leq 1, \quad \underline{\lim_{N\to\infty}} \frac{S_N}{(2\tau^2 W\rho^N \log N)^{1/2}} \geq -1,$$

with the inequalities in (1.5) _replaced by equalities if for some_ $k < \infty$ Y _depends on_ $x_0, \ldots, x_k$ _only_ (i.e. is $\mathfrak{F}_k$-measurable).

2. INTERLUDE ON THE JORDAN CANONICAL FORM

In contrast to this rather simple extension of the one-dimensional case, the analysis of $\hat{x}_N[\xi]$ with $\Phi^*[\xi] = 0$ shows up essentially new features. It is natural first to ask for the behaviour of the mean, i.e. of $M^N\xi$, where positive regularity alone only gives $M^N\xi = o(\rho^N)$. In case of a finite set of types (or more generally, if ξ belongs to a finite dimensional M-invariant subspace), the answer is given by the Jordan canonical form. In order to proceed in a similar manner in the general case, we set up

2.1. CONDITION. *For any μ with $0 < \mu < \rho$ there exists a decomposition of $\mathfrak{L}^1_{\Phi^*}$ as an algebraically direct sum $\mathfrak{L}^1_{\Phi^*} = \mathfrak{C}_1 + \mathfrak{C}_2$ of M-invariant real subspaces $\mathfrak{C}_1$, $\mathfrak{C}_2$, where: $\mathfrak{C}_1$ is finite dimensional and any eigenvalue ρ_ν of M restricted to $\mathfrak{C}_1$ satisfies $|\rho_\nu| \geq \mu$; for some constant c with $0 \leq c < \infty$ it holds for all $N = 1,2,\ldots$ and all $\eta \in \mathfrak{C}_2$ that*

$$(2.1)\qquad -c\mu^N\varphi(x)\Phi^*[|\eta|] \leq M^N\eta(x) \leq c\mu^N\varphi(x)\Phi^*[|\eta|].$$

The motivation for 2.1 derives, apart from the fact that 2.1 is trivial in case of a finite set of types, from

2.2. THEOREM. *Condition 2.1 holds for any discrete skeleton of a branching diffusion satisfying the assumptions of V.3*

The proof is contained in the proof of (M) in the cases where (M) could be obtained using a full spectral decomposition, that is, in the one-dimensional connected case or the symmetric n-dimensional case. In the general n-dimensional case a minor extension of the proof is needed.

We shall need to write M restricted to $\mathfrak{C}_1$ on the Jordan canonical form. To this end, we first note that it is apparently no restriction to allow $\mathfrak{L}^1_{\Phi^*}$, $\mathfrak{C}_1$, $\mathfrak{C}_2$ to be complex vector spaces.

2.3. PROPOSITION. *Suppose 2.1 holds and let $\rho_\nu \in \mathbb{C}$; $\nu = 1,\ldots,\bar{\nu}$ be the eigenvalues of M restricted to $\mathfrak{C}_1$, $\varphi_{\nu,j} \in \mathfrak{C}_1$; $\nu = 1,\ldots,\bar{\nu}$; $j = 1,\ldots,\bar{j}_\nu$ the corresponding generalized eigenvectors, i.e.*

$$(2.2)\qquad M\varphi_{\nu,1} = \rho_\nu\varphi_{\nu,1},\ M\varphi_{\nu,j} = \varphi_{\nu,j-1} + \rho_\nu\varphi_{\nu,j}\quad j > 1,$$

and $\Phi^*_{\nu,j}\colon \mathfrak{L}^1_{\varphi^*} \to \mathbb{C}$ the associated projectors. Then each $\eta \in \mathfrak{L}^1_{\Phi^*}$ has a unique expansion

$$(2.3) \qquad \eta = \sum_{\nu=1}^{\bar\nu} \sum_{j=1}^{\bar J_\nu} \Phi^*_{\nu,j}[\eta]\varphi_{\nu,j} + \eta_2$$

with $\eta_2 \in \mathcal{E}_2$. Furthermore,

$$(2.4) \qquad M^N\varphi_{\nu,j} = \sum_{i=1}^{j} \rho_\nu^{N-j+i}\binom{N}{j-i}\varphi_{\nu,i},$$

$$(2.5) \qquad |\varphi_{\nu,j}| \le c_{\nu,j}\varphi \quad \text{for some} \quad c_{\nu,j} \in (0,\infty).$$

PROOF. The first part of 2.3 is just a parametrization of the Jordan canonical form of M restricted to $\mathcal{E}_1$, (2.4) follows from (2.2) and (2.5) is an immediate consequence of (M) and (2.2). □

We write $\rho_\nu := \lambda_\nu e^{i\theta_\nu}$ with $\lambda_\nu \ge 0$, $\theta_\nu \in [0,2\pi)$.

2.4. REMARK. Of course, there is a unique ν such that $\rho_\nu = \lambda_\nu = \rho$, $\bar J_\nu = 1$, $\varphi_{\nu,1} = \varphi$, $\Phi^*_{\nu,1} = \Phi^*$. For all other ν, $\lambda_\nu < \rho$.

2.5. REMARK. The η with which we are concerned are real, while expressions like (2.4) formally contain complex terms. If η is real, the complex terms of course cancel. The complex set-up is therefore to some extent just a convenient notation and everything could be rewritten in real form. We shall not carry this out in full, but only give a few examples. Note, however, that whenever ρ_ν is eigenvalue, the complex conjugate $\bar\rho_\nu$ must be so, hence have the form $\bar\rho_\nu = \rho_{\nu'}$, and it is possible and often convenient to assume the ν to fall into such couples ν, ν' with the additional property $\bar\varphi_\nu = \varphi_{\nu'}$, $\bar\Phi^*_\nu = \Phi^*_{\nu'}$. This will be assumed throughout in the following and will make a number of complex-looking expressions real.

We now define two basic parameters λ, γ associated with ξ. They will be defined in terms of the parameters of 2.1, 2.3 which in general can be chosen in a number of ways (even for the same μ), but the expressions (2.6), (2.8) below show that the particular choice is immaterial. If $\xi \in \mathcal{E}_2$ for any decomposition of type 2.1, we define $\lambda := \lambda(\xi) := 0$. Otherwise, choose a decomposition as in 2.1 with $\xi \notin \mathcal{E}_2$ and define

$$\lambda := \lambda(\xi) := \max\{\lambda_\nu : \Phi^*_{\nu,j}[\xi] \ne 0 \quad \text{for some} \quad j\}$$

$$\gamma := \gamma(\xi) := \max\{j : \Phi^*_{\nu,j}[\xi] \ne 0 \quad \text{for some} \quad \nu \quad \text{with} \quad \lambda_\nu = \lambda\}.$$

Combining (2.3) and (2.4) with (2.5) and (2.1) then yields

$$(2.6) \quad M^N\xi = \lambda^N\binom{N}{\gamma-1}\xi_N + \nu_N(\xi), \quad \text{where } \xi_N := \sum_\nu^{(\xi)} \rho_\nu^{-(\gamma-1)}\Phi^*_{\nu,\gamma}[\xi]e^{i\theta_\nu N}\varphi_{\nu,1},$$

$$(2.7)\quad |\nu_N(\xi)| \leq c_N\varphi,\ c_N = \begin{cases} O(\lambda^N N^{\gamma-2}) \text{ if } \gamma > 1 \\ O(\lambda_1^N) \text{ for some } \lambda_1 < \lambda \text{ if } \gamma = 1 \end{cases}.$$

Here and in the following $\Sigma_\nu^{(\xi)}$ is shorthand for summation over $\{\nu:\lambda_\nu = \lambda(\xi),\ \overline{J}_\nu \geq \gamma\}$. If $\lambda = 0$, one has to replace (2.6), (2.7) by

$$(2.8)\qquad |M^N\xi| \leq c_N\varphi,\ c_N = O(\mu^N) \text{ for all } \mu > 0.$$

If the set of types is finite, 0 must then be eigenvalue and $M^N\xi = 0$, $M^{N-1}\xi \neq 0$ for some N, and one could define $\gamma := \gamma(\xi) := N$. In the general case, we leave γ undefined so that the case $\lambda = 0$ sometimes requires a slightly different formulation. We shall not carry this out since the modifications are always trivial.

2.6. REMARK. The magnitude of ξ_N does not depend on N in the sense that ξ_N is in

$$\{\Sigma_\nu^{(\xi)}\alpha_\nu\varphi_{\nu,1}:\ |\alpha_\nu| = |\Phi^*_{\nu,\gamma}[\xi]|\lambda^{-(\gamma-1)}\}$$

which is a compact surface in a finite dimensional space, not containing 0 by independence of the $\varphi_{\nu,1}$. Appealing to 2.5, ξ_N (and hence $\nu_n(\xi)$) is real and could be written (with $\rho_\nu^{-(\gamma-1)}\Phi^*_{\nu,1}[\xi]\varphi_{\nu,1} := \eta'_\nu + i\eta''_\nu$) $\xi_N = \Sigma_\nu^{(\xi)}\{\cos N\theta_\nu\eta'_\nu - \sin N\theta_\nu\eta''_\nu\}$.

3. INTRODUCTION. II

After having introduced the basic parameters λ, γ in the preceding section, we can now state the main results concerning the limiting behaviour of $\hat{x}_N[\xi]$ in the case $\Phi^*[\xi] = 0$. Condition 2.1 is throughout supposed to be in force.

3.1. THEOREM. *Consider the supercritical case* $\rho > 1$, *suppose that* $\Phi^*[\xi] = 0$ *and that the process has finite variance w.r.t.* φ *and* ξ.

1° *If* $\lambda^2 < \rho$, *then:*

$$\sigma^2 := \lim_{N\to\infty} \rho^{-N}\Phi^*[\mathrm{Var}^{\langle\cdot\rangle}\hat{x}_N[\xi]]$$

exists and is finite. If $\sigma^2 > 0$, *then the limiting distribution w.r.t.* P_∞ *of* $\hat{x}_N[\xi]/(\sigma^2 W\rho^N)^{1/2}$ *is standard normal and a.s. on* $\{W > 0\}$, *it holds that*

$$(3.1)\quad \overline{\lim_{N\to\infty}} \frac{\hat{x}_N[\xi]}{(2\sigma^2 W\rho^N \log N)^{1/2}} = 1, \quad \underline{\lim_{N\to\infty}} \frac{\hat{x}_N[\xi]}{(2\sigma^2 W\rho^N \log N)^{1/2}} = -1.$$

2° *If* $\lambda^2 = \rho$, *then:*

$$\sigma^2 := \lim_{N\to\infty} (N^{2\gamma-1}\rho^N)^{-1/2}\Phi^*[\mathrm{Var}^{\langle\cdot\rangle}\hat{x}_N[\xi]]$$

exists and is finite. If $\sigma^2 > 0$, *then the limiting distribution w.r.t.* P_∞ *of* $\hat{x}_N[\xi]/(\sigma^2 W N^{2\gamma-1}\rho^N)^{1/2}$ *is standard normal and a.s. on* $\{W > 0\}$, *it holds* (writing $\log_2 = \log\log$) *that*

$$(3.2)\quad \overline{\lim_{N\to\infty}} \frac{\hat{x}_N[\xi]}{(2\sigma^2 W N^{2\gamma-1}\rho^N \log_2 N)^{1/2}} = 1, \quad \underline{\lim_{N\to\infty}} \frac{\hat{x}_N[\xi]}{(2\sigma^2 W N^{2\gamma-1}\rho^N \log_2 N)^{1/2}} = -1.$$

3° *If* $\lambda^2 > \rho$, *then a.s.*

$$(3.3)\quad \lim_{N\to\infty}\{(N^{\gamma-1}\lambda^N)^{-1}\hat{x}_N[\xi] - \sum_{\nu}{}^{(\xi)}\frac{\rho_\nu^{-(\gamma-1)}}{(\gamma-1)!}\Phi^*_{\nu,\gamma}[\xi]e^{i\theta_\nu N}W^\nu\} = 0,$$

where $W^\nu := \lim_{N\to\infty} W_N^\nu := \lim_{N\to\infty} \rho_\nu^{-N}\hat{x}_N[\varphi_{\nu,1}]$ *exists since* $\{W_N^\nu\}$ *is a* $\mathcal{L}^2$-*bounded* (possibly complex) *martingale*.

We next give some discussion of 3.1.

The quantities governing the rate of growth of $\hat{x}_N[\xi]$ and the mean $M^N\xi$ are the same, viz. λ and γ. But the rate of growth is only the same if $\lambda^2 > \rho$, while otherwise the standard deviation dominates the mean and becomes the proper norming (for the weak laws and similarly with an extra factor for the strong laws). The quantitative sizes of λ and γ do not even influence the normalizing constants in the range

$0 \leq \lambda^2 < \rho$.

The results are not satisfactory if $\sigma^2 = 0$ in 1° and 2° and if in 3°

$$(3.4) \qquad \lim_{N\to\infty} \Sigma_\nu^{(\xi)} \rho_\nu^{-(\gamma-1)} \Phi_{\nu,\gamma}^*[\xi] e^{iN\theta_\nu} W^\nu = 0.$$

The precise criteria for such phenomena to occur are given in the next section. Essentially, these are special dependencies such that the distributions of certain linear functionals have to degenerate. Since this can be argued to be quite special, the (easy) modifications needed in such situations will not be spelled out.

A remark similar to 2.6 applies to the real form of (3.3) which becomes

$$(3.5) \quad \lim_{N\to\infty} \{(N^{\gamma-1}\lambda^N)^{-1}\hat{x}_N[\xi] - \Sigma_\nu^{(\xi)}\{\cos N\theta_\nu U_\nu' - \sin N\theta_\nu U_\nu''\}\} = 0$$

where $\Phi_{\nu,\gamma}^*[\xi]W^\nu := U_\nu' + iU_\nu''$.

Formally, (3.3) of course also holds for $\xi = \varphi$ (with $\lambda = \rho$, $\gamma = 1$, $W^\nu = W$) and in that case, a finer limit result (giving the rate of convergence) is in 1.1. Similar refinements could presumably be given in the general case. Some remarks on this are in Section 10. It follows by the way from Section 11 that the assumption on finite variance could be somewhat weakened for part 3° of 3.1.

As an application of 3.1, consider the problem raised in Section 1 of studying $\Delta_N := \rho^{-N}\hat{x}_N[\eta] - \Phi^*[\eta]W$, also if $\Phi^*[\eta] \neq 0$. Letting $\xi := \Phi^*[\eta]\varphi - \eta$, we see from 1.1 and 3.1 that (with non-degenerate variances) $\rho^{-N}\hat{x}_N[\xi]$ dominates $\Phi^*[\eta](\rho^{-N}\hat{x}_N[\varphi]-W)$ if $\lambda^2 \geq \rho$, so that 3.1 applies. If $\lambda^2 < \rho$, the two terms are of the same magnitude, but easy modifications of the proofs will yield

3.2. COROLLARY. *If $\lambda^2 < \rho$, then the limiting distribution w.r.t. P_∞ of $\rho^{N/2}\Delta_N/W^{1/2}$ is normal with mean zero and variance $\omega^2 := \Phi^*[\eta]^2\tau^2+\sigma^2$* [with τ^2 and σ^2 as in 1.1, 3.1 1°]. *Furthermore a.s. on $\{W > 0\}$,*

$$(3.6) \quad \overline{\lim_{N\to\infty}}\left(\frac{\rho^N}{2\omega^2 W \log N}\right)^{1/2}\Delta_N = 1, \quad \underline{\lim_{N\to\infty}}\left(\frac{\rho^N}{2\omega^2 W \log N}\right)^{1/2}\Delta_N = -1.$$

The proof will be omitted. Another application (to statistics) of 3.1 is in Section 11.

In the critical case, always $\lambda < \rho = 1$ so that $\lambda^2 < \rho$, and the results take a less complex form:

3.3. THEOREM. *Suppose that $\rho = 1$, $\Phi^*[\xi] = 0$ and that the process has finite variance w.r.t. ξ and φ. Then*

$$\sigma^2 := \lim_{N\to\infty} \Phi^*[\mathrm{Var}^{\langle\cdot\rangle}\hat{x}_N[\xi]]$$

exists and is finite. If $\sigma^2 > 0$ and the usual non-degeneracy condition $0 < \mu := \frac{1}{2}\Phi^*[\mathrm{Var}^{\langle\cdot\rangle}\hat{x}_1[\varphi]] < \infty$ holds, then for all $y \in \mathbb{R}$

$$(3.7) \qquad \lim_{N\to\infty} P_N\left(\frac{\hat{x}_N[\xi]}{(\sigma^2\hat{x}_N[\varphi])^{1/2}} \leq y\right) = \Phi(y),$$

$$(3.8) \qquad \lim_{N\to\infty} P_N\left(\frac{\hat{x}_N[\xi]}{(\sigma^2 N)^{1/2}} \leq y\right) = \int_{-\infty}^{y} \tfrac{1}{2}\varkappa e^{-\varkappa|s|}ds$$

where $\varkappa := (2/\mu)^{1/2}$.

The Laplacian limit distribution in (3.8) enters essentially as a simple corollary of (3.7) and the exponential limit law for $\hat{x}_N[\varphi]$.

The rest of this chapter is devoted to the highly technical proofs of the above results as well as also some generalizations and refinements are given. The plan is the following. In Section 4 we collect a number of the (elementary but sometimes lengthy) calculations needed to estimate second moments. In particular, we prove the asymptotic expressions stated above and study the non-degeneracy problem. In Section 5, we exploit the additivity property. That is, we give the proof of 1.2 and the applications, viz. the proof of 1.1, the CLT's of 3.1 and 3.2 in the case $\lambda^2 < \rho$, the weak laws in 3.3 for the critical case and, in the particular case where the set of types is finite, of the LIL's of 3.1 and 3.2 in the case $\lambda^2 < \rho$. Sections 6-10 give the proofs of the remaining results in full generality (and reprove most of the above results), using more refined methods involving martingales indexed by the total set of all individuals ever alive. Section 11 gives an example from asymptotic estimation theory, showing that the scope of the preceding analysis is somewhat wider than the results for linear functionals. Section 12 then discusses the extension to continuous time. Essentially all CLT's carry over as an immediate consequence of the Croft-Kingman lemma, while a gap is left in the LIL's of 3.1 if the set of types is infinite. Finally the analysis of Section 13 indicates (in a somewhat more restricted set-up than above) the consequences of dropping the assumption on finite variance.

4. INTERLUDE ON SECOND MOMENTS

The basic identity used here and in a number of subsequent proofs is the expansion

$$(4.1)\quad \hat{x}_N[\xi] = \hat{x}_0[M^N\xi] + \sum_{n=0}^{N-1}\{\hat{x}_{n+1}[M^{N-n-1}\xi]-\hat{x}_n[M^{N-n}\xi]\}$$

of $\hat{x}_N[\xi]$ in a sum of martingale increments. It follows at once from (4.1) and $\mathcal{L}^2$-orthogonality that

$$(4.2)\quad \begin{aligned} \mathrm{Var}^{\langle x\rangle}\hat{x}_N[\xi] &= \sum_{n=0}^{N-1}\mathrm{Var}^{\langle x\rangle}(\hat{x}_{n+1}[M^{N-n-1}\xi]-\hat{x}_n[M^{N-n}\xi]) \\ &= \sum_{n=0}^{N-1}E^{\langle x\rangle}\mathrm{Var}(\hat{x}_{n+1}[M^{N-n-1}\xi\,|\mathfrak{F}_n) \\ &= \sum_{n=0}^{N-1}E^{\langle x\rangle}\hat{x}_n[\mathrm{Var}^{\langle\cdot\rangle}\hat{x}_1[M^{N-n-1}\xi]] \end{aligned}$$

$$(4.3)\quad \begin{aligned} \Phi^*[\mathrm{Var}^{\langle\cdot\rangle}\hat{x}_N[\xi]] &= \sum_{n=0}^{N-1}\rho^n\Phi^*[\mathrm{Var}^{\langle\cdot\rangle}\hat{x}_1[M^{N-n-1}\xi]] \\ &= \rho^N\sum_{n=1}^{N}\rho^{-n}\Phi^*[\mathrm{Var}^{\langle\cdot\rangle}\hat{x}_1[M^{n-1}\xi]]. \end{aligned}$$

The formula in 1.1 for $\mathrm{Var}\,W$ is an immediate corollary. Indeed, letting $\xi = \varphi$ yields

$$\begin{aligned} \Phi^*[\mathrm{Var}^{\langle\cdot\rangle}W_N] &= \rho^{-2N}\Phi^*[\mathrm{Var}^{\langle\cdot\rangle}\hat{x}_N[\varphi]] = \rho^{-N}\sum_{n=1}^{N}\rho^{n-2}\Phi^*[\mathrm{Var}^{\langle\cdot\rangle}\hat{x}_1[\varphi]] \\ &= \frac{\rho^N-1}{\rho^{N+1}(\rho-1)}\,\Phi^*[\mathrm{Var}^{\langle\cdot\rangle}\hat{x}_1[\varphi]]. \end{aligned}$$

As $N\to\infty$, $\mathrm{Var}^{\langle\cdot\rangle}W_N \uparrow \mathrm{Var}^{\langle\cdot\rangle}W$ by the martingale property and convexity.

If $\Phi^*[\mathrm{Var}^{\langle\cdot\rangle}U] = 0$, we say that U degenerates $[\Phi^*]$. This is the natural degeneracy criterion in most situations, as can be motivated by the observation that $\Phi^*[\mathrm{Var}^{\langle\cdot\rangle}U] = 0$ if and only if $\mathrm{Var}^{\langle x\rangle}U = 0$ for a.a. (and with weak continuity assumptions or if the set of types is finite, for all) x.

4.1. PROPOSITION. Suppose that $\rho \geq 1$, that $\lambda^2 < \rho$ [e.g. if $\rho = 1$, $\Phi^*[\xi] = 0$] and that the process has finite variance w.r.t. ξ and φ. Then $\sigma^2(\xi) := \lim_{N\to\infty}\rho^{-N}\Phi^*[\mathrm{Var}^{\langle\cdot\rangle}\hat{x}_N[\xi]]$ exists, is finite and is given by

$$(4.4)\quad \sigma^2(\xi) = \sum_{n=1}^{\infty}\rho^{-n}\Phi^*[\mathrm{Var}^{\langle\cdot\rangle}\hat{x}_1[M^{n-1}\xi]].$$

In particular, $\sigma^2(\xi) = 0$ if and only if $x_1[\xi], x_1[M\xi], x_1[M^2\xi], \ldots$ all degenerate $[\Phi^*]$. Furthermore

(4.5) $$\sigma^2(M^k\xi) = o(\rho^k).$$

PROOF. The first part of 4.1 is immediate from (4.3) and

(4.6) $$\Phi^*[\mathrm{Var}^{\langle\cdot\rangle}\hat{x}_1[M^n\xi]] \le \Phi^*[E^{\langle\cdot\rangle}\hat{x}_1[M^n\xi]^2]$$
$$\le O(\lambda^n n^{\gamma-1})\,\Phi^*[E^{\langle\cdot\rangle}\hat{x}_1[\varphi]^2] = O(\lambda^n n^{\gamma-1}).$$

Finally (4.5) follows from (4.4) since $\rho^{-k}\sigma^2(M^k\xi)$ is the sum of the terms for $n = k+1, k+2, \ldots$ □

4.2. PROPOSITION. Suppose that $\rho > 1$, $\lambda^2 = \rho$ and that the process has finite variance w.r.t. ξ and φ. Then: (i) $\sigma^2(\xi) := \lim_{N\to\infty} \sigma_N^2(\xi)$ $:= \lim_{N\to\infty}(\rho^N N^{2\gamma-1})^{-1}\Phi^*[\mathrm{Var}^{\langle\cdot\rangle}\hat{x}_N[\xi]]$ exists and is finite. Furthermore, $\sigma^2(\xi) = 0$ if and only if $\hat{x}_1[\xi_N]$ (with ξ_N the leading term in $(\lambda^N\binom{N}{\gamma-1})^{-1}M^N\xi$ as in (2.6)) degenerates $[\Phi^*]$ for all n; (ii) for any sequence $\{n(N)\}$ such that $N/n(N) \to \theta > 1$,

$$\lim_{N\to\infty}(\rho^N N^{2\gamma-1})^{-1}\Phi^*[\mathrm{Var}^{\langle\cdot\rangle}(\hat{x}_N[\xi]-\hat{x}_{n(N)}[M^{N-n(N)}\xi]] = \sigma^2(1-\tfrac{1}{\theta})^{2\gamma-1};$$

(iii) $\sigma^2(M^k\xi) = \rho^k\sigma^2(\xi)$.

PROOF. The basic analytic observation for the proof is $\Sigma_0^N n^{2\gamma-2} \cong N^{2\gamma-1}/(2\gamma-1)$. From (4.3) and (4.6) we have at once that

$$\overline{\lim_{N\to\infty}}\,\sigma_N^2(\xi) = \overline{\lim_{N\to\infty}}\,N^{-(2\gamma-1)}\sum_{n=1}^{N} O(\binom{n-1}{\gamma-1}^2) < \infty.$$

In order to obtain the more precise assertions of 4.2, a more careful study of the terms is needed. Letting ξ_N be the leading term in $(\lambda^N\binom{N}{\gamma-1})^{-1}M^N\xi$ as in (2.6) and $\tau_n^2 := \rho^{-1}\Phi^*[\mathrm{Var}^{\langle\cdot\rangle}\hat{x}_1[\xi_{n-1}]$, it is clear from (2.7) that

(4.7) $$\sigma_N^2 = N^{-(2\gamma-1)}\sum_{n=\gamma}^{N}\binom{n-1}{\gamma-1}^2\tau_n^2 + o(1) = N^{-(2\gamma-1)}\sum_{n=\gamma}^{N} n^{2\gamma-2}\tau_n^2 + o(1),$$

noting the fact that $\{\tau_n^2\}$ is bounded by 2.6. We now need

4.3. LEMMA. If $\vartheta_1, \ldots, \vartheta_r$ are arbitrary in $[0, 2\pi)$, then for each $\epsilon > 0$ the sequence of k for which

(4.8) $$\sum_{j=1}^{r} |e^{ik\vartheta_j} - 1| \le \epsilon$$

has positive density. More precisely, there is a $\ell = \ell(\epsilon)$ such that for any s (4.8) holds for some k in $[s, s+\ell]$.

For the proof, see e.g. van der Corput (1931), in particular pg. 421.

It is clear from the explicit expression for ξ_N that to each $\delta > 0$ there is a $\varepsilon > 0$ such that $|\tau_n^2 - \tau_{n+k}^2| \leq \delta$ for all n if (4.8) holds, with the ϑ_j as the θ_ν with $\lambda_\nu^2 = \rho$. Choose ℓ as in 4.3, fix M and let $t(N+1)$ be the first k after $t(N) + M$ such that (4.8) holds, $t(0) := \gamma$. Then $t(N+1)-t(N)-M \leq \ell$, and $|\tau_n^2 - \tau_{t(N)+n}^2| \leq \delta$ for all n and N. Write

$$\sum_{n=\gamma}^{t(N+1)} n^{2\gamma-2}\tau_n^2 = \sum_{i=0}^{N} \sum_{n=t(i)}^{t(i)+M} n^{2\gamma-2}\tau_n^2 + \sum_{i=0}^{N} \sum_{n=t(i)+M+1}^{t(i+1)} n^{2\gamma-2}\tau_n^2$$

and note first that the second term can be bounded by $K\ell C_N$, where $K := \sup \tau_n^2$, $C_N := \Sigma_0^N t(i+1)^{2\gamma-2}$. The first term is, up to $o(t(N)^{2\gamma-2})$,

$$\sum_{i=0}^{N} t(i+1)^{2\gamma-2} \sum_{n=t(i)}^{t(i)+M} \tau_n^2 = C_N \sum_{n=t(0)}^{t(0)+M} \tau_n^2 + A_N,$$

where $|A_N| \leq (M+1)\delta\, C_N$. Noting that $iM \leq t(i)-\gamma \leq i(M+\ell)$ implies that

$$\frac{1}{2\gamma-1} \cdot \frac{M^{2\gamma-2}}{(M+\ell)^{2\gamma-1}} \leq \varliminf_{N\to\infty} \frac{C_N}{t(N+1)^{2\gamma-1}} \leq \varlimsup_{N\to\infty} \frac{C_N}{t(N+1)^{2\gamma-1}} \leq \frac{1}{2\gamma-1} \frac{(M+\ell)^{2\gamma-2}}{M^{2\gamma-1}}$$

and inserting in (4.7), it follows that

$$\varlimsup_{n\to\infty} \sigma_n^2(\xi) = \varlimsup_{N\to\infty} \sigma_{t(N+1)}^2(\xi) \leq \frac{1}{2\gamma-1} \frac{(M+\ell)^{2\gamma-2}}{M^{2\gamma-1}} \Big(\sum_{n=t(0)}^{t(0)+M} \tau_n^2 + (M+1)\delta + K\ell \Big),$$

$$\varliminf_{n\to\infty} \sigma_n^2(\xi) = \varliminf_{N\to\infty} \sigma_{t(N+1)}^2(\xi) \geq \frac{1}{2\gamma-1} \frac{M^{2\gamma-2}}{(M+\ell)^{2\gamma-1}} \Big(\sum_{n=t(0)}^{t(0)+M} \tau_n^2 - (M+1)\delta \Big).$$

Bounding $\Sigma_{t(0)}^{t(0)+M} \tau_n^2$ by $(M+1)K$ and letting $M \to \infty$ shows that the difference between the $\varlimsup$ and the $\varliminf$ cannot exceed $2\delta/(2\gamma-1)$. Let $\delta \downarrow 0$ to obtain the existence of $\sigma^2(\xi) = \lim_{N\to\infty} \sigma_N^2(\xi)$.

It is clear at once from the above expressions that $\sigma^2(\xi) = 0$ if all $\hat{x}_1[\xi_N]$ degenerate $[\Phi^*]$, i.e. if all $\tau_n^2 = 0$. If $\tau_n^2 \geq 2\varepsilon$ (say) for some n, then a similar application of 4.3 as above shows that the sequence of n with $\tau_n^2 \geq \varepsilon$ has positive density so that $\varliminf_{M\to\infty} M^{-1}\Sigma_{t(0)}^{t(0)+M} \tau_n^2 > 0$. Inserting in the expression for $\varliminf \sigma_n^2(\xi)$ and letting first $M \to \infty$ and next $\delta \downarrow 0$ shows that indeed $\sigma^2(\xi) > 0$.

Part (ii) is a consequence of part (i) and (4.3) since

$$\Phi^*[\mathrm{Var}^{\langle\cdot\rangle}\{\hat{x}_N[\xi] - \hat{x}_{n(N)}[M^{N-n(N)}\xi]\}] =$$

$$\sum_{n=n(N)}^{N-1} \rho^{n}\Phi^{*}[\mathrm{Var}^{\langle\cdot\rangle}\hat{x}_1[M^{N-n-1}\xi]] =$$

$$\rho^{n(N)}\Phi^{*}[\mathrm{Var}^{\langle\cdot\rangle}\hat{x}_{N-n(N)}[\xi]] \cong \rho^{N}(N-n(N))^{2\gamma-1}\sigma^{2}(\xi).$$

For part (iii), note first that the leading term in $(\lambda^{N}\binom{N}{\gamma-1})^{-1}M^{N}(M^{k}\xi)$ is $\lambda^{k}\xi_{N+k} + o(1)$. Hence, replacing ξ by $M^{k}\xi$ in (4.7),

$$\sigma_N^2(M^k\xi) = N^{-(2\gamma-1)}\sum_{n=\gamma}^{N}\lambda^{2k}n^{2\gamma-2}\tau_{n+k}^2 + o(1)$$

$$\cong \rho^{k}N^{-(2\gamma-1)}\sum_{n=\gamma}^{N+k} n^{2\gamma-2}\tau_n^2 = \rho^{k}\sigma^{2}(\xi). \qquad \square$$

Though nice expressions for variances only obtain after integration w.r.t. Φ^{*}, one often needs also some information on $\mathrm{Var}\ x_N[\xi]$ with more general initial distributions. The basic identity used here is

$$(4.9)\quad \mathrm{Var}\ \hat{x}_{N+k}[\xi] = E(\mathrm{Var}(\hat{x}_{N+k}[\xi]\,|\,\mathfrak{F}_k)) + \mathrm{Var}(E(\hat{x}_{N+k}[\xi]\,|\,\mathfrak{F}_k))$$

$$= E\hat{x}_k[\mathrm{Var}^{\langle\cdot\rangle}\hat{x}_N[\xi]] + \mathrm{Var}\ \hat{x}_k[M^{N}\xi].$$

Note that $\mathrm{Var}\ \hat{x}_N[\xi]$ may be infinite without second moment assumptions on $\hat{x}_0$ (take $k = 0$ in (4.9), $\mathrm{Var}\ \hat{x}_0[M^N\xi] = \infty$). To avoid this problem, one simply conditions on $\mathfrak{F}_0$ or consider only $P = P^{\langle x\rangle}$. We collect some useful and/or interesting consequences of (4.9) in

4.4. LEMMA. *Suppose* $\lambda^2 \leq \rho$, $\sigma^2 > 0$. *Then*

$$(4.10)\quad \lim_{N\to\infty}\frac{\mathrm{Var}^{\langle x\rangle}\hat{x}_N[\xi]}{\Phi^{*}[\mathrm{Var}^{\langle\cdot\rangle}\hat{x}_N[\xi]]} = \varphi(x)\quad \forall x\in X,$$

and for some $\zeta \in \mathcal{L}^1_{\Phi^*}$ (depending on ξ) *and all* N,

$$(4.11)\quad \mathrm{Var}^{\langle x\rangle}\hat{x}_N[\xi] \leq \begin{cases} \rho^N\zeta(x) & \text{if } \lambda^2 < \rho \\ \rho^N N^{2\gamma-1}\zeta(x) & \text{if } \lambda^2 = \rho \end{cases}.$$

Finally if $\lambda^2 < \rho$, *then*

$$(4.12)\quad \alpha_k := \varlimsup_{N\to\infty}\rho^{-N}E(\hat{x}_N[M^k\xi]^2\,|\,\mathfrak{F}_0) = o(\rho^k).$$

PROOF. Let $E = E^{\langle x\rangle}$ and let $N \to \infty$ with k fixed in (4.9). The last term on the r.h.s. is at most

$$E^{\langle x\rangle}\hat{x}_k[M^N\xi]^2 \leq O(\lambda^{2N}N^{2\gamma-2})E^{\langle x\rangle}\hat{x}_k[\varphi]^2 = o(\Phi^*[\mathrm{Var}^{\langle\cdot\rangle}\hat{x}_{N+k}[\xi]])$$

while the first, up to a factor tending to one as $k \to \infty$, is $\rho^k\varphi(x)\Phi^*[\mathrm{Var}^{\langle\cdot\rangle}\hat{x}_N[\xi]]$. Inserting the explicit form of $\Phi^*[\mathrm{Var}^{\langle\cdot\rangle}\hat{x}_N[\xi]]$ yields (4.10) and a minor change in the argument yields (4.11). For (4.12), it is clear that

$$\alpha_k = \{\lim_{N\to\infty} \rho^{-N}\mathrm{Var}(\hat{x}_N[M^k\xi]|\mathfrak{F}_0) + \rho^{-N}E(\hat{x}_N[M^k\xi]|\mathfrak{F}_0)^2\}$$

$$= \hat{x}_0[\varphi]\sigma^2(M^k\xi) + 0 = o(\rho^k) \quad \text{by (4.5)}. \quad \Box$$

4.5. PROPOSITION. *Suppose that* $\lambda_\nu^2 > \rho$ *and that the process has finite variance w.r.t.* φ [*and hence both* $\mathrm{Re}\,\varphi_{\nu,1}$ *and* $\mathrm{Im}\,\varphi_{\nu,1}$]. *Then* $\{W_N^\nu\} := \{\rho_\nu^{-N}\hat{x}_N[\varphi_{\nu,1}]\}$ *is a* $\mathfrak{L}^2$-*bounded martingale* [*i.e.,* $\{\mathrm{Re}\,W_N^\nu\}$ *and* $\{\mathrm{Im}\,W_N^\nu\}$ *are* $\mathfrak{L}^2$-*bounded real martingales*], *and therefore has an a.s. limit* W^ν. *If* ξ *is real with* $\lambda^2 = \lambda^2(\xi) > \rho$ *and the process has finite variance w.r.t.* φ, *then for any* k

$$U_k := \sum_\nu^{(\xi)} \rho_\nu^{-(\gamma-1)}\Phi^*_{\nu,\gamma}[\xi]e^{i\theta_\nu k}W^\nu$$

degenerates $[\Phi^*]$ *if and only if* $\hat{x}_1[\xi_N]$ (*with* ξ_N *defined by* (2.6)) *degenerates* $[\Phi^*]$ *for all* N. *Otherwise* $\Phi^*[P^{\langle\cdot\rangle}(\underline{\lim}_{k\to\infty}|U_k - E^{\langle\cdot\rangle}U_k|>0)]>0$.

PROOF. The martingale property follows immediately from the eigenvector property $M\varphi_{\nu,1} = \rho_\nu\varphi_{\nu,1}$ and $\mathfrak{L}^2$-boundedness for (say) $\{\mathrm{Re}\,W_N^\nu\}$ from

$$\mathrm{Var}(\mathrm{Re}\,W_n^\nu|\mathfrak{F}_{n-1}) = \hat{x}_{n-1}[\mathrm{Var}^{\langle\cdot\rangle}\hat{x}_1[\mathrm{Re}\,\rho_\nu^{-n}\varphi_{\nu,1}]]$$

$$\leq \hat{x}_{n-1}[E^{\langle\cdot\rangle}\hat{x}_1[\mathrm{Re}\,\rho_\nu^{-n}\varphi_{\nu,1}]^2]$$

$$\leq c_{\nu,1}^2\lambda_\nu^{-2n}\hat{x}_{n-1}[E^{\langle\cdot\rangle}\hat{x}_1[\varphi]^2],$$

$$\sup_N \mathrm{Var}\,W_n^\nu \leq c_{\nu,1}^2 \sum_{n=1}^\infty \lambda_\nu^{-2n}E\hat{x}_{n-1}[E^{\langle\cdot\rangle}\hat{x}_1[\varphi]^2] < \infty,$$

using $\mathfrak{L}^2$-orthogonality of the increments. Now U_k is again a martingale limit,

$$U_k = \lim_{N\to\infty} \hat{x}_N[\sum_\nu^{(\xi)} \rho_\nu^{-(\gamma-1)}\Phi^*_{\nu,\gamma}[\xi]e^{i\theta_\nu k}\rho_\nu^{-N}\varphi_{\nu,1}]$$

and, noting that by 2.6 everything is real, it follows as above that

$\Phi^*[\mathrm{Var}^{\langle\cdot\rangle} U_k]$

$$= \sum_{n=1}^{\infty} \Phi^*[E^{\langle\cdot\rangle}\hat{x}_{n-1}[\mathrm{Var}^{\langle\cdot\rangle}\hat{x}_1[\Sigma_\nu^{(\xi)}\rho_\nu^{-(\gamma-1)}\Phi^*_{\nu,\gamma}[\xi]e^{i\theta_\nu k}\rho_\nu^{-n}\varphi_{\nu,1}]]]$$
$$= \sum_{n=1}^{\infty} \rho^{n-1}\Phi^*[\mathrm{Var}^{\langle\cdot\rangle}(\lambda^{-n}\hat{x}_1[\Sigma_\nu^{(\xi)}\rho_\nu^{-(\gamma-1)}\Phi^*_{\nu,\gamma}[\xi]e^{i\theta_\nu(k-n)}\varphi_{\nu,1}]]$$
$$= \sum_{n=1}^{\infty} \frac{\rho^{n-1}}{\lambda^{2n}}\Phi^*[\mathrm{Var}^{\langle\cdot\rangle}\hat{x}_1[\xi_{k-n}]].$$

The degeneracy criterion asserted becomes clear by noting that by 4.3 either $\Phi^*[\mathrm{Var}^{\langle\cdot\rangle}\hat{x}_1[\xi_N]] \neq 0$ for no N or for infinitely many. Finally if $\Phi^*[\mathrm{Var}^{\langle\cdot\rangle}U_k] > 0$, then $\Phi^*[P^{\langle\cdot\rangle}A_k] > 0$, where $A_k := \{U_k \neq E^{\langle\cdot\rangle}U_k\}$. But on A_k, $\underline{\lim}_{N\to\infty}|U_{k+n}-E^{\langle\cdot\rangle}U_{k+n}| > 0$ by 4.3. □

For a simple example with $\hat{x}_1[\xi]$ degenerate $[\Phi^*]$, take the set of types as $\{1,2\}$, $\xi(x) := I(x=1) - I(x=2)$ and suppose that any particle produces the same number of offsprings of types 1 and 2, respectively. Then $\hat{x}_0[\xi] = \hat{x}_1[\xi] = \hat{x}_2[\xi] = \dots$ for any initial distribution. The example is, of course, somewhat special, and in general the situation of $\hat{x}_1[\xi]$ to degenerate $[\Phi^*]$ requires offspring dependencies of a special structure. For example, we have

4.6. PROPOSITION. *Suppose* $\{\hat{x}_t\}_{t\geq 0}$ *is a branching diffusion on a simply connected domain, with termination density* $k(x)$ *and a local branching law specified by* $\{p_n(x)\}$. *Suppose that* (M) *is satisfied, that* $\{p_n(x)\}$ *is not degenerate at* 1 *for a.a.* x *and that* $\xi(x) \neq 0$ *for a.a.* x. *Then* $\Phi^*[\mathrm{Var}^{\langle\cdot\rangle}\hat{x}_t[\xi]] > 0$ *for all* $t > 0$.

The proof is in Section 12.

5. EXPLOITING THE ADDITIVITY PROPERTY

We first give an elementary lemma formalizing some common steps in the proofs.

5.1. LEMMA. *Let* $\rho \geq 1$ *and let* $\{F_N\}$ *be any sequence of events. Then:* (a) *For any* k,

$$\overline{\lim_{N\to\infty}} P_{N-k}F_N = \overline{\lim_{N\to\infty}} P_N F_N, \quad \underline{\lim_{N\to\infty}} P_{N-k}F_N = \underline{\lim_{N\to\infty}} P_N F_N, \tag{5.1}$$

$$\overline{\lim_{N\to\infty}} P_N F_N = \overline{\lim_{N\to\infty}} P_\infty F_N, \quad \underline{\lim_{N\to\infty}} P_N F_N = \underline{\lim_{N\to\infty}} P_\infty F_N \quad \text{if} \quad \rho > 1; \tag{5.2}$$

(b) *If* $\lim_{N\to\infty} P_N(|P(F_N|\mathfrak{F}_N) - \alpha| > \varepsilon) = 0 \quad \forall\varepsilon > 0$, *then* $\lim_{N\to\infty} P_N F_N = \alpha$, *and if* $\rho > 1$, *then also* $\lim_{N\to\infty} P_\infty F_N = \alpha$.

(c) *If* A_N, $B_{N,k}$, $C_{N,k}$ *are* *r.v. such that* $A_N = B_{N,k} + C_{N,k}$ $\forall k$ *and that*

$$q_{k,\varepsilon} := \overline{\lim_{N\to\infty}} P_{N-k}(|C_{N,k}| > \varepsilon) \to 0 \quad \text{as} \quad k \to \infty \quad \forall\varepsilon > 0, \tag{5.3}$$

$$\lim_{N\to\infty} P_{N-k}(B_{N,k} \leq \tau_k y) = \Phi(y) \quad \forall y \in \mathbb{R} \quad \forall k \quad \text{or} \tag{5.4)'}$$

$$\lim_{N\to\infty} P_{N-k}(|P(B_{N,k} \leq \tau_k y|\mathfrak{F}_{N-k}) - \Phi(y)| > \varepsilon) = 0 \quad \forall k, \varepsilon, y \tag{5.4)''}$$

then if $\tau_k^2 \to \tau^2 \in (0,\infty)$ *as* $k \to \infty$,

$$\lim_{N\to\infty} P_N(A_N \leq \tau y) = \Phi(y) \quad \forall y \in \mathbb{R} \tag{5.5}$$

$$\lim_{N\to\infty} P_\infty(A_N \leq \tau y) = \Phi(y) \quad \forall y \in \mathbb{R} \quad \text{if} \quad \rho > 1. \tag{5.6}$$

PROOF. (a) We recall that if $\rho = 1$, then

$$\lim_{N\to\infty} NP(\hat{x}_N \neq 0) := \varkappa \tag{5.7}$$

exists (with $\varkappa \in (0,\infty)$ depending on initial conditions). Therefore

$$\overline{\lim_{N\to\infty}} P_{N-k}F_N = \overline{\lim_{N\to\infty}} \frac{P(\hat{x}_N \neq 0)}{P(\hat{x}_{N-k} \neq 0)} \frac{P(F_N, \hat{x}_{N-k} \neq 0)}{P(\hat{x}_N \neq 0)} =$$

$$\overline{\lim_{N\to\infty}} \frac{P(F_N, \hat{x}_N \neq 0) + P(F_N, \hat{x}_N = 0, \hat{x}_{N-k} \neq 0)}{P(\hat{x}_N \neq 0)} =$$

$$\overline{\lim_{N\to\infty}}\{P_N F_N + O(\frac{P(\hat{x}_{N-k}\neq 0) - P(\hat{x}_N\neq 0)}{P(\hat{x}_N \neq 0)})\} = \overline{\lim_{N\to\infty}}\{P_N F_N + O(\frac{1}{N})\} = \overline{\lim_{N\to\infty}} P_N F_N .$$

The $\underline{\lim}$ part of (5.1) is similar and (5.2) follows in quite the same way from the expressions for $P(x_N \neq 0)$ in the supercritical case. Part (b) follows from arguments of the type

$$\overline{\lim_{N\to\infty}} P_N F_N = \overline{\lim_{N\to\infty}} P_N(F_N, |P(F_N|\mathfrak{F}_N) - \alpha| \leq \varepsilon)$$

$$= \overline{\lim_{N\to\infty}} \frac{EI(\hat{x}_N \neq 0, |P(F_N|\mathfrak{F}_N)-\alpha| \leq \varepsilon)P(F_N|\mathfrak{F}_N)}{P(\hat{x}_N \neq 0)}$$

$$\leq (\alpha+\varepsilon)\overline{\lim_{N\to\infty}} P_N(|P(F_N|\mathfrak{F}_N)-\alpha| \leq \varepsilon) \leq \alpha+\varepsilon .$$

For part (c), note first that (5.4)" implies (5.4)' because of part (b). If (5.4)' holds, then

$$\overline{\lim_{N\to\infty}} P_N(A_N \leq \tau y) = \overline{\lim_{N\to\infty}} P_{N-k}(A_N \leq \tau y)$$

$$\leq \overline{\lim_{N\to\infty}}\{P_{N-k}(B_{N,k} \leq \tau y + \varepsilon) + P_{N-k}(C_{N,k} < -\varepsilon)\} = \Phi(\frac{\tau y+\varepsilon}{\tau_k}) + q_{k,\varepsilon} .$$

The $\overline{\lim} \leq$ part follows by letting first $k \to \infty$ and next $\varepsilon \to 0$. $\underline{\lim} \geq$ is similar. □

We now turn to the

PROOF OF 1.2. Define

$$F^{\langle x\rangle}(y) := P^{\langle x\rangle}(Y \leq y), \quad F(y) := \Phi^*[F^{\langle\cdot\rangle}(y)],$$

$$\omega_N^2 := \mathrm{Var}(S_N|\mathfrak{F}_N) = \hat{x}_N[\int y^2 dF^{\langle\cdot\rangle}(y)],$$

$$\delta_N := \omega_N^{-2} \sum_{i=1}^{|\hat{x}_N|} E(Y_{N,i}^2 I(|Y_{N,i}| > \varepsilon\omega_N)|\mathfrak{F}_N) = \omega_N^{-2}\hat{x}_N[\int_{|y|>\varepsilon\omega_N} y^2 dF^{\langle\cdot\rangle}(y)] .$$

Then a.s.,

$$\lim_{N\to\infty} \rho^{-N}\omega_N^2 = W\Phi^*[\int y^2 dF^{\langle\cdot\rangle}(y)] = W\tau^2, \tag{5.8}$$

$$\lim_{N\to\infty} \rho^{-N}\hat{x}_N[\int_{|y|>A} y^2 dF^{\langle\cdot\rangle}(y)] = W\Phi^*[\int_{|y|>A} y^2 dF^{\langle\cdot\rangle}(y)] \tag{5.9}$$

and hence $\delta_N \to 0$ a.s. on $\{W > 0\}$ so that (1.4) is immediate from the Lindeberg-Feller criterion and the branching property. From part (b) of 5.1, (1.4) implies $P_\infty(S_N \leq W^{1/2}\tau y) = \Phi(y)$ $\forall y$, completing the proof of the CLT.

The proof of the LIL is a more or less routine extension of the proof of II.3.2, given the ideas from Chapter VII, but is given for the sake of completeness. Write for notational convenience $EY_{N,i}$ instead of $E(Y_{N,i}|\mathfrak{F}_N)$ etc. and define

$$Y'_{N,i} := Y_{N,i} I(|Y_{N,i}| \leq \rho^{N/2}), \quad \tilde{Y}_{N,i} := Y'_{N,i} - EY'_{N,i},$$

$$\tilde{S}_N := \sum_{i=1}^{|\hat{x}_N|} \tilde{Y}_{N,i}, \quad \tilde{\omega}_N^2 := \operatorname{Var}(\tilde{S}_N|\mathfrak{F}_N), \quad T_N := \tilde{S}_N/\tilde{\omega}_N.$$

By a standard moment inequality,

$$E|\tilde{Y}_{N,i}|^3 \leq E|Y'_{N,i}|^3 + 3E|Y'_{N,i}|(E|Y'_{N,i}|)^2 + 3E|Y'_{N,i}|EY'^2_{N,i} + E|Y'_{N,i}|^3$$
$$\leq 8E|Y'_{N,i}|^3 = 8\int_{|y|\leq\rho^{N/2}} |y|^3 dF^{\langle x_i\rangle}(y).$$

Letting C be the Berry-Esseen constant, we get on $\{\tilde{\omega}_N > 0\}$

$$\Delta_N := \sup_{-\infty<y<\infty} |P(T_N \leq y|\mathfrak{F}_N) - \Phi(y)|$$
$$\leq 8C\tilde{\omega}_N^{-3}\hat{x}_N\left[\int_{|y|\leq\rho^{N/2}} |y|^3 dF^{\langle\cdot\rangle}(y)\right].$$

Since one can easily check that

$$\lim_{N\to\infty} \rho^{-N}\tilde{\omega}_N^2 = W\tau^2, \tag{5.10}$$

it follows that the assumption A(7.3) $\Sigma_0^\infty \Delta_N < \infty$ of A 7.2 holds since

$$\sum_{N=1}^{\infty} \rho^{-3N/2} E\hat{x}_N\left[\int_{|y|\leq\rho^{N/2}} |y|^3 dF^{\langle\cdot\rangle}(y)\right]$$
$$\leq C\sum_{N=1}^{\infty} \rho^{-N/2}\int_{|y|\leq\rho^{N/2}} |y|^3 dF(y) = \int_{-\infty}^{\infty} O(y^2)dF(y) < \infty.$$

Thus by A7.2 it suffices to verify

$$\overline{\lim_{N\to\infty}}\, S_N/(2\tau^2 W\rho^N \log N)^{1/2} = \overline{\lim_{N\to\infty}}\, T_N/(2\log N)^{1/2}. \tag{5.11}$$

Recalling (5.10) and the explicit definitions of $Y'_{N,i}$, $\tilde{Y}_{N,i}$, T_N, it suffices that

$$\sum_{i=1}^{|\hat{x}_N|} \{Y_{N,i} - Y'_{N,i}\} = o((\rho^N \log N)^{1/2}),$$

$$\sum_{i=1}^{|\hat{x}_N|} EY'_{N,i} = o((\rho^N \log N)^{1/2}),$$

or, appealing to Kronecker's lemma, that

$$\sum_{N=0}^{\infty} (\rho^N \log N)^{-1/2} \sum_{i=1}^{|\hat{x}_N|} |Y_{N,i} - Y'_{n,i}| < \infty,$$

(5.12)

$$\sum_{N=0}^{\infty} (\rho^N \log N)^{-1/2} \sum_{i=1}^{|\hat{x}_N|} |EY'_{n,i}| < \infty.$$

Noting that

$$|EY'_{N,i}| = |E(Y'_{N,i} - Y_{N,i})| \leq E|Y'_{N,i} - Y_{N,i}|$$

$$= E|Y_{N,i}|I(|Y_{N,i}| > \rho^{N/2}) = \int_{|y|>\rho^{N/2}} |y| dF^{\langle x_i \rangle}(y),$$

it suffices for both assertions of (5.12) that

$$\sum_{N=0}^{\infty} (\rho^N \log N)^{-1/2} \hat{x}_N [\int_{|y|>\rho^{N/2}} |y| dF^{\langle \cdot \rangle}(y)] < \infty \tag{5.13}$$

(for the first, take $E(\cdot|\mathfrak{F}_N)$ of the Nth term and use A 1.2). And (5.13) certainly holds because the mean of the l.h.s. is finite since even

$$\sum_{N=0}^{\infty} \rho^{N/2} \Phi^*[\int_{|y|\geq\rho^{N/2}} |y| dF^{\langle \cdot \rangle}(y)] = \sum_{N=0}^{\infty} \rho^{N/2} \int_{|y|>\rho^{N/2}} |y| dF(y)$$

$$= \int_{-\infty}^{\infty} O(y^2) dF(y) < \infty. \qquad \square$$

PROOF OF 1.1. Let $Y: = W - W_0 = W - \hat{x}_0[\varphi]$ in 1.2. Then $S_N = \rho^N(W-W_N)$ and therefore the CLT is immediate from 1.2 as well as

$$\overline{\lim_{N\to\infty}}\, C_N(W-W_N) \leq 1, \quad \underline{\lim_{N\to\infty}}\, C_N(W-W_N) \geq -1, \tag{5.14}$$

where $C_N: = (\rho^N/2\tau^2 W \log N)^{1/2}$. Let next $Y: = W_k - W_0$, $\tau_k^2: = \Phi^*[\mathrm{Var}^{\langle \cdot \rangle} W_k]$. Then $S_N = \rho^N(W_{N+k}-W_N)$ and equality holds in (1.5). Thus

$$C_N(W-W_N) = C_{N+k}(W-W_{N+k})\frac{C_N}{C_{N+k}} + \frac{S_N}{(2\tau_k^2 W \rho^N \log N)^{1/2}} \frac{\tau_k}{\tau},$$

$$\overline{\lim_{N\to\infty}}\, C_N(W-W_N) \geq -\lim_{N\to\infty} \frac{C_N}{C_{N+k}} + \frac{\tau_k}{\tau} = -\rho^{-k/2} + \frac{\tau_k}{\tau} .$$

As $k \uparrow \infty$, $\overline{\lim} = 1$ follows by combining with (5.14). □

We next turn to the critical case where we shall need

5.2. LEMMA. Under the assumptions of 3.3, it holds for all $\eta \in \mathcal{L}^1_{\Phi^*}$, $k = 0,1,2,\ldots$ and $\varepsilon > 0$ that

$$(5.15) \qquad \lim_{N\to\infty} P_{N-k}\left(\left|\frac{\hat{x}_{N-k}[\eta]}{\hat{x}_N[\varphi]} - \Phi^*[\eta]\right| > \varepsilon\right) = 0.$$

PROOF. By VI.3.14 , (5.15) holds if $k = 0$, i.e.

$$(5.16) \qquad \lim_{N\to\infty} P_{N-k}\left(\left|\frac{\hat{x}_{N-k}[\eta]}{\hat{x}_{N-k}[\varphi]} - \Phi^*[\eta]\right| > \varepsilon\right) = 0,$$

so that it suffices to consider (5.15) for $\eta = \varphi$, where the statement is equivalent to

$$(5.17) \qquad \lim_{N\to\infty} P_{N-k}\left(\left|\frac{\hat{x}_N[\varphi]}{\hat{x}_{N-k}[\varphi]} - 1\right| > \varepsilon\right) = 0.$$

Now

$$P\left(\left|\frac{\hat{x}_N[\varphi]}{\hat{x}_{N-k}[\varphi]} - 1\right| > \varepsilon \,\middle|\, \mathfrak{F}_{N-k}\right) \leq \frac{\mathrm{Var}(\hat{x}_N[\varphi]\,|\,\mathfrak{F}_{N-k})}{\varepsilon^2 \hat{x}_{N-k}[\varphi]^2} = \frac{\hat{x}_{N-k}[\mathrm{Var}^{\langle\cdot\rangle}\hat{x}_k[\varphi]]}{\varepsilon^2 \hat{x}_{N-k}[\varphi]^2} .$$

Appealing to (5.16) with $\eta := \mathrm{Var}^{\langle\cdot\rangle}\hat{x}_k[\varphi]$ and to $P_{N-k}(\hat{x}_{N-k}[\varphi] \leq y) \to 0 \quad \forall y$,

$$\lim_{N\to\infty} P_{N-k}\left(\frac{\hat{x}_{N-k}[\mathrm{Var}^{\langle\cdot\rangle}\hat{x}_k[\varphi]]}{\hat{x}_{N-k}[\varphi]^2} > \delta\right) = 0 .$$

Hence (5.17) follows by part (b) of 5.1. □

PROOF OF 3.3. Define

$$\vartheta_k^2(x) := \mathrm{Var}^{\langle x\rangle}\hat{x}_k[\xi], \quad \sigma_k^2 := \Phi^*[\vartheta_k^2], \quad \tau_k := \sigma_k/\sigma,$$

$$A_N := \frac{\hat{x}_N[\xi]}{(\sigma^2\hat{x}_N[\varphi])^{1/2}}, \quad \omega_{N,k}^2 := \mathrm{Var}(\hat{x}_N[\xi]\,|\,\mathfrak{F}_{N-k}) = \hat{x}_{N-k}[\vartheta_k^2]$$

$$B_{N,k} := B'_{N,k}\cdot B''_{N,k} := \frac{\hat{x}_N[\xi]-\hat{x}_{N-k}[M^k\xi]}{\omega_{N,k}} \cdot \frac{\omega_{N,k}}{(\sigma^2\hat{x}_N[\varphi])^{1/2}}$$

$$C_{N,k}:=C'_{N,k}\cdot C''_{N,k}:=\frac{\hat{x}_{N-k}[M^k\xi]}{(N-k)^{1/2}}\cdot\left(\frac{N-k}{\sigma^2\hat{x}_N[\varphi]}\right)^{1/2}.$$

We prove the CLT (3.7) by verifying the conditions of part (c) of 5.1. By 4.1, $\tau_k\to 1$ as $k\to\infty$, and by 5.2, $P_{N-k}(|B''_{N,k}-\tau_k|>\varepsilon)\to 0$ as $N\to\infty$ so that in (5.4)" we can replace $B_{N,k}$ by $B'_{N,k}$ and τ_k by one. Letting

$$\delta_{N,k}:=\omega_N^{-2}\hat{x}_{N-k}[E^{\langle\cdot\rangle}\hat{x}_k[\xi]^2I(|\hat{x}_k[\xi]>\varepsilon\omega_N)],$$

it suffices for (5.4)" by the Lindeberg-Feller criterion that $P_{N-k}(\delta_{N,k}>\delta)\to 0$ as $N\to\infty$. But since $P_{N-k}(\omega^2_{N,k}>(N-k)\sigma_k y)\to e^{-y/\mu}$ as $N\to\infty$, cf. VI.3.13, this follows exactly as in the proof of the CLT in 1.2 (replacing ρ^N by $N-k$). For (5.3), note first that $C''_{N,k}$ has a limit law (w.r.t. the P_{N-k}) in $(0,\infty)$, so that we can replace $C_{N,k}$ by $C'_{N,k}$. Suppose without loss of generality $\hat{x}_0$ to be degenerate and define α_k by (4.12). We get

$$q_{k,\varepsilon}\le\varepsilon^{-2}\overline{\lim_{N\to\infty}}\,E_{N-k}C'^{\,2}_{N-k}=\varepsilon^{-2}\overline{\lim_{N\to\infty}}\,\frac{EC'^{\,2}_{N,k}}{P(\hat{x}_{N-k}\neq 0)}=\frac{\alpha_k}{\varkappa\varepsilon^2},$$

cf. (5.7). Now (4.12) proves (5.3).

The proof of the Laplacian limit law (3.8) is a minor modification. Define

$$A_N:=\frac{\hat{x}_N[\xi]}{\sigma N^{1/2}},\quad C_{N,k}:=\frac{\hat{x}_{N-k}[M^k\xi]}{\sigma N^{1/2}},$$

$$B_{N,k}:=B'_{N,k}\cdot B'''_{N,k}:=\frac{\hat{x}_N[\xi]-\hat{x}_{N-k}[M^k\xi]}{\omega_{N,k}}\cdot\frac{\omega_{N,k}}{\sigma N^{1/2}}.$$

The estimate (5.3) follows exactly as before (even simpler) and we also have from the above that

$$\lim_{N\to\infty}P_{N-k}(|P(B'_{N,k}\le y|\mathfrak{F}_{N-k})-\Phi(y)|>\varepsilon)=0,$$

$$\lim_{N\to\infty}P_{N-k}(B'''^{\,2}_{N,k}>y)=e^{-y/\mu\tau_k^2}.$$

Hence

$$\lim_{N\to\infty}P_N(B_{N,k}\le y)=\lim_{N\to\infty}P_{N-k}(B_{N,k}\le y)=P(\tau_k U^{1/2}Y\le y),$$

where U,Y are independent, Y is standard normal and $P(U>u)=e^{-u/\mu}$.

As in 5.1 (c), this implies $\lim_{N\to\infty} P_N(A_N \le y) = P(U^{1/2}Y \le y)$. The proof is therefore completed by elementary calculations showing that the distribution of $U^{1/2}Y$ is Laplacian with the parameter asserted. □

PROOF OF THE CLT IN THE CASE $1^\circ(\lambda^2 < \rho)$ OF 3.1. The argument is parallel to the critical case, with the main exception that the Lindeberg condition has already been checked in 1.2. Define

$$A_N := \frac{\hat{x}_N[\xi]}{(\sigma^2 W\rho^N)^{1/2}}, \quad C_{N,k} := C'_{N,k}\cdot C''_{N,k} := \frac{\hat{x}_{N-k}[M^k\xi]}{\rho^{N/2}} \cdot \frac{1}{W^{1/2}\sigma},$$

$$B_{N,k} := B'_{N,k}\cdot\tau_k := \frac{\hat{x}_N[\xi]-\hat{x}_{N-k}[M^k\xi]}{(W\Phi^*[\mathrm{Var}^{\langle\cdot\rangle}\hat{x}_k[\xi]]\rho^{N-k})^{1/2}} \cdot \left(\frac{\rho^{-k}\Phi^*[\mathrm{Var}^{\langle\cdot\rangle}\hat{x}_k[\xi]]}{\sigma^2}\right)^{1/2}.$$

Then (5.4)' follows at once from 1.2 with $Y := \hat{x}_k[\xi] - \hat{x}_0[M^k\xi]$, $S_N = \hat{x}_{N+k}[\xi] - \hat{x}_N[M^k\xi]$ and (5.3) follows from $P_\infty(0 < W < \infty) = 1$ and

$$\overline{\lim_{N\to\infty}}\, P(|C'_{N,k}| > \epsilon) \le \alpha_k/\epsilon^2\rho^k,$$

cf. (4.12). The proof is complete since $\tau_k \to 1$. □

We next turn to the LIL.

5.3. LEMMA. <u>A sufficient condition for the LIL in part 1° of 3.1 is that</u>

$$D_k := \overline{\lim_{N\to\infty}} \left|\frac{\hat{x}_N[M^k\xi]}{(2\sigma^2 W\rho^N \log N)^{1/2}}\right| = o(\rho^{k/2}) \quad \text{as} \quad k\to\infty. \tag{5.18}$$

PROOF. Using the notation in the proof of the CLT above and applying 1.2, we get

$$\overline{\lim_{N\to\infty}} \frac{A_N}{(2\log N)^{1/2}} \le \overline{\lim_{N\to\infty}} \frac{B_N}{(2\log N)^{1/2}} + \rho^{-k/2}D_k = \tau_k + \rho^{-k/2}D_k,$$

$$\overline{\lim_{N\to\infty}} \frac{A_N}{(2\log N)^{1/2}} \ge \overline{\lim_{N\to\infty}} \frac{B_N}{(2\log N)^{1/2}} - \rho^{-k/2}D_k = \tau_k - \rho^{-k/2}D_k.$$

As $k\to\infty$, $\overline{\lim} = 1$ follows since $\tau_k \to 1$. □

PROOF OF THE LIL IN THE CASE $1^\circ(\lambda^2 < \rho)$ OF 3.1 AND OF A FINITE SET OF TYPES. Choosing $\{\eta_1,\eta_2,\ldots,\eta_s\}$ as a base for $\mathrm{span}\{\xi, M\xi, M^2\xi,\ldots\}$, we can write $M^k\xi := \alpha_1^k\eta_1 + \ldots + \alpha_s^k\eta_s$, where $\alpha^k := |\alpha_1^k| + \ldots + |\alpha_s^k|$ satisfies

$$\alpha^k = O(\lambda^k\gamma^{k-1}) \quad \text{and hence} \quad \sum_{n=0}^{\infty} \frac{\alpha_{n+k}}{\rho^{n/2}} = O(\lambda^k\gamma^{k-1}). \tag{5.19}$$

Using 1.2 as above, it follows that

$$E_i := \sup_{n=0,1,2,\ldots} \left|\frac{\hat{x}_n[\eta_i]-\hat{x}_{n-1}[M\eta_i]}{(2\sigma^2 W\rho^n \log n)^{1/2}}\right| < \infty, \quad E := E_1 \ldots E_s < \infty.$$

Replacing ξ by $M^k\xi$ in the usual expansion (4.1), we get

$$\left|\frac{\hat{x}_N[M^k\xi]}{(2\sigma^2 W\rho^N \log N)^{1/2}}\right| \leq \frac{\hat{x}_0[M^{N+k}\xi]}{(2\sigma^2 W\rho^N \log N)^{1/2}} + \sum_{n=0}^{N-1}\left|\frac{\hat{x}_{n+1}[M^{N+k-n-1}\xi]-\hat{x}_n[M^{N+k-n}\xi]}{(2\sigma^2 W\rho^N \log N)^{1/2}}\right|$$

$$\leq O\left(\frac{\alpha^{N+k}}{(\rho^N \log N)^{1/2}}\right) + E\sum_{n=0}^{N-1}\frac{\alpha^{N+k-n-1}}{\rho^{(N-n-1)/2}} = O\left(\frac{\alpha^{N+k}}{(\rho^N \log N)^{1/2}}\right) + E\sum_{n=0}^{N-1}\frac{\alpha^{n+k}}{\rho^{n/2}} .$$

(5.18) follows by (5.19). □

6. EXPLOITING MARTINGALES: A CLT FOR TRIANGULAR ARRAYS

The identity (4.1) gives an expansion of $\hat{x}_N[\xi]$ in a sum of martingale increments $U_n^N := \hat{x}_{n+1}[M^{N-n-1}\xi] - \hat{x}_n[M^{N-n}\xi]$ and suggests therefore to approach the CLT via the martingale CLT A 8.1. Some problems come up, however: The norming in the CLT's in 3.1 involves a r.v. W in addition to constants and the Lindeberg condition corresponding to (4.1) would lead to computation of quantities like $E(U_n^{N2} I(|U_n^N| > c_n^N)|\mathfrak{F}_n)$, which is not straightforward since, given $\mathfrak{F}_n$, U_n^N is a sum

$$U_n^N := \sum_{i=1}^{|\hat{x}_n|} V_{n,i}^N := \sum_{i=1}^{|\hat{x}_n|} \{x_{n+1}^{n,i}[M^{N-n-1}\xi] - E(x_{n+1}^{n,i}[M^{N-n-1}\xi]|\mathfrak{F}_n)\}.$$

In contrast, $E(V_{n,i}^N I(|V_{n,i}^N| > c_n^N)|\mathfrak{F}_n)$ is a simple function of the type x_i of the ith particle alive at time n and we shall therefore consider martingales indexed not by physical time but instead by the set $\{n,i: n = 0,1,2,\ldots; i = 1,\ldots,|\hat{x}_n|\}$ of all individuals ever alive. This turns out also to be an appropriate approach for the LIL.

It will be convenient to order the n,i linearly. To this end, let $\tau_N := |\hat{x}_0|+\ldots+|\hat{x}_N|$ and assume the nth generation represented as $\{k \in \mathbb{N}: \tau_{n-1} < k \leq \tau_n\}$. When needed, $n = n(k)$ denotes the generation of k (the dependence on k is suppressed for brevity) and we let y_k be the offspring population produced by k (i.e. $y_k = x_{n+1}^{n,i}$ with $i = k - \tau_{n-1}$), $\mathfrak{G}_k = \sigma(y_\ell;\ \ell \leq k)$ so that $\mathfrak{F}_{N+1} = \mathfrak{G}_{\tau_N}$.

The main result 6.1 of the present section is an adaptation of A 8.1 to this setting. Applications to linear functionals are in the next section but the scope seems substantially wider. Somewhat related applications are given as examples in Section 11, but one could easily think of more remote examples. Also the CLT for $W - W_N$ in 1.1 is contained in 6.1 in a similar manner as in the one-dimensional case (Section 3 of Chapter II).

6.1. THEOREM. *Let $\{X_{k,N}\}$ $(N = 0,1,2,\ldots,k=0,1,2,\ldots)$ be a martingale difference triangular array w.r.t. $\{\mathfrak{G}_k\}$ and define*

$$s_{k,N}^2 := \sum_{\ell=1}^{k} E(X_{\ell,N}^2|\mathfrak{G}_{\ell-1}),\quad s_N^2 := \sum_{\ell=1}^{\infty} E(X_{\ell,N}^2|\mathfrak{G}_{\ell-1}) = \sup_k s_{k,N}^2.$$

Suppose

(6.1) For any N, $S_N := \sum_{k=0}^{\infty} X_{k,N}$ converges in $\mathfrak{L}^2$, i.e. $Es_N^2 < \infty$,

(6.2) $$s_N^2 \xrightarrow{P} W,$$

$$(6.3)\qquad L_N := \sum_{k=0}^{\infty} E(X_{k,N}^2 I(|X_{k,N}| > \epsilon | \mathcal{G}_{k-1}) \overset{P}{\to} 0 \quad \forall \epsilon > 0.$$

Then the limiting distribution of $W^{-1/2}S_N$ *w.r.t.* P_∞ *is standard normal.*

In the proof, we need

6.2. LEMMA. *Let* $\{X_{k,N}\}$ $(N = 0,1,2,\ldots,k = 1,\ldots,k(N))$ $(k(N) \leq \infty)$ *be a martingale difference triangular array w.r.t.* $\{\mathcal{F}_{k,N}\}$ *and define*

$$s_{k,N}^2 := \sum_{\ell=1}^{k} E(X_{\ell,N}^2 | \mathcal{F}_{k,N}), \quad s_N^2 := s_{k(N),N}^2,$$

$$S_{k,N} := \sum_{\ell=1}^{k} X_{\ell,N}, \quad S_N := S_{k(N),N}$$

[if $k(N) = \infty$, then cf. A2.4]. *Then* $s_N^2 \overset{P}{\to} 0$ *implies that* $S_N \overset{P}{\to} 0$ *as well.*

PROOF. Define $\epsilon > 0$

$$\nu(N) := \inf\{k: 1 \leq k < k(N),\ s_{k+1,N}^2 > \epsilon\},$$

$\nu(N) := k(N)$ if no such k exists. Then

$$P(\nu(N) = k(N)) \to 1, \quad \epsilon \geq E\, s_{k(N)\wedge\nu(N),N}^2 = E\, S_{k(N)\wedge\nu(N),N}^2$$

[the last identity requires an argument as given in Neveu (1972), pg. 148]. Hence

$$\overline{\lim_{N\to\infty}}\, P(|S_N| > \delta) = \overline{\lim_{N\to\infty}}\, P(|S_{k(N)\wedge\nu(N),N}| > \epsilon) \leq \frac{\epsilon}{\delta^2}. \quad \text{Let } \epsilon \downarrow 0. \ \square$$

PROOF OF 6.1. We approximate $W^{-1/2}S_N$ by

$$T_N := \sum_{k=\underline{k}(N)}^{\overline{k}(N)} \tilde{X}_{k,N}$$

where $\tilde{X}_{k,N} := W_{R(N)}^{-1/2} X_{k,N}$ and $R(N)$, $\underline{k}(N) = \tau_{R(N)} + 1$, $\overline{k}(N) = \underline{k}(N) + \Delta(N)$ are to be determined later such that $R(N)$ and $\Delta(N)$ are non-random,

$$(6.4)\qquad R(N) \uparrow \infty,$$

$$(6.5)\qquad s_{\underline{k}(N)-1,N}^2 \overset{P}{\to} 0,$$

$$(6.6)\qquad s_N^2 - s_{\overline{k}(N),N}^2 \overset{P}{\to} 0.$$

Note that $W_{R(N)} = \rho^{-R(N)} x_{R(N)}[\varphi]$ is $\mathfrak{G}_{\underline{k}(N)}$-measurable and that thus the $\tilde{X}_{k,N}$ are again martingale differences. By 6.1, (6.5) and (6.6) ensure that the limiting distribution of $W^{-1/2} S_N$ is that of T_N, and to show that this in fact is standard normal it suffices by A8 that

$$(6.7) \qquad \sum_{k=\underline{k}(N)}^{\overline{k}(N)} E(\tilde{X}_{k,N}^2 | \mathfrak{G}_{k-1}) = W_{R(N)}^{-1} (s^2_{\overline{k}(N),N} - s^2_{\underline{k}(N)-1,N}) \overset{P}{\to} 1,$$

$$(6.8) \qquad \sum_{k=\underline{k}(N)}^{\overline{k}(N)} E(\tilde{X}_{k,N}^2 I(|\tilde{X}_{k,N}| > \varepsilon) | \mathfrak{G}_{k-1}) \overset{P}{\to} 0 \quad \forall \varepsilon > 0.$$

But (6.8) is an easy consequence of (6.3) and (6.7) follows from (6.2), (6.4)-(6.6). Thus we have only to specify $R(N)$, $\Delta(N)$ such that (6.4)-(6.6) hold. Recalling that $\underline{k}(N) - 1 = \tau_{R(N)}$, define for fixed $R = 0, 1, 2, \ldots$ and $\varepsilon > 0$

$$A_{R,N} := \sum_{k=0}^{\tau_R} E(X_{k,N}^2 I(|X_{k,N}| \leqq \varepsilon) | \mathfrak{G}_{k-1}), \; B_{R,N} = \sum_{k=0}^{\tau_R} E(X_{k,N}^2 I(|X_{k,N}| > \varepsilon) | \mathfrak{G}_{k-1}).$$

Let $N \to \infty$ with R fixed. Then by (6.3), $B_{R,N} \overset{P}{\to} 0$ and since ε is arbitrary and $A_{R,N} \leqq \varepsilon^2 \tau_R$,

$$s^2_{\tau_R, N} = A_{R,N} + B_{R,N} \overset{P}{\to} 0$$

as well. Thus if $R(N) \uparrow \infty$ sufficiently slowly, (6.5) will hold. By (6.1), we can choose $\Delta(N)$ such that

$$\sum_{k=\Delta(N)}^{\infty} EX_{k,N}^2 \to 0$$

and the proof is completed by observing that (6.6) follows from

$$s_N^2 - s^2_{\overline{k}(N),N} \leqq \sum_{k=\Delta(N)}^{\infty} E(X_{k,N}^2 | \mathfrak{G}_{k-1}) \overset{P}{\to} 0. \qquad \square$$

7. EXPLOITING MARTINGALES: THE INCREASING PROCESS AND THE CLT FOR LINEAR FUNCTIONALS

The CLT's in 3.1 are easy corollaries of the slightly more general

7.1. THEOREM. *Let* $\{\eta_n\}$ *be a sequence in* $\mathcal{L}^1_{\Phi^*}$ *and* $\{\gamma_n\}$ *a sequence of real constants with* $\gamma_n > 0$. *Define* $\sigma_n^2 := \gamma_n^{-2}\Phi^*[\mathrm{Var}^{\langle\cdot\rangle}\hat{x}_1[\eta_n]]$,

$$\beta_N^2 := \sum_{n=1}^{N} \rho^{-n}\gamma_n^2\sigma_n^2, \quad \alpha_N^2 := \sum_{n=1}^{N} \rho^{-n}\gamma_n^2,$$

$$S_N := \left(\sum_{n=0}^{N-1} \{\hat{x}_{n+1}[\eta_{N-n}] - \hat{x}_n[M\eta_{N-n}]\}\right)/\rho^{N/2}\beta_N$$

and suppose that

(7.1) *There is a* $\eta \in \mathcal{L}^1_{\Phi^*}$ *such that the process has finite variance w.v.t.* η *and that* $|\eta_n|/\gamma_n \leq \eta$ *for all* n,

$$\varliminf_{N\to\infty} \beta_N^2/\alpha_N^2 > 0, \tag{7.2}$$

$$\gamma_N^2 = o(\rho^N\alpha_N^2). \tag{7.3}$$

Then the limiting distributions w.v.t. P_∞ *of* $S_N/W^{1/2}$ *is standard normal*.

The one-dimensional analogue of 7.1 was studied at the end of Section 3 in Chapter II, where also some discussion of the minimality of the conditions was given. The proof proceeds via 6.1. Define

$$\vartheta_n^2(x) := \mathrm{Var}^{\langle x\rangle}\hat{x}_1[\eta_n/\gamma_n] \quad (\text{so that } \sigma_n^2 = \Phi^*[\vartheta_n^2]),$$

and, in the linear ordering k of the n, i of Section 6,

$$X_{k,N} := \{\hat{y}_k[\eta_{N-n}] - E(\hat{y}_k[\eta_{N-n}]\,|\,\mathfrak{F}_n)\}/\rho^{N/2}\beta_N \quad n < N,$$

$X_{k,N} := 0$ $n \geq N$. Then the S_N defined in 6.1 and 7.1 coincide and s_N^2 defined in 6.1 is

$$s_N^2 = (\rho^N\beta_N^2)^{-1}\sum_{n=0}^{N-1} \gamma_{N-n}^2\hat{x}_n[\vartheta_{N-n}^2]. \tag{7.4}$$

VERIFICATION OF THE LINDEBERG CONDITION (6.3). Defining

$$Z := \hat{x}_1[\eta] + \hat{x}_0[M\eta],\ F(x) := \Phi^*[P^{\langle\cdot\rangle}(Z \leq x)],\ Z_k := \hat{y}_k[\eta] + E(\hat{y}_k[\eta]\,|\,\mathfrak{G}_{k-1}),$$

we have from (7.1) that $|X_{k,N}| \leq \gamma_{N-n}Z_k/\rho^{N/2}\beta_N$,

$$L_N \leq (\rho^N\beta_N^2)^{-1}\sum_{k=1}^{\tau_{N-1}} \gamma_{N-n}^2 E\left(Z_k^2 I\left(Z_k > \frac{\rho^{N/2}\beta_N\epsilon}{\gamma_{N-n}}\right)\Big|\,\mathfrak{F}_n\right),$$

$$EL_N \leq c\beta_N^{-2} \sum_{n=0}^{N-1} \frac{\gamma_{N-n}^2}{\rho^{N-n}} \int_{\frac{\rho^{N/2}\beta_N\epsilon}{\gamma_{N-n}}}^{\infty} x^2 dF(x).$$

Given $\delta > 0$, use (7.1),(7.2) to choose N_1 with $\gamma_N^2 \leq \delta^2\rho^N\beta_N^2 \quad N \geq N_1$ and hence (since $\beta_N^2 \uparrow$) $\gamma_{N-n}^2 \leq \delta^2\rho^N\beta_N^2 \quad N-n \geq N_1$. For $N \geq N_2$ also $\gamma_i^2 \leq \delta^2\rho^N\beta_N^2$ for $i = 1,\ldots,N_1$ and hence all $i = 1,2,\ldots$ so that

$$EL_N \leq c \int_{c\delta^{-1}}^{\infty} x^2 dF(x) \quad \beta_N^{-2} \sum_{n=0}^{N-1} \rho^{-n}\gamma_n^2 .$$

Take first $\overline{\lim}_{N\to\infty}$ and let next $\delta \downarrow 0$ to obtain $EL_N \to 0$, $L_N \xrightarrow{P} 0$. □

PROOF OF $s_N^2 \xrightarrow{a.s.} W$ IN THE CASE OF A FINITE SET OF TYPES. We can write $\rho^{-n}\hat{x}_n \leq \tilde{W}_m \Phi^*$ when $n \geq m$, where $\tilde{W}_m \to W$ as $m \to \infty$. Therefore

$$\begin{aligned}
\overline{\lim_{N\to\infty}} s_N^2 &= \overline{\lim_{N\to\infty}} (\rho^N\beta_N^2)^{-1} \sum_{n=m}^{N-1} \gamma_{N-n}^2\hat{x}_n[\vartheta_{N-n}^2] \\
&\leq \tilde{W}_m \overline{\lim_{N\to\infty}} (\rho^N\beta_N^2)^{-1} \sum_{n=m}^{N-1} \gamma_{N-n}^2\rho^n\Phi^*[\vartheta_{N-n}^2] \\
&= \tilde{W}_m \overline{\lim_{N\to\infty}} \beta_N^{-2} \sum_{n=m}^{N-1} \gamma_{N-n}^2\rho^{n-N}\sigma_{N-n}^2 = \tilde{W}_m .
\end{aligned}$$

Here the first equality follows from $\sup_n \vartheta_n^2 < \infty$ combined with (7.2), (7.3) in the form

(7.5) $\gamma_{N-n}^2 = o(\rho^{N-n}\beta_N^2)$ as $N \to \infty$ with n fixed

and the last equality is similar. Letting $m \to \infty$ yields $\overline{\lim}\, s_N^2 \leq W$ and $\underline{\lim} \geq$ is similar. □

The general case seems much more involved, even what regards convergence in probability:

PROOF OF $s_N^2 \xrightarrow{P} W$ IN THE GENERAL CASE. Define

$$p_N^m := (\rho^N\beta_N^2)^{-1} \sum_{n=0}^{m-1} \gamma_{N-n}^2\hat{x}_n[\vartheta_{N-n}^2]$$

$$q_N^m := (\rho^N\beta_N^2)^{-1} \sum_{n=m}^{N-1} \gamma_{N-n}^2\hat{x}_{n-m}[M^m\vartheta_{N-n}^2]$$

$$r_N^m := (\rho^N\beta_N^2)^{-1} \sum_{n=m}^{N-1} \gamma_{N-n}^2 \sum_{i=1}^{|\hat{x}_{n-m}|} \{\hat{x}_n^{n-m,i}[\vartheta_{N-n}^2] - E(\hat{x}_n^{n-m,i}[\vartheta_{N-n}^2] | \mathfrak{F}_{n-m})\},$$

and note that $s_N^2 = p_N^m + q_N^m + r_N^m$. The proof will be accomplished by showing that

$$(7.6) \qquad \lim_{N\to\infty} p_N^m = 0 \text{ a.s. } \forall m,$$

$$(7.7) \qquad c_m^- W \leq \underline{\lim}_{N\to\infty} q_N^m \leq \overline{\lim}_{N\to\infty} q_N^m \leq c_m^+ W \text{ a.s. where } c_m^- \to 1,\ c_m^+ \to 1\ m \to \infty,$$

$$(7.8) \qquad r_N^m \xrightarrow[N\to\infty]{P} 0 \ \forall\ m.$$

Here (7.6) follows at once from (7.5) and the consequence

$$(7.9) \qquad \vartheta^2 := \sup_n \vartheta_n^2 \in \mathcal{L}^1_{\Phi^*}$$

of (7.1). Also (7.7) is fairly straightforward. Indeed,

$$M^m \vartheta_n^2 \leq c_m^+ \rho^m \varphi \Phi^* [\vartheta_n^2] = c_m^+ \rho^m \varphi \sigma_n^2 ,$$

$$\overline{\lim}_{N\to\infty} q_N^m \leq \overline{\lim}_{N\to\infty} (\rho^N \beta_N^2)^{-1} \sum_{n=m}^{N-1} \gamma_{N-n}^2 c_m^+ \rho^n W_{n-m} \sigma_{N-n}^2$$

$$\leq c_m^+ \sup_{k\geq k_0} W_k \overline{\lim}_{N\to\infty} \beta_N^{-2} \sum_{n=m+k_0}^{N-1} \gamma_{N-n}^2 \rho^{N-n} \sigma_{N-n}^2 = c_m^+ \sup_{k\geq k_0} W_k,$$

using (7.5) to show that the terms with $n \leq m+k_0$ is negligible. As $k_0 \uparrow \infty$, the r.h. inequality in (7.7) follows, and the l.h. is similar.

In the proof of (7.8), define ϑ^2 by (7.9) and let

$$Z := \hat{x}_m[\vartheta^2] + \hat{x}_0[M^m \vartheta^2],\ F(x) := \Phi^*[P^{\langle\cdot\rangle}(Z \leq x)],$$

$$Y_\ell := \hat{x}_m[\vartheta_\ell^2] - E\hat{x}_m[\vartheta_\ell^2],$$

and let as usual $Y_\ell^{n,i}$ be Y_ℓ evaluated in the line of descent initiated by the ith particle alive at time n so that

$$r_N^m = (\rho^N \beta_N^2)^{-1} \sum_{n=m}^{N-1} \gamma_{N-n}^2 \sum_{i=1}^{|\hat{x}_{n-m}|} Y_{N-n}^{n-m,i} .$$

Define $\tilde{r}_N^m$ as r_N^m but replacing $Y_{N-n}^{n-m,i}$ by $Y_{N-n}^{n-m,i} I(Z^{n-m,i} \leq \rho^n) - E(Y_{N-n}^{n-m,i} I(Z^{n-m,i} \leq \rho^n) | \mathfrak{F}_{n-m})$. Then, since $|Y_\ell| \leq Z$ for all ℓ,

$$\Delta_N := |r_N^m - \tilde{r}_N^m| =$$

$$(\rho^N \beta_N^2)^{-1} \left| \sum_{n=m}^{N-1} \gamma_{N-n}^2 \sum_{i=1}^{|\hat{x}_{n-m}|} E(Y_{N-n}^{n-m,i} I(Z^{n-m,i} > \rho^n) | \mathfrak{F}_{n-m}) \right|$$

$$\leq (\rho^N \beta_N^2)^{-1} \sum_{n=m}^{N-1} \gamma_{N-n}^2 \hat{x}_{n-m}[E^{\langle\cdot\rangle} Z I(Z > \rho^n)],$$

$$\overline{\lim_{N\to\infty}} \Delta_N \leq \overline{\lim_{n\to\infty}} \rho^{-n} x_{n-m}[E^{\langle\cdot\rangle} ZI(Z > \rho^n)] \cdot \overline{\lim_{N\to\infty}} \beta_N^{-2} \sum_{n=m}^{N-1} \frac{\gamma_{N-n}^2}{\rho^{N-n}} ,$$

$$\leq \rho^m W \int_{\rho^k}^{\infty} x \, dF(x) \cdot \overline{\lim_{N\to\infty}} \alpha_N^2/\beta_N^2 ,$$

for any $k \geq m$, appealing once more to (7.9), (7.5) and (7.2). As $k \uparrow \infty$, it follows that $\Delta_N \to 0$ a.s.

The final step of the proof is to show that $\operatorname{Var} \tilde{r}_N^m \to 0$ as $N \to \infty$. If $m = 1$, then the terms for different n in the definition of $\tilde{r}_N^m$ are orthogonal and hence

$$(7.10) \qquad \operatorname{Var} \tilde{r}_N^m = (\rho^N \beta_N^2)^{-2} \sum_{n=m}^{N-1} \gamma_{N-n}^4 E\hat{x}_{n-m}[\operatorname{Var}^{\langle\cdot\rangle} Y_{N-n} I(Z \leq \rho^n)]$$

$$\leq (\rho^N \beta_N^2)^{-2} \sum_{n=m}^{N-1} \gamma_{N-n}^4 E\hat{x}_{n-m}[E^{\langle\cdot\rangle} Z^2 I(Z \leq \rho^n)]$$

$$\leq c\rho^{-m} \beta_N^{-4} \sum_{n=m}^{N-1} \frac{\gamma_{N-n}^4}{\rho^{2(N-n)}} \rho^{-n} \int_0^{\rho^n} x^2 dF(x) .$$

Since by (7.5) $\gamma_k^2/\rho^k \beta_{N+k}^2 \leq c_1$ for all N, k, it follows that

$$\overline{\lim_{N\to\infty}} \operatorname{Var} \tilde{r}_N^m \leq c_2 \overline{\lim_{N\to\infty}} \beta_N^{-2} \sum_{n=m}^{N-1} \frac{\gamma_{N-n}^2}{\rho^{N-n}} \rho^{-n} \int_0^{\rho^n} x^2 dF(x)$$

$$\leq c_2 \overline{\lim_{n\to\infty}} \rho^{-n} \int_0^{\rho^n} x^2 dF(x) \overline{\lim_{N\to\infty}} \alpha_N^2/\beta_N^2 = 0,$$

using $\int_0^\infty x \, dF(x) < \infty$. If $m > 1$, one can split $\tilde{r}_N^m$ up into m sums, each having orthogonal increments, which can be treated in a similar manner. $\square$

The proofs of the CLT's in parts 1°, 2° in 3.1 (where 1° has been treated in Section 5 by a different method) are now completed by the following lemma, which also yields the a.s. convergence of s_N^2 needed for the LIL's:

7.2. LEMMA. *Let $\eta_n := M^{n-1}\xi$ with $\lambda = \lambda(\xi) \leq \rho^{1/2}$. Then under the assumptions of 3.1 (including $\sigma^2 > 0$), there exists γ_n such that the conditions of 7.1 hold and in fact, the CLT in 7.1 is the one asserted in 3.1. Furthermore (with s_N^2 defined by (7.4)), $s_N^2 \to W$ a.s. and if $\lambda^2 = \rho$, $N/n(N) \to \theta > 1$, then a.s.*

$$\lim_{N\to\infty} (\rho^N N^{2\gamma-1})^{-1} \sum_{n=n(N)}^{N-1} \hat{x}_n[\operatorname{Var}^{\langle\cdot\rangle} \hat{x}_1[M^{N-n-1}\xi]] = W\sigma^2 (1 - \tfrac{1}{\theta})^{2\gamma-1}.$$

PROOF. By definition of S_N, β_N^2 we have

$$(\rho^N\beta_N^2)^{1/2}S_N = \hat{x}_N[\xi]-\hat{x}_0[M^N\xi], \quad \rho^{-N}\Phi^*[\mathrm{Var}^{\langle\cdot\rangle}x_N[\xi]] = \beta_N^2 .$$

Let $\gamma_n := \lambda^n n^{\gamma-1}$. Then $\alpha_N \uparrow \alpha_\infty < \infty$ if $\lambda^2 < \rho$, while $\alpha_N/N^{2\gamma-1}$ has a limit in $(0,\infty)$ if $\lambda^2 = \rho$. Hence the condition $\sigma^2 > 0$ in 3.1 is equivalent to (7.2) and also (7.3) follows. Finally (7.1) follows from $\eta_1 = \xi$ and $|\eta_n| \leq c\gamma_n\varphi$, $n > 1$.

To prove that $s_N^2 \to W$ a.s., it suffices by (7.6), (7.7) and $\Delta_N \to 0$ a.s. to prove that $\sum_N \mathrm{Var}\, \tilde{r}_N^m < \infty$ (implying $\tilde{r}_N^m \to 0$ a.s.). But assume as above $m = 1$. Then by (4.10),

$$\sum_{N=m+1}^{\infty} \mathrm{Var}\, \tilde{r}_N^m \leq c \sum_{n=m}^{\infty} \rho^{-n} \int_0^{\rho^n} x^2 dF(x) \sum_{N=n+1}^{\infty} \frac{\gamma_{N-n}^2}{\beta_N^4 \rho^{2(N-n)}}$$

$$\leq \int_0^{\infty} O(x)dF(x)\beta_m^{-4} \sum_{n=1}^{\infty} \frac{\lambda^{2n}n^{2\gamma-2}}{\rho^{2n}} < \infty .$$

Finally the last assertion of 7.2 follows by exactly the same type of arguments. The heuristic motivation is (using the last part of 4.3)

$$\sum_{n=n(N)}^{N-1} \hat{x}_n[\mathrm{Var}^{\langle\cdot\rangle}\hat{x}_1[M^{N-n-1}\xi]] \doteq \sum_{n=n(N)}^{N-1} \gamma_{N-n}^2 \hat{x}_n[\vartheta_{N-n}^2]$$

$$\cong W \sum_{n=n(N)}^{N-1} \gamma_{N-n}^2\sigma_{N-n}^2 = W\Phi^*[\mathrm{Var}^{\langle\cdot\rangle}(\hat{x}_N[\xi]-\hat{x}_{n(N)}[M^{N-n(N)}\xi])]$$

$$\cong W\rho^N N^{2\gamma-1}\sigma^2(1-\tfrac{1}{\theta})^{2\gamma-1}. \qquad \square$$

8. EXPLOITING MARTINGALE EXPONENTIAL INEQUALITIES

We are now prepared for the main step in the proof of the LIL,

8.1. LEMMA. 1°. *If* $\lambda^2 < \rho$, *let* $C_N := (2\sigma^2(\xi)W\rho^N \log N)^{1/2}$. *Then* $\overline{\lim}_{N\to\infty} |\hat{x}_N[\xi]|/C_N \leq 1$.

2°. *If* $\lambda^2 = \rho$, *let* $C_N := (2\sigma^2(\xi)W\rho^N N^{2\gamma-1}\log_2 N)^{1/2}$. *Then* $\overline{\lim}_{i\to\infty} \hat{x}_{t(i)}[\xi]/C_{t(i)} \leq 1$,

$$\overline{\lim_{i\to\infty}} \frac{\hat{x}_{t(i)}[\xi]-\hat{x}_{t(i-1)}[M^{t(i)-t(i-1)}\xi]}{C_{t(i)}} \geq (1-\tfrac{1}{\theta})^{\gamma-\frac{1}{2}}$$

for any sequence $\{t(i)\}$ *such that* $t(i)/\theta^i \to 1$ *as* $i \to \infty$, *with* $1 < \theta < \infty$.

Part 1° of 8.1 completes the proof of the LIL in the case $\lambda^2 < \rho$. Indeed, replacing ξ in 8.1 by $M^k\xi$ and defining D_k as in 5.3, it follows from 8.1 that $D_k \leq \sigma(M^k\xi)/\sigma(\xi)$. Thus the assumption (5.18) $D_k = o(\rho^{k/2})$ of 5.3 and hence the LIL holds by (4.5).

If $\lambda^2 = \rho$, some additional arguments are needed, in particular to bound $\hat{x}_N[\xi]$ when $t(i-1) \leq N \leq t(i)$. The details are given in Section 9.

If $\lambda^2 < \rho$, write (for notational conformity) $t(i) := i$. In the proof of 8.1, we choose an integer $n(i) < t(i)$ and approximate the r.v. in the denumerator by

$$\hat{x}_{t(i)}[\xi]-\hat{x}_{n(i)}[M^{\Delta(i)}\xi] = \sum_{k=\underline{k}(i)}^{\overline{k}(i)} \{U_{k,i}-E(U_{k,i}|\mathfrak{F}_n)\} \tag{8.1}$$

where, in the notation of Section 6, $U_{k,i} := \hat{y}_k[M^{t(i)-n-1}\xi]$, $\overline{k}(i) := \tau_{t(i-1)}$, $\underline{k}(i) := \tau_{n(i)-1}+1$, $\Delta_i := t(i)-n(i)$. In the $\overline{\lim} \geq$ part of 2° we let $n(i) := t(i-1)$, while in the $\overline{\lim} \leq$ parts of 1°, 2° we choose $n(i)$ increasing so slowly to infinity that

$$\sum_{i=1}^{\infty} \frac{E\hat{x}_{n(i)}[M^{\Delta(i)}\xi]^2}{\rho^{t(i)}} < \infty \quad \text{if} \quad \lambda^2 < \rho \, , \tag{8.2}$$

$$\sum_{i=1}^{\infty} \frac{E\hat{x}_{n(i)}[M^{\Delta(i)}\xi]^2}{t(i)^{2\gamma-1}\rho^{t(i)}} < \infty \quad \text{if} \quad \lambda^2 = \rho \tag{8.3}$$

(to see that this is possible, use e.g. for (8.2) a similar argument as in the proof of (4.12) to show that

$$E\hat{x}_{n(i)}[M^{\Delta(i)}\xi]^2 = O(\rho^{n(i)}\lambda^{2\Delta(i)}\Delta(i)^{2\gamma-2})).$$

The expression (8.1) will be studied by means of the conditional Borel-Cantelli lemma and the martingale exponential inequalities A 7.5 .

To this end, we need to truncate. We first recall the existence of an ϑ such that, letting $F(y) := \Phi^*[P^{\langle\cdot\rangle}(|\hat{x}_1[\vartheta]| \leq y)]$,

$$(8.4) \qquad |M^N \xi| \leq \lambda^N N^{\gamma-1} \vartheta \ \forall N, \int_0^\infty y^2 dF(y) < \infty.$$

If $\lambda^2 < \rho$, choose $\{K_n\}$, $\{b_n\}$ as in A 7.3. Define

$$D_i := (\sigma^2 \hat{x}_{n(i)}[\varphi]\rho^{t(i)-n(i)})^{1/2}, \quad c_n := b_{[\rho^n]} \quad \text{if } \lambda^2 < \rho,$$

$$D_i := (\sigma^2 \hat{x}_{n(i)}[\varphi]\rho^{t(i)-n(i)} t(i)^{2\gamma-1})^{1/2}, \quad c_n := \rho^{n/2} \quad \text{if } \lambda^2 = \rho.$$

Choose a double sequence $\{B_{k,i}\}$ of r.v. with $P(B_{k,i} = 1) = P(B_{k,i} = -1) = 1/2$ which are mutually independent and independent of the branching process, and define

$$\tilde{U}_{k,i} := U_{k,i} I(\hat{y}_k[\vartheta] \leq c_n),$$

$$Y_{k,i} := \begin{cases} \{\tilde{U}_{k,i} - E(\tilde{U}_{k,i} | \mathfrak{F}_n)\}/D_i & k \leq \overline{k}(i) \\ B_{k,i}/D_i & k > \overline{k}(i) \end{cases}$$

$$X_{k,i} := \sum_{j=\underline{k}(i)}^{k} Y_{k,i}, \quad s^2_{k,i} := \sum_{\ell=\underline{k}(i)}^{\overline{k}(i)} \operatorname{Var}(Y_{\ell,i} | \mathfrak{G}_{\ell-1})$$

(the $B_{k,i}$ and $s^2_{k,i}$ will first come into the proof after a while).

8.2. LEMMA. *In the* $\overline{\lim} \leq$ *parts of* 8.1,

$$\overline{\lim_{i\to\infty}} \frac{\hat{x}_{t(i)}[\xi]}{c_{t(i)}} = \overline{\lim_{i\to\infty}} \frac{X_{\overline{k}(i),i}}{(2 \log i)^{1/2}},$$

while in the $\overline{\lim} \geq$ *part of* 2°,

$$\overline{\lim_{i\to\infty}} \frac{\hat{x}_{t(i)}[\xi] - \hat{x}_{n(i)}[M^{\Delta(i)}\xi]}{c_{t(i)}} = \overline{\lim_{i\to\infty}} \frac{X_{\overline{k}(i),i}}{(2 \log i)^{1/2}}.$$

PROOF. Clearly, $D_i(2 \log i)^{1/2} \cong c_{t(i)}$. Write

$$V_{k,i} := U_{k,i} - \tilde{U}_{k(i)} - E(U_{k,i} - \tilde{U}_{k,i} | \mathfrak{F}_n)$$

so that

$$\hat{x}_{t(i)}[\xi] - \hat{x}_{n(i)}[M^{\Delta(i)}\xi] - X_{\overline{k}(i),i} = \sum_{k=\underline{k}(i)}^{\overline{k}(i)} V_{k,i}.$$

Since in the $\overline{\lim} \leq$ parts, $\hat{x}_{n(i)}[M^{\Delta(i)}\xi]/c_{t(i)} \to 0$ by (8.2), (8.3), it thus suffices that

$$A_i := \sum_{k=\underline{k}(i)}^{\overline{k}(i)} |V_{k,i}| = o(C_{t(i)}) \text{ a.s.}$$

Recalling (8.4),

$$\sum_{k=\tau_{n-1}+1}^{\tau_n} |V_{k,i}| \leq \lambda^{t(i)-n-1}(t(i)-n-1)^{\gamma-1} Z_n$$

where

$$Z_n := \sum_{k=\tau_{n-1}+1}^{\tau_n} \{\hat{y}_k[\vartheta] I(\hat{y}_k[\vartheta] > c_n) + E(\hat{y}_k[\vartheta] I(\hat{y}_k[\vartheta] > c_n) | \mathfrak{F}_n)\}, \text{ so that}$$

$$A_i \leq \sum_{n=0}^{t(i)-1} \lambda^{t(i)-n-1}(t(i)-n-1)^{\gamma-1} Z_n .$$

If $\lambda^2 < \rho$, we assert that

$$\sum_{n=0}^{\infty} \frac{Z_n}{(\rho^n \log n)^{1/2}} < \infty . \tag{8.5}$$

To this end, note first that

$$\sum_{n=0}^{\infty} \frac{EZ_n}{(\rho^n \log n)^{1/2}} = 2 \sum_{n=0}^{\infty} \frac{E\hat{x}_n[E^{\langle\cdot\rangle} \hat{x}_1[\vartheta] I(\hat{x}_1[\vartheta] > c_n)]}{(\rho^n \log n)^{1/2}} \tag{8.6}$$

$$\leq c \sum_{n=0}^{\infty} (\rho^n/\log n)^{1/2} \int_{c_n}^{\infty} x \, dF(x).$$

We recall the notation of A 7.3 . Define $N^*(m) := \sup\{n : [c_n] \leq m\}$ ([ ] denoting integer part of). Substituting $k = [\rho^n]$ yields

$$N^*(m) \leq \sup\{\frac{\log(k+1)}{\log \rho} : k \in \{[\rho^n]\}, [b_k] \leq m\}$$

$$\leq \sup\{\frac{\log(k+1)}{\log \rho} : k \in N, [b_k] \leq m\} \leq \frac{\log(N(m)+1)}{\log \rho} ,$$

$$\frac{\rho^{N^*(m)}}{\log N^*(m)} = O(\frac{N(m)}{\log_2 N(m)}) = O(\frac{m^2 \log_2 m / K_m^2}{\log_2(m^2 \log_2 m)}) = O(\frac{m^2}{K_m}) ,$$

so that the finiteness of (8.5), (8.6) follows from

$$\sum_{n=0}^{\infty} (\rho^n/\log n)^{1/2} I(x > c_n) = O((\rho^n/\log n)^{1/2} \Big|_{n=N^*([x])})$$

$$= O(x/K_{[x]})$$

and A 7.3 . Therefore indeed

$$A_i \leq \sum_{n=0}^{t(i)} \lambda^{t(i)-n}(t(i)-n)^{\gamma-1} o((\rho^n \log n)^{1/2})$$

$$= o((\rho^{t(i)}\log t(i))^{1/2} \sum_{n=0}^{t(i)} (\frac{\lambda}{\rho^{1/2}})^n n^{\gamma-1}$$

$$= o((\rho^{t(i)}\log t(i))^{1/2}) = o(C_{t(i)}).$$

The case $\lambda^2 = \rho$ requires less care. Instead of (8.5), we assert that $\Sigma_0^\infty Z_n/\rho^{n/2} < \infty$ a.s. Recalling $c_n = \rho^{n/2}$, this follows as above from

$$\sum_{n=0}^{\infty} \rho^{n/2} \int_{\rho^{n/2}}^{\infty} x\, dF(x) = \int_0^\infty O(x^2)dF(x) < \infty.$$

Thus

$$A_i \leq \rho^{t(i)/2} t(i)^{\gamma-1} \sum_{n=0}^{t(i)} Z_n/\rho^{n/2}$$

$$= O(\rho^{t(i)/2} t(i)^{\gamma-1}) = o(C_{t(i)}). \quad \square$$

8.3. LEMMA. *In the* $\overline{\lim} \leq$ *parts of* 8.1, $\lim_{i\to\infty} s^2_{\overline{k}(i),i} = 1$ *a.s., while in the* $\overline{\lim} \geq$ *part of* 2°, $\lim_{i\to\infty} s^2_{\overline{k}(i),i} = (1-1/\theta)^{2\gamma-1}$ *a.s.*

PROOF. For $\underline{n} \leq N$, $\Delta := N-\underline{n}$ define

$$\omega^2(\eta;\underline{n},N) := \sum_{n=\underline{n}}^{N-1} \hat{x}_n[\mathrm{Var}^{\langle\cdot\rangle}\hat{x}_1[M^{N-n-1}\eta]]$$

$$= \omega^2(\eta;0,N)-\omega^2(M^\Delta\eta;0,\underline{n}) .$$

The proof consists in making precise the obvious heuristic estimate $s^2_{\overline{k}(i),i} \cong \omega^2(\boldsymbol{\xi};n(i),t(i))/D_i^2$ and applying the a.s. estimates for the ω^2 in 7.2. To this end, check first that if U, Z are r.v. such that $|U| \leq \alpha Z$ and we let $\tilde{U} := UI(Z \leq c)$, then

$$|\mathrm{Var}\, U-\mathrm{Var}\, \tilde{U}| \leq 2\alpha^2(EZ^2I(Z> c)+EZ\cdot EZI(Z> c)).$$

Letting $U := U_{k,i}, Z := \hat{y}_k[\vartheta], \alpha := \lambda^{t(i)-n}(t(i)-n)^{\gamma-1}, c := c_n$,

$$\zeta_n(x) := E^{\langle x\rangle}\hat{x}_1[\vartheta]^2 I(x_1[\vartheta] > c_n)+E^{\langle x\rangle}\hat{x}_1[\vartheta]E^{\langle x\rangle}\hat{x}_1[\vartheta]I(\hat{x}_1[\vartheta] > c_n),$$

it follows from (8.4) that $\infty > \Phi^*[\zeta_n] \downarrow 0$ and thus

$$|D_i^2 s^2_{\overline{k}(i),i}-\omega^2(\boldsymbol{\xi};n(i),t(i))| \leq \sum_{n=n(i)}^{t(i)-1} \lambda^{2(t(i)-n-1)}(t(i)-n-1)^{2\gamma-2}\hat{x}_n[\zeta_n]$$

$$= \rho^{t(i)} \sum_{n=n(i)}^{t(i)-1} \lambda^{2(t(i)-n-1)}(t(i)-n-1)^{2\gamma-2}\rho^{-t(i)}o(\rho^n).$$

Inserting the explicit form of D_i shows that the r.h.s. is $o(D_i^2)$ so that we can consider the limit of the $\omega^2(\xi;n(i),t(i))/D_i^2$ rather then the $s^2_{\overline{k}(i),i}$. Now from 7.2 and $\rho^{-n(i)}\hat{x}_{n(i)}[\varphi] \to W$ we have that

$$\lim_{i\to\infty} \frac{\omega^2(\xi;0,t(i))}{D_i^2} = 1 \text{ a.s.}$$

and the proof in the $\overline{\lim} \leq$ part is completed by observing that by (8.2), (8.3)

$$\lim_{i\to\infty} \frac{\omega^2(M^{\Delta(i)};0,n(i))}{D_i^2} = 0 \text{ a.s.}$$

since $E\omega^2(M^{\Delta(i)};0,n(i)) = \text{Var } \hat{x}_{n(i)}[M^{\Delta(i)}\xi]$. The $\overline{\lim} \geq$ part of 2° is immediate from the last statement of 7.2. □

8.4. REMARK. Also this proof admits for some simplifications in the case of a finite set of types. We omit the details.

Appealing to 8.2, the proof of 8.1 is completed by

8.5. LEMMA. *In the $\overline{\lim} \leq$ parts of 8.1,*

$$\overline{\lim_{i\to\infty}} \frac{X_{\overline{k}(i),i}}{(2 \log i)^{1/2}} = 1 \text{ a.s.}$$

while in the $\overline{\lim} \geq$ part of 2°,

$$\overline{\lim_{i\to\infty}} \frac{X_{\overline{k}(i),i}}{(2 \log i)^{1/2}} \geq (1 - \tfrac{1}{\theta})^{\gamma - \frac{1}{2}} .$$

PROOF. In order to make the exponential inequalities available, we must first check the bounds on the increments $Y_{k,i}$ of the martingale $\{X_{k,i}\}_{k \geq \underline{k}(i)}$. We claim that there are d_i such that d_i is measurable w.r.t. $\mathfrak{F}_{n(i)}$ and

(8.7) $|Y_{k,i}| \leq d_i$ for all $k \geq \underline{k}(i)$, $\lim\limits_{i\to\infty} (\log i)^{1/2} d_i = 0$.

Since $|Y_{k,i}| = D_i^{-1}$ for $k > \overline{k}(i)$ and clearly $(\log i)^{1/2}D_i \to 0$, we need only to consider the $k \leq \overline{k}(i)$. Using (8.4),

$$|Y_{k,i}| \leq \frac{2\lambda^{t(i)-n-1}(t(i)-n-1)^{\gamma-1}c_n}{D_i} .$$

Let first $\lambda^2 < \rho$. Then for some constant b, $c_n \leq bK_{[\rho^n]}(\rho^n/\log n)^{1/2}$ so that $|Y_{k,i}|$ cannot exceed

$$\frac{1}{(\log i)^{1/2}} \frac{2b}{(2\sigma^2\rho^{-n(i)}x_{n(i)}[\varphi])^{1/2}} \cdot K_{[\rho^n]} \cdot \frac{\lambda^{i-n-1}(i-n-1)^{\gamma-1}(\log i)^{1/2}}{(\rho^{i-n}\log n)^{1/2}}$$

(recalling $t(i) = i$). Here the second factor is bounded, the third factor $K_{[\rho^n]}$ small if n is large, say $n \geq i/2$ with i large, and the last factor uniformly bounded in n,i if $n \geq i/2$ and small if $n < i/2$ and i is large (since then $m := i-n-1$ is large and thus $\lambda^m m^{\gamma-1} \log m/\rho^{m/2}$ small). Hence (8.7) holds.

If $\lambda^2 = \rho$, $c_n = \rho^{n/2}$, note simply that $|Y_{k,i}|$ cannot exceed

$$\frac{1}{(\log i)^{1/2}} \cdot \frac{2}{(2\sigma^2\rho^{-n(i)}x_{n(i)}[\varphi])^{1/2}} \cdot \frac{t(i)^{\gamma-1}}{t(i)^{\gamma-1/2}}$$

where the second factor is bounded and the last factor tends to zero.

Now define

$$T(i) := \inf\{k \geq \underline{k}(i) : s^2_{\underline{k}(i)} \geq \beta^2\}.$$

In the $\overline{\lim} \leq$ parts, we take $\beta > 1$, $\epsilon_i := \alpha(2 \log i)^{1/2}$ with $\alpha > \beta$. It follows by (8.7) that to each $\delta > 0$ with $\alpha^2/\beta^2 > 1+\delta$ we can find i_0 such that

$$\frac{\epsilon_i d_i}{\beta^2} \leq 1, \ \frac{\alpha^2}{\beta^2}(1 - \frac{\epsilon_i d_i}{2\beta}) > 1+\delta \quad \text{for all } i \geq i_0 .$$

Applying the upper exponential inequality,

$$P(X_{T(i),i} > \epsilon_i |\ _{n(i)}) \leq e^{-\frac{\epsilon_i^2}{2\beta^2}(1 - \frac{\epsilon_i d_i}{2\beta})} < i^{-(1+\delta)} \quad i \geq i_0 .$$

Combining with Levy's inequality and the way the $B_{k,i}$ have been chosen, we obtain

$$\sum_{i=1}^{\infty} P(X_{\overline{k}(i),i} > \epsilon_i, T(i) > \overline{k}(i) | \mathfrak{F}_{n(i)})$$

$$\leq 2 \sum_{i=1}^{\infty} P(X_{\overline{k}(i),i} > \epsilon_i, B_{\overline{k}(i),i} + \dots + B_{T(i),i} \geq 0, T(i) > \overline{k}(i) | \mathfrak{F}_{n(i)})$$

$$\leq 2 \sum_{i=1}^{\infty} P(X_{T(i),i} > \epsilon_i, T(i) > \overline{k}(i) | \mathfrak{F}_{n(i)})$$

$$\leq 2 \sum_{i=1}^{\infty} P(X_{T(i),i} > \epsilon_i | \mathfrak{F}_{n(i)}) < \infty.$$

Since $T(i) > \overline{k}(i)$ eventually by 8.3, it follows by the conditional Borel-Cantelli lemma that $X_{\overline{k}(i),i} \leq \epsilon_i$ eventually so that

$$\overline{\lim_{i\to\infty}} \frac{X_{\overline{k}(i),i}}{(2 \log i)^{1/2}} \leq \alpha.$$

Let $\alpha, \beta \downarrow 1$ to obtain $\overline{\lim} \leq 1$.

In the $\overline{\lim} \geq$ part of 2^0, we take $\beta^2 < (1 - \frac{1}{\theta})^{2\gamma-1}$, $\epsilon_i := \alpha(2 \log i)^{1/2}$ with $0 < \alpha < \beta$, choose γ such that $\alpha^2(1+\gamma)/\beta^2 < 1$ and define $\delta > 0$ by $1-\delta := \alpha^2(1+\gamma)/\beta^2$. Applying the lower exponential inequality,

$$(8.8)\qquad P(X_{T(i),i} > \epsilon_i | \mathfrak{F}_{n(i)}) > e^{-\frac{\epsilon_i^2}{2\beta^2}(1+\gamma)} = i^{-(1-\delta)}$$

for i sufficiently large. Now define

$$A_{k,i} := \mathrm{Var}(\overline{X}_{\overline{k}(i),i} | \mathfrak{G}_k), M_i := \max_{\underline{k}(i) \leq k \leq \overline{k}(i)} (X_{k,i} - (2A_{k,i})^{1/2})$$

and note that

$$A_{k,i} = E(s^2_{\overline{k}(i),i} - s^2_{k,i} | \mathfrak{G}_k) = E(s^2_{\overline{k}(i),i} - s^2_{k,i} | \mathfrak{F}_n)$$

$$\leq E(s^2_{\overline{k}(i),i} - s^2_{\tau_{n-1}} | \mathfrak{F}_n) = \mathrm{Var}(X_{\overline{k}(i),i} | \mathfrak{F}_n)$$

$$= \frac{\hat{x}_n[\mathrm{Var}^{\langle\cdot\rangle}\hat{x}_{t(i)-n}[\xi]]}{D_i^2} \leq \frac{\rho^{-n}\hat{x}_n[\zeta]}{\sigma^2\rho^{-n(i)}\hat{x}_{n(i)}[\varphi]},$$

with ζ as in (4.11). Therefore $\sup_{k,i} A_{k,i} < \infty$ and we get

$$(8.9)\qquad \varlimsup_{i\to\infty} \frac{M_i}{(2\log i)^{1/2}} = \varlimsup_{i\to\infty} \max_{\underline{k}(i) \leq k \leq \overline{k}(i)} \frac{X_{k,i}}{(2\log i)^{1/2}}$$

$$\geq \varlimsup_{i\to\infty} \frac{X_{T(i),i}}{(2\log i)^{1/2}} \geq \alpha.$$

Here we have in addition used $T(i) < \overline{k}(i)$ eventually, cf. 8.3, and the converse conditional Borel-Cantelli lemma (which is applicable since the $[\underline{k}(i), \overline{k}(i)]$ are disjoint) combined with (8.8) for the right equality. By a maximal inequality A. 6.3

$$2\sum_{i=1}^{\infty} P\left(\frac{X_{\overline{k}(i),i}}{(2\log i)^{1/2}} > \alpha - \epsilon | \mathfrak{F}_{n(i)}\right) \geq \sum_{i=1}^{\infty} P\left(\frac{M_i}{(2\log i)^{1/2}} > \alpha - \epsilon | \mathfrak{F}_{n(i)}\right).$$

The right-hand side being infinite by (8.9), it follows similarly that $\varlimsup_{i\to\infty} \overline{X}_{\overline{k}(i),i}/(2\log i)^{1/2} \geq \alpha - \epsilon$. Let $\epsilon \downarrow 0, \alpha, \beta \uparrow (1-1/\theta)^{\gamma-1/2}$. □

8.6. REMARKS. Letting $n(i) := i-k$ in the case $\lambda^2 < \rho$ and applying the lower exponential inequality, one could derive a $\varlimsup \geq$ part analogous to the case $\lambda^2 = \rho$, which could replace the additivity argument used in the proof of the LIL for $\hat{x}_N[\xi]$. Other variants of the proofs are possible. In the case $\lambda^2 = \rho$, one could replace the use of exponential inequalities by the remainder term estimate in the martingale CLT given by Levy (1954), pg.243. See Asmussen (1977). In order to make a similar approach possible in the case $\lambda^2 < \rho$, one would need a full generalization of the Berry-Esseen theorem to martingales. This is not available at present.

9. THE LIL IN THE CASE $\lambda^2 = \rho$

If $\lambda^2 = \rho$, we define

$$\xi_1 := \sum_{\nu:\lambda_\nu^2=\rho} \sum_{j=1}^{\bar{j}(\nu)} \Phi^*_{\nu,j}[\xi]\varphi_{\nu,j}, \quad \xi_2 := \xi - \xi_1 .$$

Then $\lambda^2(\xi_2) < \rho$, so that the order of magnitude of $\hat{x}_N[\xi_2]$ (appealing to the LIL established for the case $\lambda^2 < \rho$) is vanishing compared to the LIL asserted for $\hat{x}_N[\xi]$. Hence we may assume $\xi = \xi_1$, $\xi_2 = 0$.

We recall the convention of 2.5. Define $\mathcal{K}_j$ as the complex span of the $\varphi_{\nu,j}$ with $\lambda_\nu^2 = \rho$ and let $\mathcal{H}_j$ be the real span of the Re $\varphi_{\nu,j}$, Im $\varphi_{\nu,j}$ with $\lambda_\nu^2 = \rho$, $\mathcal{K} := \mathcal{K}_1 + \ldots + \mathcal{K}_\gamma$, $\mathcal{H} := \mathcal{H}_1 + \ldots + \mathcal{H}_\gamma$. Then $\mathcal{H}$ and $\mathcal{K}$ are finite dimensional, $\xi \in \mathcal{H}$, $\lambda(\eta) = \rho^{1/2}$ and $\gamma(\eta) = j$ when $\eta \in \mathcal{H}_j$, and $\mathcal{K}$ and $\mathcal{H}$ are M-invariant. All eigenvalues of the restriction of M to have moduli $\rho^{1/2}$ and hence M is one-one on $\mathcal{K}$ so that $M^N:\mathcal{K} \to \mathcal{K}$ exists for $N = 0,-1,-2,\ldots$. One can check that the formula (2.4) for the $M^N\varphi_{\nu,\gamma}$ remains valid for $N < 0$ and one easily gets

9.1. LEMMA. *There exists* $\eta_{k,j} \in \mathcal{H}_j$, $j = 1,\ldots,\gamma$, $k = 0,\pm 1,\pm 2,\ldots$ *such that* $\{\eta_{k,j}\}$ *is relatively compact for each* j *and that*

$$M^k\xi = \rho^{k/2} \sum_{j=1}^{\gamma} |k|^{\gamma-j}\eta_{k,j} \qquad k = 0,\pm 1,\pm 2,\ldots . \tag{9.1}$$

Let as in the preceding section $t(i)/\theta^i \to 1$ with $\theta > 1$.

9.2. LEMMA. *Let* $B \subseteq \mathcal{H}_j$ *be relatively compact for some* $j = 1,\ldots,\gamma$ *and define*

$$\sigma^2(\eta) := \lim_{N\to\infty} \frac{\Phi^*[\mathrm{Var}^{\langle\cdot\rangle}\hat{x}_N[\eta]]}{\rho^N N^{2j-1}}, \quad \eta \in \mathcal{H}_j ,$$

$$E_N := (2 \sup_{\eta\in B} \sigma^2(\eta)W\rho^N N^{2j-1}\log_2 N)^{1/2} .$$

Then $\overline{\lim}_{i\to\infty} \sup_{\eta\in B} \frac{|\hat{x}_{t(i)}[\eta]|}{E_{t(i)}} \leq 1$ *a.s.*

PROOF. Choose $\eta_1,\ldots,\eta_m$ as a basis for $\mathcal{H}_j$ and, to a given $\varepsilon > 0$, n and $\eta^1,\ldots,\eta^n \in B$ such that to any $\eta \in B$ there exist $s(\eta)$ and $c_r(\eta)$ such that $\eta = \eta^{s(\eta)} + \Sigma_1^m c_r(\eta)\eta_r$ with $\Sigma_1^m|c_r(\eta)| < \varepsilon$. Then the $\overline{\lim}$ is bounded by

$$\overline{\lim}_{i\to\infty}(\max_{s=1,\ldots,n} |\hat{x}_{t(i)}[\eta^s]| + \varepsilon \max_{r=1,\ldots,m} |\hat{x}_{t(i)}[\eta_r]|)/E_{t(i)}$$

which, according to 8.1, cannot exceed

$$1+\varepsilon \max\{\sigma(\eta_1),\ldots,\sigma(\eta_m)\}/\sup_{\eta\in B}\sigma(\eta). \qquad \Box$$

9.3. LEMMA. *Let* $\xi, \gamma=\gamma(\xi)$, $\sigma^2=\sigma^2(\xi)$ *be as in* 3.1, *define* $D_N := (2\sigma^2 W\rho^N N^{2\gamma-1}\log_2 N)^{1/2}$ *and let, for each* i, $K(i)$ *be a subset of the integers. Then*

$$\overline{\lim_{i\to\infty}}\ \sup_{k\in K(i)} \frac{\hat{x}_{t(i)}[M^k\xi]}{\rho^{k/2}D_{t(i)}} \leq 1+\sum_{j=1}^{\gamma-1} w_j\mu^{\gamma-j} \tag{9.2}$$

where $w_j^2 := \sup_{k=0,\pm1,\pm2,\ldots} \sigma^2(\eta_{k,j})/\sigma^2 < \infty$, $\mu := \overline{\lim_{i\to\infty}}\ \sup_{k\in K(i)} |k|/t(i)$.

PROOF. Expanding $M^k\xi$ as in (9.1), it follows that the l.h.s. of (9.2) is bounded by

$$\sum_{j=1}^{\gamma} \overline{\lim_{i\to\infty}}\ \sup_{k\in K(i)} \left(\frac{|k|}{t(i)}\right)^{\gamma-j} \frac{|\hat{x}_{t(i)}[\eta_{k,j}]|}{(2\sigma^2 W\rho^{t(i)}t(i)^{2\gamma-1}\log i)^{1/2}}$$

and appealing to 9.2, we need only remark that by 4.2(iii),

$$\sigma^2(\eta_{k,\gamma}) = \sigma^2(\rho^{-k/2}M^k\xi) = \sigma^2(\xi) = \sigma^2. \qquad \Box$$

We can now easily give the

PROOF OF $\overline{\lim} \geq$ IN 3.1. We take $K(i)$ as the set with $\Delta(i+1)=t(i+1)-t(i)$ as its only point. Remarking that

$$D_{t(i)} \cong \theta^{\gamma-1/2}\rho^{\Delta(i)/2}D_{t(i-1)}, \mu = \theta-1,$$

we get

$$\overline{\lim_{N\to\infty}} \frac{\hat{x}_N[\xi]}{D_N} \geq \overline{\lim_{i\to\infty}} \frac{\hat{x}_{t(i)}[\xi]}{D_{t(i)}}$$

$$\geq \overline{\lim_{i\to\infty}} \frac{\hat{x}_{t(i)}[\xi]-\hat{x}_{t(i-1)}[M^{\Delta(i)}\xi]}{D_{t(i)}} - \overline{\lim_{i\to\infty}} \frac{|\hat{x}_{t(i-1)}[M^{\Delta(i)}\xi]|}{D_{t(i)}}$$

$$\geq \left(1-\frac{1}{\theta}\right)^{\gamma-1/2} - \frac{1}{\theta^{\gamma-1/2}}\left(1+\sum_{j=1}^{\gamma-1} w_j(\theta-1)^{\gamma-j}\right),$$

using 9.3 and part 2° of 8.1. Let $\theta\uparrow\infty$. $\Box$

PROOF OF $\overline{\lim} \leq$ IN 3.1. It remains to bound $\hat{x}_N[\xi]$ when $t(i-1)\leq N \leq t(i)$. We approximate $\hat{x}_N[\xi]$ by $\hat{x}_{t(i)}[M^{N-t(i)}\xi]$, using similar methods as at the end of Section 8. More precisely, define

$$K(i) := \{0,-1,\ldots,t(i-1)-t(i)\}, M_i' := \max_{k\in K(i)} \frac{\hat{x}_{t(i)}[M^k\xi]}{\rho^{k/2}\rho^{t(i)/2}},$$

$$A_{N,i} := \mathrm{Var}(\frac{x_{t(i)}[M^{N-t(i)}\xi]}{\rho^{N/2}}|\mathfrak{F}_N), M_i'' := \max_{t(i-1)\leq N\leq t(i)} [\frac{\hat{x}_N[\xi]}{\rho^{N/2}} - (2A_{N,i})^{1/2}].$$

Let F_N be the event that

$$\hat{x}_n[\xi]/\rho^{n/2} - (2A_{n,i})^{1/2} > \epsilon$$

for $n = N$ but not for any $n = t(i-1),\ldots,N-1$. Then, using Chebycheff's inequality,

$$P(M_i' > \epsilon|\mathfrak{F}_{t(i-1)}) \geq \sum_{N=t(i-1)}^{t(i)} E[P(M_i' > \epsilon, F_N|\mathfrak{F}_N)|\mathfrak{F}_{t(i-1)}]$$

$$\geq \sum_{N=t(i-1)}^{t(i)} E(I(F_N)P(\hat{x}_{t(i)}[M^{N-t(i)}\xi] - \hat{x}_N[\xi] > -\rho^{N/2}(2A_{N,i})^{1/2}|\mathfrak{F}_N)|\mathfrak{F}_{t(i-1)})$$

$$\geq \sum_{N=t(i-1)}^{t(i)} \tfrac{1}{2}P(F_N|\mathfrak{F}_{t(i-1)}) = \tfrac{1}{2}P(M_i'' > \epsilon|\mathfrak{F}_{t(i-1)}).$$

Letting

$$E_i := (2\sigma^2\rho^{-t(i-1)}\hat{x}_{t(i-1)}[\varphi]t(i)^{2\gamma-1}\log i)^{1/2},$$

it follows by the conditional Borel-Cantelli Lemma (both ways!) that

$$\overline{\lim_{i\to\infty}} \frac{M_i''}{E_i} \leq \overline{\lim_{i\to\infty}} \frac{M_i'}{E_i} \leq 1 + \sum_{j=1}^{\gamma-1} w_j(1-\tfrac{1}{\theta})^{\gamma-j},$$

using 9.3 for the r.h. inequality. Applying (4.11) in a similar manner as in Section 8, one can show that

$$\overline{\lim_{i\to\infty}} \max_{t(i-1)\leq N\leq t(i)} A_{N,i}/E_i = 0.$$

Thus

$$\overline{\lim_{N\to\infty}} \frac{x_N[\xi]}{D_N} \leq \theta^{\gamma-1/2} \overline{\lim_{i\to\infty}} \max_{t(i-1)\leq N\leq t(i)} \frac{\hat{x}_N[\xi]}{\rho^{N/2}E_i}$$

$$= \theta^{\gamma-1/2} \overline{\lim_{i\to\infty}} \frac{M_i''}{E_i} \leq \theta^{\gamma-1/2}(1 + \sum_{j=1}^{\gamma} w_j(1-\tfrac{1}{\theta})^{\gamma-j}).$$

Let $\theta \downarrow 1$. □

10. THE CASE $\lambda^2 > \rho$

We shall take the opportunity to give also some higher order expansions (essentially 10.1 below) since this comes out of the proof of part 3° of 3.1 with little extra effort. We first introduce some notation. Write

$$\binom{N-n-1}{k} := \sum_{\delta=0}^{k} N^{k-\delta} \alpha_{\delta,k}(n),$$

where

$$\alpha_{\delta,k}(n) = \frac{(-1)^\delta}{k!} \sum_{1\leq j_1<j_2<\cdots<j_\delta\leq k} (n+j_1)(n+j_2)\cdots(n+j_\delta) = O(n^\delta),$$

$$Z_N(\delta,\nu,r) := \sum_{n=0}^{N-1} \alpha_{\delta,r+\delta}(n)\rho_\nu^{-n-1} \cdot \{\hat{x}_{n+1}[\sum_{j=r+\delta+1}^{\tilde{j}(\nu)} \Phi^*_{\nu,j}[\xi]\rho_\nu^{-r-\delta}\varphi_{\nu,j-r-\delta}]-\hat{x}_n[M\ldots]\}$$

where $\tilde{j}(\nu) := \gamma\, \overline{j}(\nu)$ and $\hat{x}_{n+1}[\eta]-\hat{x}_n[M\ldots] := \hat{x}_{n+1}[\eta]-\hat{x}_n[M\eta]$ (for notational convenience in case of a complicated η).

10.1. PROPOSITION. *Suppose* ξ *has the form*

$$\sum_{\nu:\lambda_\nu=\lambda} \sum_{j=1}^{\tilde{j}(\nu)} \Phi^*_{\nu,j}[\xi]\varphi_{\nu,j} \tag{10.1}$$

with $\lambda(\xi)^2 > \rho$. *Then*

$$\hat{x}_N[\xi]-\hat{x}_0[M^N\xi] = \lambda^N \sum_{r=0}^{\gamma-1} N^r \sum_{\nu:\lambda_\nu=\lambda} e^{i\theta_\nu N} \sum_{\delta=0}^{\gamma-1-r} Z_N(\delta,\nu,r). \tag{10.2}$$

If the process has finite variance w.r.t. φ *[and hence* ξ *and all* $\varphi_{\nu,j}$*], then* $Z_N(\delta,\nu,r)$ *has an a.s. limit* $Z(\delta,\nu,r)$ *as* $N\to\infty$. *More precisely,*

$$Z(\delta,\nu,r)-Z_N(\delta,\nu,r) = O\left(\frac{N^\delta(\rho^N\log N)^{1/2}}{\lambda^N}\right). \tag{10.3}$$

PROOF. The l.h.s. of (10.2) is

$$\sum_{n=0}^{N-1}\{\hat{x}_{n+1}[M^{N-n-1}\xi]-\hat{x}_n[M^{N-n}\xi]\}$$

$$= \sum_{n=0}^{N-1}\{\hat{x}_{n+1}[\sum_{\nu:\lambda_\nu=\lambda}\sum_{j=1}^{\tilde{j}(\nu)}\Phi^*_{\nu,j}[\xi]\sum_{k=0}^{j-1}\rho_\nu^{N-n-1-k}\binom{N-n-1}{k}\varphi_{\nu,j-k}]-\hat{x}_n[M\ldots]\}$$

$$= \lambda^N \sum_{\nu:\lambda_\nu=\lambda} e^{i\theta_\nu N}\sum_{n=0}^{N-1}\rho_\nu^{-n-1}\{\hat{x}_{n+1}[\sum_{j=1}^{\tilde{j}(\nu)}\Phi^*_{\nu,j}[\xi]\sum_{k=0}^{j-1}\rho_\nu^{-k}\varphi_{\nu,j-k}\sum_{\delta=0}^{k}N^{k-\delta}\alpha_{\delta,k}(n)]-\hat{x}_n[M\ldots]\}$$

which, letting $r := k-\delta$ and interchanging the summation, reduces to the r.h.s. of (10.2). The existence of $Z(\delta,\nu,r)$ is clear at once by noting that $\{Z_N(\delta,\nu,r)\}$ is a martingale and that

$$\sup_N \operatorname{Var} Z_N(\delta,\nu,r) = \sum_{n=0}^{\infty} O(\alpha_\delta(n)^2 \lambda^{-2n} \rho^n) < \infty .$$

Alternatively, one could use 1.2 to show that

$$Z_n(\delta,\nu,r) - Z_{n-1}(\delta,\nu,r) = \alpha_{\delta,r+\delta}(n-1)\rho_\nu^{-n-1} O((\rho^n \log n)^{1/2})$$

and this estimate produces also at once (10.3) (expanding the l.h.s. as a sum of the increments from N to ∞). □

PROOF OF PART 3° OF 3.1. Any ξ can be written as $\xi_1 + \xi_2$, where $\lambda(\xi_1)^2 \leq \rho$ and ξ_2 is a sum of finitely many functions of the form (10.1), with different $\lambda \in (\rho^{1/2}, \lambda(\xi)]$. It follows that it is no restriction to assume ξ to have the form (10.1). Then (10.2), (10.3) imply

$$\hat{x}_N[\xi] - \hat{x}_0[M^N \xi] = \lambda^N N^{\gamma-1} \sum_\nu {}^{(\xi)} e^{i\theta_\nu N} Z(0,\nu,\gamma-1) + o(\lambda^N N^{\gamma-1})$$

and the proof is complete since

$$Z(0,\nu,\gamma-1) = \frac{\rho_\nu^{-(\gamma-1)}}{(\gamma-1)!} \Phi^*_{\nu,\gamma}[\xi](W^\nu - W_0^\nu),$$

$$\begin{aligned} \hat{x}_0[M^N \xi] &= \lambda^N \frac{N^{\gamma-1}}{(\gamma-1)!} \xi_N + o(\lambda^N N^{\gamma-1}) \\ &= \lambda^N N^{\gamma-1} \sum_\nu {}^{(\xi)} \frac{\rho_N^{-(\gamma-1)}}{(\gamma-1)!} \Phi^*_{\nu,\gamma}[\xi] e^{i\theta_\nu N} W_0^\nu + o(\lambda^N N^{\gamma-1}). \end{aligned}$$ □

10.2. REMARKS. Writing ξ as the sum of a term ξ_1 of the form (10.1) and one ξ_2 with $\lambda(\xi_2) < \lambda$, it follows from 10.1 that

$$\hat{x}_N[\xi] = \hat{x}_0[M^N \xi_1] + \lambda^N \sum_{r=0}^{\gamma-1} N^r \sum_{\nu:\lambda_\nu=\lambda} e^{i\theta_\nu N} \sum_{\delta=0}^{\gamma-1-r} Z(\delta,\nu,r)$$

$$+ O(N^{\gamma-1}(\rho^N \log N)^{1/2}) + \hat{x}_N[\xi_2] .$$

If $\lambda(\xi_2)^2 < \rho$, then $O(\cdot)$ term dominates $\hat{x}_N[\xi_2]$ if $\gamma > 1$, while the two terms are of the same order of magnitude if $\gamma = 1$. If $\lambda(\xi_2)^2 > \rho$, $\hat{x}_N[\xi_2]$ dominates the $O(\cdot)$ term, while if $\lambda(\xi_2)^2 = \rho$, the size of $\gamma(\xi_2)$ becomes important. It would seem conceivable that the $O(\cdot)$ estimate could be strengthened to a LIL, but we have not looked into this. At least the proof of the LIL in 1.1 carries over verbatim to show

10.3. PROPOSITION. *Suppose that* ρ_ν *is real with* $\lambda_\nu^2 > \rho$, $0 < \sigma^2 := \Phi^*[\mathrm{Var}^{\langle\cdot\rangle} W^\nu] < \infty$. *Then a.s. on* $\{W > 0\}$,

$$\overline{\lim_{N\to\infty}} \frac{\hat{x}_N[\varphi_{\nu,1}] - \rho_\nu^N W^\nu}{(2\sigma^2 W \rho^N \log N)^{1/2}} = 1, \quad \underline{\lim_{N\to\infty}} \frac{\hat{x}_N[\varphi_{\nu,1}] - \rho_\nu^N W^\nu}{(2\sigma^2 W \rho^N \log N)^{1/2}} = -1.$$

11. AN EXAMPLE FROM ASYMPTOTIC ESTIMATION THEORY

Consider the supercritical case and the estimator

$$\tilde{\rho}_N := \frac{|\hat{x}_1| + \dots + |\hat{x}_N|}{|\hat{x}_0| + \dots + |\hat{x}_{N-1}|}$$

of ρ, generalizing the estimator $\hat{m}_N$ studied in Section 3 of Chapter 2. Though $\tilde{\rho}_N$ is certainly not the maximum likelihood estimator in this general setting and one could suggest more refined estimators of ρ, $\tilde{\rho}_N$ might be argued to be natural in situations where not all features of the process (such as the family tree or the type distribution) are observable. We shall here be concerned with the asymptotic properties of $\tilde{\rho}_N$. It is clear at once from $|\hat{x}_n| \cong W\rho^n$ that $\tilde{\rho}_N$ is strongly consistent, $\tilde{\rho}_N \to \rho$ a.s. on $\{W > 0\}$. To see where 3.1 comes in in a more refined study of $\tilde{\rho}_N$, write

$$(11.1) \quad \tilde{\rho}_N - \rho = \sum_{n=0}^{N-1} \{|\hat{x}_{n+1}| - \rho|\hat{x}_n|\}/(|\hat{x}_0| + \dots + |\hat{x}_{N-1}|)$$

$$= (S_N + T_N)/(|\hat{x}_0| + \dots + |\hat{x}_{N-1}|)$$

where

$$S_N := \sum_{n=0}^{N-1} \{\hat{x}_{n+1}[1] - \hat{x}_n[M1]\},$$

$$T_N := \sum_{n=0}^{N-1} \hat{x}_n[\eta], \quad \eta := M1 - \rho 1 = M\xi - \rho\xi$$

where $\xi := 1 - \varphi$. The convergence in distribution of $S_N/\rho^{N/2}$ is an immediate consequence of 7.1. Noting that $\lambda := \lambda(\eta) = \lambda(\xi) < \rho$, therefore 3.1 suggest that S_N and T_N are of the same magnitude if $\lambda^2 < \rho$, while otherwise T_N might dominate S_N. The first of these assertions is made precise in

11.1. THEOREM. *If* $\lambda^2 < \rho$ *and the process has finite variance w.r.t.* 1, *then the limiting distribution w.r.t.* P_∞ *of* $[W(1 + \dots + \rho^{N-1})]^{1/2}(\tilde{\rho}_N - \rho)$ *is normal, with mean zero and variance* V *specified by* (11.2) *below, and a.s. on* $\{W > 0\}$

$$\overline{\lim_{N\to\infty}} \left[\frac{W(1 + \dots + \rho^{N-1})}{2\, V \log N}\right]^{1/2} (\tilde{\rho}_N - \rho) = 1, \quad \underline{\lim_{N\to\infty}} \left[\frac{W(1 + \dots + \rho^{N-1})}{2\, V \log N}\right]^{1/2} (\tilde{\rho}_N - \rho) = -1.$$

PROOF. Define

$$\eta_1 := 1, \eta_n := 1 + \sum_{k=0}^{n-1} M^k \eta, \beta_N^2 := \sum_{n=1}^{N} \rho^{-n} \Phi^*[\mathrm{Var}^{\langle\cdot\rangle} \hat{x}_1[\eta_n]],$$

$$(11.2)\quad V := \lim_{N\to\infty} \frac{\rho^N\beta_N^2}{1+\ldots+\rho^{N-1}} = (\rho-1)\sum_{n=1}^{\infty} \rho^{-n}\Phi^*[\mathrm{Var}^{\langle\cdot\rangle}\hat{x}_1[\eta_n]],$$

$\gamma_n^2 := \mu^n$ with $\lambda < \mu < \rho^{1/2}$. Then, expanding $\hat{x}_n[\eta]$ in the usual telescope sum (4.1),

$$(11.3)\quad S_N+T_N = \sum_{n=0}^{N-1} \hat{x}_0[M^n\eta] + \sum_{n=0}^{N-1} \{\hat{x}_{n+1}[\eta_{N-n}]-\hat{x}_n[M\eta_{N-n}]\}.$$

Here the first term is $o(\rho^{N/2})$, while (in the non-trivial case $V>0$) the conditions of 7.1 are easily verified so that the limiting distribution w.r.t. P_∞ of the second term normalized by $(W\rho^N\beta_N^2)^{1/2}$ is standard normal, and the CLT follows from (11.1). The LIL is proved exactly as in Section 3 of Chapter II, using 1.2 with

$$Y := \sum_{n=0}^{k-1} \{\hat{x}_{n+1}[\eta_{k-n}]-\hat{x}_n[M\eta_{k-n}]\}$$

and k large. □

When $\lambda^2 \geq \rho$, we rewrite (11.3) somewhat. We shall use

11.2. LEMMA. *If $\lambda(\xi^2) \geq \rho$, then we can write $\xi = (M-I)\xi_1+\xi_2$ with ξ_1,ξ_2 real and $\lambda(\xi_1) = \lambda(\xi), \gamma(\xi_1) = \gamma(\xi), \lambda(\xi_2) \leq 1$.*

PROOF. In 2.1, choose $\mu > 1$ such that no eigenvalue of M has modulus in $(1,\mu]$ and let $\xi := \xi_3+\xi_2$ with $\xi_3 \in {}_1$, $\xi_2 \in {}_2$, $\xi_1 := (M-I)^{-1}\xi_3$ which exists since $M-I$ is one-one on E_1. □

Now define

$$\xi_3 := (M-\rho I)\xi_1, \vartheta_r := (M-\rho I)\{\sum_{k=0}^{r-1} M^k\xi_2-\xi_1\} .$$

Then

$$(11.4)\quad T_N = \sum_{n=0}^{N-1} \hat{x}_0[M^n\eta] + \hat{x}_{N-1}[\xi_3] - \hat{x}_0[M^{N-1}\xi_3] + \sum_{n=0}^{N-2} \{\hat{x}_{n+1}[\vartheta_{N-n}]-\hat{x}_n[M\vartheta_{N-n}]\}.$$

The last term of (11.4), normalized by $\rho^{N/2}$, can be seen to converge in distribution by 7.1 and can also be proved to be $O((\rho^N\log N)^{1/2})$ a.s. Since $\lambda(\xi_3) = \lambda$, $\gamma(\xi_3) = \gamma$, it follows that the dominant term in S_N+T_N is $\hat{x}_{N-1}[\xi_3]-\hat{x}_0[M^{N-1}\xi_3]$ and from 3.1, we obtain

11.3. COROLLARY. *Suppose $\lambda^2 = \rho$ and define $V := (1-\rho^{-1})\lim_{N\to\infty}(N^{2\gamma-1}\rho^N)^{-1}\Phi^*[\mathrm{Var}^{\langle\cdot\rangle}\hat{x}_N[\xi_3]]$. Then if $V > 0$, the limiting distribution w.r.t. P_∞ of $[W(1+\ldots+\rho^{N-1})/N^{2\gamma-1}]^{1/2}(\tilde{\rho}_N-\rho)$ is normal with mean zero and variance $V > 0$, and a.s. on $\{W > 0\}$,*

$$\overline{\lim_{N\to\infty}} \left[\frac{W(1+\ldots+\rho^{N-1})}{2VN^{2\gamma-1}\log_2 N}\right]^{1/2} (\tilde{\rho}_N-\rho) = 1, \quad \underline{\lim_{N\to\infty}} \left[\frac{W(1+\ldots+\rho^{N-1})}{2VN^{2\gamma-1}\log_2 N}\right]^{1/2} (\tilde{\rho}_N-\rho) = -1.$$

11.4. COROLLARY. *Suppose* $\lambda^2 > \rho$. *Then there exist r.v.* $\{H_N\}$ *such that*

$$\lim_{N\to\infty}\left(\frac{W(1+\ldots+\rho^{N-1})}{\lambda^{N-1}(N-1)^{\gamma-1}}(\tilde{\rho}_N-\rho)-H_{N-1}\right) = \lim_{N\to\infty}\left(\frac{\hat{x}_N[\xi_3]}{\lambda^N N^{\gamma-1}}-H_N\right) = 0.$$

Furthermore $\overline{\lim_{N\to\infty}}|H_N| < \infty$ *and, except unless certain linear functionals degenerate, cf. 4.5,* $P(\underline{\lim_{N\to\infty}}|H_N| > 0) > 0$.

12. CONTINUOUS TIME

We shall take the obvious approach, to study the continuous time process $\{\hat{x}_t\}_{t\geq 0}$ by means of the results obtained so far for discrete skeletons $\{\hat{x}_{n\delta}\}_{n\in\mathbb{N}}$.

Instead of the offspring mean operator M and its iterates M^N we are now dealing with a semigroup $\{M_t\}_{t\geq 0}$ and when studying a discrete skeleton, the Jordan canonical form is defined relative to M_δ and depends essentially on δ: Not only are the eigenvalues of M_δ not the same for different δ, but properties like $M_\delta \varphi_{\nu,j} = \varphi_{\nu,j-1} + \rho_\nu \varphi_{\nu,j}$ are not preserved when passing to different δ's. The obvious relation between the parameters for different δ's are in terms of the generator. Assume henceforth in 3.1 that $\mathcal{C}_1, \mathcal{C}_2$ are M_t-invariant for all t and that the semigroup $\{M_t^1\}_{t\geq 0}$ formed by the M_t restricted to $\mathcal{C}_1$ is continuous. Then, by standard semigroup theory, $M_t^1 = e^{At}$ for some $A:\mathcal{C}_1 \to \mathcal{C}_1$. Indeed, in the simplest examples 3.1 could be derived from the properties of a generator for the whole semigroup $\{M_t\}_{t\geq 0}$, see V.3 and Hering (1978 a).

Now let the eigenvalues of A be $\mu_\nu := a_\nu + ib_\nu$ and choose generalized eigenvectors $\psi_{\nu,j}$ and projectors $\Psi^*_{\nu,j}$ as in 2.3 so that for example (cf. (2.4))

$$A^N \psi_{\nu,j} = \sum_{i=1}^{j} \mu_\nu^{N-j+i} \binom{N}{j-i} \psi_{\nu,i}\,. \tag{12.1}$$

Then

$$M_t \psi_{\nu,j} = e^{At}\psi_{\nu,j} = \sum_{N=0}^{\infty} \frac{t^N}{N!} A^N \psi_{\nu,j} = \sum_{i=1}^{j} \psi_{\nu,i} \sum_{N=0}^{\infty} \frac{t^N \mu_\nu^{N-j+i}}{N!} \binom{N}{j-i} \tag{12.2}$$

$$= e^{t\mu_\nu} \sum_{i=1}^{j} \frac{t^{j-i}}{(j-i)!} \psi_{\nu,i}\,.$$

The eigenvalues of M_δ^1 are

$$\rho_\nu(\delta) := \lambda_\nu(\delta) e^{i\theta_\nu(\delta)} = e^{\delta\mu_\nu} = e^{\delta a_\nu} e^{i\delta b_\nu}\,.$$

We can choose the $\varphi_{\nu,j} := \varphi_{\nu,j}(\delta)$ of the form $\varphi_{\nu,j} = \Sigma_1^j c_{i,j}(\delta)\psi_{\nu,i}$, normalized by $c_{1,j}(\delta) = 1$. Indeed, the requirement (2.2) reduces (after some algebra) to the set of equations

$$e^{\delta\mu_\nu} \sum_{k=i+1}^{j} \frac{\delta^{k-i}}{(k-i)!} c_{k,j} = c_{i,(j-1)}$$

which can be solved successively for $j = 1,2,\ldots$ and, for fixed j, for $k = j, j-1, \ldots, 2$. The only explicit property of the solution which will be needed is

$$(12.3)\qquad c_{j,j} = (\delta e^{\delta\mu_\nu})^{-1} c_{(j-1),(j-1)} = \dots = (\delta e^{\delta\mu_\nu})^{-(j-1)}.$$

Now define

$$a := a(\xi) := \max\{a_\nu : \Psi^*_{\nu,j}[\xi] \neq 0 \text{ for some } j\},$$

$$\gamma := \gamma(\xi) := \max\{j : \Psi^*_{\nu,j}[\xi] \neq 0 \text{ for some } \nu \text{ with } a_\nu = a\}$$

and let $\lambda(\xi,\delta)$ be $\lambda(\xi)$ defined relative to M_δ. Then γ has the same meaning as relative to the skeleton, while $\lambda(\xi,\delta) = e^{\delta a(\xi)}$. In particular, the critical value $\lambda(\xi,\delta)^2 = \rho^\delta$ corresponds to the case $2a(\xi) = \log\rho$. The continuous time version of (2.6) becomes

$$(12.4)\qquad M_t\xi = e^{at}\frac{t^{\gamma-1}}{(\gamma-1)!}\xi_t + o(e^{at}t^{\gamma-1}),$$

where

$$(12.5)\qquad \xi_t := \sum_\nu{}^{(\xi)} e^{itb_\nu}\,\Psi^*_{\nu,\gamma}[\xi]\psi_{\nu,1}\,.$$

To see that the notation ξ_t is consistent with ξ_N in (2.6), note that if $a_\nu = a$, $\bar{j}(\nu) = \gamma$, then (considering the skeleton $\{0,\delta,2\delta,\dots\}$)

$$\sum_{j=1}^{\gamma}\Phi^*_{\nu,j}[\xi]\varphi_{\nu,j} = \sum_{j=1}^{\gamma}\Psi^*_{\nu,j}[\xi]\psi_{\nu,j}\,.$$

Inserting $\varphi_{\nu,j} = \Sigma_1^j c_{i,j}\psi_{\nu,i}$, using (12.3) and identifying the coefficients of $\psi_{\nu,\gamma}$ yields

$$(12.6)\qquad \Phi^*_{\nu,\gamma}[\xi] = (\delta e^{\delta\mu_\nu})^{\gamma-1}\Psi^*_{\nu,\gamma}[\xi]$$

$$= (\delta\rho_\nu(\delta))^{\gamma-1}\Psi^*_{\nu,\gamma}[\xi],$$

implying the consistency of the notations as asserted.

Consider next second moments. From (4.3),

$$(12.7)\qquad \Phi^*[\mathrm{Var}^{\langle\cdot\rangle}\hat{x}_{N\delta}[\xi]] = \rho^{N\delta}\sum_{n=0}^{N-1}\rho^{-(n+1)\delta}\Phi^*[\mathrm{Var}^{\langle\cdot\rangle}\hat{x}_\delta[M_{n\delta}\xi]].$$

It will typically be the case that

$$(12.8)\qquad \tau^2(\xi) := \lim_{\delta\to 0}\frac{1}{\delta}\,\Phi^*[\mathrm{Var}^{\langle\cdot\rangle}\hat{x}_\delta[\xi]]$$

exists. One can then interpret (12.7) as a Riemann sum and obtains

$$(12.9)\qquad \Phi^*[\mathrm{Var}^{\langle\cdot\rangle}\hat{x}_T[\xi]] = \rho^T\int_0^T \rho^{-t}\tau^2(M_t\xi)\,dt.$$

For example, if the set of types is finite, the termination density at x is $k(x)$, the branching law $\pi(x,\cdot)$ and one lets $\mu_i(\xi,x) := \int x[\xi]^i \pi(x,dx)$, $i = 1,2$, then one can check that

$$\tau^2(\xi) = \Phi^*[k(\mu_2(\xi,\cdot)-2\xi\mu_1(\xi,\cdot)+\xi^2)].$$

Similarly, if the process is constructed from a diffusion on a simply connected domain with generator

$$A\xi = \sum_{i,j=1}^{n} a_{ij}\frac{\partial^2\xi}{\partial x_i \partial x_j} + \sum_{i=1}^{n} b_i \frac{\partial \xi}{\partial x_i}, \quad \xi \in \mathscr{D}(A),$$

a termination density $k(x)$ and a local branching law specified by $\{p_n(x)\}$, then

$$(12.10) \qquad \tau^2(\xi) = \Phi^*[k\mu\xi^2] + \Phi^*[\sum_{i,j=1}^{n} a_{i,j}\frac{\partial\xi}{\partial x_i}\frac{\partial\xi}{\partial x_j}]$$

where $\mu(x) := \Sigma_0^\infty (n-1)^2 p_n(x)$. This formula yields the promised proof of 4.6: Note first that the last term is non-negative since $\{a_{ij}\}$ is positive definite. It follows from (12.9) and (12.10) that for all sufficiently smooth ξ (and hence by continuity for all ξ),

$$(12.11) \qquad \Phi^*[\mathrm{Var}^{\langle\cdot\rangle}\hat{x}_T[\xi]] \geq \rho^T \int_0^T \rho^{-t}\Phi^*[k\mu(M_t\xi)^2]dt .$$

Under the assumptions of 4.6, $\Phi^*[k\mu\xi^2] > 0$ so that by continuity, the r.h.s. of (12.11) is > 0.

The generalization of the CLT's to continuous time is a trivial application of the Croft-Kingman lemma and we get

12.1. THEOREM. *The CLT's of 1.1, 1.2, 3.1, 3.2, 3.3 and the Laplacian limit law (3.8) in 3.3 hold in continuous time as well* (i.e., replacing $N \in \mathbb{N}$ by $t \in [0,\infty)$) *provided that in addition to the assumptions for discrete skeletons it is also assumed that the distribution of the relevant functionals (i.e., $\hat{x}_t[\xi]$ in 3.1, 3.2, 3.3, $\hat{x}_t[\varphi]$ in 1.1 and (3.8) and Y_t in 1.2) depends continuously (in the weak sense) on* t.

The problems come up in the strong laws. We remark first that it is not apparent at all whether or why the LIL in 1.2 should hold in continuous time. The strong laws for linear functionals carry over after some work, but we have had to set up

12.2. CONDITION. ξ *belongs to a finite dimensional subspace* $\mathscr{E}$ *invariant under the mean semigroup* $\{M_t\}_{t\geq 0}$.

(Which is, of course, automatic if the set of types is finite.) As will be seen, the difficulty is to get a good bound (essentially

$O((\rho^{n\delta}\log n)^{1/2}))$ for the supremum over $t \in [n\delta,(n+1)\delta]$ of $\hat{x}_t[\xi]$ in the case $\lambda^2(\xi] < \rho$. Assuming 12.2, estimates of this type will come out from

12.3. LEMMA. *Let* t,s *vary with* n *in such a way that* $n\delta \leq t \leq (n+1)\delta$, $t+s = (n+1)\delta$, *let* ζ,ϑ_t *be averaging functions such that* $M_{-s}\vartheta_t$ *exists and let* $c_t \geq 0$ *be* $\mathfrak{F}_t$*-measurable. Define* $B_t := \hat{x}_t[\vartheta_t] - \hat{x}_{n\delta}[\zeta]$,

$$K_n := \sup_{n\delta\leq t\leq(n+1)\delta} \frac{\hat{x}_{(n+1)\delta}[M_{-s}\vartheta_t]-\hat{x}_{n\delta}[\zeta]}{c_{(n+1)\delta}},$$

$$\alpha := \overline{\lim_{t\to\infty}} \frac{c_{(n+1)\delta}}{c_t} \quad \textit{and suppose that}$$

$$A_t := \mathrm{Var}\left(\frac{\hat{x}_{(n+1)\delta}[M_{-s}\vartheta_t]}{c_{(n+1)\delta}}\middle|\mathfrak{F}_t\right) = \frac{\hat{x}_t[\mathrm{Var}^{\langle\cdot\rangle}\hat{x}_s[M_{-s}\vartheta_t]]}{c^2_{(n+1)\delta}} \to 0 \quad \text{a.s.}$$

Then $\overline{\lim_{t\to\infty}} B_t/c_t \leq \alpha \overline{\lim_{n\to\infty}} K_n$.

PROOF. The idea has already been met in Sections 8 and 9, so we just outline the proof. If $B_{t(n)}$ is large for some stopping time $t(n) \in [n\delta,(n+1)\delta]$, we bound K_n below by

$$\frac{\hat{x}_{(n+1)\delta}[M_{-s(n)}\,{}_{t(n)}]-\hat{x}_{n\delta}[\zeta]}{c_{(n+1)\delta}} = \frac{B_{t(n)}}{c_{(n+1)\delta}} + Y_n,$$

$$Y_n := \frac{\hat{x}_{(n+1)\delta}[M_{-s(n)}\vartheta_{t(n)}]-\hat{x}_{t(n)}[\vartheta_{t(n)}]}{c_{(n+1)\delta}}.$$

Since $P(Y_n \geq -(2A_{t(n)})^{1/2}|\mathfrak{F}_{t(n)}) \geq 1/2$ it follows by the conditional Borel-Cantelli lemma that $Y_n \geq -(2A_{t(n)})^{1/2}$ i.a. and $A_{t(n)}$ being negligible in the limit, the claim follows. □

We can now easily prove

12.4. THEOREM. *The LIL's of 1.1, parts* 1° *and* 2° *of 3.1 and 3.2 hold in continuous time as well provided that in 3.1 and 3.2 Condition 12.2 is assumed to be in force.*

PROOF. Since $\overline{\lim_{t\to\infty}} \geq \overline{\lim_{n\to\infty}}$, we only have to prove the $\overline{\lim_{t\to\infty}} \leq$ parts. For 1.1, let $\vartheta_t := \rho^{s-\delta}\varphi$, $\omega_\delta^2 := \Phi^*[\mathrm{Var}^{\langle\cdot\rangle}W_\delta]$, $c_t := (2\omega_\delta^2\hat{x}_t[\varphi]\log t)^{1/2}$. Then $M_{-s}\vartheta_t = \rho^{-\delta}\varphi, \alpha = \rho^{\delta/2}$,

$$A_t = \frac{\hat{x}_t[\mathrm{Var}^{\langle\cdot\rangle} x_s[\rho^{-\delta}\varphi]]}{c^2_{(n+1)\delta}} \leq \frac{\hat{x}_t[\mathrm{Var}^{\langle\cdot\rangle} W_s]}{c^2_{(n+1)\delta}} = O(\frac{1}{\log n}),$$

$$\overline{\lim_{n\to\infty}}\, K_n = \overline{\lim_{n\to\infty}}\, \frac{\rho^{-\delta}\hat{x}_{(n+1)\delta}[\varphi] - \hat{x}_{n\delta}[\varphi]}{c_{(n+1)\delta}} = \rho^{-\delta/2} \quad (\text{using } \underline{1}.\underline{2}),$$

$$\overline{\lim_{t\to\infty}}\, \frac{\rho^{s-\delta}\hat{x}_t[\varphi] - \hat{x}_{n\delta}[\varphi]}{c_t} \leq \alpha\, \overline{\lim_{n\to\infty}}\, M_n = 1 \quad (\text{using } \underline{12}.\underline{3}),$$

$$\overline{\lim_{t\to\infty}} \left| \frac{\rho^{s-\delta}\hat{x}_t[\varphi] - \hat{x}_{n\delta}[\varphi]}{c_t} \right| = \overline{\lim_{t\to\infty}}\, \frac{\rho^{n\delta}|W_t - W_{n\delta}|}{c_t} \leq 1$$

(by symmetry). Now observe that $\omega_\delta^2 \to 0$ as $\delta \to 0$ and that, using the skeleton LIL,

$$\overline{\lim_{t\to\infty}} \left(\frac{\rho^t}{2\tau^2 W \log t}\right)^{1/2} (W - W_t) = \overline{\lim_{t\to\infty}} \left(\frac{\rho^t}{2\tau^2 W \log t}\right)^{1/2} ((W - W_{n\delta}) + (W_{n\delta} - W_t))$$

$$\leq \rho^{\delta/2} + \rho^{\delta} \frac{\omega_\delta}{\tau} .$$

In part 1° of $\underline{3}.\underline{1}$, let $\vartheta_t := \xi$, $\zeta := 0$, $c_t := (2\sigma^2 \hat{x}_t[\varphi] \log t)^{1/2}$ and note that the eigenvalues of M_s restricted to $\mathfrak{C}$ are non-zero, being of the form $e^{s\mu}$ with μ an eigenvalue of the corresponding generator. Hence M_s is one-one on $\mathfrak{C}$. Let

$$\eta_\delta(x) := \sup_{0 \leq s \leq \delta} \mathrm{Var}^{\langle x \rangle} \hat{x}_s[M_{-s}\xi], \quad \sigma_\delta^2 := \Phi^*[\eta_\delta].$$

Then from finite dimensionality, it follows as in the proof of $\underline{9}.\underline{2}$ and from the LIL's for discrete skeletons that $\overline{\lim}\, K_n \leq \sigma_\delta/\sigma$, and also that $A_t \leq \hat{x}_t[\eta_\delta]/c^2_{(n+1)\delta} \to 0$. Thus by $\underline{12}.\underline{3}$, $\overline{\lim_{t\to\infty}}\, \hat{x}_t[\xi]/c_t \leq \rho^{\delta/2}\sigma_\delta/\sigma$. As $\delta \downarrow 0$, $\sigma_\delta \to \sigma$. Finally for part 2° of $\underline{3}.\underline{1}$, no additional argument is required in continuous time since the way to bound the $\hat{x}_t[\xi]$ when $t(i-1) \leq t \leq t(i)$ works equally well here. □

It only remains to consider part 3° of $\underline{3}.\underline{1}$. The formulation in $\underline{3.1}$ involves the Jordan canonical form for the discrete skeletons more explicitly than so far, so we shall first reformulate in terms of the parameters $a, \psi_{\nu,j}, \Psi^*_{\nu,j}$ etc. defined above. First note that $\underline{3}.\underline{1}$ applied to a skeleton reads

$$(12.12)\quad \lim_{N\to\infty}\{(N^{\gamma-1}e^{a\delta N})^{-1}\hat{x}_{N\delta}[\xi]-\sum_{\nu}^{(\xi)}\frac{e^{-\delta\mu_\nu(\gamma-1)}}{(\gamma-1)!}\Phi^*_{\nu,\gamma}[\xi]e^{i\delta Nb_\nu}W^\nu\} = 0$$

which, letting $t = N\delta$ and using (12.6), is seen to be the skeleton version of

$$(12.13)\quad \lim_{t\to\infty}\{(t^{\gamma-1}e^{at})^{-1}\hat{x}_t[\xi] - \sum_{\nu}^{(\xi)}\frac{1}{(\gamma-1)!}\Psi^*_{\nu,\gamma}[\xi]e^{itb_\nu}W^\nu\} = 0$$

where $W^\nu = \lim_{n\to\infty}\rho_\nu^{-n}\hat{x}_{n\delta}[\varphi_{\nu,1}] = \lim_{t\to\infty} e^{-t\mu_\nu}\hat{x}_t[\psi_{\nu,1}]$ (the latter limit exists because of the convergence theorem for $\mathcal{L}^2$-bounded continuous time martingales).

<u>12.5</u>. THEOREM. *If* $2a > \log\rho$ *and Condition 12.2 holds, then* (12.13) *is valid*.

PROOF. Define

$$\eta(\xi) := \sum_{\nu}^{(\xi)}\frac{1}{(\gamma-1)!}\Psi^*_{\nu,\gamma}[\xi]\psi_{\nu,1},\ \vartheta_t := \xi - t^{\gamma-1}\eta(\xi),\ \zeta := 0,$$

$c_t := t^{\gamma-1}e^{at}$. Replacing W^ν by $e^{-t\mu_\nu}\hat{x}_t[\psi_{\nu,1}]$ shows that (12.13) is equivalent to $x_t[\vartheta_t]/c_t \to 0$ a.s. In particular, the skeleton version of (12.13) and finite dimensionality implies that

$$\sup_{0\le s\le\delta}\frac{\hat{x}_{(n+1)\delta}[M_{-s}\xi-((n+1)\delta)^{\gamma-1}\eta(M_{-s}\xi)]}{c_{(n+1)\delta}} \to 0 \text{ a.s.},$$

$$(12.14)\quad \sup_{\substack{0\le s\le\delta\\ t+s=(n+1)\delta}}\frac{\hat{x}_{(n+1)\delta}[M_{-s}\xi-t^{\gamma-1}\eta(M_{-s}\xi)]}{c_{(n+1)\delta}} \to 0 \text{ a.s.}$$

It follows from (12.2) that $\Psi^*_{\nu,\gamma}[M_{-s}\xi] = e^{-s\mu_\nu}\Psi^*_{\nu,\gamma}[\xi], M_{-s}\psi_{\nu,1} = e^{-s\mu_\nu}\psi_{\nu,1}$ and therefore that $\eta(M_{-s}\xi) = M_{-s}\eta(\xi)$ so that the l.h.s. of (12.14) is simply K_n. It follows as above that $A_t \to 0$ a.s. Hence by <u>12.3</u>, $\overline{\lim}_{t\to\infty}\hat{x}_t[\vartheta_t]/c_t \le e^{a\delta}\cdot 0 = 0$. $\underline{\lim} \ge 0$ follows similarly. □

13. INFINITE VARIANCE

We shall now investigate how the absence of second moments in the supercritical case influences the rate of convergence of $\hat{x}_N[\xi]/\rho^N$ to $W\varphi^*[\xi]$, in particular the magnitude of $\hat{x}_N[\xi]/\rho^N$ in the case $\Phi^*[\xi] = 0$. The aim is to suggest the typical form of the results in the simplest cases rather than heading for a description as complete as in Sections 1-10. To this end, some simplifying assumptions are made. Thus we consider only discrete time and the real eigenvector case

$$M^N\xi = \rho_\nu^N\xi \quad \text{with} \quad \rho_\nu \quad \text{real} \tag{13.1}$$

(in particular the case $\xi = \varphi, \rho_\nu = \rho$). For example, one-dimensional branching diffusions on simply connected domains provide nice examples of (13.1), cf. Chapter V. The results of Section 4 of Chapter II suggest that moment conditions of the form

$$\mu_\beta := \Phi^*[E^{\langle\cdot\rangle}|\hat{x}_1[\xi]|^{1/\beta}] < \infty \quad (\tfrac{1}{2} < \beta < 1), \tag{13.2}$$

$$\Phi^*[E^{\langle\cdot\rangle}|\hat{x}_1[\xi]|(\log^+|\hat{x}_1[\xi]|)^p] < \infty \quad (p \geq 1) \tag{13.3}$$

will come up (at least for $\xi = \varphi$). We shall, however, concentrate on (13.2) and not go into (13.3) except that we throughout assume (X LOG X), i.e. (13.3) with $p = 1$, $\xi = \varphi$.

We let $\alpha := \log|\rho_\nu|/\log\rho$ and note that the cases $\alpha < 1/2$, $\alpha = 1/2$, and $\alpha > 1/2$ corresponds to the cases $\lambda_\nu^2 < \rho$, $\lambda_\nu^2 = \rho$ and $\lambda_\nu^2 > \rho$, respectively. In the absence of second moments, we need, however, essentially only to distinguish between the cases $\alpha \leq 1/2$ and $\alpha > 1/2$:

__13.1__. THEOREM. _Suppose_ $\alpha \leq 1/2$. _Then condition_ (13.2) _is sufficient for_ $\hat{x}_N[\xi] = o(\rho^{N\beta})$ _a.s. If the set of types is finite_, (13.2) _is also necessary_.

__13.2__. THEOREM. _Suppose_ $\alpha > 1/2$. _Then condition_ (13.2) _is sufficient for_ (i) $\hat{x}_N[\xi] = o(\rho^{N\beta})$ _if_ $\beta > \alpha$; (ii) _the existence of_ $W^\nu := \lim_{N\to\infty} \rho_\nu^{-N}\hat{x}_N[\xi]$ _if_ $\beta = \alpha$; (iii) $\hat{x}_N[\xi] - \rho_\nu^N W^\nu = o(\rho^{N\beta})$ _if_ $\beta < \alpha$ [of course, $W^\nu = W$ if $\rho_\nu = \rho$]. _If the set of types is finite_, (13.2) _is also necessary_.

Defining $S_N := \delta^{N+1}\{\hat{x}_{N+1}[\xi] - \rho_\nu\hat{x}_N[\xi]\}$ (with $|\delta| < \rho^{-1/2}$), the proofs of __13.1__, __13.2__ reduce essentially to the two following lemmata:

__13.3__. LEMMA. _If_ (13.2) _holds, then_ $\Sigma_0^\infty S_n$ _converges a.s. provided that_ $-\log|\delta|/\log\rho \geq \beta$.

13.4. LEMMA. If (13.2) fails and the set of types is finite, then $\overline{\lim}_{N\to\infty}|S_N|$ a.s. on $\{W>0\}$ provided that $-\log|\delta|/\log\rho \le \beta$.

PROOF OF 13.3: Routine: Let

$$\tilde{S}_N := \delta^{N+1}\Big\{\sum_{i=1}^{|\hat{x}_N|} \hat{x}_{N+1}^{N,i}[\xi]\, I(|\hat{x}_{N+1}^{N,i}[\xi]| \le |\delta|^{-N-1}) - \rho_\nu \hat{x}_N[\xi]\Big\},$$

$$\epsilon_N := -E(\tilde{S}_N|\mathfrak{F}_N) = E(S_N - \tilde{S}_N|\mathfrak{F}_N) = \delta^{N+1}\hat{x}_N[E^{\langle\cdot\rangle} x_1[\xi]\, I(|\hat{x}_1[\xi]| > |\delta|^{-N-1})],$$

$$F(x) := \Phi^*[P^{\langle\cdot\rangle}(|\hat{x}_1[\xi]| \le x)].$$

Then

$$\sum_{N=0}^{\infty} P(S_N \ne \tilde{S}_N) \le \sum_{N=0}^{\infty} E\hat{x}_N[P^{\langle\cdot\rangle}(|\hat{x}_1[\xi]| > |\delta|^{-N-1})]$$

$$\le c \sum_{N=0}^{\infty} \rho^N \int_{|\delta|^{-N-1}}^{\infty} dF(x) = c\int_0^{\infty} O(x^{-\log\rho/\log|\delta|})\, dF(x)$$

which is finite by (13.2). Similar calculations yield $\Sigma_0^{\infty} E|\epsilon_N| < \infty$, $\Sigma_0^{\infty}\mathrm{Var}\{\tilde{S}_N + \epsilon_N\} < \infty$ and hence the a.s. convergence of the sum $\Sigma_0^{\infty}\{\tilde{S}_N + \epsilon_N\}$ of martingale increments and of $\Sigma_0^{\infty}\tilde{S}_N$, $\Sigma_0^{\infty} S_N$. □

PROOF OF 13.4. Given the σ-algebra $\mathfrak{F}$ spanned by the branching process, choose random populations $\hat{y}_{n+1}^{n,i}$ which are all independent and satisfy

$$P(\hat{y}_{n+1}^{n,i} \in \hat{A}|\mathfrak{F}) = P(\hat{x}_{n+1}^{n,i} \in \hat{A}|\mathfrak{F}_n) \;\; (= P^{\langle x_i\rangle}(\hat{x}_1 \in \hat{A}))$$

for all measurable $\hat{A} \subseteq \hat{X}$. Define

$$\mathfrak{G}_{n+1} := \sigma(\mathfrak{F}_{n+1}, \hat{y}_{k+1}^{k,i} : k \le n),\quad Y_{n+1}^{n,i} := \hat{x}_{n+1}^{n,i}[\xi] - \hat{y}_{n+1}^{n,i}[\xi],$$

$$S_n' := \delta^{n+1}\Big\{\sum_{i=1}^{|\hat{x}_n|} \hat{y}_{n+1}^{n,i}[\xi] - \rho_\nu \hat{x}_n[\xi]\Big\},\quad S_n^s := S_n - S_n' = \delta^{n+1}\sum_{i=1}^{|\hat{x}_n|} Y_{n+1}^{n,i}.$$

Then

$$P(S_n' \le y|\mathfrak{G}_n) = P(S_n \le y|\mathfrak{G}_n) = P(S_n \le y|\mathfrak{F}_n), \tag{13.4}$$

$$P(Y_{n+1}^{n,i} \le y|\mathfrak{G}_n) = P(Y_{n+1}^{n,i} \ge -y|\mathfrak{G}_n). \tag{13.5}$$

Now suppose $P(\overline{\lim}_{N\to\infty}|S_N| < \infty, W > 0) > 0$, i.e., $PE := P(\overline{\lim}_{N\to\infty}|S_N| \leq K, W > 0) > 0$ for some $K < \infty$. We shall then by repeated applications of the conditional Borel-Cantelli lemma (both ways) show that (13.2) must hold with $\beta = -\log|\delta|/\log\rho$. Let the below a.s. statements be understood to hold on E and note first that by (13.4)

$$\sum_{N=0}^{\infty} P(|S_N| > K|\mathfrak{G}_N) < \infty, \sum_{N=0}^{\infty} P(|S_N'| > K|\mathfrak{G}_N) < \infty$$

so that $\overline{\lim}|S_N'| \leq K$, $\overline{\lim}|S_N^s| \leq 2K$. Define

$$M_N := \max_{i=1,\ldots,|x_N|} |Y_{N+1}^{N,i}| .$$

Then

$$\sum_{N=0}^{\infty} P(M_N > 2K\delta^{-N-1}|\mathfrak{G}_N) \leq \frac{1}{2} \sum_{n=0}^{\infty} P(|S_N^s| > 2K|\mathfrak{G}_N) < \infty$$

(using A 6.4 and the symmetry property (13.5)). Hence $\overline{\lim}\ \delta^N M_N \leq 2K$, which (using a suitable family of σ-fields indexed by the set of all individuals ever alive) is equivalent to

$$\sum_{N=0}^{\infty} \sum_{i=1}^{|\hat{x}_N|} P^{\langle x_i \rangle}(|Y_{N+1}^{N,i}| > 2K|\delta|^{-N}) < \infty.$$

But letting $F^{\langle x \rangle}$ be the symmetrized $P^{\langle x \rangle}$-distribution of $x_1[\xi]$, this sum is

$$\sum_{N=0}^{\infty} \hat{x}_N[\int_{|y|>2K|\delta|^{-N}} dF^{\langle\cdot\rangle}(y)]$$

which, since the set of types if finite, can only be finite if

$$\sum_{N=0}^{\infty} \rho^N \int_{|y|>2K|\delta|^{-N}} dF^{\langle x \rangle}(y) = \int_{-\infty}^{\infty} O(|y|^{-\frac{\log\rho}{\log|\delta|}})dF^{\langle x \rangle}(y)$$

is finite for each $x \in X$. This implies by standard symmetrization inequalities $E^{\langle x \rangle}|\hat{x}_1[\xi]|^{-\log\rho/\log|\delta|} < \infty$ and the proof is complete. □

PROOF OF 13.1, 13.2. Suppose $\rho_\nu \neq 0$ and let $\delta := |\rho_\nu|/\rho_\nu\rho^\beta$. Then if (13.2) holds, it follows by 13.3 that

$$(13.6) \qquad \sum_{n=0}^{\infty} S_n = \sum_{n=1}^{\infty} \frac{|\rho_\nu|^n}{\rho^{n\beta}} \cdot \rho_\nu^{-n}\{\hat{x}_n[\xi]-\rho_\nu\hat{x}_{n-1}[\xi]\}$$

converges. If $\alpha \leq 1/2$ or $\beta > \alpha > 1/2$, then $|\rho_\nu|^n/\rho^{\beta n} \downarrow 0$ so that by Kronecker's lemma

$$\rho_\nu^{-N}\hat{x}_N[\xi]-\hat{x}_0[\xi] = \sum_{n=1}^{N} \rho_\nu^{-n}\{\hat{x}_n[\xi]-\rho_\nu\hat{x}_{n-1}[\xi]\} = o(\frac{\rho^{N\beta}}{|\rho_\nu|^N}),$$

proving the sufficiency in 13.1 and part (i) of 13.2. The sufficiency in part (ii) comes out simply by noting that (since here $|\rho_\nu| = \rho^\beta$) the existence of W^ν is equivalent to the convergence of (13.6), while in part (iii) (where $|\rho_\nu|^n/\rho^{n\beta} \uparrow \infty$) it follows from the tail sum analogue of Kronecker's lemma II.4.2 that

$$W^\nu - \rho_\nu^{-N}\hat{x}_N[\zeta] = \sum_{n=N+1}^{\infty} \rho_\nu^{-n}\{\hat{x}_n[\zeta] - \rho_\nu \hat{x}_{n-1}[\zeta]\} = o\left(\frac{\rho^{N\beta}}{|\rho_\nu|^N}\right).$$

For necessity, note simply that all assertions imply $S_n = o(1)$ and apply 13.4. Finally if $\rho_\nu = 0$, let $\delta := \rho^{-\beta}$ and note that $\hat{x}_N[\zeta] = \rho^{N\beta}S_{N-1}$ and that the convergence of (13.6) entails $S_N = o(1)$. □

BIBLIOGRAPHICAL NOTES

1.1 is a general version of II.3.1 . Assuming the set of types to be finite, the CLT's of parts 1° and 2° of 3.1, as well as part 3° of 3.1, were proved by Kesten and Stigum (1966b) (discrete time) and Athreya (1969a, b, 1971) (continuous time), 3.3 is due to Athreya and Ney (1974), while the LIL's were given by Asmussen (1977). Section 11 is extracted from Asmussen and Keiding (1978) who also gave the method used in the proofs of a number of CLT's. Some material related to Section 11 is given by Asmussen (1982). The case of an infinite set of types is presented here for the first time. A comparison with the references will reveal that this extension sometimes requires totally different and more complicated proofs. Parts of this Chapter would therefore simplify in the finite case. Some indications of this are in the text.

PART D

RELATED MODELS

CHAPTER IX

UNBOUNDED DOMAINS

1. THE BRANCHING ORNSTEIN-UHLENBECK PROCESS

We recall that the Ornstein-Uhlenbeck process $\{y_t\}_{t\geq 0}$ is the diffusion on the line specified by the differential generator

$$Af(x) = \frac{1}{2}f''(x) - \gamma xf'(x), \quad f \in \mathcal{D}(A) \tag{1.1}$$

$(\gamma > 0)$. An alternative description is that $\{y_t\}_{t\geq 0}$ is a Gaussian Markov process with continuous paths and the distribution of y_{t+s} given $y_s = y$ Gaussian with mean $ye^{-\gamma t}$ and variance $(1-e^{-2\gamma t})/2\gamma$.

The branching mechanism is specified by the (constant) termination density $k \in (0,\infty)$ and the (space-independent) branching law given by the offspring distribution $\{p_n\}_{n\in\mathbb{N}}$ of a Galton-Watson process. Then with $m := \Sigma_0^\infty np_n$, the moment semigroup has generator

$$Lf(x) = \frac{1}{2}f''(x) - \gamma xf'(x) + k(m-1)f(x) \tag{1.2}$$

and the total number of particles $|\hat{x}_t|$ is a Markov branching process with Malthusian parameter $\lambda = k(m-1)$. We shall be concerned only with the critical case $m = 1$, $\lambda = 0$ and the supercritical case $1 < m < \infty$, $0 < \lambda < \infty$.

As is well known, the Ornstein-Uhlenbeck is ergodic, with the limiting stationary distribution the normal distribution $\tilde{\Phi} = N(0,1/2\gamma)$. This leads one to expect a similar limiting behaviour as for the models of Part C. However, the moment semigroup is not uniformly primitive and thus the theory of Part C is inapplicable. This necessitates a parallel treatment.

Before stating the results, we introduce some notation. Let $E := \{|\hat{x}_t| = 0$ eventually$\}$ denote the set of extinction, let in the supercritical case $\{\gamma_t\}$ be the constants leading to a proper normalization of $|\hat{x}_t|$ on E^c, i.e. $\gamma_t|\hat{x}_t| \overset{a.s.}{\to} W$ with $0 < W < \infty$ a.s. on E^c (cf. Ch. II, Sect. 5, Ch. III, Sect. 5, Ch. IV, Sect. 4) and write for the ease of notation $\hat{x}_t[B] := \hat{x}_t[1_B]$, $1_B(x) := I(X\in B)$. Then:

1.1. THEOREM. *If* $1 < m < \infty$, *then for any Borel set* B *with* $|\partial B| = 0$ ($|\cdot| :=$ Lebesgue measure)

$$\lim_{t\to\infty} \gamma_t \hat{x}_t[B] = W\tilde{\Phi}[B] \quad \text{a.s.,} \tag{1.3}$$

$$\lim_{t\to\infty} \frac{\hat{x}_t[B]}{|\hat{x}_t|} = \tilde{\Phi}[B] \quad \text{a.s. on } E^c. \tag{1.4}$$

1.2. THEOREM. If $m = 1$, then for any Borel set B and any $\varepsilon > 0$,

$$(1.5)\qquad \lim_{t\to\infty} P\left(\left|\frac{\hat{x}_t[B]}{|\hat{x}_t|} - \tilde{\Phi}(B)\right| > \varepsilon \,\Big|\, |\hat{x}_t| > 0\right) = 0.$$

Notice that for 1.2 we assume nothing beyond first moments. That is, (1.5) holds even if there exists no normalization leading to a proper conditional limit law for $|\hat{x}_t|$, as may be the case if the offspring variance is infinite, cf. III.3.7, III.3.8 and Ch. VI, Sect. 4.

The proofs of 1.1, 1.2 proceed from the decomposition

$$(1.6)\qquad \frac{\hat{x}_{t+s}[B]}{|\hat{x}_{t+s}|} - \tilde{\Phi}[B] = \frac{|\hat{x}_t|}{|\hat{x}_{t+s}|}(A_{t,s} + e^{\lambda s}B_{t,s}) + C_{t,s}$$

$$A_{t,s} := |\hat{x}_t|^{-1} \sum_{i=1}^{|\hat{x}_t|} [\hat{x}_{t+s}^{t,i}[B] - E(\hat{x}_{t+s}^{t,i}[B] \mid \mathfrak{F}_t)],$$

$$B_{t,s} := |\hat{x}_t|^{-1} \sum_{i=1}^{|\hat{x}_t|} [e^{-\lambda s} E(\hat{x}_{t+s}^{t,i}[B] \mid \mathfrak{F}_t) - \Phi[B]]$$

$$= |\hat{x}_t|^{-1}\hat{x}_t[P^{\cdot}(y_s \in B)] - \tilde{\Phi}[B],$$

$$C_{t,s} := \left(\frac{e^{\lambda s}|\hat{x}_t|}{|\hat{x}_{t+s}|} - 1\right)\tilde{\Phi}[B].$$

Here $A_{t,s}$ has the form considered in A10, i.e. a normalized sum of independent mean zero r.v. with a common integrabel stochastical majorant, say the distribution Q of $|\hat{x}_s| + e^{\lambda s}$. The basis for estimating $B_{t,s}$ is the ergodicity of the Ornstein-Uhlenbeck process in the form

$$(1.7)\qquad \eta_s := \sup_{-s<y<s} |P^y(y_s \in B) - \tilde{\Phi}(B)| \to 0 \quad \text{as} \quad s \to \infty.$$

Throughout we let $x^{t,i}$ denote the position of the ith individual alive at time t and first give the comparatively simple

PROOF OF 1.2. Appealing to A10.1,

$$(1.8)\qquad P(|A_{t,s}| > \varepsilon \mid |\hat{x}_t| > 0)$$

$$\leq P(|\hat{x}_t| < N \mid |\hat{x}_t| > 0) + c_1 \sup_{n\geq N} n\int_n^{\infty} dQ(x) + c_2 \sup_{n\geq N} \frac{1}{n}\int_0^n x^2 dQ(x) \to 0$$

since the two last terms tend to zero as $N \to \infty$ (using dominated convergence) and the first tends to zero as $t \to \infty$ with N fixed by Lemma 4.9 of Athreya and Kaplan (1978). Next we have

$$|B_{t,s}| \leq \eta_s + 2\,\frac{\hat{x}_t[(-s,s)^c]}{|\hat{x}_t|},$$

$$E\left(\frac{\hat{x}_t[(-s,s)^c]}{|\hat{x}_t|}\,\Big|\,|\hat{x}_t| > 0\right) \leq \max_{i=1,\ldots,|x_0|} P^{x^{0,i}}(|y_t| > s) \underset{t\to\infty}{\to} \tilde{\Phi}((-s,s)^c)$$

so that it follows that

$$\lim_{s\to\infty} \overline{\lim_{t\to\infty}}\, P(|B_{t,s}| > \varepsilon \,|\, |\hat{x}_t| > 0) = 0.$$

Finally, note that

$$P(\cdot\,|\,|\hat{x}_{t+s}| > 0) \leq P(\cdot\,|\,|\hat{x}_t| > 0) \cdot \frac{P(|\hat{x}_t|>0)}{P(|\hat{x}_{t+s}|>0)},$$

$$\frac{P(|\hat{x}_t|>0)}{P(|\hat{x}_{t+s}|>0)} \underset{t\to\infty}{\to} 1,\quad P\left(\left|\frac{|\hat{x}_t|}{|\hat{x}_{t+s}|} - 1\right| > \varepsilon \,\Big|\, |\hat{x}_t| > 0\right) \to 0 \text{ (take } B=\mathbb{R} \text{ in (1.8))}.$$

Hence $A_{t,s}$, $B_{t,s}$, $C_{t,s}$ tend to zero w.r.t. $P(\cdot\,|\,|\hat{x}_{t+s}| > 0)$. □

When estimating $B_{t,s}$ in the proof of 1.1, we shall need

1.3. LEMMA. _Define_ $\vartheta(x) := |x|$. _If_ $1 < m < \infty$, _then for any_ $\delta > 0$ $\overline{\lim}_{n\to\infty} \hat{x}_{n\delta}[\vartheta]/|\hat{x}_{n\delta}| < \infty$ _a.s. on_ E^c.

PROOF. Suppose for the ease of notation $\delta = 1$ and define $\rho := e^{\lambda} = e^{k(m-1)}$, $\eta := \rho^{-1}E^{\langle 0\rangle}\hat{x}_1[\vartheta]$. The proof is carried out by showing that a.s. on E^c

$$(1.9)\quad \rho^{-1}\,\overline{\lim_{n\to\infty}}\left\{\frac{\hat{x}_{n+1}[\vartheta]}{|\hat{x}_n|} - \rho e^{-\gamma}\frac{\hat{x}_n[\vartheta]}{|\hat{x}_n|}\right\} = \overline{\lim_{n\to\infty}}\left\{\frac{\hat{x}_{n+1}[\vartheta]}{|\hat{x}_{n+1}|} - e^{-\gamma}\frac{\hat{x}_n[\vartheta]}{|\hat{x}_n|}\right\} \leq \eta\,.$$

In fact, the claim follows from (1.9) upon iteration. To prove (1.9), define $Y_{n,i} := |\hat{x}^{n,i}_{n+1}|\,I(|\hat{x}^{n,i}_{n+1}| \leq |\hat{x}_n|)$, let the ith individual at time n have position $x^{n,i}$ and his j^{th} child position $x^{n,i}_j$, and notice that

$$(1.10)\qquad \frac{\hat{x}_{n+1}[\vartheta]}{|\hat{x}_n|} - \rho e^{-\gamma}\frac{\hat{x}_n[\vartheta]}{|\hat{x}_n|} - \rho\eta \leq a_n+b_n+c_n+d_n,$$

$$a_n := \frac{1}{|\hat{x}_n|}\sum_{i=1}^{|\hat{x}_n|}\{U_{n,i} - \rho\eta\},\quad U_{n,i} := \sum_{j=1}^{|\hat{x}_{n+1}^{n,i}|} |x_j^{n,i} - e^{-\gamma}x^{n,i}|,$$

$$b_n := \frac{1}{|\hat{x}_n|}\sum_{i=1}^{|\hat{x}_n|} |x^{n,i}|e^{-\gamma}\{\hat{x}_{n+1}^{n,i} - Y_{n,i}\},$$

$$c_n := \frac{1}{|\hat{x}_n|}\sum_{i=1}^{|\hat{x}_n|} |x^{n,i}|e^{-\gamma}\{Y_{n,i} - E(Y_{n,i}|\mathfrak{F}_n)\}$$

$$d_n := \frac{1}{|\hat{x}_n|}\sum_{i=1}^{|\hat{x}_n|} |x^{n,i}|e^{-\gamma}\{E(Y_{n,i}|\mathfrak{F}_n) - \rho\}$$

Let F be the distribution of $|\hat{x}_1|$ given one initial particle. Since $d_n \leq 0$, it suffices to show that $a_n \to 0$, $b_n \to 0$, $c_n \to 0$. Here $a_n \to 0$ is immediate from a suitable LLN for triangular arrays, see A10.2, since given $\mathfrak{F}_n$ the $U_{n,i} - \rho\eta$ are i.i.d. with mean zero. (The common distribution is the centered law of the sum of the absolute values of k normal $(0,1-e^{-2\gamma})$ r.v., k chosen at random according to F.) For $b_n \to 0$, note that

$$\sum_{n=0}^{\infty} P(b_n > 0|\mathfrak{F}_n) \leq \sum_{n=0}^{\infty} |\hat{x}_n|\int_{|\hat{x}_n|}^{\infty} dF(x) = \int_0^{\infty} O(x)dF(x) < \infty,$$

cf. II.5.3, and use the conditional Borel-Cantelli Lemma. For $c_n \to 0$, note first that

$$\sum_{n=0}^{\infty} E(c_n^2|\mathfrak{F}_n) = \sum_{n=0}^{\infty} \operatorname{Var}(c_n|\mathfrak{F}_n) \leq e^{-2\gamma}\sum_{n=0}^{\infty}\frac{1}{|\hat{x}_n|^2}\int_0^{|\hat{x}_n|} x^2 dF(x)\sum_{i=1}^{|\hat{x}_n|} {x^{n,i}}^2.$$

Given the σ-algebra $\mathfrak{G}$ spanned by the branching process (without motion), and the $x^{0,j}$, the distribution of $x^{n,i}$ is that of an Ornstein-Uhlenbeck process started at the position $x^{0,j}$ of the ancestor of i. Hence

$$E\Big(\sum_{i=1}^{|\hat{x}_n|} {x^{n,i}}^2\Big|\mathfrak{G}\Big) = \sum_{j=0}^{|\hat{x}_0|} |\hat{x}_n^{0,j}|\{e^{-2n\gamma}{x^{0,j}}^2 + 1 - e^{-2n\gamma}\} = O(|\hat{x}_n|),$$

$$E\Big(\sum_{n=0}^{\infty} E(c_n^2|\mathfrak{F}_n)\Big|\mathfrak{G}\Big) \leq \sum_{n=0}^{\infty} O\Big(\frac{1}{|\hat{x}_n|}\Big)\int_0^{|\hat{x}_n|} x^2 dF(x) = \int_0^{\infty} O(x)dF(x) < \infty$$

(using II.5.3). This implies $\Sigma_0^\infty E(c_n^2|\mathfrak{F}_n) < \infty$ and $c_n \to 0$. □

1.4. LEMMA. *For any Borel set* B *and any* $\delta > 0$,

$$\lim_{n\to\infty} \frac{\hat{x}_{n\delta}[B]}{|\hat{x}_{n\delta}|} = \tilde{\Phi}[B] \quad \text{a.s. on } E^c.$$

PROOF. Let $t = n\delta$, $s = m\delta$ in (1.6). Then for any fixed m, $\lim_{n\to\infty} C_{n\delta,m\delta} = 0$. Also $\lim_{n\to\infty} A_{n\delta,m\delta} = 0$ is immediate from A10.2 and thus the assertion follows upon letting first $n \to \infty$ and next $m \to \infty$ from

$$B_{n\delta,m\delta} \leq \eta_{m\delta} + 2\frac{\hat{x}_{n\delta}[(-m\delta,m\delta)^c]}{|\hat{x}_{n\delta}|} \leq \eta_{m\delta} + \frac{\hat{x}_{n\delta}[\theta]}{m\delta|\hat{x}_{n\delta}|}. \qquad □$$

We can now easily complete the

PROOF OF 1.1. It suffices to show that $\underline{\lim}_{t\to\infty} \hat{x}_t[B]/|\hat{x}_t| \geq \tilde{\Phi}[B]$ if B is bounded with $|\partial B| = 0$ since the general case can be reduced to this by first approximating B by $B \cap [-K,K]$ with K large and next passing to the complement. Define

$$Y^i_{n,\delta} := I(\hat{x}_t^{n\delta,i}[B] > 0 \quad \forall t \in [n\delta,(n+1)\delta]),$$

$$\mu(x) := P^{\langle x\rangle}(\hat{x}_t[B] > 0 \quad \forall t \in [0,\delta]).$$

Then by A10.2

$$\underline{\lim}_{t\to\infty} \frac{\hat{x}_t[B]}{|\hat{x}_t|} \geq \underline{\lim}_{n\to\infty} \inf_{n\delta\leq t\leq(n+1)\delta} \frac{|\hat{x}_{n\delta}|}{|\hat{x}_t|} \cdot \frac{1}{|\hat{x}_{n\delta}|} \sum_{i=1}^{|\hat{x}_{n\delta}|} Y^i_{n,\delta}$$

$$= e^{-\delta} \underline{\lim}_{n\to\infty} \frac{1}{|\hat{x}_{n\delta}|} \sum_{i=1}^{|\hat{x}_{n\delta}|} E(Y^i_{n,\delta}|\mathfrak{F}_{n\delta}) = e^{-\delta} \underline{\lim}_{n\to\infty} \frac{\hat{x}_{n\delta}[\mu]}{|\hat{x}_{n\delta}|}.$$

But if $x \in B_\epsilon := \{x \in B: \inf_{y\in\partial B} |x-y| > \epsilon\}$, then clearly $\mu(x) \geq e^{-k\delta} P^x(|y_t - x| \leq \epsilon \;\; \forall t \in [0,\delta])$ becomes close to one as $\delta \downarrow 0$, uniformly in x by boundedness. Hence

$$\varliminf_{t\to\infty} \frac{\hat{x}_t|B|}{|\hat{x}_t|} \geq \varliminf_{n\to\infty} \frac{\hat{x}_{n\delta}[B^\epsilon]}{|\hat{x}_{n\delta}|} = \tilde{\Phi}[B^\epsilon].$$

As $\epsilon \downarrow 0$, $\tilde{\Phi}[B^\epsilon] \uparrow \tilde{\Phi}[B]$. □

2. BRANCHING BROWNIAN MOTION

The model and notation is the same as in Section 1 except that $\{y_t\}_{t\geq 0}$ is now standard Brownian motion specified by the differential generator

$$Af(x) = \frac{1}{2} f''(x), \quad f \in \mathfrak{D}(A). \tag{2.1}$$

(The discussion below carries over to a velocity $\sigma^2 \neq 1$ and a drift $\mu \neq 0$ by means of straightforward transformations.)

An alternative description is that $\{y_t\}_{t\geq 0}$ is a Gaussian Markov process with continuous paths and stationary independent increments. In particular, the distribution of y_t is normal $(0,t)$ (in contrast to the ergodicity of the Ornstein-Uhlenbeck process).

In particular, if $t^{1/2}B := \{t^{1/2}b : b \in B\}$, then

$$E^{\langle 0\rangle}\hat{x}_t[t^{1/2}B] = e^{\lambda t}\Phi[B], \quad E^{\langle x\rangle}\hat{x}_t[t^{1/2}B] \simeq e^{\lambda t}\Phi[B]$$

for all $x \in \mathbb{R}$, Φ standard normal, which leads one to expect

2.1. THEOREM. *If* $1 < m < \infty$ *and* $(x \log x)$ *holds, then for any Borel set* B *with* $|\partial B| = 0$

$$\lim_{t\to\infty} \frac{\hat{x}_t[t^{1/2}B]}{e^{\lambda t}} = W\Phi[B] \quad \text{a.s.,} \tag{2.1}$$

$$\lim_{t\to\infty} \frac{\hat{x}_t[t^{1/2}B]}{|\hat{x}_t|} = \Phi[B] \quad \text{a.s. on } E^c. \tag{2.2}$$

If one drops the expansion at rate $t^{1/2}$, one has

$$E^0\hat{x}_t[B] = e^{\lambda t}\Phi[t^{-1/2}B] \simeq \frac{e^{\lambda t}}{\sqrt{2\pi t}}|B|,$$

$$E^x\hat{x}_t[B] \simeq \frac{e^{\lambda t}}{\sqrt{2\pi t}}|B|$$

and in fact

2.2. THEOREM. *If* $1 < m < \infty$ *and*

$$(2.3) \qquad \sum_{n=1}^{\infty} n(\log n)^{3/2} p_n < \infty$$

<u>then for any bounded Borel set with</u> $|\partial B| = 0$

$$(2.4) \qquad \lim_{t\to\infty} \sqrt{2\pi t}\, \frac{\hat{x}_t[B]}{|\hat{x}_t|} = |B| \quad \text{a.s. on} \quad E^c.$$

<u>2.3</u>. LEMMA. <u>Define</u> $\vartheta_2(x) := x^2$. <u>Then without conditions beyond</u> $1 < m < \infty$,

$$\overline{\lim_{n\to\infty}} \frac{\hat{x}_{n\delta}[\vartheta_2]}{n|\hat{x}_{n\delta}|} < \infty \quad \text{a.s. on} \quad E^c.$$

PROOF. It suffices to take $\delta = 1$ and prove that $\gamma_n \hat{x}_n[\vartheta_2] = O(n)$. To this end, note that $E(x_j^{n,i^2} | \mathfrak{F}_n) = 1 + x^{n,i^2}$, define $Y_{n,i} := |\hat{x}_{n+1}^{n,i}| I(|\hat{x}_{n+1}^{n,i}| \le \gamma_n)$ and write

$$\gamma_{n+1}\hat{x}_{n+1}[\vartheta_2] = a_n + \gamma_{n+1}|\hat{x}_{n+1}| + b_n + c_n + d_n$$

$$a_n := \gamma_{n+1} \sum_{i=1}^{|\hat{x}_n|} \sum_{j=1}^{|\hat{x}_{n+1}^{n,i}|} \{x_j^{n,i^2} - E(x_j^{n,i^2} | \mathfrak{F}_n)\},$$

$$b_n := \gamma_{n+1} \sum_{i=1}^{|\hat{x}_n|} x^{n,i^2} \{|\hat{x}_{n+1}^{n,i}| - Y_{n,i}\},$$

$$c_n := \gamma_{n+1} \sum_{i=1}^{|\hat{x}_n|} x^{n,i^2} \{Y_{n,i} - E(Y_{n,i} | \mathfrak{F}_n)\},$$

$$d_n := \gamma_{n+1} \sum_{i=1}^{|\hat{x}_n|} x^{n,i^2} E(Y_{n,i} | \mathfrak{F}_n) = \gamma_{n+1}\hat{x}_n[\vartheta_2] \int_0^{\gamma_n^{-1}} x\, dF(x)$$

with F the distribution of $|\hat{x}_1|$ given one initial particle. The γ_n can be chosen as the c_n^{-1} in II.<u>5</u>.<u>6</u>, and then $d_n = \gamma_n \hat{x}_n[\vartheta_2]$.

We shall show that $\Sigma_1^n\{a_k + b_k + c_k\} = o(n)$ which implies the claim since $\Sigma_1^n \gamma_k |\hat{x}_k| \cong nW$. To this end, note first that as in the proof of <u>1</u>.<u>3</u> $b_n = 0$ eventually since

$$\sum_{n=1}^{\infty} |\hat{x}_n| \int_{\gamma_n^{-1}}^{\infty} dF(x) \cong \sum_{n=1}^{\infty} \gamma_n^{-1} \int_{\gamma_n^{-1}}^{\infty} dF(x) = \int_0^{\infty} O(x)\, dF(x).$$

Furthermore, it is easily checked that if U is normal $(0,1)$, then $\mathrm{Var}((x+U)^2) = O(x^2)$. Hence, with $\mathfrak{G}$ the σ-algebra spanned by the branching process,

$$E(a_n^2|\mathfrak{F}_n) = E(E(a_n^2|\mathfrak{F}_n, \hat{x}_{n+1}^{n,i};\ i=1,\ldots,|\hat{x}_n|)|\mathfrak{F}_n)$$

$$= E(\gamma_{n+1}^2 \sum_{i=1}^{|\hat{x}_n|} |\hat{x}_{n+1}^{n,i}| O(x^{n,i^2})|\mathfrak{F}_n) = O(\gamma_n^2 \sum_{i=1}^{|\hat{x}_n|} x^{n,i^2}),$$

$$E(\sum_{n=0}^{\infty} E(a_n^2|\mathfrak{F}_n)|\mathfrak{G}) = \sum_{n=0}^{\infty} O(\gamma_n^2 n|\hat{x}_n|) < \infty$$

so that $a_n \to 0$. Thus Kronecker's lemma completes the proof if we can show that $\Sigma_0^\infty k^{-1}c_k$ converges a.s. But the terms of the series are martingale increments and it suffices to notice that

$$\sum_{n=0}^{\infty} \mathrm{Var}(\frac{c_n}{n}|\mathfrak{F}_n) \leq \sum_{n=0}^{\infty} \gamma_{n+1}^{-2} \int_0^{\gamma_n^{-1}} x^2 dF(x) \sum_{i=1}^{|\hat{x}_n|} x_j^{n,i^4}/n^2$$

is finite since

$$E(\sum_{n=0}^{\infty} \mathrm{Var}(\frac{c_n}{n}|\mathfrak{F}_n)|\mathfrak{G}) = \sum_{n=0}^{\infty} \gamma_{n+1}^2 |x_n| \int_0^{\gamma_n^{-1}} x^2 dF(x) O(1)$$

$$= \sum_{n=0}^{\infty} \cdot O(\gamma_n \int_0^{\gamma_n^{-1}} x^2 dF(x) = \int_0^{\infty} O(x) dF(x) < \infty. \qquad \square$$

PROOF OF (2.1). Define $\mu := e^\lambda = E(|\hat{x}_1|\,|\,|\hat{x}_0| = 1)$,

$$A_{n,r} := \frac{1}{\mu^n} \sum_{i=1}^{|\hat{x}_n|} \mu^{-r} [\hat{x}_{n+r}^{n,i}[(n+r)^{1/2}B] - E(\hat{x}_{n+r}^{n,i}[(n+r)^{1/2}B]|\mathfrak{F}_n)],$$

$$B_{n,r} := \frac{1}{\mu^n} \sum_{i=1}^{|\hat{x}_n|} [\mu^{-r} E(\hat{x}_{n+r}^{n,i}[(n+r)^{1/2}B]|\mathfrak{F}_n) - \Phi[B]]$$

$$= \frac{1}{\mu^n} \hat{x}_n[P^{\cdot}(y_r \in (n+r)^{1/2}B)] - \Phi[B]\frac{|\hat{x}_n|}{\mu^n}$$

so that

$$\frac{\hat{x}_{n+r}[(n+r)^{1/2}B]}{\mu^{n+r}} - \Phi[B]W \cong A_{n,r} + B_{n,r}.$$

By standard properties of weak convergence, it suffices to consider the case $B = (-\infty, y]$. Then

$$|P^x(y_r \in (n+r)^{1/2}B) - \Phi(y)| = |\Phi((1+\tfrac{n}{r})^{1/2}y - xr^{-1/2}) - \Phi(y)|$$

$$\leq \frac{|x|}{r^{1/2}} + c_{n,r}, \quad c_{n,r} := |\Phi((1+\tfrac{n}{r})^{1/2}y) - \Phi(y)|.$$

Now write $N = n(N) + r(N)$ with $n(N) := [N\epsilon]$. Then clearly $\underline{\lim}_{N\to\infty} |\hat{x}_{n(N+m)}|/|\hat{x}_{n(N)}| > 1$ if $m > \epsilon^{-1}$ and, with Q the distribution of $\sup_r |\hat{x}_r|/\mu^r + 1$, the mean of Q is finite by II.2.1 , and

$$P(\mu^{-r}|\hat{x}^{n,i}_{n+r}[(n+r)^{1/2}B] - E(\hat{x}^{n,i}_{n+r}[(n+r)^{1/2}B]|\mathfrak{F}_n)| > u) \leq 1 - Q(u).$$

Hence by A10.2, $A_{n(N),r(N)} \to 0$ (since $\mu^n \cong W^{-1}|\hat{x}_n|$). Furthermore, letting $\vartheta(x) := |x|$, 2.3 implies that $M := \overline{\lim}_{n\to\infty} \hat{x}_n[\vartheta]/n^{1/2}\mu^n < \infty$. Hence

$$\overline{\lim_{N\to\infty}} |B_{n(N),r(N)}| \leq \overline{\lim_{N\to\infty}} \{\frac{\hat{x}_{n(N)}[\vartheta]}{\mu^{n(N)}r(N)^{1/2}} + \frac{|\hat{x}_{n(N)}|}{\mu^{n(N)}} c_{N(n),r(N)}\}$$

$$\leq M \overline{\lim_{N\to\infty}} (\frac{n(N)}{r(N)})^{1/2} + W|\Phi((1+\tfrac{\epsilon}{1-\epsilon})^{1/2}y) - \Phi(y)|$$

$$\leq M(\frac{\epsilon}{1-\epsilon})^{1/2} + W|\Phi((1+\tfrac{\epsilon}{1-\epsilon})^{1/2}y) - \Phi(y)|.$$

As $\epsilon \downarrow 0$, it follows that the limit statement of (2.1) holds along the sequence $0, 1, 2, \ldots$. The same argument applies to any discrete skeleton and an argument entirely similar to that of Section 1 extends the convergence to the continuum. □

PROOF OF 2.2. Entirely analogous to that of 2.1, the main difference being that one has to replace A10.2 by A10.3. Define $\mu := e^\lambda$,

$$A_{n,r} := \frac{\sqrt{2\pi(n+r)}}{\mu^n} \sum_{i=1}^{|\hat{x}_n|} \mu^{-r}\{\hat{x}^{n,i}_{n+r}[B] - E(\hat{x}^{n,i}_{n+r}[B]|\mathfrak{F}_n)\},$$

$$B_{n,r} := \frac{1}{\mu^n} \sum_{i=1}^{|\hat{x}_n|} \left[\frac{\sqrt{2\pi(n+r)}}{\mu^r} E(\hat{x}^{n,i}_{n+r}[B] | \mathfrak{F}_n) - |B|\right]$$

$$= \frac{\sqrt{2\pi(n+r)}}{\mu^n} \hat{x}_n[P^{\cdot}(y_r \in B)] - |B| \frac{|\hat{x}_n|}{\mu^n}$$

so that

$$\sqrt{2\pi(n+r)} \frac{\hat{x}_{n+r}[B]}{\mu^{n+r}} - |B| W \cong A_{n,r} + B_{n,r}.$$

It suffices to take $B := (a,b)$. Define $n(N) := [N\epsilon]$, $r(N) := N - n(N)$, $M := \sup_n |\hat{x}_n|/\mu^n$. Then, using II.4.6, (2.3) implies that $EM(\log^+ M)^{1/2} < \infty$ and thus, using the same stochastical domination as in the proof of 2.1, it follows from A10.3 that $A_{n(N),r(N)} \to 0$ a.s. as $N \to \infty$. For $B_{n,r}$, note first that for suitable constants c_1, c_2 (independent of x, r) and some $\theta(x,y,r)$,

$$|\sqrt{2\pi r}\, P^x(y_r \in B) - |B|| \leq \int_a^b |e^{-\frac{(x-y)^2}{2r}} - 1| dy$$

$$= \int_a^b \frac{(x-y)^2}{2r} \sqrt{2\pi} |\varphi'(\theta(x,y,r))| dy \leq \frac{c_1}{r} + c_2 \frac{x^2}{r}.$$

Thus

$$\overline{\lim_{N\to\infty}} |B_{n(N),r(N)}|$$

$$= \overline{\lim_{N\to\infty}} \left\{\sqrt{\frac{N}{r(N)}} \frac{x_N[\sqrt{2\pi r(N)}\, P^{\cdot}(y_{r(N)} \in B) - |B|]}{\mu^N} + \left(\sqrt{\frac{N}{r(N)}} - 1\right) \frac{|\hat{x}_N|}{\mu^N} |B|\right\}$$

$$\leq M \left\{\sqrt{\frac{1}{1-\epsilon}} \overline{\lim_{N\to\infty}} \frac{c_1}{r(N)} + \sqrt{\frac{1}{1-\epsilon}} - 1\right\} + c_2 \sqrt{\frac{1}{1-\epsilon}} \overline{\lim_{N\to\infty}} \frac{n(N)}{r(N)} \overline{\lim_{n\to\infty}} \frac{\hat{x}_n[\vartheta_2]}{n\mu^n}$$

$$= M \sqrt{\frac{1}{1-\epsilon}} - 1 + c_2 \sqrt{\frac{1}{1-\epsilon}} \cdot \frac{1}{1-\epsilon} \overline{\lim_{n\to\infty}} \frac{\hat{x}_n[\vartheta_2]}{n\mu^n}$$

Let $\epsilon \downarrow 0$ and appeal to 2.3. □

BIBLOGRAPHICAL NOTES

The purpose is to illustrate by some examples the behaviour of branching diffusions (on the line), which do not satisfy the conditions of Part C for positive regularity.

The Ornstein-Uhlenbeck model of Section 1 was treated by Enderle and Hering (1982). They needed the condition $\Sigma_0^\infty n \log^+ \log^+ n\, p_n < \infty$ for a.s. convergence, but the present proof (using finite mean only) differs from theirs in a few details only.

The branching Brownian motion model of Section 2 is closely related to branching random walks which have recently received much attention (for directions different from the one taken here, see the series of papers by Gorostiza and Moncayo (e.g.1978), also Biggins (e.g. (1978), and Bramson (1978)). Concerning the conditions for 2.1, which is proved here with $(x \log x)$, the second moment hypotheses was dispensed with for the first time by Kaplan and Asmussen (1976), who needed $\Sigma_0^\infty n (\log^+ n)^{1+\epsilon} p_n < \infty$ for some $\epsilon > 0$. Kaplan (1977) and Athreya and Kaplan (1978) claimed to have a proof with $(x \log x)$, but were in error, as the magnitude of the integrand on pg 56 of their paper is not $O(x)$. Recently Kaplan has communicated to us a proof with $(x \log x)$, which is valid also for branching random walks. (In the present context, the 4th moment of motion comes in via 2.3). In view of 2.3, one would not exclude 2.1 (in the formulation of (2.2)) to be valid only with finite mean.

Other examples related to Chapter IX are in Watanabe (1967), Kesten (1978), Kageyama and Ogura (1980) and Ogura (1979).

Chapter X

Generalized Age-Dependence and Random Characteristics

1. INTRODUCTION

Associated with the typical individual of the process is

a) A point process ξ (the reproduction) on $[0,\infty)$ describing the instants of births. That is, $\xi(t)$ is the number of children born before or at age t, $\xi(t) = \int_0^t \xi[ds]$.

b) A r.v. τ (not necessarily finite), the lifelength.

c) A collection of $[0,\infty)$-valued stochastic processes $(\Phi(t))_{0\leq t<\infty}$, $(\Psi(t))_{0\leq t<\infty}, \ldots$, random characteristics.

No specific assumptions (e.g., concerning independence) are made on the joint distribution of $(\xi,\tau,\Phi,\Psi,\ldots)$ (thus one could even allow births to take place after death, $P(\xi(\infty)-\xi(\tau) > 0) > 0$). The collection of individuals ever born is denoted by J, the above quantities associated with $n \in J$ are denoted by $(\xi_n,\tau_n,\Phi_n,\Psi_n,\ldots)$, and the $(\xi_n,\tau_n,\Phi_n,\Psi_n,\ldots)$ are assumed to be i.i.d. For convenience, we assume the $n \in J$ ordered according to their times σ_n of births and start off with one ancestor of age 0 (so that $\sigma_1 = 0$).

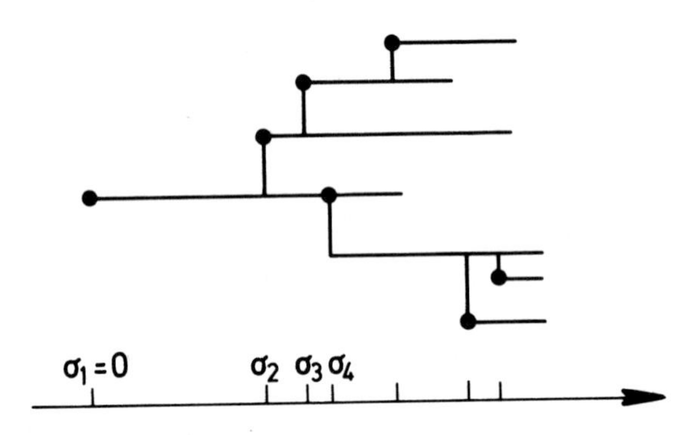

The construction of a process matching the above description is a trivial matter. We shall not go into the explosion problem since the conditions relevant for the limit theory will always exclude explosion.

The objects of study are of the form

$$Z^{\Phi}(t) := \sum_{n\in J:\sigma_n\leq t} \Phi_n(t-\sigma_n).$$

1.1. EXAMPLE, $\Phi(t) := 1$. $Z^{\Phi}(t)$ is the total number of births before t. This quantity plays a predominant role in the following and is denoted by $T(t)$.

1.2. EXAMPLE, $\Phi(t) := I(\tau > t)$. $Z^{\Phi}(t)$ is the total number of individuals alive at time t. Note that since this particular characteristic describes the lifelength, one could formally omit the lifelength b) in the above description a), b), c) of the process.

1.3. EXAMPLE, $\Phi(t) := I(\tau \wedge a > t)$. $Z^{\Phi}(t)$ is the number of particles which at time t are alive and of age strictly less than a.

1.4. EXAMPLE, $\Phi(t) := a(t \wedge \tau)$. If the individuals are bacteria, which during the life produce toxin at the linear rate a, then $Z^{\Phi}(t)$ is the total amount of toxin produced before t.

1.5. EXAMPLE, $\Phi(t) := I(\xi(t) = k, \tau > t)$. In demographic terms, $Z^{\Phi}(t)$ is the number of individuals of parity k alive at time t.

1.6. EXAMPLE. In the ususal cell kinetic model the life of a cell has has 4 biologically well defined phases G_1, S, G_2, M (e.g., M is the mitotic phase), and at the end of its life the cell splits into 2 cells (which need not survive). Thus ξ has 0, 1, or 2 epochs, all at τ. If Φ is the indicator function of the cell being, say, in the mitotic phase, then $Z^{\Phi}(t)$ is the total number of cells in mitosis at time t.

2. RENEWAL TECHNIQUES

The reproduction measure μ is defined as the intensity measure (in general point process language) of ξ, i.e., $\mu(t) := E\xi(t)$, and we let $m^{\Phi}(t) := EZ^{\Phi}(t)$. In order to compute $m^{\Phi}(t)$, we note that the expected contribution to $m^{\Phi}(t)$ due to the line of descent initiated by an individual of the first generation born at time $u \leq t$ is simply $m^{\Phi}(t-u)$. The u being distributed in the mean according to $\mu[du]$ and the expected contribution from the ancestor being $E\Phi(t)$, we arrive at the renewal equation

$$(2.1) \qquad m^{\Phi}(t) = E\Phi(t) + \int_0^t m^{\Phi}(t-u)\mu[du].$$

Hence renewal theory will come up. The basic facts used below without further reference can be found in treatments of the standard theory such as Feller (1971) Ch.XI, Karlin and Taylor (1975) Ch.5, Cinlar (1975) Ch.9 or Asmussen (1978).

In the following, multiplication by $e^{-\alpha t}$ will be denoted by a left lower subscript. E.g. ${}_{\alpha}\mu[dt]$ is the measure with density $e^{-\alpha t}$ w.r.t. $\mu[dt]$ and ${}_{\alpha}Z^{\Phi}(t) := e^{-\alpha t}Z^{\Phi}(t)$. We assume throughout that $\mu(0) = 0$, and that μ is non-lattice and Malthusian.
By this we mean that the Malthusian parameter, defined as the (necessarily unique) solution α of the equation ${}_{\alpha}\mu(\infty) = 1$, exists and that furthermore $\varkappa := \int_0^{\infty} u\,{}_{\alpha}\mu[du] < \infty$. Regarding this assumption, we remark that it is automatic in the supercritical case $\alpha > 0$, if $1 < \mu(\infty) < \infty$, that it amounts to $\int_0^{\infty} u\mu[du] < \infty$ in the critical or proper case, while in the subcritical or defective case it involves a somewhat more restrictive bound on the tail of μ.

Now multiplication by $e^{-\alpha t}$ shows that (2.1) is equivalent to

$$(2.2) \qquad {}_{\alpha}m^{\Phi}(t) = E_{\alpha}\Phi(t) + \int_0^t {}_{\alpha}m^{\Phi}(t-u)\,{}_{\alpha}\mu[du]$$

where by definition ${}_{\alpha}\mu$ is a probability measure. We let ${}_{\alpha}U := \Sigma_0^{\infty}\, {}_{\alpha}\mu^{*n}$ be the renewal measure associated with ${}_{\alpha}\mu$.

2.1. PROPOSITION. *If $E_{\alpha}\Phi(t)$ is directly Riemann integrable, then*

$${}_{\alpha}m^{\Phi}(t) = E_{\alpha}\Phi * {}_{\alpha}U(t) = \int_0^t E_{\alpha}\Phi(t-s)\,{}_{\alpha}U[ds] \underset{t\to\infty}{\to} {}_{\alpha}m^{\Phi}(\infty) := \varkappa^{-1}\int_0^{\infty} E_{\alpha}\Phi(t)dt.$$

PROOF. By reference to the key renewal theorem, the only thing to prove is ${}_\alpha m^\Phi(t)$ being bounded on finite intervals. Now $E_\alpha\Phi(t)$, being d.R.i., is bounded, say by c_1 so that with $c_2 = \sup_{0\leq s\leq t} e^{\alpha s}$

$$ {}_\alpha m^\Phi(t) = e^{-\alpha t} E \sum_{n:\sigma_n\leq t} \Phi_n(t-\sigma_n) \leq c_2 E \sum_{n:\sigma_n\leq t} E_\alpha\Phi(t-\sigma_n) \leq c_1c_2 ET(t) $$

and it suffices to show $ET(t) < \infty$. But since the intensity measure of the birth times of nth generation individuals is μ^{*n}, this follows by standard renewal theory from $ET(t) = \Sigma_0^\infty \mu^{*n}(t)$. □

2.2. EXAMPLE. Let $\alpha > 0$, $\Phi(t) := I(\tau \wedge a > t)$, $\Psi(t) := I(\tau > t)$. Then the d.R.i. of $E_\alpha\Phi$, $E_\alpha\Psi$ follows by monotonicity and it follows from 2.1 that, with $G(t) := P(\tau \leq t)$ the lifelength distribution,

$$ (2.3) \qquad \frac{m^\Phi(t)}{m^\Psi(t)} = \frac{{}_\alpha m^\Phi(t)}{{}_\alpha m^\Psi(t)} \to A(a) := \frac{{}_\alpha m^\Phi(\infty)}{{}_\alpha m^\Psi(\infty)} = \frac{\int_0^a e^{-\alpha t}(1-G(t))dt}{\int_0^\infty e^{-\alpha t}(1-G(t))dt} . $$

Thus the proportion of the expected number of individuals of age at most a to the expected total population size has a limit. Due to (2.3) (and the stochastic analogue

$$ (2.4) \qquad Z^\Phi(t)/Z^\Psi(t) \to A(a) $$

to be shown in Section 5), A is called the limiting age-distribution.

3. AGE-DEPENDENCE, THE STABLE AGE-DISTRIBUTION AND THE REPRODUCTIVE VALUE

We first consider some examples. G throughout denotes the lifelength distribution $G(t) = P(\tau \leq t)$.

3.1. EXAMPLE, the Bellman-Harris process. Reproduction takes place at the time τ of death and independent of τ. That is, at time τ, k offspring are produced w.p. p_k so that with $\nu := \Sigma_0^\infty kp_k$ the offspring mean, the reproduction measure is given by $\mu = \nu G$ and the Malthusian parameter α by $\int_0^\infty e^{-\alpha t} dG(t) = \nu^{-1}$.

3.2. EXAMPLE, the Sevastyanov process. Reproduction takes place at the time τ of death, but not necessarily independently of τ. That is, at time τ k offspring are produced w.p. $p_k(\tau)$ $(\Sigma_0^\infty p_k(t) = 1\ \forall t)$ so that, with $\nu(t) := \Sigma_0^\infty kp_k(t)$ the offspring mean given $\{\tau = t\}$, the reproduction measure is given by $d\mu(t)/dG(t) = \nu(t)$ and the Malthusian parameter by $\int_0^\infty e^{-\alpha t}\nu(t)\,dG(t) = 1$.

3.3. EXAMPLE, the age-dependent birth-death process. An individual of age a at time t gives birth to a child (and survives) in $[t,t+s]$ w.p. $\lambda(a)s+o(s)$ and dies w.p. $\mu(a)s+o(s)$. The lifelength distribution is given by $1-G(t) = P(\tau > t) = \exp(-\int_0^t \mu(a)du)$ and the reproduction measure by $d\mu(t)/dt = \lambda(t)\exp(-\int_0^t \mu(a)du)$.

The salient feature of the above examples is, of course, that the future reproduction of an individual depends only on its age a and not on the reproduction in $[0,a)$. Put in different terms, the process is Markovian in the age-structure. That is, if we identify types with ages, the age-chart $\hat{x}_t = \langle a_1,\ldots,a_n\rangle$ of the individuals alive constitutes a branching Markov process, the non-branching part being simply a linear motion on the age-scale $[0,\infty)$ (or $(0,a)$ if G is supported by $[0,a)$). A process with this property is called age-dependent and we may recall some of the discussion from the Markov branching point of view offered in Chapter V.3. That is, we consider only the age-structure, i.e., characteristics of the form $\Phi(t) = I(\tau > t)\eta(t)$ for which $Z^\Phi(t) = \hat{x}_t[\eta]$, $m^\Phi(t) = M_t\eta(0)$ in the notation of Part C.

It follows from 3.1 below that in the age-dependent case the limiting age-distribution A defined in 2.2 is in fact stable, i.e.,

$AM_t = e^{\alpha t}A$. Let $\mu^a, \mu^A := \int_0^\infty \mu^a dA(a)$ be the reproductive measures of an individual whose age is a, respectively distributed according to the stable age-distribution, i.e.,

$$(3.1) \qquad \mu^a[ds] = \frac{\mu[a+ds]}{1-G(a)}, \quad \mu^a(t) = \frac{\mu(a+t)-\mu(a)}{1-G(a)}$$

and define the <u>reproductive value</u> V by

$$V(a) := {}_\alpha\mu^a(\infty) = \int_0^\infty e^{-\alpha s}\mu^a[ds] = \frac{e^{\alpha a}}{1-G(a)}\int_a^\infty e^{-\alpha s}\mu[ds].$$

We then have the following parallel of (M) (assuming $\alpha > 0$):

3.1. THEOREM. *In the age-dependent case,* A *and* V *are eigenfunctions of* $\{M_t\}_{t\geq 0}$, $AM_t = e^{\alpha t}A$, $M_t V = e^{\alpha t}V$. *Furthermore for any* x *and any bounded, a. e. continuous* η,

$$(3.2) \qquad M_t\eta(x) = \frac{1}{\beta\varkappa}e^{\alpha t}V(x)A[\eta] + o(e^{\alpha t}), \quad \beta^{-1} := \int_0^\infty e^{-\alpha t}(1 - G(t))dt.$$

3.2. REMARK. Of course, it is possible to reformulate the more general model of the preceding sections as a branching Markov process and formulate analogues of 3.1, but the set of types, having to incorporate not only the age a but also the reproduction before a, would be very large. In connection with Part C, note that the remainder term estimates in (3.2) cannot in general be strengthened to that of (M).

PROOF OF 3.1. We first note that

$$\mu^A(t) = \int_0^\infty \mu^a(t)dA(a) = \beta\int_0^\infty e^{-\alpha a}(\mu(a+t)-\mu(a))da$$

$$= \beta e^{\alpha t}\int_t^\infty e^{-\alpha a}\mu(a)da - \beta\int_0^\infty e^{-\alpha a}\mu(a)da,$$

$$(3.3) \qquad \frac{d\,{}_\alpha\mu^A(t)}{dt} = e^{-\alpha t}\frac{d\mu^A}{dt} = \beta e^{-\alpha t}(\alpha e^{\alpha t}\int_t^\infty e^{-\alpha a}\mu(a)da - \mu(t))$$

$$= \beta\int_t^\infty \alpha e^{-\alpha a}(\mu(a)-\mu(t))da = \beta\int_t^\infty e^{-\alpha s}\mu[ds] = \beta(1-{}_\alpha\mu(t)),$$

$$(3.4) \qquad {}_\alpha\mu^A * {}_\alpha U[du] = \beta\, du \quad \text{(Lebesgue measure)},$$

(3.4) being a consequence of (3.3) either by a direct calculation or by reference to stationary renewal processes. Next, we apply the usual renewal argument to get

$$M_t\eta(x) = \eta(x+t)\frac{1-G(x+t)}{1-G(x)} + \int_0^t M_{t-s}\eta(0)\mu^x[ds]. \tag{3.5}$$

Now let $\eta(x) := I(x < a)$. Then (3.5) with $x = 0$ is a renewal equation with solution ${}_\alpha M_t\eta(0) = {}_\alpha z * {}_\alpha U(t)$, $z(t) := I(t < a)(1-G(t))$. Integrating (3.5) w.r.t. $A[dx]$ thus yields for $t < a$

$$\begin{aligned} e^{-\alpha t}AM_t[\eta] &= \int_0^\infty e^{-\alpha t}\eta(x+t)\frac{1-G(x+t)}{1-G(x)}A[dx] + \int_0^t {}_\alpha M_{t-s}\eta(0)\,{}_\alpha\mu^A[ds] \\ &= \beta\int_t^a e^{-\alpha u}(1-G(u))du + {}_\alpha z * {}_\alpha U * {}_\alpha\mu^A(t) \\ &= \beta\int_t^a e^{-\alpha u}(1-G(u))du + \int_0^t {}_\alpha z(t-s)ds \\ &= \beta\int_t^a e^{-\alpha u}(1-G(u))du + \beta\int_0^t e^{-\alpha s}(1-G(s))ds = A(a) = A[\eta]. \end{aligned}$$

Iteration shows that this holds for all t, establishing the eigenfunction property of A. For that of V, let again first $x=0$, $\eta=V$ so that (3.5) becomes a renewal equation with solution ${}_\alpha M_tV(0) = h * {}_\alpha U(t)$, with $h(t) := e^{-\alpha t}V(t)(1-G(t)) = \int_t^\infty e^{-\alpha s}\mu[dx] = 1-{}_\alpha\mu(t)$. Hence by (3.4) ${}_\alpha M_tV(0) = 1$ for all t. For a general x, we thus get from (3.5)

$$\begin{aligned} {}_\alpha M_tV(x) &= e^{-\alpha t}V(x+t)\frac{1-G(x+t)}{1-G(x)} + {}_\alpha\mu^x(t) \\ &= \frac{e^{\alpha x}}{1-G(x)}\int_{x+t}^\infty e^{-\alpha s}\mu[ds] + \frac{e^{\alpha x}}{1-G(x)}\int_x^{x+t} e^{-\alpha s}\mu[ds] = V(x). \end{aligned}$$

If $x = 0$, then (3.2) is an immediate consequence of (3.5) and the renewal theorem. For a general x, the first term in (3.5) (normalized by $e^{\alpha t}$) tends to zero and the second (using dominated convergence justified by the boundedness of ${}_\alpha M_t\eta(0)$) to $\int_0^\infty A[\eta]\,{}_\alpha\mu^x[ds] = V(x)A[\eta]$.

□

3.3. REMARK. A result similar to 3.1 holds also in the critical case $\alpha = 0$, $\mu(\infty) = 1$ subject to slightly stronger conditions (the proof is almost the same except that one has to proceed slightly differently for the analogue of (3.4)). More precisely, one has to assume that the lifetime distribution has finite mean in order for A and β to be defined and to add the assumption of η being directly Riemann integrable (d.R.i.) or the analogue of (3.2).

We conclude this section by discussing some demographic interpretations and examples [within the spirit of the classical deterministic Sharpe-Lotka theory]. Note first that the number of births in $[0,dt]$ is $(\mu(x+dt)-\mu(x))/(1-G(x))$ if the age is x and

$$\int_0^\infty \frac{\mu(x+dt)-\mu(x)}{1-G(x)}\, \beta e^{-\alpha x}(1-G(x))dx$$

$$= \int_0^\infty \beta e^{-\alpha x}dx \int_x^{x+dt} \mu[du] = \beta \int_0^\infty \mu[du] \int_{u-dt}^{u} e^{-\alpha x}dx$$

$$= \int_0^\infty \beta e^{-\alpha u}dt\ \mu[du] = \beta dt$$

if x is sampled according to A. Hence β is the birth rate in the stable population (which could also be derived from (3.3)). The contribution βdt due to the age group x is $\beta e^{-\alpha x}(\mu(x+dt)-\mu(x)) = \beta e^{-\alpha x}\mu[dx]dt$ and for this reason ${}_{\alpha}\mu$ is called the distribution of the age at childbearing in the stable population and $\varkappa = \int_0^\infty x\,{}_{\alpha}\mu[dx]$ the average age at childbearing in the stable population. In contrast, the average age at childbearing at the individuum level is $\tilde{\varkappa} := \int_0^\infty x\mu[dx]/\mu(\infty)$.

3.4. EXAMPLE. Suppose that in a stable supercritical population births are suddenly reduced (say due to a contraceptional program) so as to make the process critical. That is, the new reproduction measure is $\mu_1[dx] := \mu[dx]/\mu(\infty)$ (the lifetime distribution remains unchanged). Then the expected population size will reach an ultimate size as $t \to \infty$, viz. the initial one multiplied by

$$\gamma := \frac{A_1[1]}{\beta_1 \varkappa_1} \int_0^\infty V_1(a)A[da],$$

with subscript 1 referring to the changed parameters. The constant γ describes the momentum of population growth and reduces to an expression involving only quantities of direct demographic interpretation.

Indeed, note first that $\varkappa_1 = \tilde{\varkappa}$ is the average age of childbearing (at the individuum level) before and after the change and that the complete expectation of life is

$$\mathring{e} = \int_0^\infty (1-G(x))dx = \beta_1^{-1}$$

so that

$$\gamma = \frac{\beta}{\beta_1 \varkappa_1} \int_0^\infty \frac{1}{1-G(a)} \int_a^\infty \mu_1[ds] e^{-\alpha a} (1-G(a))da$$

$$= \frac{\beta \mathring{e}}{\mu(\infty)\tilde{\varkappa}} \int_0^\infty \mu[ds] \int_0^s e^{-\alpha a} da = \frac{\beta \mathring{e}}{\alpha \tilde{\varkappa}} \left(\frac{\mu(\infty)-1}{\mu(\alpha)}\right) .$$

4. MARTINGALES AND A THEOREM OF KESTEN-STIGUM TYPE

We return to the general model and define

$$\mathfrak{F}_n := \sigma((\xi_k, \tau_k, \Phi_k, \Psi_k, \dots): 1 \leq k \leq n),$$

$$\mathfrak{G}_t := \mathfrak{F}_{T(t)} := \sigma((\xi_k, \tau_k, \Phi_k, \Psi_k, \dots): 1 \leq k \leq T(t))$$

(noting that σ_k is $\mathfrak{F}_{k-1}$-measurable and hence the $T(t)$ are stopping times w.r.t. $\{\mathfrak{F}_n\}$),

$$Y_k := {}_\alpha\xi_k(\infty) - E_\alpha\xi_k(\infty) = {}_\alpha\xi_k(\infty) - 1, \quad R_1 := {}_\alpha\xi_1(\infty),$$

$$R_n := 1 + \sum_{k=1}^{n} e^{-\alpha\sigma_k} Y_k, \quad W(t) := R_{T(t)} \, .$$

Thus R_n is a weighted sum of children $\ell > n$ of the first n individuals, weight $e^{-\alpha\sigma_\ell}$ for ℓ,

$$(4.1) \qquad R_n = \sum_{\ell=n+1}^{\infty} e^{-\alpha\sigma_\ell} I(\ell \text{ is child of one of } 1,\dots,n)$$

(to see this, note that if $2 \leq k \leq n$, then k is child of one of $1,\dots,k-1$, say i, and hence the $-e^{-\alpha\sigma_k}$ term in the definition is cancelled by a term in $e^{-\alpha\sigma_i} Y_i$). Note also that $W(t)$ is of the form ${}_\alpha Z^\Phi(t)$, with $\Phi(t) := e^{\alpha t}({}_\alpha\xi(\infty) - {}_\alpha\xi(t)) = e^{\alpha t}\int_t^\infty e^{-\alpha s}\xi[ds]$.

We shall assume for the limit results that ${}_\gamma\mu(\infty) < \infty$ for some $\gamma < \alpha$ (though this assumption could be weakened for many partial results). E.g., $\mu(\infty) < \infty$ (which is automatic say in the Bellman-Harris case) suffices for this. The main result of the present section is then

4.1. THEOREM. (i) *$\{R_n\}_{n=1,2,\dots}$ and $\{W(t)\}_{t\geq 0}$ are non-negative martingales w.r.t. $\{\mathfrak{F}_n\}$, resp. $\{\mathfrak{G}_t\}$.* (ii) *Denote $W := \lim_{t\to\infty} W(t) = \lim_{n\to\infty} R_n$ and suppose $\alpha > 0$. Then $\{W > 0\} = \{T(\infty) = \infty\}$ if and only if*

$$(X \text{ LOG } X) \qquad E_\alpha\xi(\infty)\log^+_\alpha\xi(\infty) = \infty$$

while $W = 0$ a.s. otherwise.

PROOF OF PART (i). Non-negativity is immediate from (4.1) and the martingale property of $\{R_n\}$ from

$$E(R_k - R_{k-1} | \mathfrak{F}_{k-1}) = E(e^{-\alpha\sigma_k} Y_k | \mathfrak{F}_{k-1}) = e^{-\alpha\sigma_k} EY_k = 0.$$

Since the $T(t)$ are stopping times, a variant of the optional sampling theorem (e.g. Neveu (1972), Th.II-2-13) shows that $\{W(t)\}$ is a supermartingale and it suffices to show $EW(t) = EW(0) = 1$. But $ET(t) < \infty$ (as shown in Section 2) and

$$E(|R_n - R_{n-1}| | \mathfrak{F}_{n-1}) = e^{-\alpha\sigma_n} E|Y_n| \leq 2$$

implies that $EW(t) = ER_{T(t)} = ER_1 = 1$, cf e.g. Breiman (1968), Prop. 5.33. □

The proof of part (ii) is based on the third proof of the Kesten-Stigum theorem given in Section 5 of Chapter II. That is, we write

$$R_N = R_1 \prod_{n=2}^{N} \frac{R_n}{R_{n-1}} = R_1 \prod_{n=2}^{N} (1+A_n) \tag{4.2}$$

$$A_n := \frac{R_n - R_{n-1}}{R_{n-1}} = \beta_n Y_n, \quad \beta_n := \frac{e^{-\alpha\sigma_n}}{R_{n-1}}$$

and study $\Sigma_2^\infty A_n$. We need some a.s. estimates of β_n, which we combine with some auxiliary results for the next section. We let $J(t,c)$ denote the individuals born in $(t,t+c]$ as children of some of the individuals $1,\ldots,T(t)$ born before t and $N(t,c) := |J(t,c)|$ the number of elements of $J(t,c)$.

4.2. LEMMA. *Let* c *be chosen with* $\mu(c) > 1$. *Then it holds a.s. on* $\{T(\infty)=\infty\}$ *that*: (i) $\underline{\lim}_{t\to\infty} N(t,c)/T(t) > 0$; (ii) $\underline{\lim}_{t\to\infty} T(t+c)/T(t) > 1$; (iii) $\underline{\lim}_{t\to\infty} N(t+s,c)/N(t,c) > 1$ *for all large enough* s; (iv) $\sup_t e^{-\alpha t} T(t)/W(t) < \infty$; (v) $\sup_n n\beta_n < \infty$; (vi) $\inf_n n\beta_n > 0$.

PROOF. (i) Let $S_n := \Sigma_1^n \xi_k(c)$ be the total number of children born before age c to the first n individuals. Then if $1 < \eta < \mu(c)$, it follows by the LLN that $S_n \geq n\eta$ for all large enough n, say $n \geq N$. Thus if $T(t) \geq N$,

$$N(t,c) \geq S_{T(t)} - T(t) \geq T(t)(\eta - 1).$$

(ii) follows from (i) and $T(t+c) \geq N(t,c) + T(t)$. For (iii), let $0 < \epsilon < \underline{\lim}\, N(t,c)/T(t)$ and choose (justified by (ii)) s with $\underline{\lim}\, T(t+s)/T(t+c)$

$> \epsilon^{-1}$. Then

$$\varliminf_{t\to\infty} \frac{N(s+t,c)}{N(t,c)} > \varliminf \epsilon \frac{T(t+s)}{T(t+c)} \cdot \frac{T(t+c)}{N(t,c)} \geq \epsilon \cdot \epsilon^{-1} \cdot 1 = 1.$$

For (iv), note that $W(t) \geq e^{-\alpha(t+c)} N(t,c)$ and apply (i). (v) is a special case of (iv) (take $t = \sigma_n$ so that $T(t) = n$, $\beta_n = e^{-\alpha t}/W(t-0)$). Finally (vi) is a corollary of 5.1 of the next section (proved without reference to 4.1). A short proof is possible if $\mu(\infty) < \infty$. Indeed, let $S_n := \Sigma_1^n \xi_k(\infty)$ be the total number of children born to $1,\ldots,n$. Then $R_{n-1} \leq e^{-\alpha\sigma_n} S_{n-1}$, $\beta_n \geq S_{n-1}^{-1}$ and by the LLN, $n/S_n^{-1} \to 1/\mu(\infty)$. □

4.3. LEMMA. *Without moment conditions beyond* $_\gamma\mu(\infty) < \infty$, $\Sigma_2^\infty A_n^2 < \infty$. *If* (X LOG X) *holds, then* $\Sigma_2^\infty A_n$ *converges a.s. Otherwise* $\Sigma_2^\infty A_n = -\infty$ *on* $\{T(\infty) = \infty\}$.

4.3 completes the proof of 4.1. Indeed, since $\Sigma_2^\infty A_n^2 < \infty$, a Taylor expansion of $\log(1+A_n)$ yields $R_N = D_N \exp(\Sigma_2^N A_n)$ where D_N has an a.s. limit in $(0,\infty)$. Hence $W = \lim R_N$ is 0 or > 0 according to whether $\Sigma_2^\infty A_n = -\infty$ or $> -\infty$.

PROOF OF 4.3. Define

$$F(x) := P(Y_n \leq x), \quad \tilde{Y}_n := Y_n I(Y_n \leq n), \quad \tilde{A}_n := \beta_n \tilde{Y}_n,$$

$$\epsilon_n := -E(\tilde{A}_n \mid \mathfrak{F}_{n-1}) = \beta_n \int_n^\infty x\, dF(x).$$

Then the usual calculations yield

$$\sum_{n=2}^{\infty} P(\tilde{Y}_n \neq Y_n) = \sum_{n=2}^{\infty} \int_n^\infty dF(x) = \int_2^\infty O(x)\,dF(x) < \infty, \tag{4.3}$$

$$\sum_{n=2}^{\infty} \operatorname{Var}(\tilde{A}_n + \epsilon_n \mid \mathfrak{F}_{n-1}) = \sum_{n=2}^{\infty} \beta_n^2 \operatorname{Var}(\tilde{Y}_n \mid \mathfrak{F}_{n-1}) \leq \sum_{n=2}^{\infty} O(n^{-2}) \int_{-1}^{n} x^2\,dF(x) \tag{4.4}$$

$$= \int_{-1}^{\infty} O(x)\,dF(x) < \infty,$$

using part (v) of 4.2. Invoking also (vi), it follows similarly that on $\{T(\infty) = \infty\}$

$$\sum_{n=2}^{\infty} \epsilon_n < \infty \iff \sum_{n=2}^{\infty} \frac{1}{n} \int_n^\infty x\,dF(x) = \int_2^\infty O(x \log x)\,dF(x) < \infty \tag{4.5}$$

$$\iff \text{(X LOG X)}.$$

Now since the $\tilde{A}_n+\epsilon_n$ are martingale increments, $\Sigma_2^\infty\{\tilde{A}_n+\epsilon_n\}$ converges a.s. by (4.4). Hence $\Sigma_2^\infty\{A_n+\epsilon_n\}$ does so by (4.3) and (4.5) completes the proof of the last part of the lemma. For the first, note that the same calculation as in (4.4) yields $\Sigma_2^\infty E(\tilde{A}_n^2|\mathfrak{F}_{n-1}) < \infty$, hence $\Sigma\tilde{A}_n^2 < \infty$, and by (4.3), $\Sigma A_n^2 < \infty$. $\square$

5. EMPIRICAL RATIO LIMIT THEOREMS IN THE SUPERCRITICAL CASE

The main result is:

5.1. THEOREM. *Let* Φ, Ψ *be characteristics with sample functions which are right continuous and satisfy* $E \sup_{t\geq 0} {}_\beta\Phi(t) < \infty$, $E \sup_{t\geq 0} {}_\beta\Psi(t) < \infty$ *for some* $0 \leq \beta < \alpha$, *and suppose that* $\alpha > 0$, ${}_\gamma\mu(\infty) < \infty$ *for some* $0 \leq \gamma < \alpha$ [e.g. $\mu(\infty) < \infty$ suffices for this]. *Then a.s. on* $\{T(\infty) = \infty\}$

$$(5.1) \qquad \frac{Z^\Phi(t)}{Z^\Psi(t)} \to \frac{{}_\alpha m^\Phi(\infty)}{{}_\alpha m^\Psi(\infty)} .$$

5.2. EXAMPLES. The assumption on ${}_\beta\Phi$ is automatic whenever Φ is bounded. Thus 5.1 contains as corollary the fact (2.4) that, subject to the assumption ${}_\gamma\mu(\infty) < \infty$, the empirical age-distribution tends to the limiting age-distribution A derived in 2.2, the continuity assumptions being trivial. Letting $\Psi(t) := e^{\alpha t}\int_t^\infty e^{-\alpha s}\xi[ds]$, it is also easy to see that ${}_\gamma\mu(\infty) < \infty$ for some $\gamma < \alpha$ is equivalent to $E \sup_{t\geq 0} {}_\beta\Psi(t) < \infty$ for some $\beta < \alpha$. Again, the continuity is automatic and noting that ${}_\alpha Z^\Psi(t) = W(t)$, ${}_\alpha m^\Psi(\infty) = 1$, we get the following corollary (useful mainly when (X LOG X) holds).

5.3. COROLLARY. *Under the conditions of 5.1,* $\lim_{t\to\infty} {}_\alpha Z^\Phi(t) = W_\alpha m^\Phi(\infty)$ *a.s.*

Note also that with $\Phi(t) := 1$, ${}_\alpha Z^\Psi(t) := W(t)$ it follows that ${}_\alpha T(t)/W(t)$ and hence ${}_\alpha T(t)/W(t-0)$ has a non-zero limit. Let $t := \sigma_n$ to obtain part (vi) of 4.2.

The proof of 5.1 is based upon discrete skeletons and the inequality

$$(5.2) \qquad \frac{Z^{\Phi_\delta}(n\delta)}{Z^{\Psi^\delta}((n+1)\delta)} \leq \frac{Z^\Phi(t)}{Z^\Psi(t)} \leq \frac{Z^{\Phi^\delta}((n+1)\delta)}{Z^{\Psi_\delta}(n\delta)} \qquad n\delta \leq t \leq (n+1)\delta,$$

where $\Phi^\delta(t) := \sup_{t-\delta\leq s\leq t} \Phi(s)$, $\Phi_\delta(t) := \inf_{t\leq s\leq t+\delta} \Phi(s)$. We need

5.4. LEMMA. *The assumptions on* Φ *in 5.1 imply that* $E \sup_{t\geq 0} {}_\beta\Phi^\delta(t) < \infty$, $E \sup_{t\geq 0} {}_\beta\Phi_\delta(t) < \infty$, *that* $E_\alpha\Phi$, $E_\alpha\Phi^\delta$, $E_\alpha\Phi_\delta$ *are all d.R.i. and that* $\lim_{\delta\downarrow 0} {}_\alpha m^{\Phi^\delta}(\infty) = \lim_{\delta\downarrow 0} {}_\alpha m^{\Phi_\delta}(\infty) = {}_\alpha m^\Phi(\infty)$.

5.5. LEMMA. $\lim_{n\to\infty} \frac{W((n+1)\delta)}{W(n\delta)} = 1$ *a.s.* $\forall \delta > 0$.

5.6. LEMMA. Under the assumptions of 5.1,

$$\lim_{n\to\infty} \frac{{}_\alpha Z^\Phi(n\delta)}{W(n\delta)} = {}_\alpha m^\Phi(\infty) \quad \text{a.s. on } \{T(\infty)=\infty\} \quad \forall\delta > 0.$$

These three lemmata complete the proof of 5.1. Indeed, appealing to 5.4 we can let $\Phi = \Phi^\delta$ or $\Phi = \Phi_\delta$ in 5.6 so that from (5.2)

$$\overline{\lim_{t\to\infty}} \frac{Z^\Phi(t)}{Z^\Psi(t)} \le e^{\alpha\delta} \overline{\lim_{n\to\infty}} \frac{{}_\alpha Z^{\Phi^\delta}((n+1)\delta)}{W((n+1)\delta} \cdot \frac{W(n\delta)}{{}_\alpha Z^{\Psi_\delta}(n\delta)} = e^{\alpha\delta} \frac{{}_\alpha m^{\Phi^\delta}(\infty)}{{}_\alpha m^{\Psi_\delta}(\infty)}$$

which tends to the asserted limit ${}_\alpha m^\Phi(\infty)/{}_\alpha m^\Psi(\infty)$ as $\delta \downarrow 0$. The proof of $\underline{\lim} \ge$ is similar.

PROOF OF 5.4. The first assertion is obvious by noting that $\sup_{t\ge 0} {}_\beta\Phi^\delta(t) \le \sup_{t\ge 0} {}_\beta\Phi(t)$. For the second, note that right continuity of paths and domination of ${}_\alpha\Phi(t)$ by $e^{-(\alpha-\beta)t} \sup_{s\ge 0} {}_\beta\Phi(s)$ implies that $E_\alpha\Phi(t)$ is right continuous and dominated by a d.R.i. function. Hence $E_\alpha\Phi(t)$ is d.R.i., right continuity implying continuity a.e. and hence R.i. on finite intervals. Similar remarks apply to show first the d.R.i. of $E_\alpha\Phi^\delta(t)$, $E_\alpha\Phi_\delta(t)$, next that these functions tend to $E_\alpha\Phi(t)$ a.e. and finally that the integrals indeed converge. □

Define

$$W(t,c) := \sum_{k=T(t)+1}^{\infty} e^{-\alpha\sigma_k} I(\sigma_k > t+c,\ k \text{ is child of one of } 1,\dots,T(t))$$

as the contibution to $W(t)$ of individuals born after $t+c$ and note the identity

$$W(t+c) = \sum_{k\in J(t,c)} e^{-\alpha\sigma_k} W_k(t+c-\sigma_k) + W(t,c). \tag{5.3}$$

Here $W_k(s)$ is the $W(s)$ functional evaluated in the line of descent initiated by k. Similar notations are used in the following.

PROOF OF 5.5. It suffices to show that $W(n\delta+c)/W(n\delta) \to 1$ for all large enough c. (Consider $c = r\delta$ and $c = (r+1)\delta$). We take c with $\mu(c) > 1$ and use (5.3) to write

$$\frac{W(n\delta+c)}{W(n\delta)} = \frac{e^{-\alpha n\delta}N(n\delta,c)}{W(n\delta)} S_n + 1 \tag{5.4}$$

$$S_n := \frac{\sum_{k\in J(n\delta,c)} Y_{k,n}}{N(n\delta,c)}, \quad Y_{k,n} := e^{-\alpha(\sigma_k - n\delta)}\{W_k(n\delta+c-\sigma_k)-1\}.$$

Given $\mathfrak{G}_{n\delta}$, the $Y_{k,n}$ are independent with mean zero and appealing to part (iii) of 4.2, it holds for all large enough m that $\underline{\lim}_{n\to\infty} N((n+m)\delta,c)/N(n\delta,c) > 1$. Hence A10.2 will apply to show that $S_n \to 0$ a.s. (completing the proof since in (5.4) $W(n\delta) \geq e^{-\alpha(n\delta+c)}N(n\delta,c)$), if the $|Y_{k,n}|$ can be shown to be stochastically bounded by a r.v. $Y \geq 0$ with $EY < \infty$. We take $Y := 1 + \Sigma_1^{T(c)}\,{}_\alpha\xi_k(\infty)$ and note that $EY = 1 + ET(c)\cdot E\,{}_\alpha\xi_k(\infty) < \infty$, $W(t) \leq Y$ when $0 \leq t \leq c$, and finally that $0 \leq n\delta+c-\sigma_k$ when $k \in J(n\delta,c)$. □

5.7. LEMMA. *Suppose* Φ *satisfies the assumptions of* 5.1 *and define* ${}^c\Phi(t) := \Phi(t)I(t > c)$. *Then a.s. on* $\{T(\infty) = \infty\}$,

$$\text{(5.5)}\qquad H(c) := \overline{\lim_{n\to\infty}} \frac{{}_\alpha Z^{{}^c\Phi}(n\delta+c)}{W(n\delta)} = \overline{\lim_{n\to\infty}} \frac{{}_\alpha Z^{{}^c\Phi}(n\delta)}{W(n\delta)} \to 0 \qquad c \to \infty.$$

PROOF. Define $V := \sup_{s\geq 0} {}_\beta\Phi(s)$, $\Psi(t) := VI(t \leq \delta)$. Then

$$B := \sup_j \frac{{}_\alpha Z^{\Psi}(j\delta)}{W(j\delta)} \leq \sup_j \frac{e^{-\alpha j\delta}T(j\delta)}{W(j\delta)} \cdot \frac{1}{T(j\delta)} \sum_{k=1}^{T(j\delta)} V_k < \infty,$$

estimating $e^{-\alpha j\delta}T(j\delta)/W(j\delta)$ by part (iv) of 4.2 and the next factor by A10.2 (justified by part (ii) of 4.2). Furthermore, the contribution to $Z^{{}^c\Phi}(n\delta+c)$ due to individuals born in $((j-1)\delta, j\delta]$ cannot exceed $e^{\beta((n-j+1)\delta+c)}Z^{\Psi}(j\delta)$ so that

$$\begin{aligned}
H(c) &\leq \overline{\lim_{n\to\infty}} \frac{e^{-\alpha(n\delta+c)}}{W(n\delta)} \sum_{j=1}^{n} e^{\beta((n-j+1)\delta+c)}Z^{\Psi}(j\delta) \\
&\leq Be^{-(\alpha-\beta)c}\overline{\lim_{n\to\infty}} \sum_{j=1}^{n} e^{-(\alpha-\beta)(n-j+1)\delta}\frac{W(j\delta)}{W(n\delta)} \\
&= Be^{-(\alpha-\beta)c}\overline{\lim_{n\to\infty}} \sum_{i=1}^{n} e^{-(\alpha-\beta)i\delta}\frac{W((n-i+1)\delta)}{W(n\delta)}
\end{aligned}$$

But using 5.5, for any $\varepsilon > 0$ there is a C_ε such that $W(m\delta) \leq C_\varepsilon e^{\varepsilon(n-m)}W(n\delta)$ for all n, m with $n \leq m$. Taking $\varepsilon < (\alpha-\beta)\delta$, dominated convergence shows that

$$\begin{aligned}
H(c) &\leq Be^{-(\alpha-\beta)c} \sum_{i=1}^{\infty} \lim_{n\to\infty} e^{-(\alpha-\beta)i\delta}\frac{W(n-i)\delta)}{W(n\delta)} \\
&= Be^{-(\alpha-\beta)c} \sum_{i=1}^{\infty} e^{-(\alpha-\beta)i\delta}. \qquad \square
\end{aligned}$$

5.8. COROLLARY. $M(c) := \overline{\lim_{n\to\infty}} \frac{W(n\delta,c)}{W(n\delta)} \to 0$ as $c \to \infty$.

PROOF. If $\Phi(t) := e^{\alpha t}({}_\alpha\xi(\infty) - {}_\alpha\xi(t))$, then ${}_\alpha Z^{c_\Phi}(t) = W(t,c)$. Appeal to 5.2. □

PROOF OF 5.6. We can decompose $\Phi = (\Phi - {}^c\Phi) + {}^c\Phi$ and choose c so large that $H(c)$ is arbitrarily small and ${}_\alpha m^{c_\Phi}(\infty)$ arbitrarily close to ${}_\alpha m^\Phi(\infty)$. Hence it is no loss of generality to assume ${}^c\Phi = 0$, i.e. $\Phi(t) = 0 \quad t > c$. Appealing to 5.5, it suffices to prove

$$(5.6) \qquad \lim_{r\to\infty} \overline{\lim_{n\to\infty}} \left| \frac{{}_\alpha Z^\Phi((n+r)\delta)}{W(n\delta)} - {}_\alpha m^\Phi(\infty) \right| = 0.$$

Let c be fixed with $\mu(c) > 1$ and let r be so large that $|{}_\alpha m^\Phi(s) - {}_\alpha m^\Phi(\infty)| \leq \epsilon,\ s \geq r\delta - c,\ r\delta \geq c$. Then individuals born before $n\delta$ only contibute to $Z^\Phi((n+r)\delta)$ through the lines of descent of their children born in $(n\delta,(n+r)\delta]$ so that

$$Z^\Phi((n+r)\delta) = \sum_{k\in J(n\delta,r\delta)} Z_k^\Phi((n+r)\delta - \sigma_k),$$

with $Z_k^\Phi(t)$ the $Z^\Phi(t)$ functional evaluated in the line of descent of k. Thus

$$\frac{{}_\alpha Z^\Phi((n+r)\delta)}{W(n\delta)} = S_n^1 + S_n^2 + S_n^3 + S_n^4 \quad \text{where}$$

$$S_n^1 := \frac{1}{W(n\delta)} \sum_{k\in J(n\delta,r\delta)} e^{-\alpha\sigma_k}\{{}_\alpha Z_k^\Phi((n+r)\delta-\sigma_k) - {}_\alpha m^\Phi((n+r)\delta-\sigma_k)\}$$

$$S_n^2 := \frac{1}{W(n\delta)} \sum_{k\in J(n\delta,c)} e^{-\alpha\sigma_k}\{{}_\alpha m^\Phi((n+r)\delta-\sigma_k) - {}_\alpha m^\Phi(\infty)\},$$

$$S_n^3 := \frac{{}_\alpha m^\Phi(\infty)}{W(n\delta)} \sum_{k\in J(n\delta,c)} e^{-\alpha\sigma_k} = {}_\alpha m^\Phi(\infty)\left(1 - \frac{W(n\delta,c)}{W(n\delta)}\right),$$

$$S_n^4 := \frac{1}{W(n\delta)} \sum_{k\in J(n\delta,r\delta)\setminus J(n\delta,c)} e^{-\alpha\sigma_k}\, {}_\alpha m^\Phi((n+r)\delta-\sigma_k).$$

One can easily cheek that $E \sup_{t\leq r\delta} {}_\alpha Z^\Phi(t) < \infty$. Hence $S_n^1 \to 0$ follows as above from A 10.2 and it follows that

$$\overline{\lim_{n\to\infty}} \left| \frac{{}_\alpha Z^{\Phi}((n+r)\delta)}{W(n\delta)} - {}_\alpha m^{\Phi}(\infty) \right|$$

$$\leq 0 + \overline{\lim_{n\to\infty}} |S_n^2| + M(c) + \overline{\lim_{n\to\infty}} |S_n^4|$$

$$\leq \sup_{s\geq r\delta-c} |{}_\alpha m^{\Phi}(s) - {}_\alpha m^{\Phi}(\infty)| + M(c) + \sup_{s\geq 0} {}_\alpha m^{\Phi}(s)\cdot M(c).$$

Letting first $r \to \infty$, the first term vanishes so that (5.6) follows as next $c \to \infty$. □

6. THE SUBCRITICAL CASE

Let $J_n(t)$ be the set of children of individual n born in $[\sigma_n, t]$. Then

$$Z^{\Phi}(t) = \sum_{n \in J(t)} \Phi_n(t-\sigma_n)$$

$$= \Phi_1(t) + \sum_{j \in J_1(t)} Z_j^{\Phi}(t-\sigma_j),$$

where Z^{Φ} and the Z_j^{Φ} are i.i.d. Introducing the generating functions

$$F^{\Phi}(s,t) := E\, s^{Z^{\Phi}(t)}, \quad s \leq 1,\ t \geq 0,$$

we immediately get

(6.1) $$F^{\Phi}(s,t) = E\, s^{\Phi_1(t)} \prod_{j \in J_1(t)} F(s, t-\sigma_j),$$

which again leads to (2.1) and (2.2)

As before we assume throughout that μ is non-lattice and Malthusian,

$$\varkappa = \int_0^\infty u\,{}_\alpha\mu[du] < \infty,$$

that $\Phi(t)(\omega)$ is measurable as a function of (t,ω), and $E_\alpha \Phi(t)$ directly Riemann integrable (d.R.i.). According to 2.1, the latter guarantees

(6.2) $${}_\alpha m^{\Phi}(t) \to {}_\alpha m^{\Phi}(\infty) := \varkappa^{-1}\int_0^\infty E_\alpha \Phi(t)dt, \quad t \to \infty,$$

and to avoid trivialities, we assume this limit to be positive.

We are now interested in the case $\mu(\infty) < 1$, i.e., $\alpha < 0$.

6.1. THEOREM. *If* $\alpha < 0$, *then for each* $s \in [0,1]$

(6.3) $$e^{-\alpha t}(1-F^{\Phi}(s,t)) \to \gamma^{\Phi}(s), \quad t \to \infty,$$

where $\gamma^{\Phi}(s)$ *is finite. If in addition*

(X LOG X) $\quad E_\alpha \xi(\infty) \log^+ {}_\alpha\xi(\infty) < \infty,$

then $\gamma^\Phi(s) > 0$ _for all_ $s \in [0,1)$, _while otherwise_ $\gamma^\Phi \equiv 0$.

_6.2_. COROLLARY. _If_ $\alpha < 0$, _then_

$$(6.4) \qquad e^{-\alpha t} P(Z^\Phi(t) > 0) \to \gamma^\Phi(0) < \infty, \quad t \to \infty,$$

where $\gamma^\Phi(0) > 0$ _if and only if_ (X LOG X) _is satisfied, and if the latter is the case, then in the sense of weak convergence of distribution functions_

$$(6.5) \qquad P(Z^\Phi(t) \le \lambda | Z^\Phi(t) > 0) \to D^\Phi(\lambda), \quad t \to \infty,$$

where D^Φ _is a proper d.f. with_

$$(6.6) \qquad \int_0^\infty \lambda dD^\Phi(\lambda) = {}_\alpha m^\Phi(\infty)/\gamma^\Phi(0).$$

_6.3_. REMARK. Let

$$\Psi = I(\tau > t)$$

and suppose that $E_\alpha \Psi(t)$ is integrable and

$$(*) \qquad P(\Phi(t) = 0 | \tau \le t) = 1.$$

Then

$$E(s^{\Phi(t)} | Z^\Psi(t) > 0) = 1 - \frac{1-F^\Phi(s,t)}{1-F^\Psi(0,t)}.$$

That is, if (*) and the conditions for (6.5) and (6.6) are satisfied, then

$$(6.5^*) \qquad P(Z^\Phi(t) \le \lambda | Z^\Psi(t) > 0) \xrightarrow{D} D_0^\Phi(t), \quad t \to \infty,$$

where D_0^Φ is a proper d.f. with

$$(6.6^*) \qquad \int_0^\infty \lambda dD_0^\Phi(\lambda) = {}_\alpha m^\Phi(\infty)/\gamma^\Psi(0).$$

We split the proof of _6.1_ and _6.2_ into several lemmata. Omitting the superscript Φ, set

$$H(s,t) := 1 - F(s,t)$$

and rewrite (6.1) as

$$(6.7)\qquad H(s,t) = h(s,t) + \int_0^t H(s,t-u)\mu[du],$$

where $h(s,t) = h_1(s,t) - h_2(s,t)$,

$$h_1(s,t) := E\{(1-s^{\Phi_1(t)}) \prod_{j\in J_1(t)} F(s,t-\sigma_j)\},$$

$$h_2(s,t) := E \prod_{j\in J_1(t)} F(s,t-\sigma_j) - 1 + \int_0^t H(s,t-u)\mu[du].$$

Assume throughout that $\alpha < 0$.

6.4. LEMMA. *For each fixed* $s \in [0,1)$, $e^{-\alpha t}h(s,t)$ *is d.R.i.*

Given 6.4, it follows from (6.7) that for each $s \in [0,1)$

$$e^{-\alpha t}H(s,t) \to \kappa^{-1}\int_0^\infty e^{-\alpha t}h(s,t)dt < \infty, \quad t \to \infty,$$

cf. 2.1.

PROOF OF 6.4. Since $0 \le h_1(s,t) \le E\Phi(t)$, direct Riemann integrability of $e^{-\alpha t}h_1(s,t)$ is immediate from the assumption that $E_\alpha\Phi(t)$ is d.R.i.

Now notice that

$$(6.8)\qquad h_2(s,t) = E\{\sum_{j\in J_1(t)}[1-F(s,t-\sigma_j)] - [1-\prod_{j\in J_1(t)}F(s,t-\sigma_j)]\},$$

which is an antitone functional of $F(s,t-\cdot)$. In particular,

$$0 \le h_2(s,t) \le h_2(0,t).$$

From (6.7)

$${}_\alpha h_2(0,t) = \int_0^t {}_\alpha H(0,t-u)\,{}_\alpha\mu[du] + {}_\alpha h_1(0,t) - {}_\alpha H(0,t).$$

But

$${}_\alpha H(0,t) \le {}_\alpha m(t) \to {}_\alpha m(\infty) < \infty,$$

so that ${}_\alpha H(0,t)$ is bounded. Hence, ${}_\alpha h_2(0,t)$ is bounded, its Laplace transform ${}_\alpha\hat{h}_2(0,\lambda)$ exists, and

$${}_\alpha\hat{h}_2(0,\lambda) = {}_\alpha\hat{h}_1(0,\lambda) - {}_\alpha\hat{H}(0,\lambda)[1 - {}_\alpha\hat{\mu}(\lambda)]$$

in obvious notation. Thus

$$ {}_{\alpha}\hat{h}_2(0,\lambda) \leq {}_{\alpha}\hat{h}_1(0,\lambda) $$

and, letting $\lambda \downarrow 0$,

$$ \int_0^\infty {}_{\alpha}h_2(0,t)dt \leq \int_0^\infty {}_{\alpha}h_1(0,t)dt. $$

Unfortunately, ${}_{\alpha}h_2(0,t)$ is not monotone. However, since $F(0,t)\uparrow$, it follows from (6.8) that for $n \leq t < n+1$

$$ h_2(0,n+1) - (\mu(\infty)-\mu(n)) \leq h_2(0,t) \leq h_2(0,n) + \mu(\infty) - \mu(n), $$

thus

$$ 0 \leq {}_{\alpha}h_2(0,t) \leq e^{-\alpha(n+1)}(h_2(0,n) + \mu(\infty) - \mu(n)), $$

and all that remains to be seen is summability of the r.h.s. But

$$ \sum_{n=0}^{\infty} e^{-\alpha n}(\mu(\infty) - \mu(n+1)) \leq \int_0^\infty e^{-\alpha t}(\mu(\infty) - \mu(t))dt $$

$$ \leq |\alpha|^{-1}(\int_0^\infty e^{-\alpha t}\mu[dt] - \mu(\infty)) < \infty, $$

$$ \sum_{n=0}^{\infty} e^{-\alpha n}h_2(0,n+1) \leq \sum_{n=0}^{\infty} e^{-\alpha n}(\mu(\infty) - \mu(n)) $$

$$ + \int_0^\infty e^{-\alpha t}h_2(0,t)dt < \infty. \qquad \square $$

Recall that for any r.v. $X \geq 0$ with finite mean, $E\, X \log^+ X < \infty$ if and only if for any $a > 0$

$$ \int_0^a v^{-2}\, E(e^{-vX}-1+vX)dv < \infty, \tag{6.9} $$

cf.proof of II.1.8. Here $X := {}_{\alpha}\xi(\infty)$.

6.5. LEMMA. *If (6.9) is satisfied,*

$$ \lim_{s\uparrow 1} (1-s)^{-1}\gamma(s) = {}_{\alpha}m(\infty). \tag{6.10} $$

Given (6.10), we have $\gamma(s) > 0$ for $s < 1$ sufficiently close to 1, and thus, since $\gamma(s)\downarrow$, $\gamma(0) > 0$. Moreover, $\gamma(1-) = 0$, which proves (6.5). Finally,

$$ \int_0^\infty \lambda dD(\lambda) = \lim_{s\uparrow 1} (1-s)^{-1}\gamma(s)/\gamma(0) = {}_{\alpha}m(\infty)/\gamma(0). $$

PROOF OF 6.5. Clearly,

$$\lim_{s\uparrow 1}(1-s)^{-1}\int_0^\infty e^{-\alpha t}h_1(s,t)dt = \varkappa_\alpha m(\infty),$$

so that it suffices to show

$$\lim_{s\uparrow 1}(1-s)^{-1}\int_0^\infty e^{-\alpha t}h_2(s,t)dt = 0. \tag{6.11}$$

Since $1-F(s,t) \leq m(t)(1-s)$ and $e^{-\alpha t}m(t)$ is bounded, there exist constants $c_1, c_2 > 0$ such that

$$F(s,t) \geq 1-c_1(1-s)e^{\alpha t} \geq \exp\{-c_2(1-s)e^{\alpha t}\}$$

for all s sufficiently close to 1. Setting

$$\beta := \beta(s,t) := c_2(1-s)e^{\alpha t},$$

$$y_j := e^{-\alpha\sigma_j}, \quad Y := \sum_{j\in J_1(\infty)} y_j = {}_\alpha\xi(\infty)$$

we get

$$h_2(s,t) \leq E\{\sum_{j\in J_1(t)}(1-e^{-\beta y_j}) - (1-\prod_{j\in J_1(t)} e^{-\beta y_j})\}$$

$$= E\{\sum_{j\in J_1(\infty)}(1-e^{-\beta y_j}) - (1-e^{-\beta Y})\}$$

$$\leq E\{e^{-\beta Y} - 1 + \beta Y\} =: E\Delta(\beta Y)$$

and from this

$$(1-s)^{-1}e^{-\alpha t}h_2(s,t) \leq c_2\beta^{-1}E\Delta(\beta Y) \downarrow 0, \quad s\uparrow 1.$$

Also, for all $s > 1$ sufficiently close to 1,

$$(1-s)^{-1}e^{-\alpha t}h_2(s,t) \leq c\beta(0,t)^{-1}E\Delta(\beta(0,t)Y)$$

with some constant c, and by (6.9),

$$\int_0^\infty \beta(0,t)^{-1}E\Delta(\beta(0,t)Y) = -\alpha^{-1}\int_0^{c_2} v^{-2}E\Delta(vY)dv < \infty.$$

Hence, using dominated convergence, we indeed have (6.11). □

The following lemma completes the proof of 6.1 and 6.2.

6.6. LEMMA. *If* $\gamma(s) > 0$ *for some* $s \in [0,1)$, *then* (6.9) *is satisfied*.

PROOF. If $\gamma(s) > 0$ for some s then $\gamma(0) > 0$, so that $e^{-\alpha t}(1-F(0,t)) = c_3 > 0$ and thus

$$F(0,t) \leq 1-c_3e^{\alpha t} \leq \exp\{-c_3e^{\alpha t}\}, \quad t \geq 0.$$

Define

$$\eta := \eta(t) := c_3e^{\alpha t},$$

$$Y(t) := \sum_{j\in J_1(t)} y_j = {}_\alpha\xi(t)$$

Y_j and Δ as before. Then

$$h_2(0,t) \geq E\{\sum_{j\in J_1(t)}(1-e^{-\eta y_j}) - (1-\prod_{j\in J_1(t)} e^{-\eta y_j})\}$$

$$= E\{\Delta(\eta Y(t)) - \sum_{j\in J_1(t)} \Delta(\eta y_j)\}$$

and from this

$$c_3\int_0^\infty \eta^{-1}E\Delta(\eta Y(t))dt \leq I_1 + I_2,$$

$$I_1 := \int_0^\infty e^{-\alpha t}h_2(0,t)dt,$$

$$I_2 := \int_0^\infty e^{-\alpha t}E\sum_{j\in J_1(t)} \Delta(\eta y_j)dt.$$

Now notice that

$$\int_0^{c_3} v^{-2}E\Delta(vY)dv = -\alpha\int_0^\infty \eta^{-1}E\Delta(\eta Y)dt$$

$$\leq -\alpha\int_0^\infty \eta^{-1}E\Delta(\eta Y(t))dt + I_3,$$

$$I_3 := -\alpha\int_0^\infty E\{Y-Y(t)\}dt.$$

But $I_1 < \infty$, as we have seen before, further

$$I_2 \leq -\alpha^{-1}c_3\int_0^{c_3} \eta^{-2}E\sum_{j\in J_1(\infty)} \Delta(\eta y_j)d\eta$$

$$= -\alpha^{-1}c_3E\sum_{j\in J_1(\infty)} y_j(\int_0^{c_3} d\eta/2 + \int_{c_3}^{c_3y_j} d\eta/\eta)$$

$$\leq -\alpha^{-1} c_3 E \sum_{j\in J_1(\infty)} y_j(c_3/2 + \log y_j)$$
$$= -\alpha^{-1} c_3 (c_3/2 - \alpha\varkappa) < \infty,$$

$$I_3 = -\alpha\int_0^\infty\int_t^\infty {}_\alpha\mu[du]dt = \int_0^\infty u_\alpha\mu[du] = \quad < \infty,$$

and we are done. □

6.7. REMARK. Notice that

$$E_\alpha\xi(\infty)\log^+\xi(\infty) \leq E_\alpha\xi(\infty)\log^+{}_\alpha\xi(\infty)$$
$$\leq E\{\sum_{j\in J_1(\infty)} e^{-\alpha\sigma_j}\log^+\sum_{j\in J_1(\infty)} e^{-\alpha\sigma_j}\}$$
$$\leq \sum_{j\in J_1(\infty)} e^{-\alpha\sigma_j}(-\alpha\sigma_j + \log^+\xi(\infty))$$
$$= -\alpha\varkappa + E_\alpha\xi(\infty)\log^+\xi(\infty)$$

and, with $c = -\varepsilon/2\alpha$ and some constant C,

$$E_\alpha\xi(\infty)\log^+\varepsilon(\infty)$$
$$= E\int_0^{c\,\log^+\varepsilon(\infty)} {}_\alpha\xi[du]\log^+\xi(\infty) + E\int_{c\,\log^+\xi(\infty)}^\infty {}_\alpha\xi[du]\log^+\xi(\infty)$$
$$\leq Ee^{-\alpha c\,\log^+\xi(\infty)}\xi(\infty)\log^+\xi(\infty) + c^{-1}E\int_{c\,\log^+\xi(\infty)}^\infty y_\alpha\xi[du]$$
$$\leq E\xi(\infty)^{1+\varepsilon/2}C\xi(\infty)^{\varepsilon/2} + c^{-1}\int_0^\infty u_\alpha\mu[du]$$
$$= CE\xi(\infty)^{1+\varepsilon} + c^{-1} \quad .$$

That is, given $\varkappa < \infty$, we have

$$E_\alpha\xi(\infty)\log^+{}_\alpha\xi(\infty) < \infty \iff E_\alpha\xi(\infty)\log^+\xi(\infty) < \infty$$

and

$$E\xi(\infty)^{1+\varepsilon} \implies E_\alpha\xi(\infty)\log^+\xi(\infty) < \infty.$$

7. THE CRITICAL CASE

Assume throughout that μ is non-lattice Malthusian,

$$\alpha = 0,$$

$$\varkappa := \int_0^\infty u\mu[du] < \infty,$$

$$\sigma^2 := \operatorname{Var} \xi(\infty) < \infty.$$

To exclude the trivial case, suppose $\mu(0) < 1$. Let us first consider only the special characteristic

$$\Psi(t) := I(\tau > t).$$

7.1. THEOREM. *If*

$$\lim_{t\to\infty} t^2(1-\mu(t)) = \lim_{t\to\infty} t^2 P(\tau > t) = 0,$$

then

$$\lim_{t\to\infty} tP(Z^\Psi(t) > 0) = 2\varkappa/\sigma^2.$$

We divide the proof into several steps. Define

$$\zeta_n := \text{size of n'th generation},$$

$$q_n := P(\zeta_n = 0),$$

$$\mu^{*n} := \text{n'th convolutive power of } \mu,$$

$$G(t) := P(\tau \leq t).$$

7.2. LEMMA. *For* $n \in \mathbb{N}$ *and* $t \geq 0$

$$-\mu^{*n}(t) \leq q_n - P(Z^\Psi(t) = 0) \leq (1-G)*\sum_{k=0}^{n-1} \mu^{*k}(t).$$

PROOF. We perform four simple inductions. Define

$$F_0(s,t) := s,$$

$$F_n(s,t) := E\{s^{\Psi_1(t)} \prod_{j\in J_1(t)} F_{n-1}(s,t-\sigma_j)\}, \quad n \in \mathbb{N}.$$

The first claim is that

$$(7.1) \qquad F(0,t) - F_n(0,t) \leq \mu^{*n}(t).$$

Indeed,

$$F(0,t) - F_0(0,t) = F(0,t) \leq 1 = \mu^{*0}(t),$$

and assuming (7.1),

$$F(0,t) - F_{n+1}(0,t) \leq E\{\prod_{j\in J_1(t)} F(0,t-\sigma_j) - \prod_{j\in J_1(t)} F_n(0,t-\sigma_j)\}$$

$$\leq E \sum_{j\in J_1(t)} (F(0,t-\sigma_j) - F_n(0,t-\sigma_j)) \leq E \sum_{j\in J_1(t)} \mu^{*n}(t-\sigma_j)$$

$$= E\int_0^t \mu^{*n}(t-u)\xi[du] \leq \mu^{*(n+1)}(t).$$

Next, we convince ourselves of

$$(7.2) \qquad F_n(0,t) \leq q_n.$$

In fact,

$$F_0(0,t) = 0 \leq q_0,$$

and assuming (7.2),

$$F_{n+1}(0,t) = E\{I(\tau_1 \leq t) \prod_{j\in J_1(t)} F_n(0,t-\sigma_j)\}$$

$$\leq E\{I(\tau \leq t) q_n^{\xi(t)}\}$$

$$\leq E\{I(\tau \leq t) q_n^{\xi(\infty)}\} \leq q_{n+1}.$$

Combining (7.1) and (7.2) yields the first half of 7.2.

We now show that

$$(7.3) \qquad q_n - F_n(0,t) \leq (1-G) * \sum_{k=0}^{n-1} \mu^{*k}(t).$$

Clearly,

$$q_1 - F_1(0,t) \leq 1 - G(t).$$

Assume (7.3). Then

$$q_{n+1} - F_{n+1}(0,t)$$
$$\leq E\{q_n^{\xi_1(t)} - \prod_{j\in J_1(t)} F_n(0,t-\sigma_j\} + q_n^{\xi(\infty)} - P(\Psi(t)=0)$$
$$\leq E \sum_{j\in J_1(t)} (q_n - F_n(0,t-\sigma_j) + q_n^{\xi(\infty)} - P(\Psi(t)=0)$$
$$\leq (1-G)*\sum_{k=1}^{n} \mu^{*k}(t) + 1-G(t) = (1-G)*\sum_{k=0}^{n} \mu^{*k}(t).$$

Finally, we shall see that

$$(7.4) \qquad F(0,t) \geq F_n(0,t).$$

By (7.1), $F_n(0,t) \to F(0,t)$, as $n \to \infty$. So it suffices to prove $F_n(0,t)\uparrow$. But

$$F_1(0,t) \geq 0 = F_0(0,t)$$

and

$$F_n(0,t) - F_{n+1}(0,t) \leq E[\prod_{j\in J_1(t)} F_{n-1}(0_1 t-\sigma_j) - \prod_{j\in J_1(t)} F_n(0,t-\sigma_j);\tau \leq t]$$
$$\leq E[\sum_{j\in J_1(t)} (F_{n-1}(0,t-\sigma_j) - F_n(0,t-\sigma_j);\tau \leq t],$$

so that (7.4) follows by induction. Combining (7.3) and (7.4) gives the second half of _7.2_. □

_7.3_. LEMMA. _Let_ $F(t)$ _be a d.f. on_ $[0,\infty)$,

$$0 < a := \int_0^\infty t dF(t),$$
$$\lim_{t\to\infty} t^2(1-F(t)) = 0.$$

Then

$$(7.5) \qquad \lim_{t\to\infty} tF^{*[(1+\epsilon t/a]}(t) = 0,$$

$$(7.6) \qquad \lim_{t\to\infty} t(1-F^{*[(1-\epsilon)t/a]}(t)) = 0$$

for any $\epsilon \in (0,1)$.

This is a corollary of a result on convergence rates in the law of large numbers, see A12.

7.4. LEMMA. Let $F(t)$ and $G(t)$ be d.f.'s on $[0,\infty)$,

$$0 < a := \int_0^\infty t\,dF(t),$$

$$\lim_{t\to\infty} t^2(1-F(t)) = \lim_{t\to\infty} t^2(1-G(t)) = 0. \tag{7.7}$$

Then

$$\lim_{t\to\infty} t(1-G)*\sum_{n=0}^{[t(1-\varepsilon)/a]-1} F^{*n}(t) = 0.$$

PROOF. Note that (7.7) implies $a < \infty$. Fix $\varepsilon > 0$, and set

$$\delta := (1-\varepsilon)/(1-\varepsilon/2),\ \vartheta := a/2,\ n := [t(1-\varepsilon)/a],$$

and write

$$(1-G)*\sum_{j=0}^{n-1} F^{*j}(t) = S_1+S_2+S_3,$$

$$S_1 := \sum_{j=0}^{n-1}\int_{0-}^{\delta t+} (1-G(t-y))dF^{*j}(y),$$

$$S_2 := \sum_{j=1}^{n-1}\int_{\delta t+}^{t-\vartheta+} (1-G(t-y))dF^{*j}(y),$$

$$S_3 := \sum_{j=1}^{n-1}\int_{t-\vartheta+}^{t+} (1-G(t-y))dF^{*j}(y).$$

By (7.7)

$$S_1 \leq n\{1-G((1-\delta)t-)\} = o(t^{-1}),\quad t\to\infty.$$

To estimate S_2, S_3, we need

$$\begin{aligned} 1 - F^{*k}(t) &= \sum_{j=0}^{k-1}\int_{0-}^{t+} (1-F(t-y))dF^{*j}(y) \\ &\geq \sum_{j=0}^{k-1}\int_{t-\vartheta+}^{t+} (1-F(t-y))dF^{*j}(y) \\ &\geq (1-F(\vartheta))\sum_{j=0}^{k-1}\int_{t-\vartheta+}^{t+} dF^{*j}(y) \\ &= (1-F(\vartheta))\sum_{j=0}^{k-1}\{F^{*j}(t) - F^{*j}(t-\vartheta)\}. \end{aligned} \tag{7.8}$$

From this and (7.6)

$$S_3 \leq \sum_{j=0}^{n-1} \int_{t-\vartheta+}^{t+} dF^{*j}(y) \leq (1-F^{*n}(t))/(1-F(\vartheta)) = o(t^{-1}), \quad t \to \infty.$$

By (7.7)

$$\eta := \int_0^\infty t^{3/2} dG(t) < \infty,$$

and by Markov-Chebychev

$$1 - G(t-y) \leq \eta(t-y)^{-3/2}.$$

Using this and once again (7.8),

$$S_2 \leq \eta \sum_{j=1}^{n-1} \sum_{r-1}^{k} \int_{t-(r+1)\vartheta+}^{t-r\vartheta+} (t-y)^{-3/2} dF^{*j}(y)$$

$$\leq \eta \sum_{j=1}^{n-1} \sum_{r=1}^{k} (r\vartheta)^{-3/2} \{F^{*j}(t-r\vartheta) - F^{*j}(t-(r+1)\vartheta)\}$$

$$= \eta \sum_{r=1}^{k} (r\vartheta)^{-3/2} \sum_{j=1}^{n-1} \{(F^{*j}(t-r\vartheta) - F^{*j}(t-r\vartheta-\vartheta)\}$$

$$\leq \eta \sum_{r=1}^{k} (r\vartheta)^{-3/2} (1 - F^{*n}(t-r\vartheta))/(1-F(\vartheta))$$

$$\leq \eta \sum_{r=1}^{k} (r\vartheta)^{-3/2} (1 - F^{*n}(\delta t))/(1 - F(\vartheta))$$

$$= o(t^{-1}), \quad t \to \infty. \qquad \square$$

PROOF OF 7.1. Set

$$q(t) := P(\psi(t) = 0), \quad \gamma = \sigma^2/2a.$$

By 7.2

$$\gamma t(1-q_n) - \gamma t \mu^{*n}(t) \leq \gamma t(1-q(t)), \tag{7.9}$$

$$\gamma t(1-q(t)) \leq \gamma t(1-q_n) + \gamma t(1-G) * \sum_{j=0}^{n-1} \mu^{*j}(t). \tag{7.10}$$

Take $n = [(1+\varepsilon)t/a]$ in (7.9) and $n = [(t-\varepsilon)t/a]$ in (7.10). Then, using (7.5), 7.4 with $F \equiv \mu$, and the fact that $a\gamma n(1-q_n) \to 1$, $n \to \infty$, (cf. III.3),

$$(1+\varepsilon)^{-1} \leq \liminf_{t\to\infty} \gamma t(1-q(t))$$

$$\leq \limsup_{t\to\infty} \gamma t(1-q(t)) \leq (1-\varepsilon)^{-1}.$$

But $\varepsilon > 0$ was arbitrary. □

Our next aim is the exponential limit theorem. Here Φ is any random characteristic. However, we assume throughout that $\Phi(t)(\omega)$ is measurable in (t,ω) with $E\Phi(t)$ directly Riemann integrable. As a first consequence,

$$m(t) \to m(\infty) := \varkappa^{-1}\int_0^\infty E\Phi(t)dt, \quad t \to \infty.$$

7.5. THEOREM. *If $E\Phi(t)\xi(t)$ is bounded and $E\Phi(t)\xi(t) \to 0$, $t \to \infty$, then*

$$\lim_{t\to\infty} t[1-F^\Phi(e^{-\lambda/t},t)] = \frac{m(\infty)\lambda}{1+(\sigma^2 m(\infty)/2\varkappa)\lambda}, \quad \lambda \geq 0.$$

Since by Cauchy-Schwarz

$$(E\Phi(t)\xi(t))^2 \leq E\Phi(t)^2 E\xi(t)^2 \leq \sigma^2 E\Phi(t)^2,$$

the assumption on $E\Phi(t)\xi(t)$ can be replaced by the analogous assumption on $E\Phi(t)^2$.

Given 7.1 and 7.5, the following is immediate:

7.6. COROLLARY. *In addition to the assumptions underlying 7.1 and 7.5 let the condition*

$$P(\Phi(t)|\tau \leq t) = 1$$

be satisfied. Then

$$\lim_{t\to\infty} P(Z^\Phi(t) \geq \lambda | Z^\Psi(t) > 0) = e^{-2\varkappa\lambda/\sigma^2 m(\infty)}, \quad \lambda \geq 0.$$

PROOF OF 7.5. We omit the superscript Φ. As in the preceding section

$$H(s,t) := 1 - F(s,t).$$

Then

$$H(s,t) = h(s,t) + \int_0^t H(s,t-n)\mu[du],$$

$$h(s,t) = h_1(s,t) + h_3(s,t) - h_4(s,t),$$

where as before

$$h_1(s,t) := E\{(1-s^{\Phi(t)}) \prod_{j\in J_1(t)} F(s,t-\sigma_j)\},$$

$$h_3(s,t) := 1 - E \prod_{j\in J_1(t)} F(s,t-\sigma_j) - \int_0^t H(s,t-u)\mu[du]$$

$$+ \frac{1}{2}\iint_{[0,t]\times[0,t]} H(s,t-u)H(s,t-v)\mu^{(2)}[du,dv],$$

$$h_4(s,t) := \frac{1}{2}\iint_{[0,t]\times[0,t]} H(s,t-u)H(s,t-v)\mu^{(2)}[du,dv]$$

with

$$\mu^{(2)}(u,v) := E\xi(u)\xi(v) - E\xi(u\wedge v).$$

By the renewal theorem

$$H(s,t) = \int_0^t h(s,t-u)U[du],$$

$$U(t) = \sum_{n=0}^{\infty} \mu^{*n}(t).$$

Set

$$K(\lambda,t) := (t/\lambda)H(e^{-\lambda/t},t), \quad \lambda > 0,$$

$$k_\nu(\lambda,t) = (1/\lambda)h_\nu(e^{-\lambda},t), \quad \nu = 1,3,4.$$

Then

$$K(\lambda,t) = \sum_{\nu=1,3,4} \int_0^t k_\nu(\lambda/t,t-y)U[du]. \tag{7.11}$$

Let us estimate the first summand. For sufficiently large t

$$(t/\lambda)\int_0^t E\{(1-e^{-(\lambda/t)\Phi(t-u)})(1-(\lambda/t)\sum_{j\in J_1(t-u)} m(t-u-\sigma_j))\}U[du]$$

$$\leq \int_0^t k_1(\lambda/t,t-u)U[du]$$

$$\leq (t/\lambda)\int_0^t E(1-e^{-(\lambda/t)\Phi(t-u)})U[du].$$

Since for $0 \leq x \leq y$ and $a > 0$

$$x(1-e^{-a/x}) \leq y(1-e^{-a/y}) \leq a,$$

the renewal theorem gives

$$\lim_{t\to\infty} (t/\lambda)\int_0^t E(1-e^{-(\lambda/t)\Phi(t-u)})U[du] = m(\infty).$$

Setting

$$m := \sup_t m(t)$$

we have

$$(t/\lambda)\int_0^t E\{(1-e^{-(\lambda/t)\Phi(t-u)})(\lambda/t)\sum_{j\in J_1(t-u)} m(t-u-\sigma_j)\}U[du]$$

$$\leq (\lambda/t)m\int_0^t E\{\Phi(t-u)\xi(t-u)\}U[du].$$

Using our assumptions on $E\Phi(t)\xi(t)$ and the renewal theorem, the r.h.s. tends to 0, as $t \to \infty$. Summing up,

$$\text{(7.12)} \qquad \lim_{t\to\infty} \int_0^t k_1(\lambda/t, t-u)U[du] = m(\infty),$$

where the convergence is uniform on bounded λ-intervals.

Next consider the second summand on the right of (7.11). For $0 \leq x_j \leq x \leq y$

$$1 - \sum_{j=1}^n x_j + \frac{1}{2}\sum_{\substack{j,\ell=1\\ j\neq \ell}}^n x_j x_\ell - \prod_{j=1}^n (1-x_j)$$

$$\leq (x/y)(1-ny + \tfrac{1}{2}n(n-1)y^2 - (1-y)^n) =: (x/y)\gamma(n,y).$$

Thus for any $\lambda_0 > 0$, $\lambda \in (0,\lambda_0)$, and $t > \lambda_0 m$

$$\int_0^t k_3(\lambda/t, t-u)U[du] \leq (t^2/\lambda_0)E_\gamma(\xi(\infty), \lambda_0 m(t)U(t)/t.$$

But since $\sigma^2 < \infty$,

$$E_\gamma(\xi(\infty), x) = o(x^2), \quad x \to 0,$$

so that, again by the renewal theorem,

$$\text{(7.13)} \qquad \lim_{t\to\infty} \int_0^t k_3(\lambda/t, t-u)U[du] = 0$$

uniformly on bounded λ-intervals.

Now rewrite

$$\int_0^t k_4(\lambda/t, t-u)U[du]$$

$$= (\sigma^2\lambda/2\varkappa)\iint_{[0,1]\times[0,1]} K(\lambda u, tu)K(\lambda v, tv)\mu_t[du, dv]$$

with

$$\mu_t(u,v) = \mu_t[[0,u] \times [0,v]]$$

$$:= (\varkappa/\sigma^2 t)\int_{[0,t(1-u\vee v)]} \mu^{(2)}[[t(1-u)-x, t-x] \times [t(1-v)-x, t-x]]U[dx].$$

For simplicity assume for the moment that time and random characteristics are rescaled so that $m(\infty) = 1$ and $\sigma^2 = 2\varkappa$. Then notice that

$$\mu_2(u,v) \to u \quad v, \quad t \to \infty,$$

and thus, as $t \to \infty$,

$$\iint_{[0,1]\times[0,1]} \{(1+\lambda v)^{-1}K(\lambda u, tu) - (1+\lambda u)^{-1}K(\lambda v, tv)\}\mu_t[du, dv] \to 0,$$

$$\lambda\iint_{[0,1]\times[0,1]} (1+\lambda u)^{-1}(1+\lambda v)^{-1}\mu_t[du, dv] \to \lambda\int_0^1 (1+\lambda v)^{-2}dv$$

uniformly on bounded λ-intervals. Hence, by (7.12) and (7.13),

$$K(\lambda, t) - (1+\lambda)^{-1}$$

$$= R(\lambda, t) - \lambda\iint_{[0,1]\times[0,1]} (K(\lambda u, tu) - (1+\lambda u)^{-1})$$

$$\times\ (K(\lambda v, tv) - (1+\lambda v)^{-1})\mu_t[du, dv])$$

where

$$F(\lambda, t) \to 0, \quad t \to \infty,$$

uniformly on bounded λ-intervals.

Since $K(\lambda, t) \le m < \infty$, there exists a constant $c < \infty$ such that for $\lambda > 0$, $t > 0$, and $v \in (0,1]$

$$|K(\lambda v, tv)| + |(1+\lambda v)^{-1}| \le c,$$

so that for any $\delta \in (0,1)$

$$|K(\lambda,t) - (1+\lambda)^{-1}|$$
$$\leq |K(\lambda,t)|+\lambda c^2\mu_t(\delta,1)+\iint_{[\delta,1]\times[0,1]}|K(\lambda u,tu)-(1+\lambda u)^{-1}|\mu_t[du,dv].$$

Fix $\lambda_1 \in (0,1/c)$, let $\lambda_n := n\lambda_1$, $n \in \mathbb{N}$, and set

$$\Delta_n(t) := \sup_{\substack{0<\lambda<\lambda_n \\ t'>t}} |K(\lambda,t') - (1+\lambda)^{-1}|,$$

$$F_n(t) := \sup_{\substack{0<\lambda<\lambda_n \\ t'>t}} |R(\lambda,t)|.$$

Then

$$\Delta_1(t) \leq R_1(t)+\lambda_1 c^2 \sup_{t'>t}\mu_{t'}(\delta,1)+\lambda_1 c \sup_{t'>t} \mu_{t'}(1,1)\Delta_1(t\delta).$$

As $R_1(t) \to 0$, $\mu_t(\delta,1) \to \delta$, $\mu_t(1,1) \to 1$, $\lambda_1 c < 1$, and $\delta \in (0,1)$ can be chosen arbitrarily small, it follows that

$$(7.14) \qquad K(\lambda,t) \to (1+\lambda)^{-1}, \quad t \to \infty,$$

uniformly in $\lambda \in (0,\lambda_1)$. Next assume that (7.14) holds uniformly in $\lambda \in (0,\lambda_n)$, i.e.,

$$\Delta_n(t) \to 0, \quad t \to \infty,$$

and notice that for $\delta \in (0,n/(n+1))$

$$\begin{aligned}\Delta_{n+1}(t) \leq R_{n+1}(t) &+ \lambda_{n+1}c^2 \sup_{t'>t} \mu_t(\delta,1) \\ &+ \lambda_{n+1}c\Delta_n(t\delta)\iint_{[\delta,n/(n+1)]\times[0,1]}\mu_t[du,dv] \\ &+ \lambda_{n+1}c \sup_{t'>t} [\mu_t(1,1)-\mu_t(n/(n+1),1)]\Delta_{n+1}(t\delta),\end{aligned}$$

$$\lambda_{n+1}c \sup_{t'>t} [\mu_t(1,1)-\mu_t(n/(n+1),1)] \to \lambda_1 c < 1, \quad t \to \infty,$$

so that also

$$\Delta_{n+1}(t) \to 0, \quad t \to \infty.$$

Hence, by induction, (7.14) is true for all λ, uniformly on bounded intervals. Now return to the old scale. □

REMARK. In the ordinary Bellman-Harris case Kesten (unpublished) has shown that the condition that $t^2(1-G(t)) \to 0$, as $t \to \infty$, is also necessary for the existence of a non-degenerate normalized conditioned limit law.

REMARK. For Bellman-Harris processes there exists an analogue of Slack's theory for critical BGW processes with infinite variance, cf. Goldstein and Hoppe (1977, 1978).

8. MULTITYPE GENERALIZATIONS.

Instead of one type we now allow for a finite number p of types $i = 1,\ldots,p$. The construction of the process is a minor modification of the one-type case, the main point being that the reproduction ξ is now a (multivariate) point process on $(0,\infty)^r$, the jth coordinate ξ^j describing the birth times of children of type j. The set J of individuals is still ordered according to the birth times $\sigma_1,\sigma_2,\ldots,$ the type of n is denoted by $i(n)$ and the $(\xi_n,\tau_n,\Phi_n,\Psi_n,\ldots)$ are all independent with the distribution of $(\xi_n,\tau_n,\Phi_n,\Psi_n,\ldots)$ depending only on $i(n)$. The $Z^\Phi(t)$ are defined as in the one-type case.

The main extension needed of the considerations of the preceding section concerns the expected values and we shall only give a detailed treatment of this point. We shall need the extension of the key renewal theorem to systems of coupled renewal equations. For this version of the Markov renewal theorem we refer to Cinlar (1971), Ch. 10 (somewhat extended), Athreya and Ney (1978) or Asmussen (1978). Let P^i, E^i etc. refer to the case where the ancestor is of type i and $\mu^{ij}(t):=E^i\xi^j$. The basic assumptions are now that (μ^{ij}) is non-lattice in the sense of the Markov renewal theorem, that $(\mu^{ij}(\infty))$ is irreducible and that (μ^{ij}) is Malthusian, i.e. that for some α the matrix $({}_\alpha\mu^{ij}(\infty))$ has spectral radius one and that

$$\varkappa := \sum_{k,j=1}^{p} \varphi^*(k)\varphi(j)\int_0^\infty u\,{}_\alpha\mu^{kj}[du] < \infty.$$

Here φ and φ^* denote the right and left eigenvectors of $({}_\alpha\mu^{ij}(\infty))$ corresponding to the eigenvalue 1.

Defining $m_i^\Phi(t) := E^iZ^\Phi(t)$, a similar renewal argument as in the one-type case now produces

$$(8.1) \qquad m_i^\Phi(t) = E^i\Phi(t) + \sum_{j=1}^{p}\int_0^t m_j^\Phi(t-s)\mu^{ij}[ds].$$

This is a Markov renewal equation and the above assumptions permit at once to conclude from the Markov renewal theorem that if the $E^i{}_\alpha\Phi(t)$ are d.R.i., then

$$(8.2) \qquad {}_\alpha m_i^\Phi(\infty) := \lim_{t\to\infty} {}_\alpha m_i^\Phi(t) = \varkappa^{-1}\varphi(i)\sum_{j=1}^{p}\varphi^*(j)\int_0^\infty E^j{}_\alpha\Phi(t)dt.$$

8.1. EXAMPLE. Let $\Phi_n(t) := I(i(n) = k,\ \tau_n \ \ a > t)$, $\Psi(t) := I(\tau > t)$ so that $Z^{\Phi}(t)$ is the number of type k individuals which at time t are alive and of age strictly less than a and $Z^{\Psi}(t)$ the total population size. Letting $G^i(t) := P^i(\tau \leq t)$, we get

$$\frac{m_i^{\Phi}(t)}{m_i^{\Psi}(t)} = \frac{{}_{\alpha}m_i^{\Phi}(t)}{{}_{\alpha}m_i^{\Psi}(t)} \to \frac{{}_{\alpha}m_i^{\Phi}(\infty)}{{}_{\alpha}m_i^{\Psi}(\infty)} = \frac{\varphi^*(k)\int_0^{\infty} e^{-\alpha t}(1-G^k(t))\,dt}{\sum_{j=1}^{r} \varphi^*(j)\int_0^{\infty} e^{-\alpha t}(1-G^j(t))\,dt},$$

showing the existence of a limiting type-age distribution which is independent of the type i of the ancestor.

The limit results and at least the outlines of their proofs more or less suggest themselves now by combining the preceding sections with the theory of multi-type BGW-processes (contained in the general theory of Chapters V to VIII). We merely state a few of these results.

8.2. THEOREM. *Suppose* $\alpha > 0$ *and define*

$$W(t) := \sum_{n=T(t)+1}^{\infty} \varphi(i(n))e^{-\alpha\sigma_n} I(n \text{ is child of one of } 1,\ldots,T(t)).$$

Then $\{W(t)\}_{t \geq 0}$ *is a non-negative martingale and hence has an a.s. limit* W. *This* W *satisfies* $\{W > 0\} = \{T(\infty) = \infty\}$ *if and only if*

$$\text{(X LOG X)} \qquad E^i\, {}_{\alpha}\xi^j(\infty)\log^+ {}_{\alpha}\xi^j(\infty) < \infty \quad \forall i,j$$

while otherwise $W = 0$ *a.s. Finally if* Φ, Ψ *satisfy 5.1 w.r.t. all* P^i, *then a.s. on* $\{T(\infty) = \infty\}$ $Z^{\Phi}(t)/Z^{\Psi}(t)$ *tends to the common value*

$$\sum_{j=1}^{r} \varphi^*(j)\int_0^{\infty} E^j\, {}_{\alpha}\Phi(t)\,dt \Big/ \sum_{j=1}^{r} \varphi^*(j)\int_0^{\infty} E^j\, {}_{\alpha}\Psi(t)\,dt$$

of the ${}_{\alpha}m_i^{\Phi}(\infty)/{}_{\alpha}m_i^{\Psi}(\infty)$.

To continue, we need some more notation. Define

$J(i,t) :=$ the set of all individuals of type i born before or at time t

$$Z^{\Phi}(t) := (Z_1^{\Phi}(t),\ldots,Z_p^{\Phi}(t)), \qquad Z_i(t) := \sum_{n \in J(i,t)} \Phi_n(t-\sigma_n),$$

$$s := (s_1, \ldots, s_p), \quad |s_j| \leq 1, \quad j = 1, \ldots, p$$

$$F^{\Phi}(s,t) := (F_1^{\Phi}(s,t), \ldots, F_p^{\Phi}(s,t)), \; F_i^{\Phi}(s,t) := E^i \prod_{j=1}^{p} s_j^{z_j^{\Phi}(t)}$$

$$\Lambda^{\Phi} := (\Lambda_1^{\Phi}, \ldots, \Lambda_p^{\Phi}), \quad \Lambda_i^{\Phi} := \varkappa^{-1} \varphi^*(i) \int_0^{\infty} E^i{}_{\alpha}\Phi(t)\,dt,$$

$$\underset{\sim}{1} := (1, \ldots, 1), \quad \underset{\sim}{0} = (0, \ldots, 0).$$

8.3. THEOREM. *If* $\alpha < 0$ *and* $E^i{}_{\alpha}\Phi(t)$ *is d.R.i. for all* j, *then for every* $s \in [0,1]^p$

$$e^{-\alpha t}(\underset{\sim}{1} - F^{\Phi}(s,t)) \to \gamma^{\Phi}(s)\varphi, \; t \to \infty,$$

where γ^{Φ} *is a non-increasing, bounded, real-valued function. If in addition* (X LOG X) *is satisfied, then* $\gamma^{\Phi}(\underset{\sim}{0}) > 0$, *otherwise* $\gamma^{\Phi}(\underset{\sim}{0}) = 0$.

If all assumptions of 8.3, including (X LOG X) are satisfied, then as a corollary

$$\forall_j \; P^j(Z^{\Phi}(t) \leq \lambda | Z^{\Phi}(t) \neq 0) \to D^{\Phi}(\lambda), \quad t \to \infty,$$

where $D^{\Phi}(\lambda) = D^{\Phi}(\lambda_1, \ldots, \lambda_p)$ is a proper distribution function independent of j, and

$$\forall_j \int_{\mathbb{R}^p} \lambda_j dD^{\Phi}(\lambda) = \Lambda_j^{\Phi} / \gamma^{\Phi}(\underset{\sim}{0}).$$

Defining

$$\Psi = I(\tau > t)$$

and assuming in addition that $E_{\alpha}\Psi(t)$ is integrable and

$$(8.3) \qquad \forall_j \; P^j(\Phi_1(t) = 0 | \Psi_1(t) = 0) = 1,$$

we get

$$\forall_j \; P^j(Z^{\Phi}(t) \leq \lambda | Z^{\Psi}(t) \neq 0) \to D_0^{\Phi}(\lambda), \quad t \to \infty,$$

when again $D_0^{\Phi}(\lambda) = D_0^{\Phi}(\lambda_1, \ldots, \lambda_p)$ is a proper d.f. independent of j, and

$$\forall_j \int_{\mathbb{R}^p} \lambda_j dD_0^{\Phi}(\lambda) = \Lambda_i^{\Phi} / \gamma^{\Psi}(0).$$

Turning to the critical case, define

$$G_j(t) := P^j(\tau_1 \leq t),$$

$$e^{-\lambda} := (e^{-\lambda_1},\ldots,e^{-\lambda_p}), \quad \lambda = (\lambda_1,\ldots,\lambda_p),$$

and let $(\cdot,\cdot)$ denote the scalar product of two p-dimensional vectors. Set

$$\mu_{(2)} := \frac{1}{2}\sum_j \varphi^*(j)E^j\sum_i\sum_k(\xi^i(\infty)\xi^k(\infty) - \delta_{ik}\xi^i(\infty))\varphi(i)\varphi(k).$$

8.4. THEOREM. Suppose $\alpha = 0$,

$$\forall_i \; t^2(1-G_i(t)) \to 0, \quad t \to \infty,$$

$$\forall_{i,j} \; t^2(\mu^{ij}(\infty) - \mu^{ij}(\infty)) \to 0, \quad t \to \infty.$$

Then

$$\forall_j \; tP^j(Z^\Psi(t) \neq 0) \to \frac{\varkappa}{\mu(2)}\varphi(j).$$

8.5. THEOREM. Suppose $\alpha = 0$, $E^j\Phi(t)$ is d.R.i. $\forall_j$, and $E^i\Phi(t)\xi^j(t)$ (or $E^i\Phi(t)^2$) is bounded and tends to 0, as $t \to \infty$, $\forall_{i,j}$. Then

$$t(\underset{\sim}{1} - F^\Phi(e^{-\lambda/t},t)) \to \frac{(\Lambda^\Phi,\lambda)}{1+\varkappa^{-1}\mu_{(2)}(\Lambda^\Phi,\lambda)}\varphi .$$

If (8.3) and all assumptions of 8.4 and 8.5 are satisfied, then as a corollary

$$\forall_j \; P^j(Z^\Phi(t) \geq \lambda t \,|\, Z^\Psi(t) \neq 0) \to \exp\{-\frac{\varkappa}{\mu_{(2)}}\max_j \frac{\lambda_j}{\Lambda_j^\Phi}\}.$$

BIBLIOGRAPHICAL NOTES

A source for the model of Chapter X is the textbook by Jagers (1975). Demographic aspects are also treated by Keiding and Hoem (1976). The material of Section 2 is standard, and essentially the same is true for Section 3, 3.4 being from Keyfitz (1971).

The basis for the treatment of the supercritical case is the thesis of Nerman (1979), who suggested the martingales in Section 4 and gave an analysis similar to Section 5. The main differences from Nerman (1979) are the proof of 4.2 and the incorporation of a proof of the Kesten-Stigum type theorem. Nerman has shown us a simple extension of this proof, which yields the slightly more general result by Doney (1972).

The treatment of the subcritical case is a straightforward extension of Doney (1976a), while that of the critical case is essentially a combination of Holte (1974) and Green (1977), as to the latter slightly modified and completed. However, 6.7 and 7.2 have been taken from Jagers (1975), 7.3 from Goldstein (1971).

Details for the supercritical multi-type case can be found in Doney (1976b) and Nerman (1979). In the subcritical case nothing explicit beyond Ryan's (1968) treatment of the multi-type Bellman-Harris processes seems to be in the literature. Proofs for the results on critical multi-type processes can be found in Holte's paper (preprint).

Chapter XI

Two-Sex Models

1. MODELS AND EXAMPLES

The state of the theory of models for populations with sex interaction is much less satisfying than what regards one-sex branching processes. Presumably this is one of the main reasons for the use of branching process models also in sexual populations. However, a vague justification would be an intuitive feeling that in large populations fluctuations in sex ratio are quite small and hence the branching process approximation quite good.

Making this feeling precise would be one of the goals of developing a theory of two-sex models. Another would be studying the effects of an initial sex ratio different from the typical stable value of the population (a historically important example is here the excess of women after a war). And finally the question of how at all to model sex interaction seems worth pursuing since few (or no one at all) models suggest themselves as do branching processes in one-sex situations.

We shall concentrate on a study of sex interaction in its purest form and disregard the structure of the population according to age, parity group, social groups, location, etc. though unquestionably this leads to models too crude to be useful say in practical population projection problems in demography. Different sexual patterns are, however, possible (monogamy, polygamy, promiscuity, etc.) and also it is of interest to be able to deal both with overlapping and non-overlapping generations.

The non-overlapping generations (as one would typically have in some species of plants and insects) are most naturally modelled in discrete time. Let the M_n males and F_n females present at time n form C_n couples according to some distribution H_{M_n,F_n} depending only on M_n, F_n and suppose that couples reproduce independently and all according to the same law G on $\mathbb{N}^2$ so that

$$(1.1)\qquad (M_{n+1},F_{n+1}) = \sum_{i=1}^{C_n} (M_{n,i},F_{n,i})$$

where the $(M_{n,i},F_{n,i})$ are independent with common distribution G. Thus the population development proceeds in two stages:

$$(M_0,F_0) \to C_0 \to (M_1,F_1) \to C_1 \to (M_2,F_2) \to \cdots$$

The parameters of the model are G and the $H_{M,F}$.

In continuous time, consider first as an example a reasonably general model for human reproduction. At time t, M_t single males, F_t single females and C_t married couples are present. All births takes place from married couples, with rate μ_m for male offspring and μ_f for female offspring, males die at rate ν_m and females at rate ν_f, divorces take place at rate λ and finally couples are formed at rate $R(M_t,F_t,C_t)$ with R left unspecified at the moment. In deterministic formulation, this leads to a system

$$
(1.2)\qquad
\begin{aligned}
\dot{M} &= (\mu_m+\lambda+\nu_f)C - \nu_m M - R(M,F,C)\\
\dot{F} &= (\mu_f+\lambda+\nu_m)C - \nu_f F - R(M,F,C)\\
\dot{C} &= R(M,F,C) - (\lambda+\nu_m+\nu_f)C
\end{aligned}
$$

of deterministic differential equations. Here $M_t,F_t,C_t \in [0,\infty)$ and $\dot{M} := \frac{d}{dt} M_t$ etc. An obvious stochastic analogue is a Markov jump process (with $M_t,F_t,C_t \in \mathbb{N}$) with the intensities for the possible transitions from (M,F,C) given by the following list:

(1.3)

$(M+1,F,C)$	$\mu_m C$	(a male born to a couple)
$(M,F+1,C)$	$\mu_f C$	(a female born to a couple)
$(M-1,F,C)$	$\nu_m M$	(death of a single male)
$(M,F+1,C-1)$	$\nu_m C$	(death of a married male)
$(M,F-1,C)$	$\nu_f F$	(death of a single female)
$(M+1,F,C-1)$	$\nu_f C$	(death of a married female)
$(M+1,F+1,C-1)$	λC	(divorce)
$(M-1,C-1,C+1)$	$R(M,F,C)$	(marriage)

A simpler-minded model (with promiscuity and without death) is in deterministic formulation

$$
(1.4)\qquad
\begin{aligned}
\dot{M} &= mR(M,F)\\
\dot{F} &= fR(M,F)
\end{aligned}
$$

and in stochastic formulation a pure birth process with birth rate $mR(M,F)$ for males and $fR(M,F)$ for females. For simplicity, we discuss this model rather than (1.2), (1.3). The state of the population is therefore as in discrete time given only by M, F or, equivalently, by the total population size $N := M+F$ and the sex ratio, which we represent by $X := M/(M+F)$.

The above models are too unspecified to allow for any general type of behaviour. The crucial problem is to derive a reasonable set of assumptions on the mating mechanism, i.e. on R in (1.2), (1.3), (1.4) and on the way the $H_{M,F}$ depend on M,F in the discrete time model.

The function MF, representing the number of all possible contacts between individuals of opposite sex, plays historically an important role. Models based on MF (sometimes called the _random mating_ function) exhibit, however, the controversial feature of exploding within finite time. Consider as a simple example (1.4) with $R(M,F) = MF$, $m = f = 2$, $M_0 = F_0$. Then $M_t = F_t$ for all t and, recalling $N = M+F$, (1.4) yields $\dot{N} = N^2$ which has the solution $N_t = 1/(1/N_0-t)$ tending to infinity as $t \uparrow 1/N_0$.

The main objection to the random mating type of model is that the number of theoretical possible contacts between individuals of opposite sex is an irrelevant quantity when considering large populations within a limited period of time. Rather, any fixed individual interacts only with its own local environment and not the entire population, so that the number of actual contacts with the opposite sex is limited by the size of the environment and could not increase arbitrarily with the number of individuals of the opposite sex. If this idea is combined with the assumption of the sex ratio being an overall population quantity not varying too much from environment to environment, we arrive, however, at a base sufficient for setting up other types of models. The ideas can be illustrated by the following drawing. Position represents environmental variable and the environment of an individual is the circle disc with radius δ around it. If the population is split into two according to a cut , then only individuals at most δ from the cut are affected, and if the population is large, they represent a vanishing part of the population. The overall sex ratio being the same in both parts, we see that _a population of_ $N = N_1 + N_2$ _individuals_ (with both N_1 and N_2 large) _should behave as a_ (independent in stochastic models) _sum of two populations of size_ N_1, N_2 _and the same sex ratio_.

```
                         F M F M F M M F F M F F F F
                       M M F F M F M F M M M F M F
                      F M M M M M M F F M M F F
                   F F M F M F M F M M M M F M
                 M F F M F M F M F M F F F
               F F F M F F M M F M M M M
             M M M M M M M M M F M F F
           M F M M F F M F F F M M M
         F F F M F F F M M F M F F
       F M F M F F M F M M M M M
       F F F M M M M M M F F M
     M F M M M F M M F M F M
   M F F F F F F M M F F F
   M F F M F M M M F M F
```

In (1.4), this leads immediately to the requirement

(1.5) $\quad R(M,F) \cong Nh(x) \quad$ as $\quad N \to \infty,\ X \to x$

for some function h on $(0,1)$. In the discrete time model, we arrive at

(1.6) $\quad H_{kM,\,kF} \cong H^{*k}_{M,\,F} \quad$ as $\quad N \to \infty,\ X \to x \quad \forall k$

It still remains, of course, to indicate the precise meaning of $\cong$ in (1.5), (1.6) (in particular (1.6) is rather vague and unprecise). This will be put down as technical conditions in the respective theorems.

(1.5) can be rewritten in ways seen from the standpoint of one sex only (say the female sex) and which lead to somewhat different interpretations. With $k(x) := h(x)/(1-x)$, (1.5) is equivalent to

(1.7) $\quad R(M,F) \cong Fk(x) \quad$ as $\quad M,F \to \infty,\ X \to x.$

If exactly $R(M,F) = Fk(X)$, then the female line in the corresponding pure birth process becomes a (one-sex) pure birth process in random environment (interacting with the process), environment defined as sex ratio (the $\cong$ in (1.7) is a small population adjustment). One could even arrive at explicit forms for k and h. Suppose that with probability $c_1 t + o(t)$ each female gets in contact with another individual (which then w.p. X is a male) within a time interval at length t, that a given male and female accept each others as partners and produce offspring w.p. c_2, and that the offspring is a male w.p. $\frac{m}{m+f}$ and a female w.p. $\frac{f}{m+f}$. Then $k(x) = (m+f)c_1c_2x$, corresponding to $R(M,F)$ being proportional to the harmonic mean $MF/(M+F)$. The less restrictive requirement $R(M,F) = Fk(X)$ could be obtained by arguing that c_2 (and maybe even c_1) should depend on X.

An explicit form suggested in the literature for $H_{M,F}$ (in monogamy) is $H_{M,F}$ being degenerate at $M \wedge F = N\{X \wedge (1-X)\}$ (i.e. $C_n = M_n \wedge F_n$). It is clearly compatible with (1.6), but hardly with the idea of a local environment. A similar model in polygamy is $C_n = (dM_n) \wedge F_n$. It corresponds to allowing each male to have at most d partners, each female only one, and defining a couple as a female having a partner. Another example is discussed in Section 2.

2. LIMIT THEOREMS WITH NON-OVERLAPPING GENERATIONS

The model is the two-stage chain

$$\dots (M_n, F_n) \to C_n \to (M_{n+1}, F_{n+1}) \to C_{n+1} \to \dots$$

of Section 1, specified by G and the $H_{M,F}(h) = P(C_0 \leq h | M_0 = M, F_0 = F)$. The aim is the investigation of the relation of the limiting behaviour of this process compared to the one-sex Galton-Watson process, subject to conditions motivated by (1.6). Obvious questions are whether the process either becomes extinct or grows exponentially, and the form of the extinction criteria.

Before stating the theorems, we introduce some notation. We recall that $N := M+F$, $X := M/N$ and define $E := \{C_n = 0 \text{ eventually}\}$,

$$a := E(M_1 | C_0 = 1) = \int_0^\infty \int_0^\infty x \, dG(x,y), \quad b := E(F_1 | C_0 = 1) = \int_0^\infty \int_0^\infty y \, dG(x,y),$$

$$\nu := a+b, \quad z := a/(a+b),$$

$$\nu_2 := E(N_1^2 | C_0 = 1) = \int_0^\infty \int_0^\infty (x+y)^2 dG(x,y)$$

$$\mu(M,F) := \frac{E(C_0 | M_0 = M, F_0 = F)}{N} = \frac{\int_0^\infty h \, dH_{M,F}(h)}{N}.$$

2.1. THEOREM. _Suppose that_

$$(2.1) \qquad P(\sup_{n=0,1,2,\dots} C_n > K | C_0 = c) > 0 \quad \text{for all} \quad K < \infty,\ c > 0,$$

$$(2.2) \qquad \mu := \lim_{N\to\infty, X\to z} \mu(M,F) \quad \text{exists}$$

$$(2.3) \qquad \overline{\lim_{N\to\infty, X\to z}} \frac{\mathrm{Var}(C_0 | N_0 = N, X_0 = X)}{N^\eta} < \infty \quad \text{for some} \quad \eta \in (0,2),$$

and that $\nu_2 < \infty$, $m := \mu\nu > 1$. _Then a.s. on_ E^c, $X_n \to z$ _and_ $C_{n+1}/C_n \to m$, $N_{n+1}/N_n \to m$, $N_{n+1}/C_n \to \nu$. _In particular_, $PE + P(N_n \to \infty) = 1$ _and for all_ $\epsilon > 0$, $N_n/(m+\epsilon)^n \to 0$, $N_n/(m-\epsilon)^n \to \infty$ _a.s. on_ E^c [and correspondingly for C_n, M_n, F_n].

2.2. THEOREM. _If, in addition to the conditions of 2.1,_ (2.2) _is_

<u>sharpened to</u>

(2.4) $$\mu(M,F) = \mu+O(|\log(X-z)|^{-1-\varkappa}+(\log N)^{-1-\varkappa})$$

<u>for some</u> $\varkappa > 0$ <u>as</u> $N \to \infty$, $X \to z$, <u>then also</u> $W := \lim_{n\to\infty} C_n/m^n$ <u>exists</u> <u>a.s. and</u> $0 < W < \infty$ <u>on</u> E^c, $N_n/m^n \to W/\mu$.

We note that, provided the moments exist at all, (1.6) would lead to the conditions

(2.5) $$\mu(x) := \lim_{N\to\infty, X\to x} \mu(M,F) \quad \text{exists} \quad \forall\, x$$

(2.6) $$\tau^2(x) := \lim_{N\to\infty, X\to x} \frac{\mathrm{Var}(C_0|N_0 = N, X_0 = X)}{N} \quad \text{exists} \quad \forall\, x$$

more stringent than (2.2), (2.3). The condition (2.4) might be interpreted as a requirement on the rate of convergence 1) of $\mu(x)$ to $\mu(z)$ as $x \to z$ and 2) in (2.5). The proofs will reveal that conditions of this type are indispensable for the conclusion of <u>2.2</u>. Condition (2.1) requires some sort of irreducibility of the Markov chain $C_0, C_1, \ldots$ and is not inherent in the model even with $m > 1$. For a simple example where (2.1) is violated but all other conditions hold, take G degenerate at (2,2) and $P(1 \leq C_0 \leq \frac{1}{2}(M_0 \; F_0) | 1 \leq M_0 \; F_0) \leq a) = 1$, $P(C_0 = M_0 \; F_0 | M_0 \; F_0 > a) = 1$. Then $\{0\}$, $\{1,2,\ldots,a\}$, $\{a+1, a+2, \ldots\}$ are absorbing classes for $\{C_n\}$.

The conditions (2.1)-(2.4) are obvious for the example $C = (dM) \; F$ $(d = 1,2,\ldots)$ considered at the end of Section 1. Before giving the proofs of <u>2.1</u>, <u>2.2</u>, we consider one more example. Suppose that the couples are formed in two steps. First each male chooses a female at random and next each female chooses one of the males having selected her. Thus the number of couples is the number of females being chosen by some male. The probability of a given female not being selected is $(\frac{F-1}{F})^M$. Hence

$$E(C_0|M_0 = M, F_0 = F) = F(1 - (\tfrac{F-1}{F})^M).$$

As $N \to \infty$, $X \to x$, this behaves as

$$N(1-x)(1-(1-\frac{1}{N(1-x)})^{Nx}) \cong N(1-x)(1-e^{-x/(1-x)}),$$

yielding (2.2) and (2.5). Similarly, the probability of both of two given distinct females being selected is

$$1-2\left(\frac{F-1}{F}\right)^M\left(1-\left(\frac{F-2}{F-1}\right)^M\right)-\left(\frac{F-2}{F}\right)^M$$

so that

$$E(C_0^2 \mid M_0 = M, F_0 = F) = F\left(1-\left(\frac{F-1}{F}\right)^M\right)+F(F-1)\left(1-2\left(\frac{F-1}{F}\right)^M\left(1-\left(\frac{F-2}{F-1}\right)^M-\left(\frac{F-2}{F}\right)^M\right)\right).$$

(2.3), (2.4) follow by elementary (though somewhat lengthy) calculus. Also this model is hardly compatible with the idea of a local environment, where one, however, could have a similar mechanism locally.

The main step in the proof of 2.1 is

2.3. LEMMA. *Let* ϵ *be given with* $0 < \epsilon < m-1$, *and define* $B_n := \{|\frac{C_{n+1}}{C_n} - m| \leq \epsilon\}$, $A := \bigcap_{n=0}^{\infty} B_n$. *Then* $P(A|C_0 = k) \to 1$ *as* $k \to \infty$.

PROOF. Define $\mathfrak{F}_N := \sigma(M_n, F_n, C_n : n \leq N)$, $D_n := B_0 \cdots B_{n-1}$. Then $PA^c = \Sigma_0^{\infty} P(D_n B_n^c)$, $D_n \in \mathfrak{F}_n$,

$$(2.7) \qquad P(B_n^c|\mathfrak{F}_n) \leq P(|X_{n+1}-z| > \epsilon_1|\mathfrak{F}_n) + P(|\frac{N_{n+1}}{C_n} - \nu| > \epsilon_2|\mathfrak{F}_n)$$

$$+P(|\frac{C_{n+1}}{C_n} - m| > \epsilon, |X_{n+1}-z| \leq \epsilon_1, |\frac{N_{n+1}}{C_n} - \nu| \leq \epsilon_2|\mathfrak{F}_n)$$

The estimation of the two first terms on the r.h.s. is straightforward, using

$$(2.8) \qquad P(|\frac{M_{n+1}}{C_n} - a| > \delta|\mathfrak{F}_n) \leq \frac{\nu_2}{\delta^2 C_n}, \quad P(|\frac{F_{n+1}}{C_n} - b| > \delta|\mathfrak{F}_n) \leq \frac{\nu_2}{\delta^2 C_n}$$

$$(2.9) \qquad P(|\frac{N_{n+1}}{C_n} - \nu| > \epsilon_2^2|\mathfrak{F}_n) \leq \frac{\nu_2}{\epsilon_2^2 C_n}, \quad P(|X_{n+1}-z| > \epsilon_1|\mathfrak{F}_n) \leq \frac{\gamma}{\epsilon_1^2 C_n}.$$

Here the three first inequalities are just Chebycheff, while the last, valid for some γ and all ϵ_1 small enough, follows from (2.8) by Taylor expanding $x/(x+y)$. To estimate the last term in (2.7), note first that given $\epsilon_3 > 0$, there is a K, an ϵ_1 and a κ such that if $N \geq K$, $|X-z| \leq \epsilon_1$, then (using (2.1), (2.2)) $|\mu(M,F)-\mu| \leq \epsilon_3/2$,

$$(2.10)\quad P(|\frac{C_0}{N_0}-\mu|>\epsilon_3|M_0=M,F_0=F)\le P(|\frac{C_0-EC_0}{N_0}|>\frac{\epsilon_3}{2}|M_0=M,F_0=F)$$

$$\le\frac{\mathrm{Var}(C_0|M_0=M,F_0=F)}{(\epsilon_3/2)^2N^2}\le\frac{\varkappa}{N^{2-\eta}}\ .$$

Apparently we may assume (2.10) to hold for all $N=1,2,\ldots$ and $|X-z|\le\epsilon_1$. Choose now ϵ_2, ϵ_3 so small that $(\nu+\epsilon_2)(\mu+\epsilon_3)<m+\epsilon$, $(\nu-\epsilon_2)(\mu-\epsilon_3)>m-\epsilon$. Then the last term in (2.7) cannot exceed

$$(2.11)\quad P(|\frac{C_{n+1}}{N_{n+1}}-\mu|>\epsilon_3,|X_{n+1}-z|\le\epsilon_1,|\frac{N_{n+1}}{C_n}-\nu|\le\epsilon_2|\mathfrak{F}_n)$$

$$\le E(I(|X_{n+1}-z|\le\epsilon_1,|\frac{N_{n+1}}{C_n}-\nu|\le\epsilon_2)\frac{\varkappa}{N_{n+1}^{2-\eta}}|\mathfrak{F}_n)$$

$$\le\frac{\varkappa}{((\nu-\epsilon_2)C_n)^{2-\eta}}\ .$$

Combining (2.7), (2.9), (2.11) gives that for some constant λ and $\xi:=1\wedge(2-\eta)>0$,

$$P(B_n^c|\mathfrak{F}_n)\le\frac{\lambda}{C_n^{\xi}}\le\frac{\lambda}{C_0^{\xi}(m-\epsilon)^{n\xi}}\quad\text{on }D_n\ ,$$

$$P(A|C_0=k)=\sum_{n=0}^{\infty}E(I(D_n)P(B_n^c|\mathfrak{F}_n)|C_0=k)$$

$$\le\frac{\lambda}{k^{\xi}}\sum_{n=0}^{\infty}\frac{1}{(m-\epsilon)^{n\xi}}\ .\quad\text{Let } k\uparrow\infty.\quad\square$$

PROOF OF 2.1. Let $T(K):=\inf\{n:C_n\ge K\}$ so that by (2.1), $P(T(K)<\infty|C_0=c)>0$, $c=1,\ldots,K-1$. It follows by standard Markov chain arguments that $T(K)<\infty$ a.s. on $E^c=\{C_n>0\ \text{i.o.}\}$. Hence

$$P(C_n>0,|\frac{C_{n+1}}{C_n}-m|\le\epsilon\ \text{ eventually})$$

$$\ge P(C_n>0,|\frac{C_{n+1}}{C_n}-m|\le\epsilon,\ n\ge T(K))$$

$$\ge\inf_{k\ge K}P(A|C_0=k)P(T(K)<\infty)\ge\inf_{k\ge K}P(A|C_0=k)PE^c\ .$$

As $K\to\infty$, the r.h.s. tends to PE^c and as $\epsilon\downarrow 0$, it follows that $C_{n+1}/C_n\to m$ on E^c. Hence $\Sigma_0^{\infty}C_n^{-1}<\infty$ so that the sum of the expressions in (2.9) is finite, proving $N_{n+1}/C_n \to \nu$, $X_{n+1} \to z$ on E^c. □

PROOF OF 2.2. Let $0 < \delta < 1/2$ and $\epsilon_1 := \epsilon_2 := C_n^{-\delta}$ in (2.9). Then on E^c,

$$(2.11) \qquad \sum_{n=0}^{\infty} P(|\frac{N_{n+1}}{C_n} - \nu| > C_n^{-\delta} | \mathfrak{F}_n) \leq \sum_{n=0}^{\infty} \frac{\nu_2}{C_n^{1-2\delta}} < \infty, \quad \frac{N_{n+1}}{C_n} = \nu + O(C_n^{-\delta}) .$$

Similarly, $|X_{n+1} - z| = O(C_n^{-\delta})$ so that by (2.4) $\mu(M_n, F_n) = \mu + O(n^{-(1+\varkappa)})$, and

$$\sum_{n=0}^{\infty} P(|\frac{C_{n+1}}{N_{n+1}} - \mu(M_{n+1}, F_{n+1})| > N_{n+1}^{-\delta(2-\eta)} | M_{n+1}, F_{n+1}, \mathfrak{F}_n)$$

$$\leq \sum_{n=0}^{\infty} \frac{Var(C_{n+1} | \mathfrak{F}_n)}{N_{n+1}^{2-2\delta(2-\eta)}} = \sum_{n=0}^{\infty} O(\frac{1}{N_{n+1}^{(2-\eta)(1-2\delta)}}) < \infty,$$

$C_{n+1}/N_{n+1} = \mu(M_{n+1}, F_{n+1}) + O(N_n^{-\delta(2-\eta)})$. Combining these estimates yields $C_{n+1}/C_n = m(1+A_n)$ with $A_n = O(n^{-(1+\varkappa)})$. Hence $0 < W := C_0 \Pi_0^{\infty}(1+A_n) < \infty$ on E^c and

$$\frac{C_N}{m^n} = C_0 \prod_{n=0}^{N-1} \frac{C_{n+1}}{mC_n} \to W. \qquad □$$

2.3. REMARK. Using standard truncation techniques (see e.g. Section 5 of Chapter II) one can weaken the second moment assumption $\nu_2 < \infty$ on G to the first moment condition $\nu < \infty$ in 2.1 and to a (X LOG X) condition in 2.2. Also (2.3) could be somewhat relaxed.

The above results leave the case $m \leq 1$ open. The state of the theory is here somewhat less satisfying than in the case $m > 1$. We shall, however, give some results on the extinction problem.

2.4. PROPOSITION. <u>A sufficient condition for a.s. extinction,</u> $P(E | M_0 = M, F_0 = F) = 1$ <u>for all</u> M, F, <u>is that</u>

$$(2.12) \qquad P(E | C_0 = k) > 0 \quad \forall k,$$

(2.13) There is a K such that $E(C_1 | C_0 = k) \leq k$, $k \geq K$.

Under the condition (2.2) one would expect (2.13) to be close to $m := \mu\nu \leq 1$. Indeed, if $m < 1$ we have

2.5. PROPOSITION <u>Suppose that conditions</u> (2.2), (2.3) <u>hold, that</u> $m < 1$

and that $\mu(M,F) \le \varkappa < \infty$. Then (2.13) holds. More precisely, $\overline{\lim}_{k\to\infty} E(\frac{C_1}{k} | C_0 = k) \le m$.

The situation is less clear cut if $m = 1$. We shall be satisfied by remarking that here (2.13) is obvious in the $C_n = M_n \wedge F_n$ example which is hence quite completely described by 2.1, 2.2, 2.4, 2.5. Indeed, here $m = 1$ amounts to $a\ b = 1$ and we get

$$E(C_{n+1}|C_n) = E(M_{n+1} \wedge F_{n+1}|C_n) \le E(M_{n+1}|C_n) \wedge E(F_{n+1}|C_n) = aC_n \wedge bC_n = C_n$$

(more generally, $\{m^{-n}C_n\}$ is a non-negative supermartingale no matter the value of m in this example).

PROOF OF 2.4. For any N, consider $A_n^N := \prod_{k=N}^{n} C_{k+1}/C_k I(C_k \ge K)$. Then it follows immediately from (2.13) that $\{A_n^N\}_{n=N,N+1,\ldots}$ is a non-negative supermartingale and hence has an a.s. limit $A^N < \infty$. On $B_N := \{C_n \to \infty, C_k \ge K, k = N, N+1, \ldots\}$ we have $A^N = \infty$. Hence $PB_N = 0$ for all N, implying that $P(C_n \to \infty) = 0$. It follows however by standard Markov chain arguments that (2.12) forces the number of visits to any state $k > 0$ to be finite. Hence $C_n \to \infty$ on E^c so that $PE^c = 0$. □

PROOF OF 2.5. Let $m < \gamma < 1$. Then by (2.2), there is a ϵ and a $\underline{N}$ such that $\mu(M,F) \le \gamma/\nu$ if $N \ge \underline{N}$, $|X-z| \le \epsilon$. Define $A := \{N_1 \ge \underline{N}, |X_1-z| \le \epsilon\}$. Then conditioning on N_1, X_1 yields

$$\frac{E(C_1|C_0=k)}{k} = \frac{E(C_1 I(A)|C_0=k)}{k} + \frac{E(C_1 I(A^c)|C_0=k)}{k}$$

$$\le \frac{\gamma}{\nu} E(\frac{N_1}{k} I(A)|C_0=k) + \varkappa E(\frac{N_1}{k} I(A^c)|C_0=k)$$

$$\le \gamma + \varkappa E(\frac{N_1}{k} I(A^c)|C_0=k) .$$

From the structure of N_1 as a sum of C_0 i.i.d. r.v. with finite mean, it follows by A 3.4 that given η, there is a δ independent of k such that $P(B|C_0=k) \le \delta$ implies $E(\frac{N_1}{k} I(B)|C_0=k) \le \eta$. But for k large, $P(A^c|C_0=k) \le \delta$ by the LLN so that the $\overline{\lim}$ is at most $\gamma+\varkappa\eta$. Let $\eta \downarrow 0$, $\gamma \uparrow m$. □

3. LIMIT THEOREMS WITH OVERLAPPING GENERATIONS: THE DETERMINISTIC DIFFERENTIAL EQUATIONS

The model is (1.4). For simplicity, we ignore the small population effects in the requirement (1.5) motivated in Section 1 so that (1.5) is strengthened to

$$R(M,F) = Nh(X) \quad (= Fk(X),\ k(x) := \frac{h(x)}{1-x}) . \tag{3.1}$$

Explicit forms suggested for $R(M,F)$ compatible with (3.1) are: M (_male marriage dominance_, $h(x) = x$); F (_female marriage dominance_, $h(x) = 1-x$); $(M+F)/2$ (_arithmetic mean_, $h(x) = 1/2$); $\sqrt{MF}$ (_geometric mean_, $h(x) = \sqrt{x(1-x)}$); $MF/(M+F)$ (_harmonic mean_, cf. Section 1, $h(x) = x(1-x)$; and $M\wedge F$ (_minimum_, $h(x) = x\wedge(1-x)$). The three first $(M, F, (M+F)/2)$ of these examples incorporate, however, no genuine sex interaction since, e.g., both sexes need not be present in order that births take place.

The precise conditions on h are stated in the respective theorems. They are little restrictive, essentially smoothness conditions like Hölder continuity on suitable intervals $I \subseteq (0,1)$,

$$|h(x_1) - h(x_2)| \leq c|x_1 - x_2|^p \quad x_1,x_2 \in I \tag{3.2}$$

(with $0 < p \leq 1$). In order to avoid trivialities, we also need

$$h(x) > 0,\ 0 < x < 1. \tag{3.3}$$

Further axiomatic discussions of rules for marriage would limit the class of functions h somewhat, however. For example, it would not seem unreasonable to require that

$$R(M_1,F) \geq R(M_2,F),\ M_1 \geq M_2, \tag{3.4}$$

and/or, appealing to the discussion following (1.7),

$$\lim_{M\to\infty} R(M,F) = cF \quad \text{with} \quad 0 < c < \infty. \tag{3.5}$$

Note that in the $Fk(X)$-formulation, these axioms correspond to $k(x)\uparrow c$, $x\uparrow 1$, implying

$$h(x_2) \geq h(x_1)\frac{1-x_2}{1-x_1} \quad \text{when} \quad 0 < x_1 < x_2,\ h(1-y) \leq cy = -h'(1)y, \tag{3.6}$$

$$(3.7) \qquad h(x_2) \leq h(x_1)\frac{x_2}{x_1} \quad \text{when} \quad 0 < x_1 < x_2, \; h(x) \leq dx = h'(0)x,$$

(with $0 < d < \infty$). Here formula (3.7) is derived by interchanging the role of males and females. Beyond the highly unrealistic models corresponding to arithmetic mean or one of the sexes being marriage dominant, this would exclude also the geometric mean model. However, these models are formally included in what follows since none of the conditions (3.4)-(3.7) come up.

We first study the deterministic differential equations (1.4). The existence and uniqueness of a set (M_t, F_t) of solutions, given the initial values (M_0, F_0), is well-known assuming Lipschitz type conditions ((3.2) with $p = 1$). In order to obtain the asymptotic behaviour of the solutions, we rewrite (1.4) as

$$(3.8) \qquad \dot{N} = Nh(X)(m+f)$$

$$(3.9) \qquad \dot{X} = (z-X)h(X)(m+f)$$

It follows from (1.4) that the derivative of $fM_t - mF_t$ vanishes so that $fM_t - mF_t = fM_0 - mF_0$, which is equivalent to

$$(3.10) \qquad N_t(z-X_t) = N_0(z-X_0).$$

Therefore, if X_t is known, N_t can be computed from (3.8) or (3.10). In this manner, the investigation of (1.4) reduces to the study of (3.9).

Assume without loss of generality that $0 < X_0 < z$. Then, by (3.9), $X_0 \leq X_t \leq z$ for all t.

Choosing $\beta_1 \geq \beta_2 > 0$ such that $\beta_1 \geq h(x)(m+f) \geq \beta_2$ when $X_0 \leq x \leq z$, we get

$$\beta_1(z-X) \geq \dot{X} \geq \beta_2(z-X), \quad (z-X_0)e^{-\beta_1 t} \leq z-X_t \leq (z-X_0)e^{-\beta_2 t}.$$

Thus, $X_t \uparrow z$ and $X_t < z$ for all $t < \infty$. Define

$$k(y) := h(z)\frac{1}{z-y}\left(\frac{1}{h(z)} - \frac{1}{h(y)}\right), \quad h^*(x) := e^{\int_x^z k(y)dy}.$$

Note that (3.2) with $p = 1$ ensures the integrability of k at z. We can rewrite (3.9) as

$$\lambda = \dot{X}\,\frac{h(z)}{h(X)}\,\frac{1}{z-X} = \frac{\dot{X}}{z-X} - \dot{X}k(X) \quad \text{with} \quad \lambda := h(z)(m+f)$$

and integration from 0 to t yields

$$\lambda t = -\log(z-X_t) + \log(z-X_0) + \log h^*(X_t) - \log h^*(X_0),$$

$$(3.11) \qquad z-X_t = (z-X_0)\frac{h^*(X_t)}{h^*(X_0)}e^{-\lambda t},$$

$$(3.12) \qquad N_t = N_0 \frac{h^*(X_0)}{h^*(X_t)} e^{\lambda t}.$$

Here (3.12) is obtained by combining (3.11) with (3.10). Note that since $X_t \to z$, we also have $h^*(X_t) \to h^*(z) = 1$.

Noting that (3.11) and (3.12) follow by symmetry if $z < X_0 < 1$ and are trivial if $X_0 = z$, we have proved the first part of the following result.

3.1. THEOREM. *Assume that* $0 < X_0 < 1$ *and that conditions* (3.1), (3.3) *and* (3.2) *with* $p = 1$ *and* I *the closed interval with endpoints* X_0 *and* z *hold. Let* (N_t, X_t) *be solutions of* (3.8), (3.9) *corresponding to a set* (M_t, F_t) *of solutions to* (1.4). *Then* $X_t \to x$ *monotonically and* $e^{-\lambda t}N_t \to N_0 h^*(X_0)$. *More precisely,*

$$(3.13) \qquad X_t = z + \frac{X_0 - z}{h^*(X_0)}e^{-\lambda t} + O(e^{-2\lambda t}), \quad N_t = N_0 h^*(X_0)e^{\lambda t} + O(1).$$

Furthermore, if h *has a derivative* $h'(z)$ *at* z, *then*

$$(3.14) \qquad X_t = z + \frac{X_0 - z}{h^*(X_0)}e^{-\lambda t} + \left(\frac{X_0 - z}{h^*(X_0)}\right)^2 \frac{h'(z)}{h(z)}e^{-2\lambda t} + o(e^{-2\lambda t}),$$

$$(3.15) \qquad N_t = N_0 h^*(X_0)e^{\lambda t} - N_0(X_0 - z)\frac{h'(z)}{h(z)} + o(1).$$

To complete the proof, note first that (3.13) follows immediately from (3.11), (3.12) once we observe that as $x, y \to z$, then $k(y) = O(1)$, $h^*(x) = 1 + O(x-z)$. If $h'(z)$ exists, these estimates can be strengthened to $k(y) = -h'(z)/h(z) + o(1)$, $h^*(x) = 1 + (x-z)h'(z)/h(z) + o(x-z)$ and (3.14), (3.15) follow.

REMARK. Of course, further assumptions on well-behaviour of h at z will yield further refinements of (3.13), (3.14), (3.15). In connection with the minimum model with $z = \frac{1}{2}$, we note also that existence of one-sided derivatives at z suffices for (3.14), (3.15), if one replaces $h'(z)$ by the left derivative for $X_0 < z$ and the right derivative for $X_0 > z$.

In demographic terms, λ is the Malthusian parameter, while (motivated from the second part of (3.13)) $Nh^*(X)$ is the reproductive value of a population of M males and F females. The following table gives h^* in the explicit examples above, with $z = 1/2$,

$R(M,F)$	M	F	$\frac{M+F}{2}$	$\sqrt{MF}$	$\frac{MF}{M+F}$	$M \wedge F$
$h(x)$	x	$1-x$	$\frac{1}{2}$	$\sqrt{x(1-x)}$	$x(1-x)$	$x\wedge(1-x)$
$h^*(x)$	$2x$	$2(1-x)$	1	$\frac{1}{2}+\sqrt{x(1-x)}$	$2\sqrt{x(1-x)}$	$2[x\wedge(1-x)]$

while the next figure shows (in the harmonic mean example) the effects of different values of z.

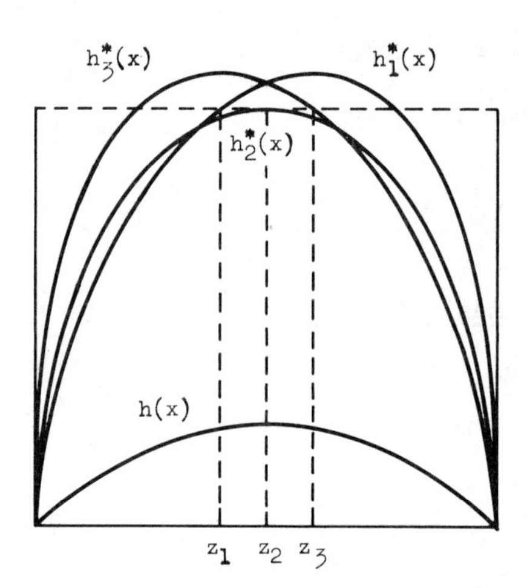

Even if no explicit form of h is assumed, some information may still be obtained concerning the properties of h^*. Of particular interest is the behaviour of h^* at one of the boundaries, say at 0. As was argued earlier, the typical case is (3.7). If, furthermore, $h(x) = dx + o(x^2)$ as $x \downarrow 0$, then $h^*(x) \cong \beta x^\alpha$, where $\alpha := h(z)/dz$. Note that by (3.7), $\alpha \leq 1$, with $\alpha = 1$ if and only if h is linear on $[0,z]$.

As is seen in the geometric mean example, $h^*(x)$ may have a non-zero limit as $x \downarrow 0$ if (3.7) is violated. This type of behaviour does not correspond nicely to intuition and it will occur if and only if $\int_0^\epsilon 1/h(y)dy < \infty$. Also the (typically unique) point y at which h^* attains its maximum, has a simple description as a solution of $h(y) = h(z)$. The situation is illustrated in the following figure.

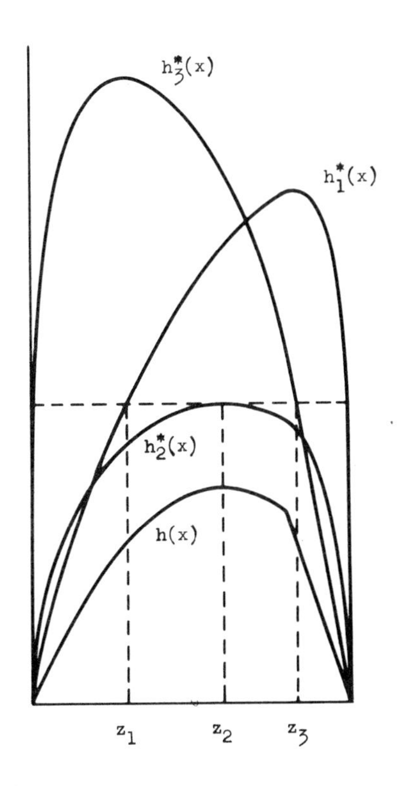

For the same h, we have taken three values z_1, z_2, z_3 of z and plotted the corresponding h^*-functions h_1^*, h_2^*, h_3^*. Note that the z_i have been chosen such that $h(z_1) = h(z_2)$ and that h attains its maximum at z_3.

We finally remark that the above analysis suggests how in particular cases to obtain exlicit solutions, viz. to compute h^*, solve (3.11) for X_t and insert in (3.10). For example, in the harmonic mean model, (3.11) yields a quadratic equation for X_t. More elegant methods may, of course, exist in this and other specific examples.

4. LIMIT THEOREMS WITH OVERLAPPING GENERATIONS: THE PURE BIRTH PROCESS

The analysis is based on the method of split times, cf. Section 4 of Chapter IV. The process $(M_t,F_t)_{t\geq 0}$ in question is a time-homogenous continuous time Markov process with state space $\{1,2,\dots\}\times\{1,2,\dots\}$, where the only possible transitions from state (M,F) are to $(M+1,F)$ or $(M,F+1)$, with intensities $mR(M,F)$, respectively $fR(M,F)$. We let $\tau(n)$ be the time of the nth birth (male or female) and $\tau(0)=0$. The process is then completely described by two independent sequences $Y_1',Y_2',\dots,V_0,V_1,\dots$ of random variables, where the Y_k' are i.i.d. 0-1 variables with $P(Y_k'=1)=z$ and the V_k are i.i.d. with $P(V_k>v)=e^{-v}$, in the following way: $M_{\tau(0)},M_{\tau(1)},\dots$ is a random walk, i.e., $M_{\tau(n)}=M_0+Y_1'+\dots+Y_n'$. Also, $N_{\tau(n)}=N_0+n$, $F_{\tau(n)}=F_0+n-Y_1'-\dots-Y_n'$, and the sojourn times $U_k:=\tau(k+1)-\tau(k)$ are given by

$$U_k=\mu_kV_k,\ \text{where}\ \ \mu_k:=\frac{1}{(m+f)R(M_{\tau(k)},F_{\tau(k)})}\,.$$

Note that conditional upon $\mathfrak{G}:=\sigma(Y_1',Y_2',\dots)$, the U_k are independent and exponentially distributed with $E(U_k|\mathfrak{G})=\mu_k$. It will be convenient to consider the centered variables $Y_k=Y_k'-z$ instead of the Y_k' themselves. Then

$$(4.1)\qquad X_{\tau(n)}=\frac{M_0+Y_1+\dots+Y_n+nz}{N_0+n}=z+\frac{Y_1+\dots+Y_n}{N_0+n}+\frac{M_0-N_0z}{N_0+n}\,,$$

$$(4.2)\qquad Y_1+\dots+Y_n=o(n^{\frac{1}{2}+\epsilon})\ \text{ for all }\ \epsilon>0,$$

using the law of the iterated logarithm for (4.2).

4.1. THEOREM. <u>Assume that conditions</u> (3.1), (3.3) <u>and</u> (3.2) <u>with</u> $p>0$ <u>and</u> I <u>containing a neighbourhood of</u> z <u>hold and let</u> $\lambda:=(m+f)h(z)$. <u>Then there exists a random variable</u> W <u>such that</u> $0<W<\infty$ <u>and</u>

$$(4.3)\qquad X_t=z+o(1),\ N_t=e^{\lambda t}W+o(e^{\lambda t})\ \text{a.s. as}\ \ t\to\infty.$$

PROOF. Combining (4.1), (4.2) and (3.2), one obtains $h(X_{\tau(k)})=h(z)+O(k^{-\delta})$, where $\delta=p(\frac{1}{2}-\epsilon)$,

$$(4.4) \qquad \mu_k = \frac{1}{(m+f)(N_0+k)h(X_{\tau(k)})} = \frac{1}{\lambda(N_0+k)} + O(k^{-1-\delta}).$$

Thus $\Sigma_0^\infty \operatorname{Var}(U_k|\mathfrak{G}) = \Sigma_0^\infty \mu_k^2$ converges a.s., and conditioning upon $\mathfrak{G}$, it follows by standard criteria for convergence of sums of independent mean zero variables that $\Sigma_0^\infty\{U_k-\mu_k\}$ converges a.s. Also from (4.4) and the well-known relation

$$(4.5) \qquad \sum_{k=1}^{n} k^{-1} = \log n + \text{Euler's constant} + O(\tfrac{1}{n}),$$

it follows that

$$\lambda \sum_{k=0}^{n} \mu_k - \log(N_0+n) = \sum_{k=N_0}^{N_0+n} k^{-1} - \log(N_0+n) + \sum_{k=0}^{n} O(k^{-1-\delta})$$

has a limit as $n \to \infty$. Therefore W is well-defined by

$$(4.6) \qquad \lambda\tau(n+1) = \lambda \sum_{k=0}^{n} \{U_k-\mu_k\} + \lambda \sum_{k=0}^{n} \mu_k = -\log W + \log(N_0+n) + o(1)$$

and $e^{-\lambda\tau(n+1)}N_{\tau(n+1)} \to W$. Choosing the paths to be right-continuous, we have $N_t = N_{\tau(n)}$ when $\tau(n) \leq t < \tau(n+1)$, so that

$$e^{-\lambda t}N_t = e^{-\lambda\tau(n)}N_{\tau(n)} \cdot e^{-\lambda(t-\tau(n))} \to W$$

as $t \to \infty$, since $0 \leq t - \tau(n) \leq \tau(n+1) - \tau(n) \to 0$ by (4.6). Similarly, $X_{\tau(n)} \to z$, as is obvious from (4.1) and (4.2); and $X_t \to z$ from $X_t = X_{\tau(n)}$ when $\tau(n) \leq t < \tau(n+1)$. □

The rate of convergence in 4.1 is described by

4.2. THEOREM. *In addition to the conditions of 4.1, suppose that*

$$(4.7) \qquad h(x) = h(z) + (x-z)h'(z) + O((x-z)^2) \quad \text{as} \quad x \to z$$

and write

$$X_t := z + \frac{A_t}{(We^{\lambda t})^{1/2}}, \quad N_t := We^{\lambda t} + (We^{\lambda t})^{1/2}B_t.$$

Then (i) *the limiting distribution of* (A_t, B_t) *exists and is the two-dimensional normal distribution with mean zero and co-variance matrix*

$$\begin{pmatrix} \gamma_1^2 & \rho \\ \rho & \gamma_2^2 \end{pmatrix} = \begin{pmatrix} z(1-z) & -z(1-z)\frac{h'(z)}{h(z)} \\ -z(1-z)\frac{h'(z)}{h(z)} & 1+2z(1-z)\left(\frac{h'(z)}{h(z)}\right)^2 \end{pmatrix}$$

and (ii) for all $(\alpha,\beta) \neq (0,0)$ and

$$C_t := \alpha A_t + \beta B_t, \quad \sigma^2 := \alpha^2\gamma_1^2 + \beta^2\gamma_2^2 + 2\alpha\beta\rho,$$

$$\overline{\lim_{t\to\infty}}\, C_t/(2\sigma^2 \log t)^{1/2} = 1, \quad \underline{\lim_{t\to\infty}}\, C_t/(2\sigma^2 \log t)^{1/2} = -1 \text{ a.s.}$$

We remark that the central limit theorem for A_t alone in (i) as well as the case $\beta = 0$ in (ii) are almost immediate from similar results on sums of independent random variables by reference to (4.1) and (4.3). The main new difficulty entering here is to obtain precise estimates of the remainder term in (4.6). The (somewhat lengthy) details are in Asmussen (1980) and will not be reproduced here, see A 11, however.

In the deterministic case, the reproductive value $V(M_0,F_0)$ of the initial population was defined as $\lim_{t\to\infty} e^{-\lambda t}N_t$ and an explicit formula $V(M_0,F_0) = N_0 h^*(X_0)$ was derived (we note that provided a Malthusian parameter λ exists at all, this factorization is a trivial consequence of the model, so that only the only computation needed was to find h^*). The obvious analogue in the stochastic model is letting $\tilde{V}(M_0,F_0) := E(W|M_0 = M, F_0 = F)$ and we get easily

4.3. PROPOSITION. *The process $e^{-\lambda t}\tilde{V}(M_t,F_t)$ is a non-negative martingale w.r.t. $\mathfrak{F}_t := \sigma(M_s,F_s : 0 \leq s \leq t)$ and $e^{-\lambda t}\tilde{V}(M_t,F_t) \to W$ a.s. Furthermore, $\tilde{V}$ solves the difference equation*

$$\lambda\tilde{V}(M,F) = (M+F)h\left(\frac{M}{M+F}\right)[m\tilde{V}(M+1,F)+f\tilde{V}(M,F+1)-(m+f)\tilde{V}(M,F)] \tag{4.8}$$

PROOF. The first assertion follows from general martingale theory since

$$E(W|\mathfrak{F}_t) = E^{M_t,F_t}e^{-\lambda t}W = E^{-\lambda t}\tilde{V}(M_t,F_t)$$

(here and in the following $E^{M,F}$ denotes expectation in a process with $M_0 = M$, $F_0 = F$). The martingale property is equivalent to $\lambda\tilde{V} = A\tilde{V}$, where A is the infinitesimal generator of the transition semigroup, and this equation is simply (4.8). □

In the deterministic case, $e^{-\lambda t}V_0(M_t,F_t)$ was constant, cf. (3.12), and the form of V_0 was derived from equation (1.4). The counterparts of these equations in the stochastic case are

$$\frac{d}{dt}EM_t = m\ ER(M_t,F_t),\quad \frac{d}{dt}EF_t = fER(M_t,F_t) \tag{4.9}$$

which cannot be reduced by the same methods. We leave it as an open question whether equations (4.8) or (4.9) are of any use for the study of V and use instead the method of split times to obtain

4.4. THEOREM. Suppose that conditions (3.1), (3.3) and (3.2) with $p > 0$ and I containing a neighbourhood of z and of $x \in (0,1)$ hold. Then $\tilde{V}(M_0,F_0)/N_0 \to h^*(x)$ (with h^* defined as in Section 3) when

$$M_0 \to \infty,\ F_0 \to \infty \quad \text{in such a way that} \quad X_0 = \frac{M_0}{M_0+F_0} \to x. \tag{4.10}$$

PROOF. We use the above notation, with the same sequence $Y_1,Y_2,\ldots$ for all M_0, F_0. The constants in the inequalities are always independent of M_0, F_0 (but may depend on x). Let

$$W_{\tau(n+1)} := N_{\tau(n+1)}e^{-\lambda\tau(n+1)} = (N_0+n+1)\prod_{k=0}^{n} e^{-\lambda\mu_k V_k}.$$

Conditioning upon $\mathfrak{G}$ yields

$$E^{M_0,F_0}(W_{\tau(n+1)}|\mathfrak{G}) = (N_0+n+1)\prod_{k=0}^{n}\left(1-\frac{1}{1/\lambda\mu_k+1}\right) \tag{4.11}$$

$$= (N_0+n+1)\prod_{k=0}^{n}\left(1-\frac{\lambda}{(N_0+k)h(X_{\tau(k)})(m+f)+\lambda}\right).$$

The idea of the proof is to observe that $W_{\tau(n)} \to W$, prove that indeed

$$E^{M_0,F_0}W = \lim_{n\to\infty} E^{M_0,F_0}\, E^{M_0,F_0}(W_{\tau(n+1)}|\mathfrak{G}) \tag{4.12}$$

and show that for large M_0, F_0, we can replace $X_{\tau(k)}$ in (4.11) by its expected value $(M_0+kz)/(N_0+k) = x_k$ (say). The asymptotic expression for $\tilde{V}$ will then come out by elementary calculus. To this end, define for some fixed $\varepsilon > 0$

$$T = \sup\{n: |Y_1+\ldots+Y_n| > n^{\frac{1}{2}+\varepsilon}\},$$

$$C_n(M_0,F_0) = \prod_{k=0}^{T\wedge n}\left(1-\frac{\lambda}{1/\mu_k+\lambda}\right),\quad C_\infty(M_0,F_0) = \prod_{k=0}^{T}\left(1-\frac{\lambda}{1/\mu_k+\lambda}\right),$$

$$D_n(M_0,F_0) = \prod_{k=T\wedge n+1}^{n}\left(1-\frac{\lambda}{1/\mu_k+\lambda}\right)$$

Note that the r.h.s. of (4.11) is $(N_0+n+1)C_n(M_0,F_0)D_n(M_0,F_0)$ and that $T < \infty$ a.s. by the law of the iterated logarithm. We shall need below the fact that even $ET^\beta < \infty$ for all $\beta > 0$. See, for example, the more general results by Strassen (1965). For C_n, the elementary estimates

(4.13) $\quad C_\infty(M_0,F_0) \leq C_n(M_0,F_0) \leq 1,\ C_\infty(M_0,F_0) \to 1$ subject to (4.10)

will suffice, while more care is needed when treating D_n. Preparing for an expansion of $\log D_n$, we first note that for $k > T$ it follows from (3.2) that

$$h(X_{\tau(k)}) = h\left(\frac{M_0+Y_1'+\ldots+Y_k'}{N_0+k}\right) = h\left(x_k+\frac{Y_1+\ldots+Y_k}{N_0+k}\right) := h(x_k)+E_k^1$$

where $|E_k^1| \leq \gamma_1 k^\delta/(N_0+k)^p$, $\delta = (\frac{1}{2}+\epsilon)p$. Also from (4.10) and (3.3), we must have $h(x_k)(m+f) \geq \xi$ for some $\xi > 0$ and all M_0, F_0, k. Without loss of generality, we can assume that $|E_k^1| + \lambda/(N_0+k) < \xi/2$ (say) for all N_0, k and it then follows for $k > T$ that

(4.14) $$\frac{\lambda}{1/\mu_k+\lambda} = \frac{\lambda}{N_0+k}\cdot\frac{1}{h(x_k)(m+f)+E_k^1+\lambda/(N_0+k)} := \frac{h(z)}{(N_0+k)h(x_k)}+E_k^2,$$

where $|E_k^2| \leq \gamma_2 k^\delta/(N_0+k)^{1+p}$. Write further

(4.15) $$\sum_{k=0}^{T\ n}\frac{h(z)}{(N_0+k)h(x_k)} := E^3,\quad \sum_{k=0}^{n}\frac{1}{N_0+k} := \log\frac{N_0+n}{N_0}+E^4.$$

Then $0 \leq E^3 \leq \gamma_3\log(T+N_0)/N_0$, $|E^4| \leq \gamma_4/N_0$. Assume without loss of generality $0 < x \leq z$, $0 \leq X_0 \leq z$ so that $x_k\uparrow$ and let $I_k := [x_{k-1},x_k)$, $\ell(y) := 1/h(z) - 1/h(y)$. The Lebesgue measure $m(I_k)$ of I_k is

$$m(I_k) = \frac{N_0(z-X_0)}{(N_0+k)(N_0+k-1)} := \frac{N_0(z-X_0)}{(N_0+k)^2}+E_k^5$$

where $|E_k^5| \leq \gamma_5 m(I_k)/N_0$. By (3.2),

$$\sup_{y_1, y_2 \in I_k} \left|\frac{\ell(y_1)}{z-y_1} - \frac{\ell(y_2)}{z-y_2}\right| \leq E^6$$

where $E^6 \leq \gamma_6/N_0^p$. Therefore

$$(4.16) \qquad \sum_{k=0}^{n} \frac{\ell(x_k)}{N_0+k} = \sum_{k=0}^{n} \frac{N_0(z-X_0)}{(N_0+k)^2} \frac{\ell(x_k)}{z-x_k} := \sum_{k=1}^{n} m(I_k)\frac{\ell(x_k)}{z-x_k} + E_8$$

$$:= \int_{x_0}^{x_n} \frac{\ell(y)}{z-y}\, dy + E^8 + E^9 := \int_{x_0}^{z} \frac{\ell(y)}{z-y}\, dy + E^{10} + E_n^{11}$$

where $|E^8| \leq \gamma_8/N_0$, $|E^9| \leq E^6$, $|E^{10}| := |E^8 + E^9| \leq \gamma_{10}/N_0^p$, $|E_n^{11}| \leq \gamma_{11}N_0/(N_0+n)$. Combining (4.15), (4.16) yields

$$\sum_{k=T\wedge n+1}^{n} \frac{\lambda}{(N_0+k)h(x_k)(m+f)} = \sum_{k=0}^{n} \frac{h(z)}{(N_0+k)h(x_k)} - E^3$$

$$= \log \frac{N_0+n}{N_0} + E^4 - h(z)\sum_{k=0}^{n} \frac{\ell(x_k)}{N_0+k} - E^3$$

$$= \log \frac{N_0+n}{N_0} - \log h^*(x_0) + E^{12} - h(z)E_n^{11} - E^3$$

where $|E^{12}| \leq \gamma_{12}/N_0^p$. Using (4.14),

$$\log D_n(M_0, F_0) \leq - \sum_{k=T\wedge n+1}^{n} \frac{\lambda}{1/\mu_k+\lambda} = - \sum_{k=T\wedge n+1}^{n} \frac{\lambda}{(N_0+k)h(x_k)(m+f)} + E^{13},$$

where $|E^{13}| \leq \Sigma_0^\infty |E_k^2| \to 0$ subject to (4.10), say by dominated convergence. Combining with $C_n < 1$, we have thus proved that

$$E^{M_0,F_0}(W_{\tau(n+1)}|\mathfrak{G}) \leq (N_0+n+1)\cdot 1\cdot \frac{N_0}{N_0+n} h^*(X_0) e^{-E^{12}+E^{13}+h(z)E_n^{11}} \cdot \left(\frac{T+N_0}{N_0}\right)^{\gamma_3},$$

$$E^{M_0,F_0}W \leq \varliminf_{n\to\infty} E^{M_0,F_0}E^{M_0,F_0}(W_{\tau(n+1)}|\mathfrak{G})$$

$$\leq N_0 h^*(X_0) e^{-E^{12}+E^{13}} E\left(\frac{T+N_0}{N_0}\right)^{\gamma_3} .$$

When (4.10) holds it follows by dominated convergence that $\overline{\lim}\ \tilde{V}(M_0,F_0)/N_0 \leq h^*(x)$.

To obtain the $\underline{\lim} \geq$ - part of 4.4, we first prove that for fixed M_0, F_0,

$$\sup_n E^{M_0,F_0} W^2_{\tau(n+1)} < \infty. \tag{4.17}$$

By uniform integrability, this is enough to ensure (4.12). Let $\check{\mu}_k = 2\mu_k$ and define $\check{C}_n$, $\check{E}^i$, $\check{\gamma}_i$ etc. as above, repeating the estimates with μ_k replaced by $\check{\mu}_k$. Then essentially one has to multiply the main terms by 2, while the order of magnitude of the E^i and the E^i are the same. We obtain

$$E^{M_0,F_0} W^2_{\tau(n+1)} = (N_0+n+1)^2 E^{M_0,F_0} \prod_{k=0}^{n} E^{M_0,F_0}(e^{-2\lambda U_k}|\mathcal{G}) <$$

$$\leq (N_0+n+1)^2 E^{M_0,F_0} \check{D}_n(M_0,F_0)$$

$$\leq (N_0+n+1)^2 \left(\frac{N_0}{N_0+n}\right)^2 h^*(X_0)^2 e^{-\check{E}^{12}+\check{E}^{13}+h(z)\check{E}^{11}_n} E\left(\frac{T+N_0}{N_0}\right)^{\gamma_3}.$$

From this and $ET^{\check{\gamma}_3} < \infty$ (4.17) is immediate and we get from (4.12), (4.13)

$$E^{M_0,F_0} W \geq \lim_{n\to\infty} (N_0+n+1) E^{M_0,F_0} C_\infty(M_0,F_0) D_n(M_0,F_0)$$

$$\geq \lim_{n\to\infty} (N_0+n+1) \frac{N_0}{N_0+n} h^*(X_0) e^{-E^{12}+E^{13}+h(z)E^{11}_n} E^{M_0,F_0} C_\infty(M_0,F_0)$$

$$= N_0 h^*(X_0) e^{-E^{12}+E^{13}} E^{M_0,F_0} C_\infty(M_0,F_0).$$

When (4.10) holds $E^{M_0,F_0} C_\infty \to 1$ by (4.13) and the $\underline{\lim} \geq$ - part of the theorem is proved. $\square$

BIBLIOGRAPHICAL NOTES

The two-sex models of Chapter XI are based on Kesten (1970) in discrete time and Asmussen (1980) in continuous time. The background for these references is the literature on population genetics (Kesten) and demography (Asmussen), and the above references may be consulted for further information. The model of Kesten is much more general than the simple case considered here. On might note, that our set of conditions for 2.1, 2.2 is used essentially to obtain estimates similar to the technical assumptions of Kesten. In order to obtain more refined limit statements like 4.2, 4.4 for the discrete time model one would need additional assumptions, which would require a reexamination of the model building. A step of the CLT stated in 4.2 is left out in Asmussen (1980) and is given here as A 10.

APPENDIX

1. THE CONDITIONAL BOREL-CANTELLI LEMMA

Let $\{\mathfrak{F}_n\}$ be an increasing sequence of σ-algebras and $\{X_n\}$, $\{Y_n\}$ sequences of random variables.

1.1. *We call* $\{X_n\}$ *adapted to* $\{\mathfrak{F}_n\}$ *if there is a* k *such that* X_{n+k} *is* $\mathfrak{F}_n$*-measurable for all* n. *A sequence of events is adapted if the indicators are so.*

Assume from now on $X_n \geq 0$. Then

1.2. $\sum_{n=1}^{\infty} X_n < \infty$ *a.s. on* $\{\sum_{n=1}^{\infty} E(X_n | \mathfrak{F}_n) < \infty\}$. *The converse holds if* $\{X_n\}$ *is adapted to* $\{\mathfrak{F}_n\}$. *I.e., then*

$$\{\sum_{n=1}^{\infty} X_n < \infty\} \overset{a.s.}{=} \{\sum_{n=1}^{\infty} E(X_n | \mathfrak{F}_n) < \infty\}$$

The case $k = 1$ is treated by Meyer (1971). If $k > 1$, consider $\{X_{nk+r}\}$ separately for each $r = 0,\dots,k-1$.

We next state some particular cases.

1.3. $Y_n \overset{a.s.}{\to} 0$ *on* $\{\sum_{n=1}^{\infty} E(|Y_n|^p | \mathfrak{F}_n) < \infty\}$ $(p > 0)$.

Indeed, 1.2 implies $\sum_{n=1}^{\infty} |Y_n|^p < \infty$ on the event considered. Taking the $\mathfrak{F}_n$ trivial yields

1.4. *If* $\sum_{n=1}^{\infty} E|Y_n|^p < \infty$ *for some* $p > 0$, *then* $Y_n \overset{a.s.}{\to} 0$.

If $\{A_n\}$ is a sequence of events and $X_n = I(A_n)$, then $\{\Sigma_1^{\infty} X_n < \infty\}$ is the event $\{A_n^c$ eventually$\}$ that only finitely many A_n occur and $\{\Sigma_1^{\infty} X_n = \infty\}$ the event $\{A_n$ i.o.$\}$ of infinitely many occurrences of A_n. Thus

1.5. $\{\sum_{n=1}^{\infty} P(A_n | \mathfrak{F}_n) < \infty\} \overset{a.s.}{\subseteq} \{A_n^c \text{ eventually}\}$.

The converse holds if $\{A_n\}$ *is adapted. I.e., then*

$$\{\sum_{n=1}^{\infty} P(A_n | \mathfrak{F}_n) < \infty\} \overset{a.s.}{=} \{A_n^c \text{ eventually}\},$$

$$\{\sum_{n=1}^{\infty} P(A_n | \mathfrak{F}_n) = \infty\} \overset{a.s.}{=} \{A_n \text{ i.o.}\}$$

2. MARTINGALE CONVERGENCE THEOREMS

A sequence $\{X_n\}$ of integrable r.v. is called a martingale w.r.t. an increasing sequence $\{\mathfrak{F}_n\}$ of σ-algebras if X_n is $\mathfrak{F}_n$-measurable and $E(X_{n+1}|\mathfrak{F}_n) = X_n$. If instead $E(X_{n+1}|\ _n) \geq X_n$ or $E(X_{n+1}|\ _n) \leq X_n$, we talk of a submartingale, resp. a supermartingale. If $\{X_n\}$ is a (sub-), (super-) martingale w.r.t. $\{\mathfrak{F}_n\}$, then $\{X_n\}$ is also a (sub-), (super-) martingale w.r.t. $\{\sigma(X_0,\dots,X_n)\}$. Hence the specification of $\{\mathfrak{F}_n\}$ is most often immaterial.

If $\{X_n\}$ is a martingale w.r.t. $\{\mathfrak{F}_n\}$ and we define $Y_n := X_n - X_{n-1}$, then clearly $E(Y_{n+1}|\mathfrak{F}_n) = 0$. R.v. with this property are called martingale increments.

2.1. Any submartingale $\{X_n\}$ satisfying

$$\sup_n EX_n^+ < \infty \tag{2.1}$$

converges a.s. to a (finite) r.v.

This is the basic convergence theorem.

(2.1) is called Doob's condition. Useful special cases are:

2.2. If $\{X_n\}$ is a non-negative martingale (i.e. $X_n \geq 0$ a.s.), then (2.1) and hence a.s. convergence is automatic. More generally, any non-negative supermartingale converges a.s.

2.3. The condition $\sup_n EX_n^2$ of $\mathfrak{L}^2$-boundedness implies (2.1).

In connection with 2.3, it is often useful to note that the increments of a $\mathfrak{L}^2$-martingale are orthogonal in $\mathfrak{L}^2$. E.g.

$$EX_n^2 = EX_0^2 + \sum_{k=0}^{n-1} E(X_{k+1} - X_k)^2 = EX_0^2 + \sum_{k=0}^{n-1} \mathrm{Var}(X_{k+1} - X_k) , \tag{2.2}$$

and thus $\Sigma_0^\infty \mathrm{Var}(X_{k+1} - X_k) < \infty$ is a sufficient condition for a.s. convergence. More generally:

2.4. Any martingale satisfying $\sum_{k=0}^{\infty} \mathrm{Var}(X_{k+1} - X_k|\mathfrak{F}_k) < \infty$ converges a.s.

An important example of 2.1 is

2.5. If X is any integrable r.v. and $\mathfrak{F}_n \uparrow \mathfrak{F}_\infty$, then $E(X|\mathfrak{F}_n) \to E(X|\mathfrak{F}_\infty)$ a.s. and in $\mathfrak{L}^1$.

Indeed, $\{E(X|\mathfrak{F}_n)\}$ is a martingale and (2.1) automatic. Additional arguments are required for $\mathcal{L}^1$-convergence (cf. A3) and to identify the limit.

Background: The above results are standard and can be found in most textbooks on probability theory. See e.g. Neveu (1972). Some typical examples of the applications in the present monograph are in Sections 1 and 2 of Chapter II.

3. UNIFORM INTEGRABILITY

The family $\{X_i\}_{i\in I}$ of r.v. is called *uniformly integrable* if

$$\lim_{a\to\infty} \sup_{i\in I} E|X_i|I(|X_i|>a) = 0 \ . \tag{3.1}$$

A sufficient condition is $\sup_{i\in I} E\varphi(|X_i|) < \infty$ where $\varphi:[0,\infty)\to[0,\infty)$ satisfies $\varphi(x)/x\to\infty$, $x\to\infty$. Examples: $\varphi(x)=x^p$ $(p>1)$, $\varphi(x)=x\log^+x$. Sometimes also the following alternative characterization is useful:

3.1. *The* X_i *are uniformly integrable if and only if*

$$\sup_{i\in I} E|X_i| < \infty \ , \tag{3.2}$$

$$\forall\varepsilon>0\ \exists\,\delta>0:\ PB\le\delta \Rightarrow E|X_i|I(B)\le\varepsilon \quad \forall i\in I \ . \tag{3.3}$$

The main importance of the concept comes from

3.2. *Let* $I=\mathbb{N}$ or $I=[0,\infty)$ [or a more general directed set] *and suppose* $X=\lim_{i\to\infty} X_i$ *exists in probability. Then* $X_i\to X$ *in* $\mathcal{L}^1$ (i.e. $E|X_i-X|\to 0$) *if and only if the* X_i *are uniformly integrable.*

Under the assumption of non-negativity a simple criterion obtains:

3.3. *If* $X_i\ge 0$, $X_i\to X$ *in probability, then* $X_i\to X$ *in* $\mathcal{L}^1$ *if and only if* $EX_i\to EX$.

For example:

3.4. *If* $Y_1,Y_2,\ldots$ *are i.i.d. with* $E|Y_1|<\infty$, *then the r.v.* $\overline{Y}_n=(Y_1+\ldots+Y_n)/n$ *are uniformly integrable. In particular*

$$\lim_{\delta\downarrow 0} \sup_{n=1,2,3,\ldots} \sup_{B:PB\le\delta} E|\overline{Y}_n|I(B) = 0$$

In fact, for the first statement it suffices to consider the case

$Y_1 \geq 0$ and appeal to A3.2, A3.3 and the WLLN. The second statement then follows from A3.1.

Reference: Neveu (1965), Section II.4, Problem II.6.5.

4. SERIES WITH INDEPENDENT TERMS

Let $X_1, X_2, \ldots$ be independent. We are interested in conditions for the a.s. convergence of $\Sigma_0^\infty X_n$, i.e. the a.s. existence of $\lim_{N\to\infty} \Sigma_0^N X_n$. The classical result is Kolmogorov's three-series criterion:

4.1. *Let $c > 0$ be any number and define $\tilde{X}_n := X_n I(|X_n| \leq c)$. Then $\Sigma_0^\infty X_n$ converges a.s. if and only if the series $\Sigma_0^\infty P(\tilde{X}_n \neq X_n) = \Sigma_0^\infty P(|X_n| > c)$, $\Sigma_0^\infty E\tilde{X}_n$, $\Sigma_0^\infty \operatorname{Var}\tilde{X}_n$ all converge.*

As standard examples of application, see e.g. Neveu (1965), pg. 152-155. The three-series criterion in the form 4.1 is used very little in the present monograph. However, the ideas behind the proof and the calculation of the series in question are very important tools. See e.g. Chapter II, Sections 2,4,5,6 and also A10. The proof of 4.1 is e.g. in Loeve (1955).

Occasionally we need the following easily derived consequence of 4.1:

4.2. *If $X_n = c_n Y_n$ with the Y_n i.i.d., $Y_n \geq 0$, $c_n \geq 0$ and $EY_n < \infty$, then $\Sigma_0^\infty c_n Y_n$ converges a.s. if and only if $\Sigma_0^\infty c_n < \infty$.*

5. SUMMATION BY PARTS

The discrete analogue of the formula $\int aB = AB - \int Ab$ $(A' = a, B' = b)$ for integration by parts is Abel's formula

$$\sum_{\nu=n+1}^{n+k} a_\nu B_\nu = A_{n+k} B_{n+k+1} - A_n B_{n+1} - \sum_{\nu=n+1}^{n+k} A_\nu b_\nu \tag{5.1}$$

for summation by parts, valid for any sequences $\{a_\nu\}$, $\{A_\nu\}$, $\{b_\nu\}$, $\{B_\nu\}$ satisfying $b_\nu = B_{\nu+1} - B_\nu$, $A_\nu - A_{\nu-1} = a_\nu$. Corollaries are:

5.1. *Let $\{\alpha_n\}$, $\{\beta_n\}$ be series of real numbers such that $0 < \beta_n \uparrow \infty$. Then*

$$\sum_{n=1}^{\infty} \frac{\alpha_n}{\beta_n} \text{ converges} \Rightarrow \sum_{n=1}^{N} \alpha_n = o(\beta_N) \ , \tag{5.2}$$

$$(5.3)\qquad \sum_{n=1}^{\infty} \alpha_n\beta_n \text{ converges} \Rightarrow \sum_{n=N}^{\infty} \alpha_n = o(1/\beta_N)$$

(5.2) is Kronecker's lemma and (5.3) the tail sum analogue stated as II.4.2. The proofs are analogous and we consider only (5.2) ((5.3) is treated in Asmussen (1976)). Define $a_\nu := \alpha_\nu/\beta_\nu$, $A_\nu := a_1 + \ldots + a_\nu$, $A_\infty := a_1 + a_2 + \ldots$, $B_\nu := \beta_\nu$, $b_\nu := B_{\nu+1} - B_\nu$. Then (5.1) with $n = 0$, $k = N$ gives

$$\sum_{\nu=1}^{N} \alpha_\nu = A_N B_{N+1} - \sum_{\nu=1}^{N} A_\nu b_\nu = -\sum_{\nu=1}^{N-1} A_\nu b_\nu + A_N\beta_N ,$$

$$\frac{1}{\beta_N}\sum_{\nu=1}^{N} \alpha_\nu = A_N - \sum_{\nu=1}^{N-1} A_\nu \frac{\beta_{\nu+1}-\beta_\nu}{\beta_N} \cong A_\infty - \sum_{\nu=1}^{N-1} A_\infty \frac{\beta_{\nu+1}-\beta_\nu}{\beta_N} =$$

$$A_\infty\{1 - \frac{\beta_N-\beta_1}{\beta_N}\} \to 0$$

References: (5.1) is trivial to check, but may be found in Knopp (1931) in conjunction with (5.2) and some related material. Se also Bromwich (1908).

6. MAXIMAL INEQUALITIES

The problem is essentially to bound the maximum of a finite number of partial sums by the last sum. A classical result in that direction is Kolmogorov's extension of Chebycheff's inequality,

6.1. *If $X_1,\ldots,X_N$ are independent with* $\sigma_n^2 := EX_n^2 < \infty$, $EX_n = 0$, *then*

$$P(\max_{n=1,\ldots,N} |X_1 + \ldots + X_n| > \varepsilon) \le \frac{\sigma_1^2 + \ldots + \sigma_N^2}{\varepsilon^2}.$$

Sharper results are in

6.2. *Let $X_1,\ldots,X_N$ be independent,* $S_n := X_1 + \ldots + X_n$, $M_N := \max_{n=1,\ldots,N} |S_n|$. *Then if* $P(X_{n+1} + \ldots + X_N \le b) \ge \eta > 0$, $P(X_{n+1} + \ldots + X_N \ge -b) \ge \eta$, $n = 1,\ldots,N$, *it holds for all* $a > 0$ *that* $P(M_N > a + b) \le \eta^{-1}P(|S_N| > a)$. *In particular:*

(6.1) (*Levy's inequality*) *If the X_n are symmetric,* $P(X_n < t) = P(X_n > -t)$, then $P(M_N > a) \le 2P(|S_N| > a)$.

(6.2) *If the X_n are i.i.d. with* $EX_n^2 < \infty$, $EX_n = 0$, *then for some* $\eta > 0$ $P(M_N > a) \le \eta^{-1}P(|S_N| > a)$.

(6.3) *If* $EX_n = 0$, $\sigma^2 := EX_1^2 + \ldots + EX_N^2 < \infty$, *then* $P(M_N > a + \sigma\sqrt{2}) \leq 2P(|S_N| > a)$.

The proof of the first part is an obvious stopping time argument, see e.g. Breiman (1968) pg. 45-46. (6.1) is obvious by taking $b = 0$, $\eta = \frac{1}{2}$, while (6.2) follows from the CLT and (6.3) from Chebycheff's inequality. A martingale version of (6.3) follows along just the same lines, see e.g. Stout (1970a) pg. 286-287, slightly adapted (a continuous time version is implicit in the proof of IV.3.4 and VIII.12.3):

6.3. *Let* $X_1,\ldots,X_N$ *be a martingale w.r.t.* $\mathfrak{F}_1,\ldots,\mathfrak{F}_N$, $s_n^2 := \mathrm{Var}(X_N|\ _n)$, $M_N := \max_{n=1,\ldots,N} \{X_n - s_n\sqrt{2}\}$. *Then*

$$P(M_N > a) \leq 2P(X_N > a) .$$

We shall also need

6.4. *Let* $X_1,\ldots,X_N$ *be independent and symmetric*, $H_N := \max_{n=1,\ldots,N} |X_n|$. *Then*

$$P(H_N > a) \leq 2P(|S_N| > a) .$$

PROOF. Define

$$A_n^+ := \{X_n > a,\ |X_k| \leq a \quad k = 1,\ldots,n-1\} ,$$

$$A_n^- := \{X_n < -a,\ |X|_k \leq a \quad k = 1,\ldots,n-1\} ,$$

$A_n := A_n^+ + A_n^-$. Then conditionally upon A_n^+, A_n^- the X_k $(k \neq n)$ are independent and symmetric (the distribution of X_k is the unconditionally distribution of X_k for $k > n$ and the distribution of X_k conditioned upon $|X_k| \leq a$ for $k < n$). Hence

$$P(|S_N| > a) \geq$$

$$\sum_{n=1}^{N} \{P(A_n^+, \sum_{k \neq n} X_k \geq 0) + P(A_n^-, \sum_{k \neq n} X_k \leq 0)\} \geq$$

$$\sum_{n=1}^{N} \tfrac{1}{2}\{PA_n^+ + PA_n^-\} = \tfrac{1}{2} \sum_{n=1}^{N} PA_n = \tfrac{1}{2}P(H_N > a) . \quad \square$$

7. RESULTS RELATED TO THE LIL

The classical version is

7.1. (The Hartman-Wintner LIL). Let $Y_1, Y_2, \ldots$ be i.i.d. with $EY_1 = 0$, $\operatorname{Var} Y_1 = 1$ and define $S_N := Y_1 + \ldots + Y_N$, $\log_2 := \log\log$. Then a.s.

$$\overline{\lim_{N\to\infty}} \frac{S_N}{(2N \log_2 N)^{\frac{1}{2}}} = 1, \quad \underline{\lim_{N\to\infty}} \frac{S_N}{(2N \log_2 N)^{\frac{1}{2}}} = -1 .$$

The normally distributed case comes up in various ways in all proofs. For the classical proof here, see e.g. Levy (1965) pg. 226-228. The basic ingredients are the Borel-Cantelli lemma (both ways), Levy's inequality and the identities

$$(7.1) \qquad (2\pi)^{\frac{1}{2}} (1 - \Phi(y)) = \int_y^\infty e^{-y^2/2} dy \simeq \frac{1}{y} e^{-y^2/2} \quad \text{as } y \to \infty$$

$$(7.2) \qquad \sum_{n=1}^{\infty} (1 - \Phi(\eta (2 \log n)^{\frac{1}{2}}) \simeq c \sum_{n=1}^{\infty} \frac{n^{-\eta}}{\log n} \quad \begin{cases} < \infty & \text{if } \eta > 1 \\ = \infty & \text{if } \eta < 1 \end{cases}$$

The more general 7.1 can be derived in at least three different ways:

1° Approximate Y_n by $Y_n := Y_n' - EY_n'$ where $Y_n' := Y_n I(|Y_n| \le n^{\frac{1}{2}})$. Study $\tilde{S}_n$ as in the normally distributed case, using the Berry-Esseen inequality to bound the deviations in distribution and replacing Levy's inequality by A(6.3).

Some of the ideas can be formalized in the following lemma (stated in greater generality than needed for 7.1):

7.2. Let $\{\mathfrak{F}_N\}$ be an increasing sequence of σ-algebras and $\{X_N\}$ a (not necessarily adapted) sequence of r.v. such that

$$(7.3) \qquad \sum_{n=0}^{\infty} \Delta_n := \sum_{n=0}^{\infty} \sup_{-\infty < y < \infty} |P(T_N \le y | \mathfrak{F}_N) - \Phi(y)| < \infty .$$

Then $\overline{\lim}_{N\to\infty} T_N/(2 \log N)^{\frac{1}{2}} \le 1$, with $\overline{\lim} = 1$ if T_N is $\mathfrak{F}_{N+k}$-measurable for some k.

For the proof, combine the conditional Borel-Cantelli lemma A1.5 applied to the events $\{T_N > \eta (2 \log N)^{\frac{1}{2}}\}$ with A(7.1), A(7.2).

2° Use the Skorohod imbedding scheme to approximate S_N a.s. by the normally distributed case. The details are in Breiman (1968), Ch. 13.

3° Approximate Y_n by $\tilde{Y}_n := Y_n' - EY_n'$ where $Y_n' := Y_n I(|Y_n| \leq b_n)$ and $\{b_n\}$ is chosen according to

7.3. If F is a distribution with finite second moment, then there are constants $\{K_n\}$ such that $K_n \to 0$,

$$b_n := K_n (n/\log_2 n)^{\frac{1}{2}} \to \infty \, , \quad \int_{-\infty}^{\infty} x^2/K_{[|x|]} \, dF(x) < \infty \, ,$$

$$N(m) := \sup\{n : [b_n] \leq m\} = o(m^2 \log_2 m/K_m^2) \, ,$$

[where $[\cdot]$ denotes integer part], with F the common distribution of the Y_n, and study $\tilde{S}_n$ by means of

7.4. (the Kolmogorov LIL) Let $Y_1, Y_2, \ldots$ be independent with $EY_n = 0$ and define $S_N := Y_1 + \ldots Y_N$, $s_N^2 := \Sigma_1^N \operatorname{Var} Y_n$. Then a.s.

$$\overline{\lim_{N\to\infty}} \frac{S_N}{(2\, s_N^2 \log_2 s_N^2)^{\frac{1}{2}}} = 1 \, , \quad \underline{\lim_{N\to\infty}} \frac{S_N}{(2\, s_N^2 \log_2 s_N^2)^{\frac{1}{2}}} = -1$$

provided there exists a sequence $\{K_n\}$ of constants such that $|Y_n| \leq K_n s_n/(\log_2 s_n^2)^{\frac{1}{2}}$, $K_n \to 0$ $n \to \infty$.

The proof of A7.3 (and the above program, in a slightly more general setting) is in Stout (1970b). See also Chow and Teicher (1973). Proofs of A7.4 are in Kolmogorov (1929) or Loeve (1955). The idea is analogous to that sketched above in the case of the normal distribution. Again A(6.3) replaces Levy's inequality, while the role of A(7.1) is taken by the exponential inequalities, which we state in the martingale case (see Stout (1970a), somewhat adapted):

7.5. Let $\beta^2, \gamma > 0$ be given constants. Then there are ε_0, d_0 with the following properties:

If $\{Y_k\}$ is any martingale difference sequence w.r.t. $\{\mathfrak{F}_k\}$ such that $|Y_k| \leq d$ for some d and that $s_n^2 := \Sigma_1^n \operatorname{Var}(Y_k | \mathfrak{F}_{k-1}) \uparrow \infty$ a.s., and we let $\tau := \inf\{n : s_n^2 > \beta^2\}$, $S := Y_1 + \ldots + Y_\tau$, then

$$(7.4) \qquad 0 \leq \frac{\varepsilon d}{\varepsilon^2} \leq 1 \Rightarrow P(S > \varepsilon) \leq e^{-\frac{\varepsilon^2}{2\beta^2}} \left(1 - \frac{\varepsilon d}{2\beta}\right)$$

$$(7.5)\qquad \varepsilon \geq \varepsilon_0,\ d \leq d_0 \Rightarrow P(S > \varepsilon) > e^{-\frac{\varepsilon^2}{2\beta^2}(1+\gamma)}$$

8. THE MARTINGALE CLT

The following martingale version of Lindeberg's theorem is used in Section 6 of Chapter VIII (see also Section 3 of Chapter II):

8.1. *Let* $\{Y_{k,N}\}_{N=1,2,\dots,k=1,\dots,k(N)}$ *be a martingale difference triangular array w.r.t.* $\{\mathfrak{F}_{k,N}\}$ (i.e., for any fixed N $\{Y_{k,N}\}_{k=1,\dots,k(N)}$ are martingale differences w.r.t. $\{\mathfrak{F}_{k,N}\}_{k=0,\dots,k(N)}$) *and suppose that*

$$s_N^2 := \sum_{k=1}^{k(N)} E(Y_{k,N}^2 | \mathfrak{F}_{k-1,N}) = \sum_{k=1}^{k(N)} \mathrm{Var}\,(Y_{k,N} | \mathfrak{F}_{k-1,N}) \overset{P}{\to} 1 \quad N \to \infty ,$$

$$\sum_{k=1}^{k(N)} E(Y_{k,N}^2 I(|Y_{k,N}| > \varepsilon) | \mathfrak{F}_{k-1,N}) \overset{P}{\to} 0 \quad N \to \infty,\ \forall \varepsilon > 0 .$$

Then the limiting distribution of $\Sigma_1^{k(N)} Y_{k,N}$ *as* $N \to \infty$ *exists and is standard normal*.

The result is due to Dvoretsky (1972). McLeish (1974) derives 8.1 as a special case of his elegant approach to martingale central limit theory. Later results close in spirit to VIII.6.1 and II.3.5 are in Hall (1977) (see in particular his corollary subject to his condition (2.12)).

9. THE CROFT-KINGMAN LEMMA

The statement is

9.1. *Let* $h: [0,\infty) \to \mathbb{R}$ *be continuous and suppose that the limit* $\gamma(\delta) := \lim_{n\to\infty} h(n\delta)$ *exists for each* $\delta > 0$. *Then* $\gamma(\delta)$ *is independent of* δ, $\gamma(\delta) = \gamma$, *and* $h(t) \to \gamma$ *as* $t \to \infty$ *(continuously)*.

The lemma is a key tool for deriving distributional limit results for continuous time stochastic processes $\{X_t\}_{t\geq 0}$ by means of their discrete skeletons $\{X_{n\delta}\}_{n\in\mathbb{N}}$. Typically, $h(t) = E\varphi(X_t)$ (with the continuity obvious). Examples in the present context come up for the first time in Section 3 of Chapter IV. It would be desirable to have a similar criterion applicable to a.s. convergence available in order to avoid lengthy technical proofs (like that of IV.3.4) of expected results. Unfortunately, at least some additional information is required since one can construct examples where $\{X_t\}_{t\geq 0}$ has continuous paths and does not converge a.s. at $t\to\infty$ though any discrete skeleton $\{X_{n\delta}\}_{n\in\mathbb{N}}$ converges a.s. as $n\to\infty$. The difficulty is that the exceptional set may depend on δ and that it does not suffice to consider a countable (say dense) subset of δ's.

Reference: Kingman (1963).

10. RESULTS RELATED TO THE LLN

10.1 *Let* $X_1,\dots,X_n$ *be independent random variables with mean* 0, *such that the* $|X_j|$ *are stochastically dominated by a distribution* Q *on* $[0,\infty)$ *with finite mean, i.e.* $P(|X_j|>t)\leq 1-Q(t)$, $\int_0^\infty t\,dQ(t)<\infty$, *and define* $\overline{X}_n := (X_1+\dots+X_n)/n$. *Then*

$$P(|\overline{X}_n|>\delta)\leq c_1\, n\int_n^\infty dQ(n) + c_2\,\frac{1}{n}\int_0^n x^2\,dQ(x) \tag{10.1}$$

where c_1,c_2 *may depend on* δ *and* Q *but not* n.

PROOF. Define

$$X_k^* := \begin{cases} -n & X_k\leq -n \\ X_k & -n<X_k<n\,, \\ n & X_k\geq n \end{cases} \qquad \varepsilon_k := EX_k^*$$

$\overline{X}_n^* = (X_1^*-\varepsilon_1+\dots+X_n^*-\varepsilon_n)/n$. Then, since $f(x)=xI(x\geq n)$ is non-decreasing on $[0,\infty)$

$$|\varepsilon_k| = |E(X_k^*-X_k)|\leq E|X_k|I(|X_k|\geq n) =$$

$$Ef(|X_k|)\leq\int_0^\infty f(x)\,dQ(x)=\int_n^\infty x\,dQ(x)\leq\delta/2$$

when n is sufficiently large, say $n>N$. Then for $n>N$ $|\varepsilon_1+\dots+\varepsilon_n|/n\leq\delta/2$ and it follows in a similar manner that

$$P(|X_n|>\delta)\leq P(X_k^*\neq X_k \text{ for some } k=1,\dots,n)+P(|\overline{X}_n^*|>\tfrac{\delta}{2})\leq$$

$$\sum_{k=1}^n P(X_k^*\neq X_k)+\frac{4\operatorname{Var}\overline{X}_n^*}{\delta^2}\leq\sum_{k=1}^n\{P(|X_k|>n)+\frac{4\,EX_k^{*2}}{n^2\delta^2}\}\leq$$

$$n\int_n^\infty dQ(x)+\frac{4}{n\delta^2}\int_0^\infty (x\wedge n)^2\,dQ(x)\,.$$

Hence for $n > N$ (10.1) holds with $c_1 = 1 + 4/\delta^2$, $c_2 = 4/\delta^2$. Adjust c_1, c_2 so that the validity remains for $n = 1,\dots,N$. □

10.2. *Let $\{\mathfrak{F}_n\}$ be an increasing sequence of σ-algebras, let $T(n)$ be $\mathfrak{F}_n$-measurable and let $X_{n,1},\dots,X_{n,T(n)}$ be r.v. which conditionally upon $\mathfrak{F}_n$ are independent with mean 0 and stochastically dominated by a distribution Q on $[0,\infty)$ with finite mean. Then, defining $\bar{X}_n := (X_{n,1} + \dots + X_{n,T(n)})/T(n)$, it holds for any m that $\bar{X}_n \to 0$ a.s. on $\{\lim_{n\to\infty} T(n+m)/T(n) > 1\}$.*

PROOF. Considering the sequence $\{\bar{X}_{nm+r}\}$ separately for each r, it follows that it suffices to consider the case $m = 1$. By A10.1

$$\sum_{n=1}^{\infty} P(|\bar{X}_n| > \delta | \mathfrak{F}_n) \le \int_0^{\infty} \{c_1 \sum_{n=1}^{\infty} T(n) I(x > T(n)) + c_2 x^2 \sum_{n=1}^{\infty} \frac{1}{T(n)} I(x \le T(n))\} dQ(x)$$

which is finite on the event considered since in view of II.5.3 the sums are $O(x)$, resp. $O(1/x)$. Apply the conditional Borel-Cantelli lemma. □

We shall occasionally need some variants:

10.3. *If in addition to the assumptions of 10.2 also $\int_0^{\infty} x(\log^+ x)^{\varepsilon}\, dQ(x) < \infty$, then $n^{\varepsilon}\bar{X}_n \to 0$ a.s. on $\{\lim_{n\to\infty} T(n)/m^n$ exists in $(0,\infty)\}$ $(1 < m < \infty)$.*

10.4. *If in addition to the assumptions of 10.2 also $\int_0^{\infty} x(\log^+ x)^{1/\beta - 1}\, dQ(x) < \infty$ $(0 < \beta < 1)$, then $\bar{X}_n \to 0$ a.s. on $\{\lim_{n\to\infty} T(n)/m^{n^{\beta}}$ exists in $(0,\infty)\}$ $(1 < m < \infty)$.*

SKETCH OF PROOFS. For 10.4, note that

$$\sum_{n=1}^{\infty} T(n) I(T(n) \le x) \simeq \sum_{n=1}^{\infty} m^{n^{\beta}} I(m^{n^{\beta}} \le x) \simeq$$

$$\int_1^{(\log_m x)^{1/\beta}} m^{y^{\beta}}\, dy = c_1 \int_1^{\log_m x} m^u u^{1/\beta - 1}\, du \simeq c_2 x(\log^+ x)^{1/\beta - 1},$$

and similarly $\sum_{n=1}^{\infty} T(n)^{-1} I(T(n) > x) \simeq c_3 \frac{1}{x}(\log^+ x)^{1/\beta - 1}$.

For 10.3, truncate at $n/(\log^+ x)^\varepsilon$ rather than n in 10.1 to obtain

$$(10.2)\qquad P(|\bar{X}_n| > \delta) \leq c_1 n \int_{n/(\log^+ n)^\varepsilon}^{\infty} dQ(x) + c_2 \frac{1}{n} \int_0^{n/(\log^+ n)^\varepsilon} x^2\, dQ(x)$$

and check that

$$\sum_{n=1}^{\infty} T(n) I(T(n)/(\log^+ T(n))^\varepsilon \leq x) \cong c_3\, x(\log^+ x)^\varepsilon$$

$$\sum_{n=1}^{\infty} T(n)^{-1} I(T(n)/(\log^+ T(n))^\varepsilon \geq x) \cong c_4\, \frac{1}{x}(\log^+ x)^\varepsilon . \quad \square$$

References: Kurtz (1972), Kaplan and Asmussen (1976), Asmussen and Kurtz (1980), Athreya and Kaplan (1978).

11. A RESULT OF ANSCOMBE-RENYI TYPE

In Asmussen (1980), it is promised to give at this place the proof of

11.1 *Let $Y_1, Y_2, \ldots$ be i.i.d. with $\sigma^2 := \mathrm{Var}Y_1 < \infty$, $EY_1 = 0$, let $V_1, V_2, \ldots$ be i.i.d. and independent of $Y_1, Y_2, \ldots$, with $\omega^2 := \mathrm{Var}\, V_1 < \infty$, $EV_1 = 0$, and define for given constants τ_1, τ_2, τ_3 satisfying $(\tau_1^2 + \tau_2^2)\sigma^2 + \tau_3^2\omega^2 = 1$*

$$C_n := n^{\frac{1}{2}}\Big[\tau_1 n^{-1} \sum_{k=1}^{n} Y_k + \tau_2 \sum_{k=n+1}^{\infty} \frac{Y_k}{k} + \tau_3 \sum_{k=n+1}^{\infty} \frac{V_k}{k}\Big] .$$

Let $n(t)$ be r.v. and $\gamma(t)$ constants such that $\gamma(t) \uparrow \infty$ and that $W := \lim_{t\to\infty} n(t)/\gamma(t)$ exists a.s., is measurable w.r.t. $\mathfrak{G} := \sigma(Y_1, Y_2, \ldots, V_1, V_2, \ldots)$ and satisfies $0 < W < \infty$. Then the limiting distribution of $C_{n(t)}$ as $t \to \infty$ exists and is standard normal.

The first result of this type is due to Anscombe (1952), who considered the case $C_n = (Y_1 + \ldots + Y_n)/\sqrt{n}$ and W constant. Renyi (1960) extended Anscombe's result to the case of a random (discrete) W. Since then a number of extensions and related results have appeared, see e.g. Guiasu (1972) or Fischler (1977) for further discussion. In view of the literature one would expect the condition that W be $\mathfrak{G}$-measurable to be immaterial and indeed, H. Rootzén has pointed out to us an alternative argument valid without this condition.

PROOF. That C_n is asymptotically standard normal as $n \to \infty$ (non-random) is straightforward to verify: Note that $\operatorname{Var} C_n \to 1$ and that the Lindeberg condition holds in view of

$$EY_1^2 I(|Y_1| \geq \varepsilon n^{\frac{1}{2}}) \to 0,$$

$$n \sum_{k=n+1}^{\infty} \frac{1}{k^2} EY_1^2 I(|Y_1| \geq \varepsilon k/n^{\frac{1}{2}}) \leq O(1) EY_1^2 I(|Y_1| \geq \varepsilon n^{\frac{1}{2}}) \to 0$$

(with a similar estimate for the last term in C_n). To deal with the random indexing, define $\mathcal{G}_r := \sigma(Y_1,\ldots,Y_r,V_1,\ldots,V_r)$, $W^{(r)} := E(W|\mathcal{G}_r)$ and let $n_r(t) := [\gamma(t)W^{(r)}]$ (integer part),

$$C_n^r := n^{\frac{1}{2}}[\tau_1 n^{-1} \sum_{k=r+1}^{n} Y_k + \tau_2 \sum_{k=n+1}^{\infty} \frac{Y_k}{k} + \tau_3 \sum_{k=n+1}^{\infty} \frac{V_k}{k}] \quad n > r .$$

The main step of the proof is to establish

$$\lim_{r\to\infty} \overline{\lim_{t\to\infty}} P(|C^r_{n(t)} - C^r_{n_r(t)}| > \delta) = 0 \qquad \forall \delta > 0 . \tag{11.1}$$

To this end, let $\theta > 1$ and define $A_{r,\theta} := \{\theta^{-1} < W/W^{(r)} < \theta\}$, $\underline{n}_r(t) := [\theta^{-1}\gamma(t)W^{(r)}]$, $\overline{n}_r(t) := [\theta\gamma(t) W^{(r)}]$. Then for all large enough t, $\underline{n}_r(t) < n(t) < \overline{n}_r(t)$ on $A_{r,\theta}$ and $\underline{n}_r(t) < n_r(t) < \overline{n}_r(t)$. Hence, with

$$M_t := \max_{\underline{n}_r(t) \leq i \leq \overline{n}_r(t)} |C_i^r - C^r_{\underline{n}_r(t)}| \quad ,$$

$$\overline{\lim_{t\to\infty}} P(|C^r_{n(t)} - C^r_{n_r(t)}| > \delta) \leq PA^c_{r,\theta} + \overline{\lim_{t\to\infty}} P(M_t > \tfrac{\delta}{2}) \quad . \tag{11.2}$$

To estimate M_t, note that for $j \leq i$

$$\begin{aligned} |C_i^r - C_j^r| &\leq i^{\frac{1}{2}}|i^{-\frac{1}{2}}C_i^r - j^{-\frac{1}{2}}C_j^r| + |1 - (\tfrac{i}{j})^{\frac{1}{2}}||C_j^r| \leq \\ &|\tau_1|\frac{i-j}{i^{\frac{1}{2}}j}|Y_{r+1} + \ldots + Y_j| + |\tau_1|\frac{i^{\frac{1}{2}}}{j}|\sum_{k=j+1}^{i} Y_k| + \\ &|\tau_2||i^{\frac{1}{2}}|\sum_{k=j+1}^{i} \frac{Y_k}{k}| + |\tau_3||i^{\frac{1}{2}}|\sum_{k=j+1}^{i} \frac{V_k}{k}| + |1 - (\tfrac{i}{j})^{\frac{1}{2}}||C_j^r| \quad . \end{aligned} \tag{11.3}$$

Letting $j = \underline{n}_r(t)$ and taking the maximum over $\underline{n}_r(t) \leq i \leq \overline{n}_r(t)$, it holds by Kolmogorov's inequality A6.1 that

$$(11.4) \qquad P\Big(\max_{\underline{n}_r(t) \le i \le \overline{n}_r(t)} \frac{i^{\frac{1}{2}}}{j} \Big| \sum_{k=j+1}^{i} Y_k \Big| > \varepsilon \,|\, \mathfrak{G}_r\Big) \le$$

$$\frac{\operatorname{Var}\Big(\sum_{k=j+1}^{\overline{n}_r(t)} Y_k \,|\, \mathfrak{G}_r\Big)}{\varepsilon^2 j^2 / \overline{n}_r(t)} \cong \frac{\sigma^2(\Theta - \Theta^{-1})\gamma(t)W^{(r)}}{\varepsilon^2 \Theta^2 \gamma(t)^2 W^{(r)^2} / \Theta\gamma(t)W^{(r)}} = \frac{\sigma^2(1-\Theta^{-2})}{\varepsilon^2} \quad .$$

Similar estimates of the other terms on the r.h.s. of (11.3) can be obtained (the first and last term can simply be estimated by Chebycheff's inequality) and in this manner one gets (after taking expectations in (11.4))

$$\overline{\lim_{t\to\infty}} P(M_t > \frac{\delta}{2}) \le \eta(\Theta) , \qquad \text{where } \eta(\Theta) \to 0 \text{ as } \Theta \downarrow 1 .$$

Hence the l.h.s. of (11.1) is at most $\lim_{r\to\infty}\{PA^c_{r,\Theta} + \eta(\Theta)\} = \eta(\Theta)$ (using $W^{(r)} \to W$ as $r \to \infty$) so that (11.1) follows as $\Theta \downarrow 1$.

The rest of the proof is now easy: Using the independence of C^r_n and $W^{(r)}$ and proceeding as when establishing the CLT for C_n one gets

$$\lim_{t\to\infty} P(C^r_{n_r(t)} < y) = \Phi(y) ,$$

$$\overline{\lim_{t\to\infty}} P(C_{n(t)} < y) = \overline{\lim_{t\to\infty}} P(C^r_{n(t)} < y) \le$$

$$\overline{\lim_{t\to\infty}} \{P(C^r_{n_r(t)} < y + \delta) + P(|C^r_{n(t)} - C^r_{n_r(t)}| > \delta)\} =$$

$$\Phi(y+\delta) + \overline{\lim_{t\to\infty}} P(|C^r_{n(t)} - C^r_{n_r(t)}| > \delta)$$

Letting first $r \to \infty$ and next $\delta \downarrow 0$ it follows that $\overline{\lim} \le \Phi(y)$. $\underline{\lim} \ge$ follows similarly or by complementation. □

12. A WEAK LLN RATE OF CONVERGENCE RESULT

Let $X_k, k=1,2,\ldots,$ be i.i.d. random variables with d.f. F,

$$S_n := \sum_{k=1}^{n} X_k .$$

Here is part of theorem 4 of Baum and Katz (1965):

12.1. For $\alpha \geq 0$ <u>the following statements are equivalent:</u>

(a) $n^{\alpha+1}P(|X_k| > n) \underset{n\to\infty}{\to} 0$ <u>and</u> $\int_{|x|<n} x\,dF(x) \underset{n\to\infty}{\to} 0$,

(b) $n^{\alpha}P(|S_n| > n\varepsilon) \underset{n\to\infty}{\to} 0$ <u>for every</u> $\varepsilon > 0$.

We only need

PROOF OF (a) ⇒ (b): First note that (a) implies the same statement for the symmetrized variables and, by the weak law of large numbers, $\text{med}\,(S_n/n) \to 0$, $n \to \infty$. Hence, by standard symmetrization inequalities, we may assume w.l.o.g. that the X_k are symmetric. Do this and define

$$X_{kn} := X_k 1_{\{|X_k| \leq n\varepsilon\}}, \quad S_n' := \sum_{k=1}^{n} X_{kn} .$$

Then

$$n^{\alpha}P(|S_n| > \varepsilon n) \leq n^{\alpha+1}P(|X_1| > \varepsilon n) + n^{\alpha}P(|S_n'| > \varepsilon n) .$$

Using Markov's inequality,

$$(*) \quad n^{\alpha}P(|S_n'| > \varepsilon n) \leq \varepsilon^{-r} n^{\alpha-r} E(S_n')^r$$

$$\leq \varepsilon^{-r} n^{\alpha-r} (n\, EX_{1n}^r + n(n-1) EX_{1n}^{r-2}\, EX_{1n}^2 + \ldots)$$

for every even integer $r > 2\alpha+1$. Let $(2\nu_1,\ldots,2\nu_m)$ be a partition of r into m positive even integers. Then the corresponding term on the right of (*) is bounded by

$$\varepsilon^{-r} n^{\alpha-r+m}\, E\, X_{1n}^{2\nu_1} \cdots E\, X_{1n}^{2\nu_m} .$$

Using (*) and integration by parts,

$$E\, X_{1n}^{2\nu_j} = \begin{cases} O(1) & ;\ 2\nu_j < \alpha+1 , \\ o(\log n) & ;\ 2\nu_j = \alpha+1 , \\ o(n^{2\nu_j-\alpha+1}) & ;\ 2\nu_j > \alpha+1 . \end{cases}$$

From this easily

$$\varepsilon^{-r} n^{\alpha-r+m}\, E\, X_{1n}^{2\nu_1} \cdots E\, X_{1n}^{2\nu_m} = o(1) ,$$

which completes the proof. □

13. SLOWLY OR REGULARLY VARYING FUNCTIONS

We compile a number of facts referred to in the text. A systematic treatment can be found in Seneta (1976).

A measurable, real-valued function $L(x)$, defined on $[a,\infty)$, is called slowly varying (at infinity), if it is positive and if, in addition, for each $\lambda > 0$

$$\lim_{x\to\infty} \frac{L(\lambda x)}{L(x)} = 1 . \tag{13.1}$$

A measurable, real-valued function $R(x)$ is called regularly varying (at infinity) with exponent α, $\alpha \in (-\infty,\infty)$, if it is of the form

$$R(x) = x^{\alpha} L(x)$$

with L slowly varying at infinity.

We say that $L(x)$ (or $R(x)$) is slowly (or regularly) varying at zero, if $L(1/x)$ (or $R(1/x)$) is slowly (or regularly) varying at infinity.

By translation of the origin we can define slow (or regular) variation at any point.

The following result was proved by Karamata (1930,1933) for continuous functions and by Korevaar, Aardenne-Ehrenfest and de Bruijn (1949) for measurable functions.

13.1. (Uniform convergence theorem) If L is slowly varying, then for every fixed $[a,b]$, $0 < a < b < \infty$, relation (13.1.) holds uniformly in $\lambda \in [a,b]$.

PROOF. Setting $f(x) := \log L(e^x)$, we have to prove uniformity of

$$f(x+\mu) - f(x) \to 0 , \quad x \to \infty ,$$

on every fixed finite μ-interval in $\mathbb{R}$.

If uniformity does not hold, there exist $\varepsilon > 0$, $x_n \uparrow \infty$, and $\mu_n \in [0,1]$, such that

$$|f(x_n+\mu) - f(x_n)| > \varepsilon \quad \forall\, n \in \mathbb{N} . \tag{*}$$

Let

$$U_n := \{\mu \in [0,1]: |f(x_m+\mu) - f(x_m)| < \varepsilon/2 \quad \forall\, m \geq n\} ,$$

$$V_n := \{\mu \in [0,1]: |f(x_m+\mu_m+\mu) - f(x_m+\mu_m)| < \varepsilon/2 \; \forall\, m \geq n\}.$$

Clearly, $U_n, V_n \uparrow [0,2]$, and since U_n, V_n are measurable, there exists an $N \in \mathbb{N}$ such that the Lebesgue measures of both U_N and V_N exceed 3/2 , so that the intersection of U_N and $V_N+\mu_N$ is nonempty. If $\mu \in U_N \cap (V_N+\mu_N)$, then

$$|f(x_N+\mu) - f(x_N)| < \varepsilon/2 ,$$

$$|(x_N+\mu_N+\mu-\mu_N) - f(x_N+\mu_N)| < \varepsilon/2$$

and thus

$$|f(x_N+\mu_N) - f(x_N)| < \varepsilon ,$$

which contradicts (*) .

Hence, we have uniformity for $\mu \in [0,1]$. For arbitrary intervals consider $f((x-a)/(b-a))$ in place of $f(x)$. □

<u>13.2.</u> <u>If</u> (13.1.) <u>holds for all</u> $\lambda \in (a,b) \subset (0,\infty)$, <u>it holds for all</u> $\lambda \in (0,\infty)$.

The proof is easy, using

$$\frac{L(cx\lambda)}{L(cx)} = \frac{L(cx\lambda)}{L(x)} \cdot \frac{L(x)}{L(cx)}$$

The next result is due to Karamata (1930,1933) and de Bruijn (1959).

<u>13.3.</u> (Representation theorem) <u>Every slowly varying function</u> L <u>defined on</u> $[a,\infty)$ <u>can be represented in the form</u>

$$L(x) = \exp\{\eta(x) + \int_b^x \frac{\varepsilon(t)}{t}\, dt\} \quad x \in [b,\infty) ,$$

<u>where</u> b <u>is some number</u> $\geq a$, $\eta(x)$ <u>bounded and measurable on</u> $[b,\infty)$, $\eta(x) \to c$, <u>as</u> $x \to \infty$, <u>and</u> $\varepsilon(x)$ <u>continuous on</u> $[b,\infty)$, $\varepsilon(x) \to 0$, as $x \to \infty$.

For a proof see, e.g., Seneta (1976).

An immediate corrollary is

<u>13.4.</u> <u>If</u> L <u>is slowly varying at infinity and</u> $\gamma > 0$, <u>then</u>

$$x^{\gamma} L(x) \to \infty , \quad x^{-\gamma} L(x) \to 0 , \quad x \to \infty .$$

PROOF. Using 13.3.,

$$x^{\gamma} L(x) = \sup \{\eta(x)+ \int_b^x \frac{\gamma+\varepsilon(t)}{t}\, dt\} .$$

Since $\gamma+\varepsilon(t) > \gamma/2$, the integral tends to ∞ , as $x \to \infty$, proving the first assertion. The proof of the second is similar. □

13.5. (Lamperti 1958) Let $a > 0$ and $f(x) > 0$ for every $x \in (0,a)$. If

$$\lim_{x \to 0+} \frac{xf'(x)}{f(x)} = \alpha , \tag{13.2.}$$

then

$$f(x) = x^{\alpha} L(x) , \tag{13.3.}$$

where L is slowly varying at zero.

Conversely, if (13.3.) holds for some α, and if f' exists and is monotone, then (13.2.) holds.

PROOF. By the mean value theorem and (13.2.),

$$\lim_{x \to 0+} \frac{\log f(cx) - \log f(x)}{\log(cx) - \log x} = \lim_{x \to 0+} \frac{xf'(x)}{f(x)} = \alpha$$

for each $c > 0$. Hence

$$\lim_{x \to 0+} \frac{f(cx)}{f(x)} = c^{\alpha} ,$$

and $x^{-\alpha} f(x)$ is slowly varying at 0.

For the converse assume (13.3.) and (w.l.o.g.) $f' \downarrow, x \downarrow 0$. Then, again by the mean value theorem, for each $c > 1$,

$$f'(x) \leq \frac{f(cx) - f(x)}{cx - x} \leq f'(cx) ,$$

and thus

$$\frac{xf'(x)}{f(x)} \leq \left[\frac{f(cx)}{f(x)} - 1\right](c-1)^{-1} \leq \frac{cxf'(cx)}{f(cx)} \cdot \frac{f(cx)}{cf(x)} .$$

Letting $x \downarrow 0$ and $c \downarrow 1$ and using (13.3.),

$$\limsup_{x \to 0+} \frac{xf'(x)}{f(x)} \leq \alpha \leq \liminf_{x \to 0+} \frac{xf'(x)}{f(x)} ,$$

which is (13.2.). □

13.6. (Rubin and Vere-Jones 1968). Let $f(x)$ be monotone increasing for $x \in (0,c)$. If

$$\lim_{\mathbb{N} \ni n \to \infty} \frac{f(\lambda \theta_n)}{f(\theta_n)} = \lambda^{\alpha} \quad \forall \lambda \in (0,1] \tag{13.4.}$$

for some $\alpha \in \mathbb{R}$ and some sequence $\{\theta\}$ of positive reals tending to 0, as $n \to \infty$, in such a way that $\theta_n/\theta_{n+1} < C$, $n \in \mathbb{N}$, $1 < C < \infty$, then $f(x)$ is regularly varying with exponent α.

PROOF. Notice that (13.4.) is upheld, if $\{\theta_n\}$ is replaced by any sequence of the form $\{\mu\theta_n\}$, $\mu \in (0,1]$, or by any "mixture" of finitely many sequences of the form $\{\mu\theta_n\}$. Hence,

$$\lim_{n\to\infty} \frac{f(\lambda\beta_n)}{f(\beta_n)} = \lambda^\alpha$$

for every sequence

$$\{\beta_n\} \subseteq B_k := \{\theta_n\, c^{-j/k} : n \in \mathbb{N} , 0 \le j \le k\}$$

such that $\beta_n \to 0$, $n \to \infty$.

For each sufficiently small $x > 0$ there exist $\alpha(x)$, $\beta(x) \in B_k$ such that

$$\alpha(x) < x \le \beta(x) , \quad \beta(x) = \alpha(x)\, c^{1/k} .$$

By monotonicity of $f(x)$

$$\frac{f(\lambda\alpha(x))}{f(\beta(x))} \le \frac{f(\lambda x)}{f(x)} \le \frac{f(\lambda\beta(x))}{f(\alpha(x))} .$$

Hence

$$\limsup_{x\to 0+} \frac{f(x\lambda)}{f(x)} \le (\lambda\, c^{1/k})^\alpha ,$$

$$\liminf_{x\to 0+} \frac{f(x\lambda)}{f(x)} \ge (\lambda\, c^{-1/k})^\alpha .$$

Now let $k\uparrow\infty$, and apply 13.2. □

<u>13.7</u>. (Slack 1972) <u>Suppose that</u> $f(t)$ <u>is monotone decreasing to</u> 0 , <u>as</u> $t \to \infty$, $f(n+1)/f(n) \to 1$, <u>as</u> $\mathbb{N} \ni n \to \infty$, <u>and for each integer</u> $k \ge 2$ <u>the limit of</u> $f(kn)/f(n)$, <u>as</u> $\mathbb{N} \ni n \to \infty$, <u>exists and is positive</u>. <u>Then</u> $f(x)$ <u>is regularly varying at infinity</u>.

PROOF. For $x \in (0,\infty)$ define $[x] \in \mathbb{Z}$ by $[x] \le x < [x]+1$. By monotonicity

$$\frac{f([kx]+1)}{f([x])} \le \frac{f(kx)}{f(x)} \le \frac{f([kx])}{f([x]+1)} , \qquad k \in \mathbb{N} .$$

For $x \to \infty$ the expressions on the extreme left and right tend to the same limit. To see this, note that $k[x]-[kx] < k$ for every $x \in [0,\infty)$ and use $f(n+1)/f(n) \to 1$, $n \to \infty$. Hence

$$\alpha(\lambda) := \lim_{x\to\infty} \frac{f(\lambda x)}{f(x)} \tag{13.5.}$$

exists for every $\lambda \in \mathbb{N}$. Since for $k,\ell \in \mathbb{N}$

$$f(xk/\ell)/f(x) \sim f(xk)/f(x\ell) , \quad x \to \infty ,$$

$\alpha(\lambda)$ exists for all positive rational λ . Moreover,

(13.6.) $\alpha(\lambda\mu) = \alpha(\lambda)\,\alpha(\mu)$, λ,μ positive rational ,

and $\alpha(\lambda)$ is positive and decreasing, as $\lambda \uparrow \infty$. Hence, with λ passing through rationals only,

$$\lim_{\lambda\downarrow 1} \alpha(\lambda) = \lim_{\lambda\downarrow 1} \alpha(\lambda^2) = \lim_{\lambda\downarrow 1} \alpha(\lambda)^2 = (\lim_{\lambda\downarrow 1} \alpha(\lambda))^2 ,$$

so that

$$\lim_{\lambda\downarrow 1} \alpha(\lambda) = 1 = \alpha(1) .$$

In view of (13.6.), this implies that $\alpha(\lambda)$ can be extended to a continuous function on $\mathbb{R}_+$, which - again by monotonicity of f - must be the limit of $f(\lambda x)/f(x)$, as $x \to \infty$, for any $\lambda \in (0,\infty)$.

It remains to be shown that $\alpha(\lambda)$ is a power of λ . In fact, by (13.6.) ,

$$\alpha(e^{m/n}) = \alpha(e)^{m/n} , \quad m,n \in \mathbb{Z} ,$$

and from this, by continuity of α ,

$$\alpha(e^{\lambda}) = \alpha(e)^{\lambda} , \quad \lambda \in \mathbb{R} ,$$

or

$$\alpha(\lambda) = \alpha(e)^{\log \lambda} = \lambda^{\log \alpha(e)} . \quad \square$$

13.8. (Kohlbecker 1958) *Let L be continuous and slowly varying at infinity , $0 < \alpha$, and u sufficiently large. Then there exists an asymptotically ($u \to \infty$) unique function $S(u)$ such that*

(13.7.) $$u = S(u)^{\alpha} L(S(u)) ,$$

(13.8.) $$S(u) = u^{1/\alpha} L^*(u) , \quad u \to \infty ,$$

where L^* is slowly varying at infinity.

PROOF. By 13.4., $x^{\alpha} L(x) \to \infty$, as $x \to \infty$, and since $L(x)$ is continuous, (13.7.) has at least one measurable solution $S(u)$ for all sufficiently large u , and every solution tends to ∞ , as $u \to \infty$. Using 13.3. , (13.7.) becomes

(13.9.) $$u = \exp\{\eta(S(u)) + \int_c^{S(u)} \frac{\alpha+\varepsilon(t)}{t}\,dt\}$$

Now let S,S^* be any two solutions of (13.7.). By (13.9.) and the fact that $\lim_{u\to\infty}\eta(S(u)) = \lim_{u\to\infty}\eta(S^*(u))$,

$$0 = \int_{S^*}^{S(u)} \frac{\alpha+\varepsilon(t)}{t}\,dt .$$

Since $\varepsilon(t) \to 0$, as $t \to \infty$, we have

$$\lim_{u\to\infty} [\log S(u) - \log S^*(u)] = 0 ,$$

i.e. asymptotic uniqueness of the solution.

Again starting with (13.9.) and arguing similarly as above, we get

(13.10.) $$|\log S(\lambda u) - \log S(u)|\ \alpha < 2|\log \lambda|$$

for every $\lambda > 0$ and all sufficiently large u. Suppose now that (13.8.) does not hold. Then, by (13.10.), there exist $\lambda, \mu > 0$ and $u_n \uparrow \infty$ such that $\mu^\alpha \neq \lambda$ and $S(u_n\lambda)/S(u_n) \to \mu$, as $n \to \infty$. However, by (13.7.) and 13.1., $\lambda = \mu^\alpha$. □

14. TAUBERIAN THEOREMS

14.1. (Karamata 1931 a,b). Let $U(x)$ be a monotone non-decreasing function on $[0,\infty)$ such that

$$w(x) := \int_0^\infty e^{-xu}\, dU(u)$$

is finite for all $x > 0$. If for some $\alpha \geq 0$

$$w(x) \sim x^{-\alpha} L(1/x), \quad x \downarrow 0,$$

L slowly varying at infinity, then

$$U(x) \sim x^\alpha \frac{L(x)}{\Gamma(\alpha+1)}, \quad x \uparrow \infty,$$

and if for some $\alpha \geq 0$

$$w(x) \sim x^{-\alpha} L(x), \quad x \uparrow \infty,$$

then

$$U(x) \sim x^\alpha \frac{L(1/x)}{\Gamma(\alpha+1)}, \quad x \downarrow 0,$$

with L varying at infinity.

PROOF. (Feller 1971). We prove the first statement. The proof of the second is analogous. For $\lambda > 0$

$$w(\lambda x)/w(x) \to \lambda^{-\alpha} = \int_0^\infty e^{-u\lambda}\, dG(u), \quad x \downarrow 0,$$

$$G(u) := u^\alpha/\Gamma(\alpha+1).$$

On the other hand,

$$w(\lambda x)/w(x) = \int_0^\infty e^{-\lambda xy} dU(y)/w(x) = \int_0^\infty e^{-\lambda u} dU(y/x)/w(x).$$

Hence, by continuity of $G(u)$ on $[0,\infty)$ and the extended continuity theorem

$$U(u/x)/w(x) \to G(u), \quad x \downarrow 0,$$

or, putting $v = u/x$,

$$U(v) \sim w(u/v)\ G(u) \sim (v/u)^\alpha\ L(v/u)\ u^\alpha/\Gamma(\alpha+1)$$
$$\sim v^\alpha\ L(v)/\Gamma(\alpha+1), \quad v \to \infty. \quad □$$

The corresponding density version is

14.2. (Landau 1916, Feller 1971). *For* $x \in [\beta,\infty)$ *let*

$$U(x) = \int_\beta^x u(y)\,dy\ ,$$

where $u(y)$ *is ultimately monotone. If for some* $\alpha \geq 0$

$$U(x) = x^\alpha L(x)\ ,$$

L *slowly varying at infinity, then*

$$x\,u(x)/U(x) \to \alpha\ ,\ x \to \infty\ .$$

PROOF. (Seneta 1973). W.l.o.g. let $u(x)$ be ultimately non-decreasing. For $b > a > 0$ and all sufficiently large x

$$\frac{U(xb)-U(xa)}{U(x)} = \int_{xa}^{xb} \frac{u(y)}{U(x)}\,dy\ .$$

By monotonicity of u,

$$\frac{x(b-a)u(bx)}{U(x)} \geq \frac{U(xb)-U(xa)}{U(x)} \geq \frac{x(b-a)u(xa)}{U(x)}\ .$$

Letting $x \to \infty$,

$$\liminf_{x\to\infty} \frac{xu(xb)}{U(x)} \geq \frac{b^\alpha - a^\alpha}{b-a} \geq \limsup_{x\to\infty} \frac{xu(xa)}{U(x)}\ .$$

Finally, letting $a \uparrow b$ in the first inequality and $b \downarrow a$ in the second,

$$\lim_{x\to\infty} \frac{xu(xc)}{U(x)} = \alpha\, c^{\alpha-1} \qquad \forall\, c > 0\ .\ \square$$

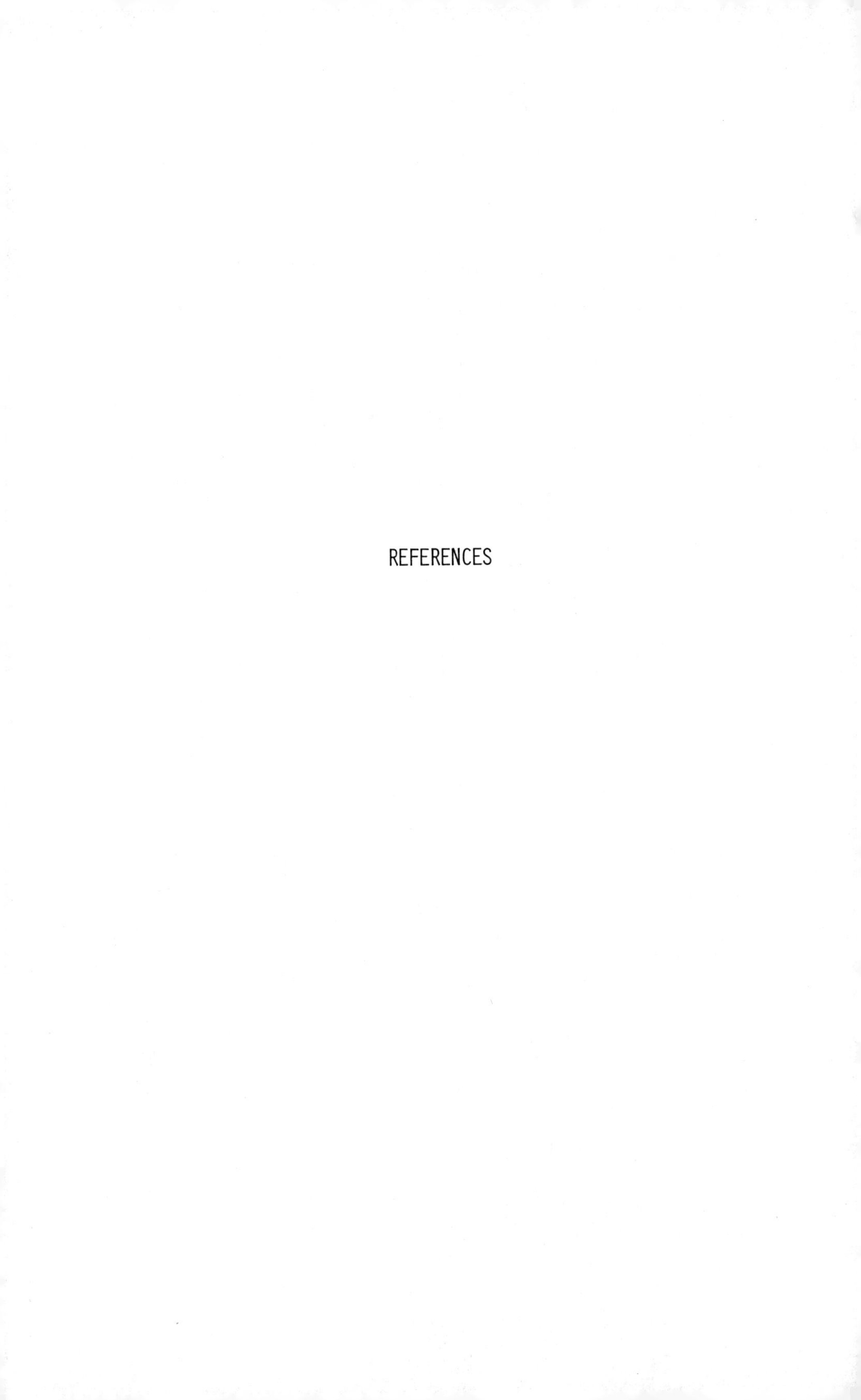

REFERENCES

REFERENCES

F.J.ANSCOMBE (1952): Large-sample theory of sequential estimation. Proc.Cambridge Philos.Soc.48, 600-607.

S.ASMUSSEN (1976): Convergence rates for branching processes. Ann. Probability 4, 139-146.

- (1977): Almost sure behavior of linear functionals of supercritical branching processes. Trans.Amer.Math.Soc. 231, 233-248.

- (1978): Some martingale methods in the limit theory of supercritical branching processes. Advances in Probability and Related Topics 5, 1-26.

- Lectures on renewal theory. Inst.Mat.Stat., University of Copenhagen.

- (1980): On some two-sex population models. Ann.Probability 8, 727-744.

- (1982): On the role of a certain eigenvalue in estimating the growth rate of a branching process. Austral.J.Statist. 24(2).

S.ASMUSSEN and H.HERING (1976a): Strong limit theorems for general supercritical branching processes with applications to branching diffusions. Z.Wahrscheinlichkeitsth.verw.Geb.36,195-212.

- (1976b): Strong limit theorems for supercritical immigration-branching processes. Math.Scand.39, 327-342.

- (1977): Some modified branching diffusion models. Math.Biosci. 35, 281-299.

S.ASMUSSEN and N.KAPLAN (1976): Branching random walks I. Stoch.Proc. Appl.4, 1-13.

S.ASMUSSEN and N.KEIDING (1978):Martingale central limit theorems and asymptotic estimation theory for multitype branching processes. Adv.Appl.Probability 10, 1o9-129.

S.ASMUSSEN and T.G.KURTZ (1980): Necessary and sufficient conditions for complete convergence in the law of the large numbers. Ann.Probability 8, 176-182.

K.B.ATHREYA (1969): Limit theorems for multitype continuous time Markov branching processes. Z.Wahrscheinlichkeitsth.verw.Geb. 12, 320-332; ibid.13, 2o4-214.

- (1971): Some refinements in the theory of supercritical multitype branching processes. Z.Wahrscheinlichkeitsth.verw.Geb.20, 47-57.

K.B.ATHREYA and N.KAPLAN (1978): Additive property and its applications in branching processes. Advances in Prob.5, 27-60.

K.B.ATHREYA and P.NEY (1970): The local limit theorem and some related aspects of super-critical branching processes. Trans.Amer. Math.Soc.152, 233-251.

- (1972): Branching Processes. Springer-Verlag, Berlin-Heidelberg-New York.

- (1974): Functionals of critical multitype branching processes. Ann.Prob. 2, 339-343.

- (1978): A Markov process approach to systems of renewal equations with applications to branching processes. Advances in Prob.5, 297-317.

K.B.ATHREYA, P.R.PARTHASARATHY, and G.SANKARANARAYANAN (1974): Supercritical age-dependent branching processes with immigration. J.Appl.Prob.11, 695-702.

L.E.BAUM and M.KATZ (1965): Convergence rates in the law of large numbers. Trans.Amer.Math.Soc. 120, 109-123.

J.D.BIGGINS (1978): The asymptotic shape of the branching random walk. Adv.Appl.Prob.10, 62-84.

M.D.BRAMSON (1978): Maximal displacement of branching Brownian motion. Comm.Pure Appl.Math.31, 531-581.

H.BRAUN (1978): Stochastic stable population theory in continuous time. Scand.Act.J., 185-203.

L.BREIMAN (1968): Probability. Addison-Wesley, Reading, Mass.

T.J. I'A BROMWICH (1908): An introduction to the theory of infinite series. MacMillan, London.

N.G. de BRUIJN (1959): Pairs of slowly oscillating functions occuring in asymptotic problems concerning the Laplace transform. Nieuw Arch.Wisk. 7, 20-26.

W.FELLER (1971): An Introduction to Probability Theory and Its Applications, vol.2, 2nd ed., Wiley, New York.

Y.S.CHOW and H.TEICHER (1973): Iterated logarithm laws for weighted averages. Z.Wahrscheinlichkeitsth.verw.Geb.26, 87-94.

E.CINLAR (1968): Introduction to stochastic processes. Prentice-Hall, N.J.

H.COHN and H.HERING (1981): Inhomogeneous Markov branching processes: Supercritical case. Stoch.Proc.Appl. - To appear.

J.G.van der CORPUT (1931): Diophantische Ungleichungen. I. Zur Gleichverteilung modulo Eins. Acta.Math.56, 373-456.

K.S.CRUMP and C.J.MODE (1968,1969): A general age-dependent branching process I,II. J.Math.Anal.Appl. 24, 494-505; 25, 8-17.

D.DAWSON and G.IVANOFF (1978): Branching diffusions and random measures. Adv.Probability 5, 61-103.

J.-P.DION and N.KEIDING (1978): Statistical inference in branching processes. Advances in Prob.5, 105-140.

R.A.DONEY (1972): A limit theorem for a class of supercritical branching processes. J.Appl.Prob.9, 707-724.

- (1976 a): A note on the subcritical generalized age-dependent branching process. J.Appl.Prob.13, 798-803.

- (1976 b): On single- and multi-type general age-dependent branching processes. J.Appl.Prob. 13, 239-246.

S.DUBUC (1970): La fonction de Green d'un processus de Galton-Watson. Studia Math. 34, 69-87.

- (1971a): Problèmes relatifs à l'iteration des fonctions suggérés par le processus en cascade. Ann.Inst.Fourier (Grenoble), 21, 171-251.

- (1971b): Processus de Galton-Watson surcritiques. Seminaire d'Analyse Moderne No.7, Departement de Mathématiques, Université de Sherbrooke.

- (1971c): La densité de loi-limite d'un processus en cascade expansif. Z.Wahrscheinlichkeitsth.verw.Geb. 19, 281-290.

- (1978): Martin boundaries of Galton-Watson processes. Adv.Prob. 5, 141-157.

S.DUBUC and E.SENETA (1976): The local limit theorem for the Galton-Watson process. Ann.Prob. 4, 490-496.

A.DVORETSKY (1972): Central limit theorems for dependent random variables. Proc.Sixth Berkeley Symp.Math.Statist.Prob., Univ. of California Press, 513-535.

E.B.DYNKIN (1965): Markov Processes I, Springer, Berlin.

K.ENDERLE and H.HERING (1982): Ratio limit theorems for branching Ornstein-Uhlenbeck processes. Stoch.Proc.Appl 13, 75-85.

D.H.FEARN (1972): Galton-Watson processes with generation dependence. Sixth Berkeley Symp.Math.Statist.Prob. IV, 159-172.

K.S.FAHADY, M.P.QUINE, and D.VERE-JONES (1971): Heavy traffic approximations for the Galton-Watson process. Adv.Appl.Prob. 3, 282-300.

W.FELLER (1971): An introduction to probability theory and its applications. II, 2nd ed. Wiley, New York.

R.FISCHLER (1976): Convergence faible avec indices aleatoires. Ann. Inst.Henri Poincaré (B) 12, 391-399.

J.FOSTER and P.NEY (1978): Limit laws for decomposable critical branching processes. Z.Wahrscheinlichkeitsth.verw.Geb. 46, 13-43.

R.T.GOETTGE (1975): Limit theorems for the supercritical Galton-Watson process in varying environment. Math.Biosci. 28, 171-190.

M.I.GOLDSTEIN (1971): Critical age-dependent branching processes: Single and multitype. Z.Wahrscheinlichkeitsth.verw.Geb. 17, 74-88.

L.G.GOROSTIZA and A.R.MONCAYO (1978): Invariance principles for the spatial distributions of branching populations. Advances in Probability 5, 159-175.

P.J.GREEN (1977): Conditional limit theorems for general branching processes. J.Appl.Prob. 14, 451-463.

D.R.GREY (1980): A new look at convergence of branching processes. Ann.Probability 8, 377-380.

S.GUIASU (1971): On the asymptotic distribution of sequences of random variables with random indices. Ann.Math.Statist. 42, 2o18-2o28.

P.HALL (1977): Martingale invariance principles. Ann.Probability 5, 875-887.

T.E.Harris (1963): The Theory of Branching Processes. Springer, Berlin.

C.R.HEATHCOTE (1966): Corrections and comments on the paper, "A branching process allowing immigration". J.Roy.Statist.Soc. 28B, 213-217.

H.HERING (1973): Asymptotic behaviour of immigration-branching processes with general set of types. I: Critical branching part. Adv.Appl.Prob. 5, 391-416.

- (1974): Limit theorem for critical branching diffusion processes with absorbing barriers, Math.Biosci. 19, 355-370.

- (1977a): Subcritical branching diffusions. Comp.Math.34,289-306.

- (1977b): Minimal moment conditions in the limit theory for general Markov branching processes. Ann.Inst. H.Poincaré, Sec.B,13, 299-319.

- (1978a): Refined positivity theorem for semigroups generated by perturbed differential operators of second order with an application to Markov branching processes. Math.Proc.Cambr.Phil. Soc.83, 253-259.

- (1978b): Uniform primitivity of semigroups generated by perturbed elliptic differential operators. Math.Proc.Cambr.Phil.Soc.83, 261-268.

- (1978c): Multigroup branching diffusions. Adv.Probability 5, 177-217.

- (1978d): The non-degenerate limit for supercritical branching diffusions. Duke Math.J., 45, 561-600.

H.HERING and F.M.HOPPE (1981): Critical branching diffusions: Proper normalization and conditional limit. Inst.H.Poincaré, Sec.B, 17, 251-274.

C.C.HEYDE (1970a): Extension of a result of Seneta for the supercritical Galton-Watson process. Ann.Math.Stat. 41, 739-742.

- (1970b): Some almost sure convergence theorems for branching processes. Z.Wahrscheinlichkeitsth.verw.Geb.20, 189-192.

C.C.HEYDE and J.R.LESLIE (1971): Improved classical limit analogues for Galton-Watson processes with or without immigration. Bull. Austral.Math.Soc. 5, 145-155.

E.HILLE and R.S.PHILIPS (1957): Functional Analysis and Semi-Groups, rev.ed., Amer.Math.Soc.Colloq.Publ. vol.31, Amer.Math.Soc., Providence, R.I.

J.M.HOLTE (1974): Extinction probability for a critical general branching process. Stoch.Proc.Appl.2, 3o3-3o9.

- Critical multitype branching processes. - Preprint.

F.M.HOPPE (1976a): Supercritical multitype branching processes, Ann. Prob. 4, 393-401.

- (1976b): On a result of Rubin and Vere-Jones concerning subcritical branching processes. J.Appl.Prob. 13, 804-808.

- (1977): Convex solutions of a Schroeder equation in several variables. Proc.Amer.Math.Soc. 64, 326-330.

F.M.HOPPE and E.SENETA (1978): Analytical methods for discrete branching processes. Advances in Probability 5, 219-261.

S.ITÔ (1957): Fundamental solution of parabolic differential equations and boundary value problems. Jap.J.Math. 27, 55-102.

P.JAGERS (1969): A general stochastic model for population development. Skand.Aktuarietidskr. 52, 84-103.

- (1974): Convergence of general branching processes and functionals thereof. J.Appl.Prob.11, 471-478.

- (1975): Branching Processes with Biological Applications. Wiley, London.

S.KAGEYAMA and Y.OGURA (1980): On a limit theorem for branching one-dimensional diffusion processes. Publ.RIMS Kyoto Univ.16, 355-376.

P.J.M.KALLENBERG (1979): Branching processes with continuous state space. Mathematical Centre Tracts, vol.117. Mathematisch Centrum, Amsterdam.

N.KAPLAN (1973): A continuous time Markov branching model with random environments. Adv.Appl.Prob.5, 37-54.

- (1974): Some results about multidimensional branching processes with random environments. Ann.Prob.2, 441-455.

- (1977): A limit theorem for a branching random walk. Preprint.

N.KAPLAN and S.ASMUSSEN (1976): Branching random walks II. Stoch.Proc. Appl.4, 15-31.

J.KARAMATA (1930): Sur un mode de croissance régulière des fonctions. Mathematica (Cluj), 4, 38-53.

- (1931a): Neuer Beweis und Verallgemeinerung der Tauberschen Sätze, welche die Laplace'sche und Stieltjesche Transformation betreffen. J.Reine Angew.Math. 164, 27-39.

- (1931b): Neuer Beweis und Verallgemeinerung einiger Tauberscher Sätze. Math.Zeitschrift 33, 294-299.

- (1933): Sur un mode de croissance régulière. Théorèmes fondamentaux. Bull.Soc.Math.France, 61, 55-62.

S.KARLIN and J.McGREGOR (1966): Spectral theory of branching processes. I. The case of discrete spectrum; II. The case of continuous spectrum. Z.Wahrscheinlichkeitsth. 5, 6-33, 34-54.

- (1968a): On the spectral representation of branching processes with mean one. J.Math.Anal.Appl. 21, 485-495.

- (1968b): Embeddability of discrete time simple branching processes into continuous time branching processes. Trans.Amer.Math.Soc. 132, 115-136.

- (1968b): Embedding iterates of analytic functions with two fixed points into continuous groups. Trans.Amer.Math.Soc. 132, 137-145.

S.KARLIN and H.M.TAYLOR (1975): A first course in stochastic processes. Academic Press, New York.

N.KEIDING and J.HOEM (1976): Stochastic stable population theory with continuous time. Scand.Act.J., 150-175.

J.G.KEMENY, J.L.SNELL, and A.W.KNAPP (1966): Denumerable Markov Chains. Van Nostrand, Princeton, N.J.

D.G.KENDALL (1948): On the role of variable generation time in the development of a stochastic birth process, Biometrika 35, 316-330.

H.KESTEN (1970): Quadratic transformations: A model for population growth. Adv.Appl.Probab. 2, 1-82, 179-228.

- (1978): Branching Brownian motion with absorption. Stoch.Proc. Appl.7, 9-47.

H.KESTEN, P.NEY, and F.SPITZER (1966): The Galton-Watson process with mean one and finite variance. Theor.Prob.Appl.11, 513-540.

H.KESTEN and B.P.STIGUM (1966a): A limit theorem for multidimensional Galton-Watson processes. Ann.Math.Statist. 37, 1211-1223.

- (1966b): Additional limit theorems for indecomposable multidimensional Galton-Watson processes. Ann.Math.Statist.37,1463-1481.

N.KEYFITZ (1971): On the momentum of population growth. Demography 8, 71-80.

K.KNOPP (1931): Theorie und Anwendungen der unendlichen Reihen. Springer, Berlin.

E.E.KOHLBECKER (1958): Weak asymptotic properties of partitions. Trans. Amer.Math.Soc.88, 346-365.

A.KOLMOGOROV (1929): Über das Gesetz des iterierten Logarithmus. Math.Ann. 101, 126-135.

H.KOREVAAR, T.van AARDENNE-EHRENFEST, and N.G.de BRUIJN (1949): A note on slowly oscillating functions. Nieuw.Arch.Wisk. 23, 77-86.

M.G.KREǏN and M.A.RUTMAN (1956): Linear operators leaving invariant a cone in a Banach space. Amer.Math.Soc.Transl. (1),26, (transl. from: Uspeki Mat.Nauk 3, (1948), 3-95.)

T.G.KURTZ (1972): Inequalities for the law of large numbers. AMS 43, 1874-1883.

\- (1978): Diffusion approximations for branching processes. Adv. Probability 5, 269-292.

J.LAMPERTI (1958): An occupation time theorem for a class of stochastic processes. Trans.Amer.Math.Soc.88, 380-387.

E.LANDAU (1916): Darstellung und Begründung einiger neuerer Ergebnisse der Funktionentheorie. Springer, Berlin.

M.LOEVE (1955): Probability Theory. Van Nostrand, Princeton.

P.LEVY (1954): Théorie de l'addition des variables aleatories. 2ième ed. Gauthier-Villars, Paris.

\- (1965): Processus stochastiques et mouvement Brownian. Gauthier-Villars, Paris.

A.J.LOTKA (1931): The extinction of families - I,II. J.Wash.Acad.Sci. 21, 377-380; 453-459.

\- (1939): Théorie analytique des associations biologiques, deuxième partie. Actualités Scientifiques et Industrielles 780, Hermann, Paris.

K.MATTHES, J.KERSTAN, and J.MECKE (1978): Infinitely Divisible Point Processes. J.Wiley & Sons, Chichester 1978.

D.L.McLEISH (1974): Dependent central limit theorems and invariance principles. Ann.Probability 2, 620-628.

P.A.MEYER (1972): Martingales and stochastic integrals I. Springer Lecture Notes in Mathematics 284, Springer, Berlin-Heidelberg-New York.

C.J.MODE (1971): Multitype Branching Processes. Elsevier, New York.

P.MONTEL (1957): Leçons sur les récurrences et leurs applications. Gauthier-Villars, Paris.

J.E.MOYAL (1962): The general theory of stochastic population processes. Acta Math. 1o8, 1-31.

T.W.MULLIKIN (1963): Limiting distributions for critical multitype branching processes with discrete time. Trans.Amer.Math.Soc. 1o6, 469-494.

O.NERMAN (1979): On the convergence of supercritical general branching processes. Ph.D.thesis, University of Gothenburg.

J.NEVEU (1965): Mathematical foundations of the calculus of probability. Holden-Day, San Francisco.

\- (1972): Martingales a temps discret. Masson, Paris.

Y.OGURA (1975): Asymptotic behavior of multitype Galton-Watson processes. J.Math.Kyoto Univ. 15, 251-302.

- (1979): A remark on a branching Brownian motion on a Riemannian manifold with negative curvature. Reports of the Faculty of Science and Engineering, Saga University 7, 13-23.

G.C.PAPANICOLAOU (1978): Boundary behaviour of branching transport processes. Stochastic Analysis, ed.A.Freedman and M.Pinsky, Academic Press, New York.

O.PERRON (1907): Zur Theorie der Matrizen. Ann.Math. 64, 248-263.

A.K.POLIN (1976): Limit theorems for decomposable critical branching processes. Math.USSR Sbornik 29, 377-392.

- (1977): Limit theorems for decomposable branching processes with finite type. Math.USSR Sbornik 33, 136-146.

A.RENYI (1960): On the central limit theorem for the sum of a random number of independent random variables. Acta Math.Acad.Sci. Hung. 11, 97-102.

H.RUBIN and D.VERE-JONES (1968): Domains of attraction for the subcritical Galton-Watson process. J.Appl.Prob.5, 216-219.

T.RYAN (1968): On age-dependent branching processes. Ph.D.Thesis. Cornell University.

- (1976): A multidimensional renewal theorem. Ann.Prob. 4, 656-661.

K.SATO and T.UENO (1965): Multidimensional diffusion and tne Markov process on the boundary. J.Math.Kyoto Univ. 4, 529-605.

T.SAVITS (1969): The explosion problem for branching Markov process. Osaka.J.Math. 6,375-395.

E.SCHROEDER (1871): Über iterierte Funktionen. Math.Ann. 3, 296-322.

H.-J. SCHUH (1982): Sums of iid random variables and an application to the explosion criterion for Markov branching processes. J.Appl.Prob. 19, 29-38.

H.-J.SCHUH and A.BARBOUR (1977): On the asymptotic behaviour of branching processes with infinite mean. Adv.Appl.Prob.9,681-723.

E.SENETA (1968): On recent theorems concerning the supercritical Galton-Watson process. Ann.Math.Statist. 39, 2o98-21o2.

- (1969): Functional equations and the Galton-Watson process. Adv.Appl.Prob.1, 1-42.

- (1970a): A note on the supercritical Galton-Watson process with immigration. Math. Biosci.6, 3o5-312.

- (1970b): On the supercritical Galton-Watson process with immigration. Math.Biosci.7,9-14.

- (1970c): An explicit-limit theorem for the critical Galton-Watson process with immigration. J.Roy.Statist.Soc. 32 B, 149-152.

- (1971): On invariant measures for simple branching processes. J.Appl.Prob.8, 43-51.

- (1973): A Tauberian theorem of E.Landau and W.Feller. Ann.Prob. 1, 1o57-1058.

- (1974): Regularly varying functions in the theory of simple branching processes. Adv.Appl.Prob. 6, 408-420.

- (1975): Characterization by functional equations of branching process limit laws. Statistical Distributions in Scientific Work., ed.G.P.Patil et al., vol.3, 249-254,D.Reidel,Dordrecht.

- (1976): Regularly Varying Functions. Lecture Notes in Maths. No.508, Springer, Berlin.

E.SENETA and D.VERE-JONES (1966): On quasi-stationary distributions in discrete time Markov chains with a denumerable infinity of states. J.Appl.Prob.3, 403-434.

B.A.SEVAST'YANOV (1959): Transient phenomena in branching stochastic processes. Theor.Prob.Appl.4, 113-128.

- (1975): Verzweigungsprozesse. R.Oldenbourg, München-Wien.

R.S.SLACK (1968): A branching process with mean one and possibly infinite variance. Z.Wahrscheinlichkeitsth.verw.Geb.9, 139-145.

- (1972): Further notes on branching processes with mean one. Z. Wahrscheinlichkeitsth.verw.Geb.25,31-38.

F.SPITZER (1967): Two explicit Martin boundary constructions. Symposium on Probability Methods in Analysis. Lecture Notes in Maths., Springer, Berlin.

W.F.STOUT (1970a): A martingale analogue of Kolmogorov's law of the iterated logarithm. Z. Wahrscheinlichkeitsth.verw.Geb.15,279-290.

- (1970b): The Hartman-Wintner law of the iterated logarithm for martingales. Ann.Math.Statist.41, 2158-2160.

D.TANNY (1976): Normalizing constants for branching processes in random environments. Stoch.Proc.Appl.6, 201-211.

- (1977): Limit theorems for branching processes in a random environment. Ann.Prob.5, 100-116.

K.UCHIYAMA (1976): Limit theorems for non-critical Galton-Watson processes with $EZ_1 \log Z_1 = \infty$. Proc.3rd Japan-USSR Symp.Prob. Theory, Taschkent 1975. Springer Lecture Notes in Mathematics 550, 646-649.

S.WATANABE (1967): Limit theorems for a class of branching processes. Markov processes and potential theory (J.Chover ed.), 205-232. Wiley, New York.

V.M.ZOLOTAREV (1957): More exact statements of several theorems in the theory of branching processes. Theor.Prob.Appl.2, 245-253.